Contrasts in Scientific Style

Contrasts in Scientific Style

Research Groups in the Chemical and Biochemical Sciences

Joseph S. Fruton

American Philosophical Society
Independence Square • Philadelphia
1990

Publication of this volume has been subsidized by the Alfred P. Sloan Foundation.

Dust jacket: Adolf von Baeyer and Frederick Gowland Hopkins.

Library of Congress Catalog
Card No: 90-55266
International Standard Book No.: 0-87169-191-4
US ISSN 0065-9738

*This book is
dedicated to the memory of
departed friends*

ALBERT CHARLES CHIBNALL

(1894–1988)

CARL FERDINAND CORI

(1896–1984)

WILLIAM HOWARD STEIN

(1911–1980)

ABRAHAM WHITE

(1908–1980)

Preface

This book is a sequel to my earlier effort to describe some aspects of the historical development of the interplay of chemistry and biology since 1800. One of the features of that development has been the emergence, during the nineteenth century, of relatively large research groups, each of them led by an individual who had attained an institutional status, such as that of an *ordentlicher Professor* at a German university, that provided an opportunity to develop a research program with the aid of predoctoral students and post-graduate assistants. My aim in this book has been to examine the manner in which several of the prominent early leaders of such groups exercised that function, and the extent to which, in their roles as teachers, they influenced through their pupils the later development of the chemical and biochemical sciences.

After an introductory chapter, the succeeding four chapters offer somewhat detailed accounts of the research groups led by these men, with special attention to their scientific attitudes, to their styles of leadership, and to the later activity of their scientific progeny. In the final chapter, I suggest that the contrasts in scientific style, evident during the nineteenth century, have persisted into the present, and I have sketched some of the research groups of the recent past in the hope that they may be examined more closely (and more critically) by historians of the chemical and biochemical sciences.

Some parts of this book have appeared in different form in previous publications of The American Philosophical Society. A section of Chapter One was taken from the introduction to Fruton (1982b), and Chapters Two and Five are extensively revised versions of Fruton (1988a) and Fruton (1985b) respectively. I am grateful to the Society for permission to reproduce this material.

A large final section of the book is devoted to appendices with brief statements about the junior members of the research groups considered in some detail in Chapters Two–Five. The appendices provide the data upon which the numerical tables and the conclusions offered in the text are based. In addition to well-known persons, there are many more who deserve to be remembered. The reader will find in some of the entries hints of human tragedy—young scientists and old, persecuted and exiled in the years after 1848 and 1933, or killed during the two World Wars of the twentieth century and in the concentration camps of Nazi Germany.

There are many quotations in the text, all in English; the translations from French, German and Russian are my own, and usually differ in detail from others that may be available. I am grateful to many authors and publishers for permission to quote or to translate copyrighted material.

In the index of personal names, those of authors of books and articles cited in the notes and bibliography are not included, nor are names which are listed in the appendices but not in the text or notes. Also, I hope that the reader will not be unduly inconvenienced by the absence of a subject index. During the period covered by this historical account, the same term was often used to denote different things, and what turned out to be the same thing was assigned different names.

The genesis of this book came during the tenure of a fellowship (1983–1984) of the John Simon Guggenheim Foundation, and the preparation of the book for publication was made possible by a personal grant (RH–20739–86) from the National Endowment for the Humanities. These two awards were decisive in encouraging me to pursue the line of study outlined above.

I am also much indebted to many individuals for their help. Valuable criticisms of early drafts of several chapters were provided by the following colleagues: Sidney Altman, John T. Edsall, Gerald L. Geison, Frederic L. Holmes, Ann Körner and Sofia Simmonds. In the collection of archival material I owe special thanks to Marie Byrne (Bancroft Library, University of California, Berkeley), Jacques d'Orléans (Archives du Bas-Rhin, Strasbourg), Clark A. Elliott (Harvard University Archives), Katherine Morton (Yale University Archives) and Inge Wojte (Staatsbibliothek Preussischer Kulturbesitz, Berlin). The search for biographical data was greatly aided by several other university archivists, notably Laetitia Boehm (Munich), Beat Glaus (Zurich) and Andreas Jacob (Erlangen), as well as many persons at state, municipal and company archives, especially Dr. Dietrich Andernacht and Karin Carl (Frankfurt/M), Archivoberrat Hecker (Munich) and Manfred Simon (Hoechst). I also owe much to Grace Epps (Kline Science Library) and Ferenc A. Gyorgyey (Medical Historical Library) at Yale for their generous assistance in obtaining printed material needed for this study.

Joseph S. Fruton

New Haven, Connecticut
January 1989

Contents

Chapter One

Research Groups in the Biochemical Sciences

Since the middle of the nineteenth century, laboratory experimentation in the chemical and medical sciences has become increasingly the collective activity of relatively large research groups. These collectives, led by scientists who achieved sufficient recognition to allow them to conduct an independent research program at a university or at a research institute, have been composed of advanced predoctoral students, postdoctoral associates and guests, and technical assistants. Through their social and intellectual interaction with the leader and with one another, the junior members of such groups have had the opportunity to learn laboratory techniques, to acquire theoretical knowledge and, by participating actively in the research program of the laboratory, to generate new scientific results and new ideas. In the formation of the character of the group, the number and caliber of the junior members, the space and research instruments available for their work, and their expectations of a later successful career in academic science, medicine, pharmacy, commerce or industry, all have been significant factors. The dominant influence, however, has been the manner in which the leader guided the efforts of his or her research group.[1]

---

[1] The term *research school* has also been used to denote what I describe as a *research group*; see Geison (1981), p.23. I prefer the latter term because *research school* has also been applied to a community of scientists, not necessarily located at a single institution, or even in the same country, who are united solely by a common interest in a

The distinctive modes of thought and action—the scientific styles—that have characterized the leadership of research groups in the chemical and medical sciences have varied widely. At one extreme, the leader has been a quasi-military director of the work of subordinates, and at the other, a senior counselor in the independent efforts of his junior associates. Apart from the scientific acumen, personal qualities and public prestige of the group leader, other factors have affected the balance between dictatorship and liberality. Foremost among these factors have been not only the leader's perception of the conceptual structure of his discipline, but also the institutional status of that discipline, as determined by official opinion regarding its relevance to medical, agricultural or industrial practice. In addition, more general social influences in different nations or institutions, and at different times, have inclined a leader to adjust his attitude toward the junior members of his group in order to promote its productivity.

The award of credit for the research achievements of a closely directed group has usually gone to the leader, as it was considered that he or she formulated the plan of attack that proved to be successful. Although some of the junior associates may have made original contributions to the group effort, they were expected to gain satisfaction from the reflected fame of their leader, and to hope that he would make these contributions known, especially to prospective employers. On the other hand, if the group leader acted principally as senior counselor to his junior associates, and his name did not appear among the authors of their research papers, his role in any success they may have achieved usually did not receive recognition, even if he may have suggested the research problem, given advice about the conduct of the work, and helped to write up the results for publication. His place in the historical record has been based largely on his personal research, usually performed at the beginning of his professional career.[2]

---

particular direction of research. For an extended discussion of the various definitions of *research school*, see Gasilov (1977). In the valuable study edited by Andrews (1979), what I have denoted as *research groups* have been termed *research units*. The term *research team*, favored by journalists, may be appropriate in some cases, but hardly descriptive of the groups led by less competitive scientists. Among the many other recent writings on the social organization of modern scientific research are: Pelz and Andrews (1966), Ravetz (1971), Merton (1973), Mikulinskij et al. (1977, 1979), Elias et al. (1982), Jagtenberg (1983) and Ziman (1983).

[2] Much has been written on the reward system in science; see Cole and Cole (1967), Merton (1967), and Blume and Sinclair (1973). Special attention has been to the award of Nobel Prizes; see Crawford and Friedman (1982), Salomon-Bayet (1982) and Craw-

Whether the group leader was an autocrat, and alone reaped the accolades if his scientific judgment proved to be sound, or an unselfish adviser whose suggestions and continued guidance may have led some junior colleagues to early success, the most important qualities of the leader of a large research group have been his or her ability to choose significant problems within the discipline, and to channel the efforts of the group in a manner that encouraged its junior members to make useful contributions to the solution of these problems. The crucial component has been, therefore, the thread of a research program in which scientific problems were clearly formulated, at least in the mind of the leader, and from which the junior members of the group learned not only particular techniques, but also something of the art of scientific research. Consequently, the effectiveness of a research group should not be measured solely in terms of the significance of its contributions to the development of knowledge within its discipline, but also in terms of the extent to which the influence of the leader may be discerned in the later independent work of the junior members of the group.

These qualities of the leader acquire special importance in relation to what Thomas Kuhn aptly termed the "essential tension" between "flexibility and open-mindedness" which encourages new theoretical and experimental contributions that challenge accepted thought and practice in his discipline, and the tendency to develop a productive research program " . . . based firmly upon a settled consensus from scientific education and reinforced by subsequent life in the profession." Indeed, as Kuhn noted, " . . . only investigations firmly rooted in the contemporary scientific tradition are likely to break that tradition and give rise to a new one. . . . To do his job the scientist must undertake a complex set of intellectual and manipulative commitments. . . . Very often the successful scientist must simultaneously display the characteristics of the traditionalist and of the iconoclast." In a footnote, Kuhn added

Strictly speaking, it is the professional group rather that the individual scientist that must display both these characteristics simultaneously. . . . Within the group some individuals may be more traditionalistic, others more iconoclastic, and their contributions may differ accordingly. Yet

ford (1984a). It should be noted here that if the leader of a closely directed research group received most of the reward for success, he was also required to accept responsibility for any mistakes or falsifications in publications bearing his name.

education, institutional norms, and the nature of the job to be done will invariably combine to insure that all group members will, to a greater or lesser degree, be pulled in both directions.[3]

Although Kuhn may have intended to use the word *group* differently from my restricted definition of a research group, there can be no doubt that the "essential tension" has long been an important feature of the styles of the scientific leadership of relatively large research groups that were directed by a single senior scientist. Thus, in bringing together young people with different aims and talents, the modern research laboratory had become a place where traditional scientific knowledge acquired coherence but also invited skepticism, whether through (or in defiance of) the authority of the leader, or through his discourse with his research assistants and students, or through the social and intellectual interactions of the junior members themselves.

In this book, I attempt to provide some historical background to the emergence of research groups in the set of scientific disciplines that may be denoted the *biochemical sciences*. I define these sciences as those in which chemical ideas and techniques, as well as the physical principles on which they are based, have been used for research on biological problems, and also those in which problems in biology, medicine and agriculture have influenced the direction of chemical investigation. Various names have been given to the large area included in this definition. During the period 1800–1914, with the parallel development of chemical and biological thought and experimentation, what Berzelius first called *organic chemistry* became *animal chemistry* (as a counterpart to *plant chemistry*) or *medical chemistry*, which in turn was replaced by *physiological chemistry* and later *biochemistry* or *biological chemistry*. In German-speaking countries, the term *Biochemie* appeared as early as 1858, in a book by Kletzinsky, but the more common term was *physiologische Chemie* while in France it was *chimie biologique*.[4] More important than the names preferred at various times is the fact that, in the development of the biochemical sciences before 1914, among the leading investigators have been people who called themselves

[3] Kuhn (1963), pp.342-343.
[4] Kletzinsky (1858). Kohler (1973), p.183, has suggested that *physiological chemistry* referred to the chemical statics and *biochemistry* to the chemical dynamics of living organisms. This is incorrect. A more plausible reason for the wider use of *biochemistry* or *biological chemistry* during the early years of the twentieth century was the wish to emphasize its separation from medical physiology.

organic chemists, physical chemists, pharmacists, physiologists, pathologists, pharmacologists, bacteriologists, clinicians, zoologists or botanists.

In more recent years, increased specialization has brought such variants of *biochemistry* as *bio-organic chemistry* or *biophysical chemistry;* some people have considered the latter to be a branch of *biophysics* in the form of *molecular biophysics.* Other kinds of *biochemistry* that have appeared in print include *cellular biochemistry* and, as an exercise in redundancy, *molecular biochemistry.* There has also been a spate of names in which the noun was derived form biology, rather than from chemistry. During the first half of this century, *chemical physiology, general physiology* or *cell biology* (along with others) were preferred by many investigators, and after 1950 the fashionable terms became *molecular biology* and its offshoots *structural biology* and *molecular genetics,* the last of which has come to public attention as *genetic engineering.* Indeed, several branches of biology and medicine now have subdivisions identified as *molecular* (for example, *molecular pharmacology*). The introduction of new names to denote areas of science has been not only a consequence of their transformation through the accretion of knowledge but also, and often more importantly, an expression of the desire of like-minded investigators to assert their separation from a prevailing tradition. When some of them have won renown, as through the award of Nobel Prizes, their declaration of a distinctive status among the sciences has received wide acceptance. It may questioned, however, whether the theories, observations, experiments and methods in what I define as the biochemical sciences can be packaged according to academic customs or fashions, and whether such manifestations of academic rivalry among leaders of factions in the biochemical sciences should be accepted uncritically, no matter how interesting such rivalry may be as a subject of historical or sociological study. There can be no doubt that the interplay of chemistry and biology since 1800 has been a continuous historical process, and has led, in this century, to the emergence of the biochemical sciences as central and interdependent links in the search for the explanation, and means of control, of the phenomena of life in terms of the specific properties and interactions of the chemical substances of which living things are composed.[5]

In this book, the research groups that are considered in some detail were active during the period from about 1830 to

---

[5] The general history of modern biochemistry has been the subject of books and articles by Lieben (1935), Fruton (1972,1976), Florkin (1972,1975,1977,1979) and Leicester (1974), among others.

about 1914. As in the recent past, the interplay of chemistry and
biology at that time was marked by competition among attitudes
and approaches derived from different parts of these established
disciplines, and the styles of biochemical research often differed
sharply. There were tensions arising from the opinion of some
leading biologically minded chemists that the research of some
leading chemically minded biologists did not meet the standards
of the chemistry of their time. Examples that will be mentioned
in later chapters are Emil Fischer's views about the contributions
of Franz Hofmeister to the study of proteins, and those of
Svante Arrhenius about the work of Paul Ehrlich in immunology.
Moreover, there were tensions within the embryonic biochemical
sciences, as a consequence of differences in methodological approach; the case of Felix Hoppe-Seyler *versus* Willy Kühne, also
to be considered in a later chapter, reflected, in part, a contrast
in their attitudes toward the application of the developing organic chemistry of their time to important physiological problems. Also, at the turn of the century, tensions arose from the
response of some leading organic chemists and biochemists to
the emergence of physical chemistry as a distinct scientific specialty. In short, it may not be an exaggeration to suggest that
during the period from 1830 to 1914, and indeed for decades
afterward, the biochemical sciences were in a state of continuous
transformation, if not revolution. Although many of the fundamental biological problems, such as the nature of metabolic processes or the material basis of heredity, had been defined, new
chemical and physical ideas and methods brought new theories
and experimental approaches to the biochemical exploration of
these problems. These evoked different responses from leading
biologists and medical scientists, as in the contrast between the
attitudes of men such as Claude Bernard and Carl Ludwig to
the use of quantitative physical and chemical methods in the
study of physiological phenomena.[6] The resulting tensions were,
in many cases, transmitted to subsequent generations by the junior members of research groups led by the principal adversaries.
Because the problems tackled by those who sought chemical ex-
planations of important biological phenomena were solved only
partially, or not at all, there continued to be many disputes
among the later participants in the interplay of chemistry and
biology. If one takes into account the enormous increment of
biochemical knowledge during the past seventy-five years, and
the striking change in the institutional status of the biochemical

---

[6] See Mendelsohn (1964).

sciences, many features of the history of the earlier groups examined in this book may suggest that although this area of science and its place in society have been transformed, the heritage of past tensions has lingered into the present.[7]

The Scope of this Book

The first of the research groups to be considered in some detail is the one led by Justus Liebig in Giessen. Although much has been written about the relatively large-scale program of chemical laboratory instruction and research he conducted there during the 1830s and 1840s, Liebig's group merits further examination. This group was a forerunner of the many important German groups that arose in response to the wider public awareness of the industrial importance of chemistry. Also, Liebig's interest in the chemical aspects of the physiology of plants and animals has conferred upon him a special place in the historiography of the biochemical sciences. After the mid-nineteenth-century acceptance of the concepts of valence, structure and stereochemistry, the most notable of the many chemical research groups at German universities were those of Adolf Baeyer at Munich and of his pupil Emil Fischer in Berlin. These groups are also examined in some detail in this book. In parallel with the formation of the large chemical groups, several German universities became important centers of organized research in the medical sciences, especially at the institutes of Rudolf Virchow in Berlin (pathology), of Carl Ludwig in Leipzig (physiology), and of Oswald Schmiedeberg in Strassburg (pharmacology). It was from such institutes that there emerged, after 1850, future leaders of sizable biochemical groups, among them Felix Hoppe-Seyler and (his successor) Franz Hofmeister at Strassburg, as well as Willy Kühne at Heidelberg; these three research groups will also be discussed in some detail.

Several considerations influenced my decision to focus attention on the groups led by the chemists Liebig, Baeyer and Fischer, and the physiological chemists Hoppe-Seyler, Hofmeister and Kühne. Obviously, no account of the emergence of relatively large research groups in chemistry would be complete without a detailed study of Liebig's group at Giessen, although it may rightly be argued that the groups of his distinguished

---

[7] See Kornberg (1987).

contemporaries Jean Baptiste Dumas, Friedrich Wöhler and
Robert Wilhelm Bunsen merit no less attention. Thus far, only
the Dumas group has been subjected to thorough examination.[8]
The choice of the Baeyer group was dictated by the pre-eminent
role it played in bringing the new theoretical insights of mid-
nineteenth-century organic chemistry into the mainstream of
chemical research. Baeyer's group was not alone in this respect,
but his scientific progeny was especially important in the later
development of the biochemical sciences, with Fischer as its most
notable representative. As for the physiological chemists, the institutional status of their discipline in Germany and elsewhere
offered fewer opportunities to establish relatively large groups.
Apart from those led by Hoppe-Seyler, Hofmeister and Kühne,
there were only a few others, such as those of Marceli Nencki at
Berne and of Olof Hammarsten at Uppsala.[9]

A second reason for choosing the six groups was that they
were all at German universities. Although there were differences
among these institutions, during the period 1830–1914 their administrative structures and procedures were quite similar, thus
allowing more valid comparisons to be made between the styles
of leadership exhibited by the men listed above. Except at times
of political stress, there was a considerable degree of academic
freedom (*Lehrfreiheit*), but state officials controlled budgets and
faculty appointments, and authority in the individual university
institutes was vested in a single *ordentlicher Professor*. In the conduct of his program of research, his most important junior associates were his personal research assistants (*Privatassistenten*).
They received a salary from government funds assigned to the
head of the institute, or from his private purse. Among these
assistants were men who held the title *Privatdozent*, by virtue of
having met the requirement for the *Habilitation* (the publication
of a scientific paper, a lecture on the subject and an oral examination by appropriate full professors). This entitled the *Privatdozent* to lecture to students, from whom he was supposed to
derive fees, but he received no salary for this service. Consequently, unless a *Privatdozent* had adequate financial means, he
was obliged, while waiting for a call to a professorship, to accept
employment as a research assistant. Nor did a promotion to *ausserordentlicher Professor* always alter the *Privatdozent's* status so far

---

[8] See Klosterman (1985).
[9] For other biochemical research groups at universities in German-speaking countries,
see Eulner (1970). The book by Szwejcerowa and Groszyńska (1956) provides an essential starting point for a detailed study of the Nencki research group. The most valuable biography of Hammarsten is by Thunberg (1933).

as a direct government salary was concerned, although such promotion usually led to improved research facilities and better opportunities to conduct an independent research program with the help of predoctoral students.[10]

A third, and the most important, reason for selecting the six groups listed above for detailed study was that, during the latter half of the nineteenth century, the organization of research in the chemical and medical sciences at German university laboratories served as a model that was emulated in other countries and, as a consequence, these laboratories attracted many foreign research students. Moreover, in times of political crisis, as in the years immediately after 1848 or 1933, many of the German members of prominent research groups emigrated abroad to avoid persecution. Thus, in examining the later scientific activity of such a student in relation to his or her membership in a German research group, some appreciation may be gained of the influence of the leader of that group on the role of the student in the development of his discipline in his home or adopted country.

In attempting to evaluate the effectiveness of each of the six research groups, the first requisite was the compilation of a list of its junior members. The purpose of such a list was not only to determine the numerical size and national composition of the group, but also to ascertain what contribution (if any) each of the members made to the research activities of the laboratory. In particular, it was necessary to identify the advanced students, research assistants and guests who actually worked in the laboratory, and to weed out the individuals who only attended lectures, did only elementary practical work or only imbibed the social and intellectual atmosphere of the laboratory (as well as the beverages of the region). As was the custom, many students matriculated at several German universities before receiving the Dr.phil. or Dr.med. degree at one of them. For the purposes of this study, the decisive criterion was whether the student had done the research for that degree as a junior member of the research group in question. Moreover, there were matriculants who already held an advanced degree, but do not appear to

---

[10] For the travail of the *Privatdozent*, see Busch (1959,1963). For an incisive account of the assistantship system in pre-World War II Germany, see Bock (1972). A valuable account of German student life has been prepared by Jarausch (1982). For a detailed compilation of data on the changes in the composition and social origins of the faculties of German universities, see Ferber (1956). From these sources it is clear why German scientists were largely drawn from the upper middle class, since the sons of titled families tended to enter military or civilian government service and, owing to the cost and uncertainty of the road to a university professorship, it was closed to sons of workers and small farmers.

have participated actively in the research program of the laboratory. Some of these individuals, especially those who achieved distinction, may have described themselves as pupils of the leader of the research group, or may have been so described by their biographers but, with a few exceptions, I did not include them in the lists that I compiled. Since somewhat different problems were encountered in the case of each list, specific details of the procedure in compiling them will be given in the individual chapters. At this point, to reduce later repetition, one general feature of the procedure may be noted. Although, in most instances, the role the co-worker was indicated by the appearance of his or her name as a co-author (or sole author) of one or more scientific publications or of a doctoral dissertation, often his contribution is only evident from an acknowledgment in a footnote or at the end of a paper by the group leader. Consequently, in addition to the use of published lists of publications from the laboratory of the group leader, and of standard bibliographic reference works (*Royal Society Catalogue of Scientific Papers, Chemisches Zentralblatt, Chemical Abstracts, Index Medicus* and others), it was also necessary to examine each of the published papers of the group leader and those of his principal junior associates.

In order to assess the influence of the group leader on the later professional activity of his junior associates, an effort was made to collect biographical data about all the individuals in the six lists.[11] It was not difficult to find such data for those who had established themselves in academic life or at least had continued to contribute articles to scientific, medical or technical journals, but those whose names disappeared from the literature presented greater problems. In this search I was greatly aided by many archivists at universities, chemical firms and municipal archives. The data are collected in a series of appendices, and were used for the preparation of tables that are presented in text. I did not succeed in finding the needed biographical information for all the persons in the lists, but I believe that this shortcoming is not sufficiently great to invalidate the use of the lists for the purposes of this inquiry.

One of these purposes was to ask whether there is a connection between the scientific renown of the leader of a research group and his role in the education of future leaders in the chemical, biological and medical sciences. Such a connection is implied in the widely reproduced genealogical table (or modifi-

---

[11] See Fruton (1982b, 1985a).

cations thereof) which depicts Liebig as the primogenitor of a scientific family tree, with emphasis on later winners of Nobel Prizes.[12] Apart from confirming the received opinion that eminent leaders in fashionable fields of science have attracted many talented students, such selective genealogy cannot be accepted at face value. In view of the large number of outstanding (now deceased) twentieth-century chemists, biologists, biochemists and medical scientists who did not receive Nobel Prizes, this criterion is not an adequate measure of relative scientific distinction in the biochemical sciences. If one wishes to gauge the role of a group leader in the later success of his scientific progeny, detailed examination of each research group is needed not only in terms of its social attributes and national composition but, more importantly, through the informed consideration of such matters as the research program of the group, the relationship of the leader to his junior associates, and the nature of their later scientific contributions to the development of their discipline.

Consequently, in this book the numerical data derived from the membership lists for the six groups only provide a background for the discussion of the scientific program of the leader and of the later independent research of his students. In recently fashionable parlance among historians of science, this inquiry has, because of its concern with aspects of the scientific life previously studied by sociologists, some of the stamp of "external" history. In its purpose, however, it cannot be separated from the search for better understanding of the "internal" development of the interplay of the chemical and biological sciences. In an earlier book,[13] I attempted to describe this interplay in terms of the historical record of theories, observations, experiments and methods, and only brief attention was paid to the personal qualities, social interactions or institutional status of the participants. In the present volume, the balance is shifted, but it would be idle to ask how the research groups were led, what the junior members did in the way of laboratory research and what success they achieved afterward, without describing, however briefly, the nature of the scientific problems studied by the groups, the theories that guided their research and the methods used in their experiments. Fuller treatments of these matters may be found in books and articles cited in the text.

---

[12] Among the appearances of this table (or variants thereof) are those in Sachtleben (1958), Dechend (1963), opposite p.33, Krebs (1967), Steiner (1977), p.138 and Fischer (1987), pp. viii-ix.

[13] Fruton (1972).

A Personal Note

To conclude this introductory chapter, I beg the reader's indulgence for some words about my reasons for wishing to do this book, and about the bias I brought to its preparation. I have been, in turn, a predoctoral student in biochemistry at Columbia University, a junior member of a chemical research group at the Rockefeller Institute for Medical Research and, at Yale, a teacher of biochemistry to undergraduate, graduate and medical students, and the leader of a biochemical research group. This is not the place for an account of my activities as a scientist; some aspects of my career have been presented elsewhere.[14] However, since this book purports to be a contribution to the history of science, it may perhaps be appropriate to offer an *apologia* for my intrusion into this field.

My interest in the history of science began during my student days at Columbia College, and developed into a hobby whose chief expression was at first the assembly of a sizable collection of historical material on the development of biochemistry. After I came to Yale in 1945, these efforts received a major impetus from the friendship of John Farquhar Fulton [1899-1960], who was largely responsible for the establishment, at the Yale School of Medicine, of the Historical Library, which I have used continuously to this day. My education in the history of science was extended under the tutelage of Fulton's successor, Lloyd Grenfell Stevenson [1918-1988] and of his junior associate Frederic Lawrence Holmes, who later returned to Yale as Professor of the History of Medicine. These men helped me to appreciate more fully the values of historical scholarship, and encouraged my efforts to write about the past development of the biochemical sciences.

As an amateur intruder, I know that professional historians of science have regarded with suspicion scientists who enter their field, and I have taken to heart the stern stricture of George Sarton, that tireless advocate of professionalism in the history of science, who took a dim view of the scientist "who has become sufficiently interested in the genesis of his knowledge to wish to investigate it, but has no idea whatever how such investigations should be conducted and is not even aware of his short-comings. . . . He generally lacks the humility of the beginner, and publishes his results with blind and fatuous

---

[14] Fruton (1982a, 1988b).

assurance."[15] It is regrettable, but true, that most working scientists have not regarded highly the search for historical understanding. Their concentration on the doing of science, and on gaining fame (perhaps even wealth), has usually been accompanied by indifference to the historical process in which they and their predecessors have been participants because they have realized that the knowledge of the history of one's discipline will not make one a more successful scientist. Also, it is undeniable that many great scientific discoveries were made by young people whose minds were not yet cluttered with the tortuous history of their field. When, near the end of a productive life, a scientist has attempted to become a historian, all too often his writings have been accounts of an unbroken series of successes or, in Butterfield's phrase, Whig history. If he has published reminiscences and reflections, too often such writings have been self-serving efforts to re-assert reputation. Also, the laudatory biographical articles and books written by admiring colleagues too often have been acts of homage, and occasionally hagiography, rather than critical evaluations of the hero's place in the history of his science. Such abuses of history are not unlike past mythology about saints, kings and warriors, not only to provide shining examples of lives worthy of emulation, but also to justify the existing religious, social and political order. In the case of modern science, the correction of such myths requires the historian, whether professional or amateur, to get his biographical facts as correct as the available record will permit and to search assiduously for additional reliable sources of such facts. Accurate information is needed not only about the names, dates and careers of individual scientists or about their social and institutional relationships, but also what they wrote about their scientific work in notebooks and professional journals or told others in letters and in recorded conversations. Such studies may lead to a re-assessment of the reputations of famous scientists as well as of those whose work had not received acclaim. Indeed, the historical books and articles written by scientists and tainted with Whiggery and hagiography are part of the historical record, for they are useful not only in providing information that needs to be checked for its historical and scientific accuracy, but also in revealing how some scientists preferred to view themselves and their colleagues, and how they perceived the role that they and their predecessors had played in the development of their discipline.

---

[15] Sarton (1938), p.469. See also Reingold (1981) and Whitaker (1984).

Although there is likely to be little disagreement about these generally accepted methods of historical inquiry, some present-day historians of science consider the social relations of science to be of primary importance and disparage efforts they describe as "internalist," "vertical" or "parochial." While I consider it essential to study the social relationships of scientists to one another as leaders and junior members of research groups, as collaborators or competitors, as the personnel of universities and research institutes, or as recipients of private or state funds for their research, I believe that at the core of any historical analysis of the development of a modern scientific discipline is the painstaking, informed and critical examination of the successive theories, observations, experiments and methods that have marked that development, including those that were later considered to be incorrect or unreliable. Many leading professional historians of modern science hold this view, and have not sought to evade the necessity of learning the scientific content of a historical problem. Indeed, how else except through a close examination of the internal history of a branch of modern science can the externalist decide whether an individual scientist, or a particular research group, discipline or institution, is of sufficient importance to be worthy of closer attention? How else can he or she learn of the work of people who achieved relatively little public fame, but whose contributions were important in the historical development of the scientific problem they studied? No doubt it is possible to write books about the social relationships among scientists without much regard to what they have done (or not done) as scientific investigators.[16] Whatever the merit of such books might be, they are not about science as a distinctive human activity, any more that a book on the relation of painters to their patrons, to art critics, to gallery owners or to museum curators would constitute a history of painting. In short, I submit that although "internal" history does not suffice to illuminate the scientific enterprise, without it the illumination is likely to be superficial and uncertain.

Nor have professional historians of modern science been immune to the tendency to see the advance of science as the result of the work and thought of a relatively few prominent individuals. This approach is implicit in the planning of such important reference works as the *Dictionary of Scientific Biography*, and is entirely justified in the case of a distinguished scientist who worked alone, or was the chief experimenter in a small research

---

[16] For the biochemical sciences, such a book is the one by Kohler (1982).

group. Although we take for granted the proposition that significant advances in the chemical and medical sciences are now made only rarely by such individuals, but rather by relatively large research groups, the criteria used by historians in the evaluation of the importance of the past leaders of such groups in the development of their discipline are often the same as those applied to the lone investigators. As a consequence, with a few notable exceptions,[17] more attention has been given to the leaders of the closely directed groups than to leaders of groups in which the junior members were encouraged to develop their talents as independent investigators. One of the aims of this book is to suggest that the exclusive focus on the research achievements attributed to the leaders of large groups has tended to obscure the fact that research groups have produced not only new scientific knowledge but also the succeeding generations of scientists. Also, by enlarging the scope of historical inquiry to include people who, as junior members of important research groups, made only a brief, though fruitful, appearance on the scientific scene, this book may help to call attention to some of the many forgotten individuals who contributed significantly to the development of the biochemical sciences, and to invite further examination of their place in the history of these sciences.

In offering this book for the consideration of colleagues in the history of the sciences, I realize that my approach to historical problems may be different from theirs. I am comforted, however, by the thought that their approaches are by no means uniform either in philosophy or in method, and that their discipline has been no less beset by tensions than the sciences whose history they have studied.[18] Although I have learned much from many historians, philosophers and sociologists, my view of the aims and values of the history of science is colored by my personal experience as a scientist. Consequently, in acknowledging the historical fallacy of "presentism," and in attempting to avoid it, I freely confess that I see the past of the biochemical sciences in terms of their present state of development, and of my hopes for their future.

---

[17] The most thorough recent historical studies of individual research groups have been those of Morrell (1972), Geison (1978), Snelders (1984) and Klosterman (1985).
[18] See Kragh (1987).

Chapter Two

The Liebig Research Group
in Giessen

Much has been written about Justus von Liebig (as he was known after his ennoblement in 1845), and the exalted place assigned to him by many historians of science reflects, in large measure, the extraordinary public esteem he won during his lifetime through his semi-popular scientific books.[1] His *Familiar Letters on Chemistry*,[2] first published in English (the first German edition appeared in 1844), went through fifty-one editions in these and nine other languages. To his admirers, Liebig was one of the greatest men of the nineteenth century[3]; among his scientific contemporaries, however, there were many notable dissenters. Before 1840, when he was thirty-seven years old, he had established himself as one of the leading research chemists of his time, through his great improvement of the method for the determination of the carbon-hydrogen content of organic compounds, as well as his important work on derivatives of ethanol,

---

[1] An extensive list of biographical books of articles about Liebig, published before about 1968, is available in Poggendorff (1971). The most important book is that of Liebig's family friend and scientific assistant Volhard (1909). The valuable book by Paolini (1968) contains references to nearly all of Liebig's personal publications, as well as a list of the then-known Liebig letters, memorial articles and selected items of secondary critical literature. More recent writings about Liebig have been discussed by Brock (1981).

[2] Liebig (1843, 1844). The chemical letters began to appear in the *Augsburger Allgemeine Zeitung* in 1841.

[3] The notice in Hermann Kolbe's *Journal für praktische Chemie* of Liebig's death in 1873 read: "Death has taken from us one of the greatest men of this century. On 18. April Justus von Liebig died in Munich in his not fully completed seventieth year of life after a brief illness involving an inflammation of the lungs."

the constitution of organic acids and (with Friedrich Wöhler) the benzoyl radical, amygdalin and uric acid. During the late 1830s, however, his interest began to turn to the applications of chemistry in agriculture and medicine; although his facile speculations stimulated new research efforts, many of them were roundly criticized by leading biologists.[4] First (in 1839) he wrote about fermentation, putrefaction and moldering (*Verwesung*), then (in 1840) on agriculture and plant physiology and (in 1842) on animal physiology and pathology.[5] By the mid-1840s, at the height of his public fame, Liebig had ceased to be a significant participant in the advances in organic chemistry that would soon lead to the concepts of structure, valence and stereochemistry. Liebig quarreled with other chemists whose work and writings he criticized severely in the pages of his *Annalen der Chemie und der Pharmacie;* there are also accounts of his readiness to mend the breaks and of the warmth and generosity of his friendships. Whatever his scientific achievements or shortcomings may have been, or whatever his personal faults or virtues, there can be little doubt that Liebig was unique among the German chemists of his time in two important respects: he was a persuasive propagandist for the institutional advancement of his discipline,[6] and an effective entrepreneur in the organization of his program of chemical studies at Giessen.[7]

In the writings about Liebig, one finds statements such as: "It is well-recognized that at the small provincial University of Giessen, in the 1820s, Justus Liebig initiated a mode of training chemists that eventually served as the principal model from which the modern teaching-research laboratory has

---

[4] There developed considerable tension between Liebig and several leading biologists. Thus, in 1844 Liebig wrote that "in chemical-physiological work, physiology is not threatened by chemists but by physiologists and physicians" (Liebig, 1874, p. 81), and four years later Emil du Bois-Reymond wrote about Liebig that "I consider his physiological fantasies to be worthless and pernicious"; see du Bois-Reymond (1927), p.19 and Cranefield (1982), p.14.

[5] Liebig's interest in fermentation was signaled by the appearance of an anonymous article (Anon., 1839), satirizing the claim of Theodor Schwann, Charles Cagniard-Latour, Friedrich Kützing and Pierre Turpin that alcoholic fermentation is caused by living organisms; see Fruton (1972), pp.44–51. Liebig's first publication on fermentation appeared in the following volume of his *Annalen* (Liebig, 1839). The first German editions of his books on agriculture and on animal chemistry were entitled *Die organische Chemie in ihrer Anwendung auf Agricultur und Physiologie* (Braunschweig 1840) and *Die organische Chemie in ihrer Anwendung auf Physiologie und Pathologie* (Braunschweig 1842) respectively.

[6] See Drössmar (1964) and Turner (1982). Fischer (1955) has compared Liebig with Wilhelm Ostwald as propagandists for the institutional advancement of their respective branches of chemistry.

[7] Detailed accounts of Liebig's program of elementary instruction have been presented by Conrad (1985) and Holmes (1986b).

descended."[8] Liebig's "students, both brilliant and mediocre, were enabled to produce reliable chemical knowledge in a systematic way on a large scale. In this sense, the era of Big Science began in Giessen in the early 1840s."[9] "Almost all the leading chemists of recent times count themselves with pride among the students of Liebig. . . . Only to have been in Giessen was already considered as a recommendation."[10] To these quotations may be added the widely reproduced genealogical table which depicts Liebig as the primogenitor of a scientific family tree, with special emphasis on later winners of Nobel Prizes.[11]

In this chapter, I attempt to evaluate Liebig's role in Giessen as the leader of a chemical research group. During the years between 1824, when he came there at the age of twenty-one, until 1852, when he went to Munich, over 700 students matriculated at Giessen in chemistry or pharmacy, the number per academic year varying between 10 and 50. In addition, matriculants in other subjects (philosophy, natural science, medicine) attended Liebig's lectures, and a sizable number of persons who worked in his laboratory are not included in the published matriculation lists. Whatever the exact total may have been, it is certain that Liebig and his assistants taught very many students, some of whom went beyond elementary lectures and laboratory work, engaged in research, received a Dr.phil. degree and were authors of one or more papers in Liebig's journal. Nor can there be any doubt that although university-level laboratory instruction in chemistry had been initiated in Europe before Liebig's advent in Giessen,[12] Liebig's entrepreneurial talent led to the establishment of a course of practical studies in which more students were taught chemical laboratory techniques at a single university than ever before. However, the statement that they "experienced systematic preparation for chemical research, and . . . were deliberately groomed for membership in a highly effective research school"[13] needs to be examined more carefully. According to Morrell, the considerable manpower was engaged in a research program centered upon Liebig's improvements of

---

[8] Holmes (1986b), p.1. I am grateful to Professor Holmes for a copy of this lecture, and for his permission to quote from it. A revised version has appeared in Holmes (1989).
[9] Morrell (1972), p.28.
[10] Roth (1898), p.172. Thomson (1885) stated that "all the eminent chemists who were young in 1845 were pupils of Liebig."
[11] See Chapter One, note 12
[12] See Morrell (1972) and Gustin (1975).
[13] Morrell (1972), p.2.

the analytical techniques for the determination of the elementary composition of organic compounds and thus permitted students

who were less than brilliant to do and to publish competent work. When these techniques were deployed on a large scale a knowledge factory was the likely result. This was characterized by the steady and systematic production of reliable experimental results by ordinary students whose scientific mediocrity had been converted into scientific competence by the acquisition of these very techniques.[14]

If this description of Liebig's research program is correct, one must ask how such routine analyses affected the abler, more imaginative, students who may have aspired to be research chemists. Did it repel them, lead them to reject Liebig's influence and to seek scientific stimulation elsewhere, or did it only train them to be highly skilled analysts and possibly inventors of new and improved techniques?

Questions also may be raised with regard to the size of Liebig's research group. There can be little doubt that Liebig's chemical achievements before 1840 and his later public renown were decisive factors in attracting many students to his laboratory. After 1839, when additional space became available, practical instruction could be given to about 30 students at a time, and in 1843 Liebig set up annex for 15 more students who worked under the direction of his senior assistant, Heinrich Will. Nor can it be questioned that Liebig's lectures on "Experimental Chemistry" provided enjoyment and stimulation to many students. As a consequence, the available laboratory spaces were readily filled, and the manpower (there is no record of female matriculants) was, by the standards of the time, quite large. But is the size of a scientific research group a valid measure of its effectiveness either in the production of significant scientific knowledge or of the next generation of scientific leaders within its discipline? Clearly, much depends not only on the qualities of the leader and of his junior associates but also on the state of theory and practice in the discipline, as well as on the opportunities for later productive research. Thus, after 1870, when organic chemistry had acquired a fruitful theoretical basis, and many hands were needed to test alternative routes to the syn-

[14] Ibid., p.5.

thesis of organic compounds, there was (in Germany, at least) an expansion of university chemical facilities, and a growing research-minded chemical industry was providing jobs for the chemical Dr.phil.'s who emerged in ever-larger numbers from German university laboratories. During the period 1830–1850, however, the theoretical state of organic chemistry was in flux and the institutional facilities for large-scale productive chemical research were more limited. One may ask therefore whether the size of Liebig's educational venture in Giessen may not have been excessive for its time. It would be idle to speculate whether Liebig would have been an even more significant contributor to the nineteenth-century development of organic chemistry if he had run a smaller show and not withdrawn from the field after 1840. Perhaps one should view the initiation of his educational effort in Giessen as an early manifestation of a drive for pre-eminence, with its later expression in a desire to supplant Berzelius as the father figure of chemistry, to be editor of the leading German chemical journal, to reform chemical education in Prussia and Austria, to put down his scientific rivals (especially the French ones) and to seek to apply the chemical knowledge of his time to the problems of agriculture and medicine, with ready proposals for their solution. It is small wonder that Liebig has fascinated his biographers, although his complex personality does not yet appear to have been fully probed. I shall not attempt to do so here, except to suggest that if he had striven for less, he might have achieved more.

Liebig's Research Program

Before 1840, Liebig's research was in the increasingly turbulent mainstream of debate on important questions of chemical theory. His careful analysis (1826) of salts of fulminic acid and cyanic acid helped resolve uncertainty about isomerism in organic chemistry. His studies during the early 1830s on derivatives of ethanol and (jointly with Wöhler) on benzaldehyde ("oil of bitter almonds") represented a major development of the organic radical theory. Later in the 1830s, their benzoyl radical theory and the work of Thomas Graham [1805–1869] on acids led Liebig to propose his "hydracid" theory, with a flurry of papers from his laboratory on the constitution of fatty acids. He also joined Wöhler in the study of the decomposition products of uric acid.

Liebig's attack on these problems was greatly aided by his "five-bulb" apparatus (1831) for carbon-hydrogen determination.[15] Analyses could be performed more rapidly and reliably, and required less chemical skill, than with the assemblies introduced by Gay-Lussac and Thenard (1810), Berzelius (1813) and Döbereiner (1816). Liebig's method was especially valuable for the continuing business of the separation of pure substances from gross mixtures, an activity in which Carl Wilhelm Scheele [1742–1786] had been pre-eminent in the eighteenth century, and which was pursued intensively by pharmaceutical chemists during the first three decades of the nineteenth century. The close tie that Liebig maintained to pharmacy, especially during his early years in Giessen, is evident not only from his service to the profession,[16] but also from the many analyses performed in his laboratory on preparations of plant alkaloids, pigments, oils and camphors. Moreover, many of Liebig's disputes arose from uncertainty about published analytical data, and the disagreements often stemmed from questions as to the purity of the material that had been subjected to analysis. Even after about 1840, when Liebig's principal research interest had shifted away from pure organic chemistry, he continued to criticize analyses published by others, and had them checked by his junior associates. Also, there were echoes in the student assignments of Liebig's earlier research on such topics as the derivatives of ethanol, the constitution of fatty acids or the nitric acid oxidation of uric acid, as well as investigations of minerals, double salts and crude industrial products.

It must be stressed, however, that the main activity of Liebig's students and assistants during 1840–1850 was to provide analytical data that Liebig wanted for his writings on fermentation, agricultural chemistry and animal physiology. Thus, his views on the role of minerals (especially phosphates) in plant nutrition obliged many students to perform ash analyses of a large variety of woody tissues. Likewise, his effort to determine the balance between the food intake and excretory output of animals led to many student reports in the *Annalen* on the analysis of dietary proteins, as well as of urine and feces. Also, Liebig's celebrated dispute with Mulder about the latter's concept of a

---

[15] The most comprehensive available account of the early history of elementary organic analysis is that by Dennstedt (1899). The subject is less well treated in the otherwise excellent book by Szabadváry (1966); for a perceptive review of this book and a valuable brief summary of the place of analytical chemistry in the development of chemical thought, see Brock (1967). See also Hickel (1979).

[16] Liebig's tie to pharmacy has been discussed by Brand (1931), Eberhard (1938) and Gustin (1975).

sulfur-free "protein" produced a series of analytical papers by
Liebig's assistants and students. However unattractive these tasks
may have been, Liebig considered the results to be important for
his theories of animal metabolism.[17] His approach, based on Lavoisier's analogy of respiration to combustion, led Liebig to the
view that muscular contraction involves protein breakdown. This
was soon disproved, but other aspects of Liebig's theories stimulated important later efforts by German and American physiologists to study animal metabolism by means of the quantitative
analysis of ingested and excreted materials and the determination of the oxygen inhaled and carbon dioxide exhaled.[18] Moreover, Liebig's emphasis on the role of muscle led not only to the
later marketing of his famous *Fleischextrakt* but also to important
studies in the Giessen laboratory on muscle lactic acid, his interest in the composition of bile to the identification of glycocholic
acid, and his views on proteins to the isolation of tyrosine.

In considering the research output of the students, assistants and guests at the Giessen laboratory during the 1840s it
should also be noted that Liebig's chief junior associate, Heinrich Will, had begun to direct the work of others. One of Will's
first contributions (with Varrentrapp) was a useful method for
the determination of the nitrogen content of organic compounds. Later work done under his direction included research
in pure organic chemistry; among the noted *Liebig-Schüler* who
did not do their Dr.phil. work with Liebig, but with Will, were
Kekulé, Erlenmeyer and Planta.

This brief sketch of Liebig's research program during 1830–
1850 may perhaps suffice to indicate that the emphasis on routine quantitative analysis in the work of the younger members of
his group was not the program itself, but a necessary adjunct to
it. Indeed, Liebig's program resembled in several respects that
of Dumas, the other rising star of organic chemistry during
the 1820s. Between 1832 and 1848, Dumas led a research group

---

[17] During the 1840s Liebig's students and assistants were required to perform routine
ash analyses of foodstuffs, wine, blood, urine and feces for his *Chemische Briefe*. In the
preface to the third edition (Heidelberg 1851), pp.vii–viii, he wrote:
> Letters 27 and 28 contain research of the past six years in the fields of animal
> chemistry and physiology. The reported analyses were performed under my close
> scrutiny by several able young chemists, Dr. Verdeil from Lausanne, Porter, Dr.
> Breed and Johnson from New York, Zedeler from Copenhagen, Lehmann from
> Dresden, Dr. Keller from Würzburg, Dr. Griepenkerl, now Professor in Göttingen,
> Dr. Stölzel in Heidelberg, Stammer from Luxembourg, Dr. Henneberg, Buchner
> and Kekulé from Darmstadt, Mr. Arzbächer, and my two assistants Dr. Strecker and
> Dr. Fleitmann, to whom I express herewith my thanks for their help.
An examination of the text of the two chapters shows that they also report analyses by
Will, Fresenius, Weber, Poleck and Geiger.
[18] See Lusk (1928), Fruton (1972), Ihde (1974) and Holmes (1987, 1988).

far smaller than the one in Giessen (Klosterman[19] lists 29 names), but no less important in the generation of new chemical knowledge or the breeding of later productive scientists. Like Liebig, Dumas made a significant contribution to analytical chemistry; his method of nitrogen determination, with later improvements, was lasting in its utility, and his research students in Paris were also well trained in other analytical techniques, including Liebig's carbon-hydrogen method.

The similarities between the research programs of Dumas and Liebig, and the successes both men achieved, are striking, and support the view that during 1830–1850 the size of an organic-chemical research group did not determine its effectiveness. Among the elements of Dumas's program was work on the "ether" theory, which paralleled Liebig's development of the organic radical theory. Liebig's work on the chlorination of ethanol was contemporaneous with Dumas's chlorination of acetic acid, and contributed to the formulation of Dumas's important substitution theory (1834). Dumas's type theory (1840), partially anticipated by Laurent and later fully developed by Gerhardt, led Dumas to study polybasic acids, whose constitution was also a part of Liebig's program during the late 1830s and, like Liebig's students, the Dumas group searched for new fatty acids in oils and waxes. And in 1841, soon after the appearance of Liebig's writings on fermentation and on plant nutrition, and before the publication of Liebig's book on animal chemistry, Dumas ventured into print on these subjects, basing his views on the researches of his colleague Boussingault as well as on many routine analyses, such as those by his student Cahours on dietary proteins. It is evident that, despite the great disparity in the laboratory manpower available to Dumas and Liebig, Dumas's dependence on a large mass of analytical data was no less than that of Liebig in the conduct of their respective research programs.

This congruence of research interest led to bitter priority disputes, briefly suspended in 1837, but resumed with greater intensity and nationalistic fervor soon afterward, except for another short-lived reconciliation in 1850. It may also be noted that, like Liebig, Dumas won the greatest public acclaim for his writings on agricultural and physiological chemistry. Moreover, before he was fifty years old, Dumas ceased to be an active participant in the development of organic chemistry, became a politician, served as Minister of Agriculture (1850–1851), was made

---

[19] See Klosterman (1985). An excellent account of the competition between Liebig and Dumas has been provided by Holmes (1973).

a senator and was influential in the professional advancement of his former students.

Who Should Be Included in the Liebig Research Group?

There are available two published lists of matriculants in chemistry and pharmacy at Giessen during the first half of the nineteenth century. In 1965–1969, Armin Wankmüller assembled a list of such students from 1800 to 1852; it appeared in installments during 1981–1982.[20] Previously, Wankmüller had published a list of Liebig's foreign students.[21] The other general list, which includes both matriculations and class inscriptions in all subjects from 1807 to 1850, was assembled by Franz Kössler,[22] who had also published lists of doctoral awards and dissertations at Giessen from 1801 to 1884.[23]

In the present study, the above sources were used for the preparation of a list of Liebig students at Giessen during about 1830–1850. These dates were chosen because before 1830 the matriculants were almost entirely in pharmacy, and because the available information about those who came after 1850 was incomplete. As was mentioned by Morrell,[24] there are uncertainties in this procedure. On comparing the Wankmüller and

---

[20] Wankmüller (1981–1982). The last installment of this series lists students through the winter semester 1850/51, but also has a name index that includes matriculants after that date. The published list goes up to No. 691 (Wankmüller's numbering); there are some duplicate entries, and regrettably Nos. 533–561 were omitted between the third and fourth installments. Also, the name index is not complete (it ends with WEIS). I am indebted to Dr. Wankmüller for sending me copies of the relevant issues of his valuable journal, as well as much other helpful information and advice.

[21] Wankmüller (1967). A modified form of this list also appeared in the book by Rossiter (1975), pp.184–190.

[22] Kössler (1976).

[23] Kössler (1970, 1971). It should be noted that many names in the list of Dr.phil. awards are absent from the list of doctoral dissertations, indicating that the submission of a dissertation was not required for the degree at Giessen during 1830–1860. Moreover, the list of Dr.phil. degrees includes the names of chemists who may not have worked in Liebig's laboratory. An example is Warren De La Rue [1815–1889] who received the degree in 1851. Although Anschütz (1929), vol.1, p.39, refers to De La Rue as a *Liebig-Schüler*, I have been unable to find in biographical writings or bibliographical sources any indication of De La Rue's association with Liebig in Giessen. They met in London during one of Liebig's visits there, and perhaps the degree was an honorary one, conferred when Liebig was dean of the philosophical faculty; see Gundel (1973). One's opinion of the criteria applied at that time in the award of Giessen Dr.phil. degrees cannot but be influenced by the fact that the degree was given in 1853 to the Englishman Albert James Bernays, who had matriculated in chemistry in 1841, for the second edition of his book *Household Chemistry: or the Rudiments of the Science Applied to Every-day Life.* I do not know whether Liebig had approved this award (he had left Giessen in 1852).

[24] Morrell (1972), p.17.

Kössler lists of matriculants, I found some names in one and not in the other, as well as many differences (and inaccuracies) in the spelling of names. Also, as I learned from other sources, some individuals studied chemistry at Giessen during Liebig's time there, but do not appear in either list. Consequently, I cannot claim accuracy or completeness for the list that I have assembled and used for a closer examination of the Liebig research group, but hope that, subject to future correction, this list may be acceptable as a basis for the conclusions offered in this chapter.[25]

The list has 718 names. For the purposes of this study it was desirable to separate them into four groups, corresponding to the time periods 1830–1835, 1836–1840, 1841–1845 and 1846–1850. In nearly all cases, the assignment was based on the reported date of initial matriculation at Giessen; for the 31 names not included in the available matriculation lists, the assignment was based on the award dates of Dr.phil. degrees, or on biographical information found in other sources. This division is arbitrary, since some of the people continued their studies at Giessen, or held assistantships there, during more than one of the four time periods mentioned above. Despite the obvious shortcomings of this procedure, I believe it to be necessary for an evaluation of the research conducted in Liebig's laboratory before 1840, when he was still productive as an organic chemist, and after 1840, when he ventured into agriculture and animal physiology.

It will be seen from Table 2-1 that over 75 percent of the matriculants in chemistry at Giessen during 1830-1850 had come after 1840. By that time, the number in chemistry had begun to exceed those in pharmacy but, as noted by Wankmüller,[21] many young men who later became successful pharmacists were enrolling in the chemistry program. Also, some students who had initially enrolled in other subjects, but attended Liebig's lectures, were attracted to chemistry, and a few later received a Dr.phil. degree for a chemical dissertation. Among them were August Wilhelm Hofmann, who began at Giessen in 1836 as a student of law, and August Kekulé, who enrolled in 1847 as a student of architecture.

Obviously, a listing in this book of all the 718 names is impracticable, and indeed unnecessary, for most of the Liebig stu-

---

[25] My uncertainty about the completeness and accuracy of the list also stems from the fact that Volhard (1909, pp.96,121) mentions Troger and Kemp, for whom I could find no record of matriculation, a degree, a publication or any other biographical data.

TABLE 2–1
Students of Chemistry and Pharmacy at Giessen 1830–1850

	1830–35	1836–40	1841–45	1846–50	Total
Initial Matriculation					
Chemistry	15	75	174	143	407
Pharmacy	53	63	74	62	252
Natural Science	1	1			2
Medicine	2		1	2	5
Philosophy	1	1	9	5	16
Public Affairs			1	1	2
Law		1		1	2
Architecture				1	1
Not in Matriculation Lists	4	7	11	9	31
TOTAL	76	148	270	224	718
Dr.phil. (Giessen) awarded to students of chem. or pharm.	10	12	49	70*	141
Postdoctoral persons in Liebig's laboratory		5	18	6	29
Persons from other than German or Austrian states					
United Kingdom	2	20	36	25	83
Switzerland	1	11	11	15	38
France	6	9	8	4	27
United States			2	14	16
Russia		2	8	3	13
Poland	1		1		2
Italy			2	3	5
Norway		1	1		2
Denmark				1	1
Netherlands			2		2
Belgium			1		1
Luxembourg			1	1	2
Spain			1		1
Mexico		1			1
TOTAL	10	44	74	66	194

*Includes degrees awarded after 1850.

dents did not proceed with their chemical studies beyond lectures and practical work at the elementary level. I therefore sought to identify the individuals who may be considered to have been junior members of Liebig's research group, as evidenced by their chemical publications or other documentary records. To begin with, I examined volumes 1-6 of the *Royal Cat-*

alogue of Scientific Papers, which cover the period 1800-1863, to learn what papers, if any, had appeared under the names of the people in the list, and whether these papers were based on work done in Liebig's laboratory.[26] Since it was his practice (at least after about 1833) to publish such papers in his journal, I examined the appropriate volumes of the *Annalen* for explicit statements by Liebig or by the author that the work had been done at Giessen. This procedure allowed a winnowing of the list and (not surprisingly) led to elimination of nearly all of the 252 people who had initially matriculated in pharmacy. Many of them were sons of pharmacists and, for those about whom I was able to find biographical data, succeeded their fathers in the ownership of the family pharmacy.[27] A significant number, however, aspired higher, switched to medicine and became physicians.[28] Of special interest for the present study are the few pharmacy students who did advanced work that led to chemical papers; they will be mentioned later in this chapter.

The proportion of the initial matriculants in chemistry who were authors of papers based on work done in the Liebig laboratory is considerably greater than that for the pharmacy students, but the list could be winnowed further after an examination of the *Royal Society Catalogue* and the available biographical data. Apart from the chemistry students who went on to become pharmacists, there was an assortment of matriculants who do not appear to have done publishable work at Giessen. Among them were scions of German titled families, as well as sons of Alsatian industrialists and British landowners, for whom a brief stay in Giessen was probably part of a pleasurable Grand

[26] The assumption that the *Royal Society Catalogue* includes citations of all the journal articles from Liebig's laboratory may have led to the omission of some papers in periodicals not covered by the *Catalogue.* I found the coverage of Liebig's *Annalen* to be complete, although not free of error. For example, the reference to a 1833 paper by Blanchet and Sell only gives Blanchet's name.

[27] It is perhaps worth noting that in 1836 the eight prizes that Liebig awarded to those of his students who had taken a special examination went to a medical student and seven pharmacists, one of whom (Ricker) received the first prize ("a laboratory knife with a platinum blade and trimmed with silver and palladium"). Apart from what the establishment of these prizes suggests about Liebig's enterprise as a teacher, the fact that he published the news of the awards in his *Annalen* indicates something of his talent as a publicity agent for his program at Giessen. See Liebig (1836).

[28] In preparing Table 1, only those Giessen medical students during 1830–1850 were included for whom there was evidence of their participation in Liebig's advanced program of chemical instruction. From Kössler's list it appears that during this period about 560 students had matriculated in medicine, surgery or veterinary medicine. It was customary in German universities for medical students to attend lectures in elementary chemistry, and at some institutions (notably Berlin) such students represented the bulk of the enrollment in these courses. Around 1840, however, few German medical students received practical chemical instruction, largely because the laboratory facilities were insufficient.

Tour before taking up their family responsibilities. Also, as was the custom, many students attended courses at several German universities before receiving the Dr.phil. degree at one of them. Consequently, the absence of a publication from the Giessen laboratory does not preclude the possibility that the student obtained the degree elsewhere. Finally, there were matriculants in chemistry who already held a Dr.phil. or Dr.med. degree, and published no papers from the Liebig laboratory; this group includes several men who subsequently gained some distinction. In later life, this varied assortment of individuals may have taken pride in having been pupils of Liebig, but since they do not appear to have contributed to the research output of his laboratory, I did not include them in his research group.

I have used the available biographical and bibliographical information to separate the *Liebig-Schüler* into two classes, with two sub-classes in each. Class I includes the students, guests and scientific assistants whose work in Liebig's laboratory led to creditable publications and who either (a) continued afterward to publish work in pure or applied chemistry (including physiological chemistry) or (b) made few, if any, further contributions in these fields because of early death or a change in professional interests. In Class II were put individuals whose stay in Liebig's laboratory led, at the most, to one or two brief reports of routine analyses and who later either (a) published worthwhile scientific papers or (b) made only slight, if any, further contributions to the scientific literature. Clearly, there are uncertainties in such an arbitrary division, based in large part on my judgment of what constitutes a "creditable" or "worthwhile" scientific paper in the context of the state of organic and physiological chemistry during 1830–1850 and in the years afterward.

In Appendix 1 are listed the 59 names (capital letters) in Class Ia and the 89 names in Class Ib; the 12 names in Class IIa (capital letters) and some (188) of those in Class IIb are listed in Appendix 2. The remaining 370 persons in Class IIb are not listed.

An examination of the list in Appendix 1 suggests little correlation between the extent or quality of the contributions of Liebig's research students or assistants to his research program and their later achievements in organic or physiological chemistry. Among those who came to Giessen before 1841, 45 (about 20 percent of the total known enrollment in chemistry and pharmacy) appear to have done worthwhile published work: Blanchet, Böckmann, Bromeis, H. Buff, Campbell, Clemm-Lennig, Crasso, Demarçay, Ettling, Fehling, Fownes, Gay-Lussac,

Gregory, Hofmann, W. Keller, Knapp, Kodweiss, Kopp, Kosmann, Marignac, Nöllner, Oppermann, Otto, Plantamour, Playfair, Redtenbacher, Richardson, Regnault, Riegel, Schoedler, Scherer, Schunck, Sell, Stein, Stenhouse, A. Strecker, Thaulow, R.D. Thomson, T. Thomson, Tilley, Varrentrapp, Voskressensky, Will, Winkelblech, Zinin. It is noteworthy that only slightly more than half of this group came from German or Austrian states. Of these 45 men, 13 (H. Buff, Fehling, Fownes, Hofmann, Kopp, Marignac, Redtenbacher, Regnault, Scherer, Schunk, A. Strecker, Will, Zinin) may be considered to have gained a high scientific reputation in their later professional careers. Three of them (H. Buff, Marignac, Regnault) left organic chemistry soon after their work in Liebig's laboratory, but the others continued to make important contributions that gave evidence of his influence. Charles Gerhardt, who turned out to have been the most distinguished organic chemist among the *Liebig-Schüler* before 1841, is listed in Appendix 1 but is not included with the above 45 men because he does not appear to have done any published experimental work at Giessen; more will be said later in this chapter about his relationship to Liebig.

After 1840, the number of matriculants and guests was much greater than in the preceding decade, but the proportion of those (102) who were authors of papers based on work at Giessen remained about the same (21 percent) as before. Thirty-six of them are listed in Appendix 1 in capital letters: Anderson, Babo, Buchner, Büchner, Erlenmeyer, Frankland, Fresenius, Genth, Gladstone, Griepenkerl, Henneberg, Horsford, Jones, Kane, Kekulé, Khodnev, Krocker, Matthiessen, Melsens, J.S. Muspratt, Nicklès, Penny, Pettenkofer, Planta, Rochleder, Rowney, Sacc, Sandberger, Schlossberger, C. Schmidt, J.L. Smith, Sobrero, Sokolov, Wetherill, Williamson, Wurtz. The presence of so many foreigners (22) in this group is of special interest. It should also be noted that several of these men were in Giessen only briefly, and made rather small experimental contributions to Liebig's research program. Thus, it may be more appropriate to consider Frankland, Genth and Matthiessen as having been members of Bunsen's research group at Marburg or Heidelberg, while Melsens and Wurtz belong more properly to Dumas's group in Paris.

A striking feature of the list of 36 men is the large proportion of those who later made contributions to agricultural chemistry (Anderson, Griepenkerl, Henneberg, Horsford, Krocker, Sacc, Wetherill) or to physiological chemistry (Buchner, Jones, Khodnev, Pettenkofer, Schlossberger, C. Schmidt), thus reflecting Liebig's interest in these fields. So far as organic chemistry was concerned, only a few (Büchner, Rochleder, Rowney, Sobrero, Sokolov) continued to work along traditional lines, mainly on the constitution of natural products, and the men later associated with the great advances in organic chemical theory broke decisively with Liebig's adherence to the views he had held since the 1830s. In this respect, the *Liebig-Schüler* who stands out most prominently is August Kekulé; because Kekulé's beginnings as a chemist provide a revealing glimpse of Liebig's role as the leader of his research group, special attention will be given later in this chapter to their relationship. The other men who later became noted organic chemists include Erlenmeyer (who owed more to Kekulé than to Liebig), Frankland (see above), Williamson and Wurtz (both of whom were more influenced by Dumas and Gerhardt). Although Liebig's group included many men who later made chemical analysis their profession, only Fresenius stands out among them, but several distinguished mineralogists (Genth, Sandberger, J.L. Smith) also reflect the emphasis on analytical chemistry during their studies at Giessen in the 1840s.

The lack of correlation between the later distinction of many of Liebig's scientific progeny and the nature of the work they did in his laboratory after 1840 is evident from the data collected in Appendix 1. Noted men such as Fehling, Hofmann, Kopp, Playfair, Redtenbacher and Will, all of whom first matriculated during the 1830s, did valuable experimental work in the Giessen laboratory, and so did some of their less well-known contemporaries (for example, Blanchet, Crasso, Ettling, Kodweiss, Plantamour, Thaulow). After about 1840, however, with only a few prominent exceptions (notably A. Strecker), most of the valuable chemical papers came from the work of men such as Bensch, Bopp, Heldt, Marsson, Merck, Schlieper, Unger and Wolff who disappeared from the later chemical literature because of early death,[29] or entry into other pursuits, especially the chemical industry.

The list in Appendix 2 is included in part because it calls attention to some of the other notable figures in nineteenth-century science who were with Liebig during 1830–1850, but

---

[29] Among the promising Liebig students who died at an early age was Friedrich Bopp, whose work on the hydrolysis of casein led him to discover tyrosine. He died soon afterward, on 12 November 1849, following his participation in the Baden revolution; see Haupt (1934). Unaccountably, in the recently published edition of Liebig's correspondence with August Wilhelm Hofmann (Brock, 1984), there are references in Liebig's letters of 8 December 1849 and 17 January 1850 to a living "Friedrich" who is identified by the editor as Friedrich Bopp. Other Liebig students who died young were Benckiser, the Englishmen Fownes and Tilley and the Norwegian Thaulow.

who do not appear to have participated in his research program. Among them were the important biochemists Ludwig Thudichum and Moritz Traube, who may have gained stimulus for their later achievements from their studies in Giessen. Of particular interest in relation to Liebig's writings on agriculture is the presence of Joseph Henry Gilbert who, together with John Bennet Lawes [1814–1900], later did decisive field experiments to test Liebig's ideas on soil nutrition. Also included in Appendix 2 are the chemists Oliver Wolcott Gibbs, William Allen Miller, Robert Angus Smith and Ludwig Ferdinand Wilhelmy, the physicist Fabian Carl Ottokar von Feilitzsch, the mineralogist Hermann von Trautschold and the biologists Willem Berlin, Francis Trevelyan Buckland and Karl Vogt. The many other names in Appendix 2 are included to indicate the variety of the later occupations (pharmacy, industry, medicine, etc.) of Liebig students whose stay at Giessen led to no publications (or at the most a report of a few routine analyses), and who do not appear to have done any chemical research afterward.[30]

Although the data in hand do not allow me to define with reasonable precision the total size of Liebig's research group during about 1830–1850, it would seem that it was between 120 and 150, with the preponderance (about 70 percent) having matriculated after 1840. Also, the portion of the group that had matriculated before 1841, although smaller in number, appears to have made more contributions of a creditable nature than the one that had enrolled afterward. It may suggested, therefore, that when Liebig's research program was in the mainstream of the organic chemistry of his time, but when he had a smaller research group, his students and assistants produced more valuable scientific knowledge than during the 1840s, when most of them were obliged to provide analytical data for Liebig's writings on agriculture and animal chemistry, or for his disputes with other chemists about the accuracy of published analyses.

In considering Liebig's effectiveness as the leader of his research group, it is not sufficient to list the names of those of his pupils whose later work may have been influenced by their association with Liebig, and who attained a measure of scientific distinction or high academic posts (or both). I have therefore

---

[30] Volhard (1909, pp.84–85) listed 64 notable chemists who had been students in the Giessen laboratory. The list is not free of error; for example, John Hall and Gladstone are given as two persons. Another list of distinguished former Giessen students appeared in Kolbe (1874), p.455; it includes 31 names, with the notable omission of Hofmann and Kekulé, among others.

TABLE 2–2
Principal Subsequent Activity of Liebig Pupils*

Initial Matriculation

	1830–35	1836–40	1841–45	1846–50	Total
Full (*ord*) prof. chem.					
University	F1	G7/F11	G5/F11	G1/F6	G13/F29
Higher techn. school	G1/F1	G4	G3/F3	G4	G12/F4
Non-univ. med. school		F1	F1	F2	F4
Full (*ord*) prof. techn. & pharm. chem.	G2	F1	G2/F1	G1	G5/F2
Other academic ranks in pure or applied chem.	G2	G2	G2/F4	G1/F1	G7/F5
Univ. prof. other subjects	G3/F1	G1/F1	G5/F3	G2/F1	G11/F6
Private research lab.		F2	G3/F3	G1/F1	G4/F6
Medical practice	G5	G1/F3	G2/F2	G5	G13/F5
Pharmacy	G11	G17/F3	G29/F2	G29/F5	G86/F10
Industrial manufacture	G4	G7/F4	G16/F4	G12/F6	G39/F14
Commercial or consulting chemists		G1/F2	G2/F4	G3/F3	G6/F9
Govt. techn. depts.		F1	F1	G4/F1	G4/F3
Other occupations (forestry, politics, landed gentry, etc.)	G1/F2	G3/F1	G6/F3	F9	G10/F15

*G = German or Austrian; F = Foreign (non-German or Austrian)

sought to determine the later professional activity, if any, of the 718 men in my complete list, but was able to find the needed biographical information for only 324 (45 percent). The data are summarized in Table 2-2. There is little doubt that the numbers for the men who entered academic life approach the actual totals more closely than the numbers for the other professional categories.

Although the statement[10] that nearly all of the leading chemists of the latter part of the nineteenth century had been students of Liebig is an exaggeration, it is significant that at least 13 later became full professors of chemistry at German or Austrian universities (Babo, Erlenmeyer, Hofmann, Hruschauer, Kekulé, Kopp, Redtenbacher, Rochleder, Scherer, Schlossberger, C. Schmidt, A. Strecker, Will), and that a similar number gained this rank at higher technical schools in Germany.[31] Of particular

[31] In 1864, Liebig (1874, p.37) recalled that

interest is the relatively large number of Liebig students from abroad who attained university professorships in chemistry; the British contingent was the largest, and included Anderson, Blyth, Brodie, Fownes, Frankland, Gregory, Kane, Miller, Playfair, Ronalds, Rowney and Williamson. More will be said later in this chapter about Liebig's foreign students and about his special relationship to the British scientific and industrial community. As is indicated in Table 2-2, 17 former Liebig pupils became university professors in disciplines other than chemistry (physics, geology, hygiene, agriculture, etc.).

The 99 men whom I have identified as having pursued academic careers represent about 14 percent of the estimated total during 1830–1850 and, as noted above, the ready availability of biographical data for this professional category in Table 2-2 makes it reasonable to consider this percentage as approaching the true figure. It is otherwise for the two largest other categories in Table 2-2, namely pharmacy (96) and industrial manufacture (53), which represent 30 and 17 percent respectively of the 324 men for whom I have found the needed biographical information. The limited data on the later professional activity of the men in the non-academic categories do not justify the application of these percentages to the total enrollment, except to suggest that the true figures are higher than the ones given above.

Despite the inadequacy of the data in Table 2-2, they allow the tentative conclusion that the main educational function of Liebig's laboratory throughout the period 1830–1850 was the training of future pharmacists and industrial chemists. Although some of them did advanced work, and published papers in the *Annalen*, nearly all their names disappeared from the scientific literature after they left that laboratory. In historical accounts of Liebig's program of instruction, little is said about this large group, probably representing over one-half of the total enrollment, and emphasis is placed on those who attained academic status or other public distinction. It is therefore necessary to note that, contrary to the statement by Morrell, the "scientific mediocrity" of the majority of Liebig's "ordinary students" had not been transformed into "scientific competence" by the acquisition of the analytical techniques taught at Giessen. There were

in 1838 and 1840 I discussed the state of chemistry in Austria and Prussia, and my unfavorable descriptions were accorded a remarkably different reception in the two countries. In the eyes of the advisers on educational matters in Berlin at that time I had committed a political crime, and the penalty was that for many years none of my students was appointed to a university in Prussia. It was the opposite in Austria: for many years hardly any candidate received a chemical teaching post in Austria who had not been educated at Giessen or had not completed his studies there.

exceptions, of course, but they do not weaken the impression that the main social function of Liebig's educational program was not appreciably different from that of the programs at other German universities, namely to produce chemical technicians for professions allied to medicine and industry.[32]

Liebig's Style of Leadership

In accounts of the manner in which Liebig developed and led his research group at Giessen, historians of science have frequently drawn upon his autobiographical sketch.[33] An oft-quoted passage is:

Actual instruction in the laboratory, with experienced assistants in charge, applied only to the beginners; my special students learned only in proportion to what they brought, I gave the assignments and supervised the execution; like the radii of a circle everybody had a common center. There was no actual guidance; every morning I received from each one a report on what he had done the day before as well as his views on what he intended to do; I either agreed or made my objections, everyone was obliged to seek his own way. In the companionship and constant intercourse, and in which each one participated in the work of all, everyone learned from the others.... We worked from daybreak until nightfall, distractions and diversions were not available in Giessen.... The memory of their stay in Giessen awakens in most of my pupils, as I have often heard, the pleasant sense of satisfaction for a well-spent time.

It is difficult to reconcile Liebig's description with the published reports of some of his pupils. According to Volhard, "If Liebig gave a student or an assistant an assignment, he sat, as the saying goes, on his neck [so sass er ihm auf dem Nacken]. His impatience could hardly be contained until the result was in hand."[34]

---

[32] For excellent discussions of the chemical instruction in mid-nineteenth-century Germany of future physicians, pharmacists and industrial personnel, see Borscheid (1976) and Burchardt (1980), as well as Turner (1982) and Gustin (1975).

[33] Liebig's autobiographical notes, probably written in about 1871, were later found by his son Georg (Liebig, 1890). They were reprinted in the *Deutsche Rundschau* 66 (1891):30–39 and in Dechend (1963), pp.13–26. An English translation by J. Campbell Brown appeared in *Chemical News* (1891):265–267,276–278. The quotation in the text is on p.828 of the first-named reference. My translation differs from those of others, and I have kept as much of Liebig's punctuation as possible.

[34] Volhard (1909), vol.1, pp.98–99.

This statement is followed by a personal recollection of Guckelberger, successively a student and assistant of Liebig during 1845–1849. To this glimpse may be added another, from the 1830s. Carl Vogt, who matriculated as a medical student in 1833, and later became a noted zoologist, was greatly attracted to Liebig and worked in his laboratory during 1834–1835. In his autobiography, Vogt wrote the Liebig "drove us incessantly: you have filtered? you have washed the precipitate? Not yet? Meanwhile you can begin another analysis! Wernekinck has given me tachylite—he wants to know whether it really contains uranium. Do this at once! *[Machen Sie sich gleich dahinter!]*."[35] These fragmentary recollections suggest that Liebig's behavior toward his assistants and "special" students may have been less liberal than was indicated in his autobiographical sketch. Moreover, the evidence they provide of his haste, impatience and dictatorial attitude is more consistent with the qualities he displayed in his public life than with those suggested in his own account.

In at least one respect Liebig was more generous toward his assistants and advanced students than was customary in nineteenth-century German chemical laboratories, by encouraging them to publish the results of some of their work in his *Annalen* under their own names. Although an examination of most of these papers reveals little more than reports of analytical data, without discussion of their significance, the appearance in print of such publications must have been a source of pleasure to the authors, apart from their possible value to Liebig in advertising the productivity of his laboratory. One's admiration of this generosity is somewhat dampened, however, by Vogt's recollection that

we were soon drawn into our master's work while at the same time he insisted that we must carry on our own. He said: "Nothing enthuses young people more than to see their names in print. The French have an entirely wrong system. Everything that is done in Paris or the provinces must go out into the world under the name of the professor. This discourages young people, apart from the fact that the professor must answer for mistakes for which he is not responsible. This Dumas, he has to defend things of which his assistants are guilty. The people who

[35] Vogt (1896), p.126. Friedrich Wernekinck [1798–1835] was professor of mineralogy at Giessen. An English translation of the section of Vogt's book dealing with Liebig appeared in the article by Good (1936).

work with me publish under their own names, even if I have helped
them. If it is something good, a part of the credit is ascribed to me and
I do not need to defend the mistakes. You understand?[36]

It would seem, therefore, that the acceptance accorded Liebig's
description of his style of leadership should be reconsidered.
Also, one should perhaps treat with reserve adulations such
those of Kolbe:

Liebig was not a teacher in the ordinary sense; to an exceptional degree productive in science and rich in chemical ideas, he shared them
with his mature students, called upon them to test his ideas experimentally, and so stirred them to think for themselves, showed them the
way and taught them the methods to solve chemical questions and
problems by means of experiment.[37]

In the light of the foregoing account of the activities of Liebig's
students both at Giessen and afterward a more appropriate picture of his role would appear to be the one offered by Williamson. He described the Giessen laboratory as "the most efficient
organization for the promotion of chemistry which had ever existed . . . a little community of which each member was fired
with enthusiasm for learning by the genius of the great master,
and of which the best energies were concentrated on the one
subject of scientific investigation."[38]

During 1830–1850, there were other German university laboratories where advanced chemistry students belonged to a research group led by a noted professor. Among these laboratories
were those of Friedrich Wöhler [1800–1882] at Göttingen and of
Robert Wilhelm Bunsen [1811–1899] at Marburg. The personal
qualities of these two men were rather different from those of
Liebig. A brief discussion of their respective styles of leadership
may therefore be of interest in connection with the theme of this
chapter.

Wöhler and Bunsen

Wöhler became *ordentlicher Professor* of Chemistry and Pharmacy
in the medical faculty at Göttingen in 1836, after having re-

[36] Vogt (1896), p.129.
[37] Kolbe (1874), p.442.
[38] Divers (1907), p.xxvi.

ceived his Dr.med. degree at Heidelberg in 1823, studied with Berzelius (1823–1824), and taught at the technical schools (*Gewerbeschulen*) in Berlin (1825–1832) and Kassel (1832–1836). In Berlin, he effected the famous conversion of ammonium cyanate to urea and also made an important contribution to the preparation of pure aluminum. This interest in both organic and inorganic chemistry continued during the 1830s when he collaborated with Liebig. By the 1840s, their scientific paths had parted, and Wöhler's personal research was increasingly devoted to problems in inorganic chemistry.[39]

A detailed study of Wöhler's program of advanced chemical instruction at Göttingen does not appear to be available. It would seem, however, that during 1843–1850 the number of laboratory students per semester was 40–50, of whom 10–14 had matriculated in pharmacy. This is similar to the enrollment at Giessen in the mid-1840s. The number at Göttingen rose to 60–70 (13–16 pharmacy students) during 1851–1855, and the proportion of advanced students to beginners reached 4:1.[40] Before 1852, there were relatively few foreign chemistry students, but after Liebig's departure from Giessen many more went to Göttingen, with a sizable contingent from the United States.[41]

In contrast to the impulsive and combative Liebig, driven by ambition for public fame, Wöhler had a stable and self-contained outlook on life, as well as an unusual measure of modesty and human kindness. Their friendship survived many trials, including Liebig's arrogation of sole credit for some of their joint research. In 1840, Wöhler was impelled to write his friend:

If in the citation of work that we have both done together only one of us is named, and especially in a journal in which both are named on the title page, about which everyone knows that you are the actual editor, and this editor allows that to happen and does not show the slight-

---

[39] See Valentin (1949) and Poggendorff (1971), pp.779–783. Much has been written about Wöhler's synthesis of urea, especially in 1928 (the centenary of his report); for critical accounts, see Lipman (1964) and Brooke (1968). The significance of Wöhler's work on aluminum has been ably summarized in the published portion of the Dr.phil. dissertation of Kunzmann (1930), pp.16–31.

[40] See Ganss (1937).

[41] See Van Klooster (1944). According to Jones (1987), it was only after 1851 that American or British students received Dr.phil. degrees at Göttingen for work in Wöhler's laboratory, and it is noteworthy that this group included 22 Americans and only 4 Britons.

est consideration to report it, then everyone will conclude that this represents an agreement between us, that the work is yours alone, and that I am a jackass.[42]

Also, Wöhler was unsuccessful in his repeated attempts to dissuade his friend from publishing in the *Annalen* Liebig's harsh criticisms of other chemists, especially Berzelius.

After his collaboration with Liebig, most of Wöhler's important contributions were in inorganic chemistry (titanium, silicon, calcium carbide, etc.), although some of the 170 papers he published during 1840–1877 dealt with organic compounds. Nearly all of these publications reported the results of his personal research; the experimental work of his many advanced students and teaching assistants was published under their own names. Although he suggested research problems, or allowed his junior associates to propose them, and helped in their work, he encouraged independent effort. A striking feature of Wöhler's own papers and those of his research group is the absence of theoretical speculation, a quality that carried over into the later writings of his scientific progeny. Among the sizable number who became full professors of chemistry were: Adolf Eduard Arppe [1818–1894], Friedrich Konrad Beilstein [1838–1906], Rudolf Fittig [1835–1910], Anton Geuther [1833–1889], Charles Anthony Goessmann [1827–1910], Hermann Kolbe [1818–1884], Heinrich Limpricht [1827–1909], Ira Remsen [1846–1927] and Georg Staedeler [1821–1871]. It is noteworthy that none of these men later specialized in inorganic chemistry; except for Goessman, who became an agricultural chemist in the United States, the others made their principal contributions in pure organic chemistry.

To conclude this brief sketch of Wöhler's style of leadership, as compared with that of Liebig, mention should be made of Wöhler's attitude to physiological chemistry, in which he had a longstanding interest. At Göttingen, as at other universities, instruction in this subject was under the jurisdiction of the professor of physiology but, in 1853, Wöhler was asked whether he would add it to his professorial responsibilities. He declined, and wrote to the university administration:

Physiological chemistry, a specialty that is not represented in Göttingen at present, is a discipline that has emerged only recently. . . . It is the chemical part of physiology, the science of the composition and forma-

---

[42] Hofmann (1888), vol.1, p.166.

tion of the substances that constitute the blood and organs of the animal body, the science of metabolism or the chemical processes whereby
nutrients are converted into the constituents of the various organs,
muscle, nerves, brain, etc., and whereby these are decomposed into excretory products.... The student learns these general relationships,
which belong, or should belong, to the basic knowledge of the physician, partly in lectures on general chemistry and partly in those on
general physiology. However, the field has become so large and so difficult that it requires of the teacher such special chemical and physiological knowledge, that neither the chemist, whose science has already
branched in many directions, nor the physiologist who has barely mas-
tered it, can do more than handle it in only fragmentary fashion. Whoever wishes to teach this subject thoroughly and successfully and to
contribute to its advances through his own researches must—it is now
clear—make it his sole lifework.[43]

As it turned out, although Wöhler's students Georg Staedeler
[1821–1871] and Carl Bödeker [1815–1895] were in charge of
teaching physiological chemistry at Göttingen during the 1850s,
the successive professors of physiology soon reasserted their control, and it was not until 1939 that the subject gained a fully
independent status there.

The other contemporary of Liebig who, during the 1840s,
attracted many aspiring young chemists was Bunsen.[44] He had
come to Marburg in 1838, after having taught at the technical
school in Kassel for two years; he remained in Marburg until
1851, when he moved to Breslau for a year before going to
Heidelberg as the successor of Leopold Gmelin.[45] The first of
his many notable scientific contributions was the investigation
during 1837–1843 of organic arsenic compounds, the so-called
cacodyl series. In this work, Bunsen showed not only exceptional
experimental skill, but also great courage, as many of these compounds are highly toxic and explosive. Bunsen's painstaking and
systematic study provided the strongest support up to that time
for the Berzelius-Liebig radical theory of the constitution of organic compounds. This investigation, however, was Bunsen's
only major effort in organic chemistry, for he then demonstrated his virtuosity in developing new methods of gasometric

---

[43] Ganss (1937), p.68.
[44] An excellent biography of Bunsen was prepared by Lockemann (1949). Among the
many obituary notices and other articles about Bunsen, those by Roscoe (1900),
Meyer (1900) and Lotze (1986) are especially noteworthy. See also Poggendorff
(1971), pp.121–126.
[45] Meinel (1978), pp.26–43. See also Debus (1901), Curtius (1906) and Rheinboldt
(1950).

analysis (1838–1846), in improving the galvanic battery and in
the use of electrochemical techniques for the preparation of
pure metals (1841–1855). These successes were followed by his
famous work with Henry Roscoe on photochemistry (1852–
1862) and with Gustav Kirchhoff on spectrochemical analysis
(1859–1875).[46] All these (and other) achievements gave evidence
of Bunsen's originality, ingenuity in the construction of scientific
instruments, analytical precision and a remarkable ability to
overcome experimental difficulties.

Bunsen and Liebig were men of entirely different make-up.
This was appreciated by Liebig's own students; in an obituary
notice for Friedrich Knapp (a pupil and brother-in-law of Liebig) it is stated:

There was a lively traffic of the younger Giessen chemists to Marburg;
they traveled as often as possible to visit the much admired Bunsen. As
Knapp reported, "Liebig and Bunsen did not understand each other
particularly well, although there was of course mutual respect. Each
went his own way, and Bunsen was a thoroughly original character, he
was no-one's pupil, and therefore was not likely to be attracted to Liebig. But we younger men had an immense admiration for Bunsen, and
I can well say that he was for me a most attractive person, perhaps the
most charming one that I had known. He spoke beautifully, so that
just to hear his voice gave me pleasure. At first he was not communi-
cative, rather reticent; but as acquaintance grew, he became intimate.[47]

To this statement should be added Roscoe's assessment:

Considerate and generous toward the opinions of others, he held
firmly to his own views, which at times he did not fail strongly to express. Simple and straightforward, he disliked assumption and hated
duplicity; single-minded and wholly devoted to his science, he ab-
horred vanity and despised publicity-hunting. Indeed, of so retiring
disposition was he, that it was difficult to get him to take part in public
proceedings, and next to impossible to induce him make any public
utterance of either a scientific or social character.[48]

---

[46] See Danzer (1972) and James (1985).
[47] Meyer (1904), pp.4781–4782. Liebig's attitude toward Bunsen may have been influenced by the fact that in 1838 he had written to the Prorector at Marburg to oppose
the appointment of Bunsen, as it involved the displacement of Winkelblech, a former
Liebig student; see Lockemann (1949), pp.72–73.
[48] Roscoe (1900), p.542.

At Marburg, and later at Heidelberg, Bunsen's lectures were much admired by his students. According to John Tyndall, who enrolled at Marburg in 1848:

Bunsen was a man of fine presence, tall, handsome, courteous, and without a trace of affectation or pedantry. He merged himself in his subject; his exposition was lucid, and his language pure; he spoke with the clear Hannoverian accent which is so pleasant to English ears; he was every inch a gentleman. After some experience of my own, I still look back on Bunsen as the nearest approach to my ideal of a university teacher.[49]

A detailed study of the composition of Bunsen's laboratory group at Marburg has not come to my attention. Apparently, between 1845 and 1850 there were, per semester, more than 30 practical students in chemistry, medicine, pharmacy or technology.[50] During Bunsen's later stay at Heidelberg, there were, on the average, about 50 such students. According to Lockemann,

Bunsen occupied himself very actively with the beginners. He showed them each manipulation and everything involved in an analysis. He strove to train the young people to be self-reliant. He also watched the more advanced students carefully and gave them good advice for their research. He warned them not to study the literature too thoroughly, and instead urged them to depend on their own observations.[51]

Meinel has stated that Bunsen never founded a "research school," and that even those of his students who had received their *Habilitation* with him did not remain in Marburg very long. It is not clear what Meinel meant by "research school," but the data provided in his valuable book[52] indicate that between 1840 and 1851 there were sixty-four publications from the Marburg chemical laboratory; forty of them bore Bunsen's name (one was a joint paper with Playfair) and the others were published under the sole authorship of his junior colleagues. Fifteen chemical Dr.phil. dissertations were submitted during this period. In the

[49] Quoted from J. Tyndall, *New Fragments* (1892), in Eve and Creasey (1945), p.23.
[50] R. Schmitz (1978), p.236. See also Meinel (1978), pp.471–472.
[51] Lockemann (1949), p.79.
[52] Meinel (1978), pp.38,480,520–523.

light of the testimony of his former assistants and students, it is probable that Bunsen encouraged his abler junior associates to embark on independent careers as soon as possible. Among them were Hermann Kolbe [1818–1884], his successor at Marburg, Heinrich Debus [1824–1915], John Tyndall [1820–1893], Edward Frankland, Friedrich August Genth and Constantin Zwenger; the last two came to Bunsen after a brief stay with Liebig. During his longer stay at Heidelberg, Bunsen's junior associates included Adolf Baeyer [1835–1917], Hans Landolt [1831–1910], Dmitri Ivanovich Mendeleev [1834–1907], Lothar Meyer [1830–1895] and Ludwig Mond [1839–1909], along with many other notable chemists of the late nineteenth century.

A prominent feature of Bunsen's scientific style was his attitude toward the chemical debates of his time. For example, even during the 1870s he did not discuss in his lectures the periodic table proposed by his former students Mendeleev and Meyer. Although he attended the 1860 Karlsruhe Congress (Liebig and Wöhler were absent), "The subject under discussion was not one in which he took much interest, and he frequently said that one new chemical fact, even an unimportant one, is worth a whole congress of discussion of matters of theory."[53] When Kolbe became editor of the *Journal für praktische Chemie* in 1870, he carried his teacher's attitude to an extreme, and even exceeded Liebig in his harsh treatment of proponents of new theories in organic chemistry.[54] It would seem, therefore, that although Liebig, Wöhler and Bunsen all discouraged chemical speculation among their junior associates, only Liebig indulged in it freely, and thus may have stimulated some of his abler students, such as Gerhardt, Kekulé and Williamson, to disregard his advice and even to formulate theories that conflicted with his views. Wöhler, on the other hand, provided no such stimulus, and Bunsen's broad tolerance allowed his students (especially in Heidelberg) to be exposed to the theories taught by the new younger leaders of organic chemistry, notably August Kekulé [1829–1896].[55]

[53] Roscoe (1906), p.68.
[54] For Kolbe as an editor, see Phillips (1966) and Rocke (1987).
[55] Bunsen's attitude toward organic chemistry has been discussed perceptively by Bodenstein (1936).

Liebig and Kekulé

In the scientific family tree stemming from Liebig, mentioned in the previous chapter,[56] the largest branch leads from August Kekulé, principally through Adolf Baeyer. It would seem, however, that Liebig did not appreciate Kekulé's chemical talent.[57] After hearing Liebig's lectures in 1848, Kekulé decided to switch from architecture to chemistry, which he studied at Giessen from 1849 until 1851. He did his practical work in analytical chemistry under the direction of Liebig's teaching assistant Theodor Fleitmann, who later reported that Liebig had reprimanded him for spending too much time with Kekulé.[58] As an advanced student in 1850, Kekulé worked in the annex under the supervision of Henrich Will on amyloxysulfates, and it was only during the winter semester 1850/51 that he was accepted into Liebig's private laboratory. There he was assigned ash analyses and the analysis of gluten and wheat bran; the data appeared in the third German edition of *Chemische Briefe*.[59] Many years later, Kekulé recalled that in 1851 Liebig said to him: "Go to Paris, you will enlarge your horizon, you will learn a new language, you will come to know a great city, but you will not learn chemistry there."[60] Indeed, it was in Paris, during 1851–1852, that Kekulé met Charles Gerhardt, and thereby enlarged his view of chemistry. When Liebig decided to go to Munich, he did not invite Kekulé to accompany him; instead he recommended Kekulé to his former students Fehling and Planta, both of whom were seeking a laboratory assistant. Kekulé chose the latter, a wealthy Swiss landowner who had set up a private laboratory in Reichenau. There Kekulé worked during 1852–1853 on the chemistry of nicotine and coniine and also performed analyses of minerals and mineral waters.

In 1853 Liebig suggested that Kekulé should become a teacher at an agricultural school, but as he wished to remain in chemical research, Kekulé accepted an assistantship with John Stenhouse (another former Liebig pupil) at St. Bartholomew's Hospital in London. Liebig had made the recommendation, but

---

[56] See Chapter I, note 12.
[57] The most important biography of Kekulé is still the two-volume work by Anschütz (1929). See also Japp (1898) and Hafner (1979). Other valuable articles on Kekulé's chemical contributions are those by Fisher (1974), by Rocke (1981), and by Wotiz and Rudofsky (1987).
[58] Anschütz (1929), vol.1, p.14.
[59] See note 17, this chapter.
[60] Anschütz (1929), vol.2, p.949.

did not write Kekulé about the terms of the appointment; instead he had his teaching assistant Wilhelm Mayer do so. As Kekulé later recalled,

I had little inclination to accept it because, if I may be permitted the expression I considered him [Stenhouse] a *Schmierchemiker*. By chance, Bunsen came to Chur to visit a brother-in-law whom I had come to know. I asked him about Stenhouse's offer, and he advised me strongly to accept it. I would learn a new language, but I would not learn chemistry. I therefore went to London, where I did not profit much from my assistantship with Stenhouse.[61]

As it turned out, Kekulé's stay in London (1853–1854), apart from routine analyses for Stenhouse, led to valuable papers on sulfur-containing acids and, most importantly, to his acquaintance with Alexander Williamson, William Odling and Edward Frankland. It was a stormy time in the history of organic chemistry, for in 1854 Kolbe launched an attack on Williamson and Gerhardt, and Kekulé aligned himself squarely with their concepts. He later said: "Originally a pupil of Liebig, I became a pupil of Dumas, Gerhardt and Williamson; I no longer belonged to any school."[62]

Liebig's attitude toward Kekulé in 1854 is evident from his failure to reply to three letters in which Kekulé asked Liebig to recommend him for a professorship at the newly established polytechnic school in Zurich. Liebig recommended Wöhler's assistant Staedeler, who received the appointment, despite strong support for Kekulé from Bunsen, Gerhardt, Williamson and A. W. Hofmann. Kekulé returned to Germany in 1855, became a *Privatdozent* in Heidelberg, and set up a small private laboratory as well as a lecture room in which the twenty-seven-year old chemist taught modern organic chemistry to a group that included Adolf Baeyer, Emil Erlenmeyer, Friedrich Beilstein, Ludwig Carius, Lothar Meyer and Henry Roscoe. In Heidelberg, Kekulé also wrote important theoretical papers and, independently of Archibald Scott Couper [1831–1892], advanced the concept of the tetravalent carbon atom.

Late in 1858, Kekulé became professor of chemistry at Ghent, as a consequence of the recommendation of both Liebig and Bunsen. His German friends deplored the fact that he was

---

[61] Ibid., vol.2, p.950.
[62] Ibid., vol.2, pp.943–944.

allowed to go to Belgium without a comparable offer from a German university, and some of the Belgians resented the appointment of a German. Despite initial difficulties, during his stay in Ghent he attracted numerous Belgian students, the Germans Baeyer, Hans Hübner [1837–1884], Wilhelm Körner [1839–1925], Albert Ladenburg [1842–1911], Hermann Wichelhaus [1842–1927], and the Englishmen James Dewar [1842–1923] and George Carey Foster [1835–1919].[63] Kekulé also wrote his uncompleted textbook of organic chemistry and proposed the holding of an international chemical congress; it was organized by his friend Carl Weltzien [1813–1870] and held in Karlsruhe on 3–5 September 1860. The historical importance of this meeting is well known, in particular because of the contribution of Stanislao Cannizzaro [1826–1910][64] At Ghent, during the 1860s, significant experimental work was done in Kekulé's laboratory, but the most notable achievements were his writings on the constitution of aromatic compounds and his famous benzene theory.[65]

In 1867, Kekulé finally received a professorship at a German university. The post in Bonn became vacant in 1863; A. W. Hofmann was the preferred candidate, and a fine new institute was promised, but he went to Berlin, where an even finer establishment was provided. Consideration was then given to Kekulé and to Kolbe; the latter was chosen, but declined the offer.[66] Kekulé lived out his years in Bonn, where his research group did extensive experimental work on aromatic compounds. An echo of his association with Liebig came in 1874, when Kekulé was offered the chair that Liebig had occupied in Munich. He declined, and proposed Adolf Baeyer instead.

---

[63] See Gillis (1966).
[64] See de Milt (1948) and Ihde (1961). A few months before the Karlsruhe Congress, Kekulé wrote: " . . . perhaps something will come of it. In any case, it should provide, in addition to serious utility, much enjoyment; a real chemists' church festival." (Letter from Kekulé to Baeyer, 13 June 1860). I am indebted to Professor Klaus Hafner of the Technische Hochschule Darmstadt for sending me photocopies of the Kekulé-Baeyer correspondence, and for his kind permission to quote passages from some of these letters.
[65] See Rocke (1985).
[66] Turner (1982, p.129) states that the Prussian government had decided "in the early 1860s . . . to call A. W. Hofmann and August Kekulé to the vacant chairs of chemistry at Berlin and Bonn and to build splendid institutes for them at a combined cost of about one million marks." This statement is not entirely accurate as regards Kekulé's appointment in Bonn.

Liebig's Influence Abroad

Liebig's international renown brought many students to Giessen from foreign countries. As noted in Table 2–1, they came principally from Great Britain, Switzerland, France, the United States and Russia, with smaller numbers for other nationalities. Apart from the chemical education they received, and the published laboratory work they may have done, the later professional careers of these *Liebig-Schüler* are of interest for an assessment of his influence on the subsequent development of chemistry in countries other than the German and Austrian states.

France. When he began at Giessen, Liebig's inspiration came chiefly from his stay in Paris during 1822–1824, and especially from his association with Gay-Lussac, whose son Jules was among Liebig's first French students.[67] Liebig's esteem for French chemistry waned with the emergence of Dumas as its leading luminary, and by the 1840s the anti-French tone of Liebig's polemical writings became more strident. This may account, in part, for the noteworthy decrease in the proportion of Frenchmen at Giessen during 1841–1850 (about 9 percent of the total foreign group) as compared with that during 1830–1840 (about 29 percent). Another feature of the French group was the preponderance of students from German-speaking Alsace; in 1830–1840, there were 10 (of 15), and in 1841–1850, 10 (of 12) Alsatians. They included several men who later achieved scientific distinction (Gerhardt, Nicklès, Oppermann, Wurtz), but most of the others for whom I found biographical data either entered family industrial firms (Dollfus, Gros, Hartmann, Koechlin, Mertzdorff),[68] or became pharmacists (Hering, Kosmann). Among the Liebig pupils, the only non-Alsatian Frenchman to gain renown as a scientist was Regnault, who left organic chemistry after his work at Giessen.

Consequently, any influence Liebig's French pupils may have exerted on the development of chemistry in their native country would most likely have come from the efforts of the four Alsatians named above. Of these, Nicklès made valuable

---

[67] As Liebig reported to Berzelius on 6 November 1832, Jules Gay-Lussac did not distinguish himself at Giessen; see Carrière (1893), p.42.

[68] During the second quarter of the nineteenth century, Alsace (especially Mulhouse) was an important center for textile dyeing and for chemical manufacture. For a comprehensive survey of the textile and chemical industry in the Mulhouse region, see Société Industrielle de Mulhouse (1902). Also see Brandt (1980).

contributions in chemistry and physics at Nancy, and Oppermann, at the pharmacy school in Strasbourg, did much to make
that school a worthy rival of the one in Paris. But the center of
scientific power was in the capital, and the two other Alsatians,
Gerhardt and Wurtz, each in his own way, did influence chemical thought there.

Gerhardt's stay in Giessen was brief (November 1836–April
1837). In charting his course, he resisted both the opposition of
his father, who wanted him to enter the family firm, and the
blandishments of Liebig. According to Gerhardt's biographers,

. . . the young man not only had won the esteem of the teacher for his
lively intelligence and astonishing scientific fervor, but also had en-
tranced him [Liebig] and his family by his [Gerhardt's] charming man-
ners, brilliant personality and spiritual qualities. It is today no longer
indiscreet to say that the impression made by the young man awakened
in the mind of the teacher, father of a young daughter, thoughts of her
union with his brilliant student. But this dream did not attract the
young Frenchman who, at the age of twenty and a half, did not intend
to tie up his future: on the pretext of an urgent summons, he took
leave of his teacher.[69]

Liebig's interest in Gerhardt had been roused by the latter's
first publication (1835) on silicates, from the Leipzig laboratory
of Otto Erdmann [1804–1869]. In bringing him to Giessen, Liebig thus enlisted the services of a promising chemist, fluent in
both French and German, and Gerhardt's principal income during the next few years came from translations of Liebig's writings. Liebig wrote in the preface to the French edition of his
treatise on organic chemistry: "Finally it is my duty to express
my special thanks to my friend and former student M. Ch. Gerhardt for his effective collaboration in the preparation of this
book. . . . It is an obligation to emphasize strongly how much I
wish his outstanding talent and thorough knowledge a happy
and successful career."[70]

After his arrival in Paris in October 1838, Gerhardt soon
showed his promise; in 1841 he was appointed *chargé de cours* at
Montpellier, and three years later he was made a professor

[69] Grimaux and Gerhardt (1900), pp.28–29. The surmise about Liebig's daughter Agnes
   may be doubted, as she was born 6 June 1829. For other biographies of Gerhardt, see
   Tiffeneau (1916), which includes a list of Gerhardt's publications, and Kahane (1968).
[70] Liebig (1840).

there. A recent biographer has asked whether "this surprisingly early nomination, made by Dumas, should be interpreted as evidence of great esteem, or of a wish to get rid of a troublesome person."[71] As was suggested earlier in this chapter, the same question was asked by Kekulé's German friends when Liebig recommended him for the professorship in Ghent. Gerhardt found Montpellier increasingly uninviting, both because of the inadequate research facilities and its distance from Paris, especially after he had begun his celebrated collaboration with Auguste Laurent [1807–1853].[72] It is perhaps not surprising that the first scientific encounter of these two rebellious spirits was marked by controversy, but after 1842 their relationship became one of close comradeship. At about that time, Gerhardt exchanged friendly letters with Liebig, who offered fatherly advice:

Do not make of science an article of commerce, as is frequently the case in France: one may get a little money, but the scientist is ruined. Give yourself a higher aim, and you will find that the money and the honors will come without your having to seek them. . . . Do not give yourself up to any kind of theoretical speculations; they will serve to satisfy only the one person whose views you support, and to gain you hundreds of enemies. Facts, particularly new facts, that is the only lasting merit; they speak more loudly, are appreciated by all minds, will bring you friends and will win the respect of your adversaries.[73]

This paternal relationship was rudely interrupted in 1845, with the appearance of a vitriolic blast by Liebig, in which he called Gerhardt a shameless liar, compared him to a highwayman, and accused him of plagiarism.[74] The outburst appears to have been triggered by Gerhardt's claim that the analytical data obtained in Liebig's laboratory for the substance the latter had named *mellon* were incorrect.[75] Liebig had previously attacked Laurent in the pages of the *Annalen,* but he now wrote him, telling him that his association with Gerhardt

[71] Kahane (1968), p.4735.
[72] See Kapoor (1973) and de Milt (1951).
[73] Letter from Liebig, 27 June 1841; Grimaux and Gerhardt (1900), pp.53–54.
[74] Liebig (1845). A French translation appeared in the *Revue Scientifique et Industrielle* 23 (1845):422–439.
[75] See Volhard (1909), vol.1, pp.277–290 and Partington (1964), pp.410–411.

would be the greatest misfortune that could befall you ... I do not like and will never like your manner in speaking of the opinions of other chemists, but I believe you to be a man of complete loyalty and honesty who, ensnared by an incredible fatality, has linked his destiny to a man without character and without morality.[76]

In 1850, there was a reconciliation initiated by Gerhardt, but his active correspondence with Liebig does not appear to have been resumed. As has been noted above, Gerhardt played a major role in the development of the type theory of the constitution of organic compounds; it is therefore of interest that after Liebig's departure from Giessen, Will lectured there on organic chemistry and presented Gerhardt's system "so clearly, convincingly and indeed so inspiringly that we all swore by Gerhardt's type theory, and revered its creator as the reformer of organic chemistry."[77]

Wurtz, who had been a schoolmate of Gerhardt, and remained a staunch friend and defender, came to Giessen in 1842 after completing his medical studies in Strasbourg. His work with Liebig dealt with the basicities of hypophosphorous and phosphorous acids, and his Strasbourg M.D. dissertation (1843) with fibrin and the albumins of egg white and serum.[78] This dual interest in pure and physiological chemistry was maintained throughout his professional career. In 1844 Wurtz went to Paris, where he worked in Dumas's laboratory until it was closed in 1848. To continue chemical research, in 1850 he joined with two former Giessen students, Charles Dollfus and François Verdeil, in establishing a private laboratory.[79] The funds were largely provided by Dollfus, the entrepreneurial talent by Verdeil, and it was Wurtz who did important research there before his appointment in 1853 as professor of chemistry at the Paris medical school. In 1874 he moved to the Sorbonne. Wurtz made outstanding experimental and theoretical contributions in organic chemistry, notably in the field of aliphatic amines (fulfilling Liebig's prediction that there might be organic compounds derived from ammonia) and of polyalcohols. Among his biochemical efforts were studies on enzymes, in particular pepsin and the

---

[76] Letter from Liebig to Laurent, 30 January 1845; see Tiffeneau (1918), p.350.
[77] Volhard (1909), vol.1, p.351.
[78] For Wurtz, see Friedel (1885) and Williamson (1885).
[79] The laboratory was located on the rue Garançière in a building that also housed the private laboratory of the biologist Charles Robin [1821–1885]. This circumstance led to the appearance of the three-volume treatise by Robin and Verdeil, *Traité de Chimie Anatomique et Physiologique, Normale et Pathologique* (Paris 1852–1853).

plant enzyme papain. Moreover, Wurtz's resolute advocacy of atomicity and the new concepts of organic chemical structure helped to counter the resistance of other French chemists to these advances in chemical theory.

Although he was more strongly influenced in his chemical thought by Dumas and Gerhardt than by Liebig, Wurtz profited from his brief stay in Giessen in several ways. Apart from his association with Dollfus and Verdeil, he won the friendship of August Wilhelm Hofmann, who later extended Wurtz's work on aliphatic amines. Also, what he learned from Liebig may have guided the style of his leadership of the effective research group he assembled after 1860. A detailed study of this group is still needed, but it is fair to say that, during 1860–1880, it rivalled in quality, if not in size, the better-known chemical research groups in Germany. Moreover, in his compaign for increased state support of university chemical laboratories, Wurtz imitated Liebig's approach. In 1864, Wurtz wrote the Minister of Education:

Since the now distant past when M. Liebig gathered at Giessen students from all nations of the world, and established a justly celebrated school, chemical studies have gained great impetus in Germany. Spacious laboratories have been constructed in Giessen, Heidelberg, Breslau, Göttingen, Karlsruhe, Greifswald. Many researches, beautiful discoveries have been the fruit of these valuable developments. Science has profited thereby; Germany is rightly proud and redoubles its efforts to put instruction in practical chemistry at the top of modern needs.[80]

The Minister was impressed, but Wurtz's chemical colleague Marcelin Berthelot [1827–1907] was not:

I have spoken often with scientific *Privatdozents:* their life is more miserable than ours, and their lack of scientific instruments is greater. The bit of independence they may have is more apparent than real, because a very poor man cannot be considered to be truly free. As for the professors, I do not know whether the supervision of 50 students subjected to a regimen of 150 almost identical manipulations is not equivalent to examinations and double functions in deadening scientific activity.[81]

---

[80] Translated from the quotation by Paul (1972), p.7.
[81] Ibid., p.9.

This was written after his visit in 1858 to Bunsen's laboratory in Heidelberg. Later, in 1882, Berthelot conceded that "it certainly is not useless, when dealing with legislative bodies and the public, to point to the advantages in the status of German professors in order to improve that of French professors, but one should not be duped by this comparison, in which one contrasts the defective aspects of our system with the shining aspects of those of our neighbors."[82]

Clearly, the full consideration of Liebig's influence on the development of chemistry in nineteenth-century France must take into account the special character of its administrative system for higher education and scientific research. This cannot be discussed here, but it may be suggested that Liebig's growing animosity after 1840 toward younger French chemists reflected not only scientific rivalry, but also the increasing political tensions that came to a head in the Franco-Prussian war.[83]

Great Britain. In no nation, including the German states, was the esteem accorded Liebig greater than in England and Scotland. His special relationship to British academic and industrial chemistry was established during the autumn of 1837, when he spent two months there. Liebig had been invited to address the Liverpool meeting of the recently founded British Association for the Advancement of Science, in the hope that he would speak about isomerism, elementary analysis and the state of organic chemistry, but his address dealt largely with the results that Wöhler and he had obtained in their studies on the products of the decomposition of uric acid. Although he visited academic colleagues (Thomas Andrews, Thomas Graham, William Gregory, Robert Kane, Thomas Thomson) during his trip to the meeting via Manchester, Dublin, Belfast, Glasgow and Edinburgh, Liebig appears to have spent most of his time in visiting factories and in enjoying the hospitality of wealthy chemical manufacturers. Among these were William Charles Henry [1804–1892] in Manchester, Walter Crum [1796–1867] in Glasgow and James Muspratt [1793–1886] in Liverpool. From the meeting, Liebig went to London (via Manchester and Newcastle) and visited Faraday, but did not find the city exciting.[84] After his

---

[82] Ibid, p.10.

[83] It is noteworthy that the long list in Poggendorff (1971), p.378, of obituary notices after Liebig's death in 1873 does not include any citations of French publications.

[84] For Henry, see Farrar et al. (1977). For Crum, see De La Rue (1868). For Muspratt, see Hardie (1955) and Stephens and Roderick (1972). Most of the information about Liebig's 1837 trip comes from letters he wrote to his wife, to Berzelius or to Wöhler.

return to Giessen, Liebig wrote to Berzelius:

I have spent some months in England, have seen an awful lot and learned little. England is not a land of science, there is only a widely practiced dilettantism, the chemists are ashamed to call themselves chemists because the pharmacists, who are despised, have assumed this name. I was very satisfied with the public, [their] courtesy and hospitality; in short, I found in them all the virtues. Scientifically, Graham is the most notable exception, he is a splendid person, also Gregory, who has replaced him in Glasgow.[85]

The favorable British response to Liebig's semi-popular scientific writings, especially the one on agriculture, which were rapidly made available in English by Lyon Playfair, William Gregory and John Gardner, led to shorter visits in 1842, 1844, 1845 and 1851. These were marked by ceremonial occasions and visits to the estates of English peers, where he met persons close to the royal entourage. Although, after 1851, the popularity of Liebig's views on agriculture had diminished in the face of the publication of the experimental data of Lawes and Gilbert, and of the criticism of his former supporter Philip Pusey [1799–1855] (President of the Royal Agricultural Society), there still remained much of the earlier adulation. It seems that the Prince Consort, who came to be admired by the British public despite his German origin, was greatly interested in Liebig, whose final visit to England was marked by an audience with the Queen.[86] By that time, however, Liebig's attitude toward Britain had soured somewhat, largely because of the rejection by British agriculturists of his theories and his artificial manures. Apart from a reply to the criticisms of Lawes and Gilbert, in 1863 there also appeared Liebig's analysis of the writings of Francis Bacon, with a denigration of the high place accorded him in history. This article did not receive a cordial response in British periodicals.[87]

Those to his wife are partially cited in Volhard (1909), vol.1, pp.132–141, who states (p.140) that Walter Crum had studied with Liebig; it was Crum's son Alexander (see Appendix 2) who later came to Giessen. This error is echoed by Morrell (1972), p.19.

[85] Carrière (1893), p.134. See Russell (1983), pp.174–192.

[86] One of Prince Albert's advisers in the planning of the 1851 Exhibition was Lyon Playfair, and Albert's personal staff included the German Ernst Becker [?–1888] who received a Dr.phil. at Giessen in 1847 (no dissertation) after matriculating there in 1843 in public affairs (Kameralwissenschaften). According to Brock (1984), p.100, Becker had been a Liebig pupil; perhaps he attended Liebig's chemical lectures, but I could find no evidence of his having worked in Liebig's laboratory.

[87] Volhard (1909), vol.2, pp.309–315.

Before Liebig's 1837 trip, there had been three British guests at his laboratory (William Gregory, William Charles Henry, Robert John Kane) and only two matriculants, Thomas Richardson and William Eatwell, the first of whom became a chemical manufacturer and the other a physician. After 1837, the British colony in Giessen grew rapidly: by 1840, among the 18 matriculants were George Fownes, Joseph Henry Gilbert, William Allen Miller, Lyon Playfair, Henry Edward Schunck and John Stenhouse. During the succeeding five years, the number of British matriculants increased markedly, and included Thomas Anderson, John Blyth, Benjamin Collins Brodie, Jr., James Sheridan Muspratt and Alexander Williamson; Frederick Penny received a Giessen Dr.phil. at that time. The flow slackened afterward; of the 20 British matriculants during 1846–1850, only John Hall Gladstone stands out and, as noted above, Edward Frankland's stay in Giessen was transitory.

A striking feature of the British contingent at Giessen during 1835–1850 is the large proportion of Scots, especially from Glasgow. Of the 83 *Liebig-Schüler* from the United Kingdom, 26 came from Scotland; they included the sons of Walter Crum and Thomas Thomson. It seems, however, that while Liebig's visit may have impressed the fathers, the stay in Giessen did not incline their offspring to a career in chemistry, as Thomas Thomson, Jr., turned to botany and Alexander Crum to commerce and politics. Indeed, many more of Liebig's British students felt the same way. As has been noted by the biographers of William Charles Henry, "Like many young Englishmen who studied at German universities during the nineteenth century, Henry came to look back upon those few months in Giessen as a brief residence in Arcadia."[88]

For the Liebig pupils who entered upon an academic career, his recommendations appear to have carried considerable weight in Britain. According to Morrell, "by the 1850s the Liebig clan had assumed most of the 'plums' available in British university chemistry."[89] However, if one examines the scientific careers of those who gained university professorships, only four of the eleven men listed by Morrell can be judged to have become outstanding research men: Anderson, Frankland, Miller and Williamson. Like Liebig's German student Strecker, Anderson pursued lines of investigation that may be considered to have reflected Liebig's influence, but, as noted above, Frankland and Williamson drew more chemical inspiration from Marburg and

---

[88] Farrar et al. (1977), p.4.
[89] Morrell (1972), p.19.

Paris than from Giessen, and it is difficult to find in Miller's brilliant spectroscopic studies evidence of Liebig's teachings. No doubt, in their lectures and in the organization of their programs of practical instruction, all the eleven men listed by Morrell reflected in varying degrees their association with Liebig. But a sustained research effort does not apear to have been led at Oxford by Brodie, at Edinburgh by either Gregory or Playfair, or for that matter at University College London by Williamson, who had withdrawn from active chemical investigation after succeeding Graham there in 1855. In assessing the later research contributions of Liebig's British pupils, an especially high place must instead be assigned to Schunck, who did his scientific work in a private laboratory.[90] Indeed, Liebig's greatest influence on the development of British academic chemistry may be said to have been exerted through the presence in London during 1845–1865 of his German pupil August Wilhelm Hofmann.[91]

Hofmann was the son of the architect Johann Phillipp Hofmann [1776–1842], who had designed the renovation and expansion of Liebig's laboratory in 1839. He began at Giessen as a student of law and philology, but switched to chemistry after attending Liebig's lectures, and soon showed his chemical promise in an investigation of the constituents of a coal tar distillate. In 1845, after serving as Liebig's assistant, Hofmann moved to Bonn, and was made *Privatdozent* and *ausserordentlicher Professor* before going to London, where Liebig had arranged for his appointment as head of the newly established Royal College of Chemistry.[92]

At Giessen, Hofmann had demonstrated the presence of aniline in coal tar and, with Auguste Laurent (who visited Liebig's laboratory briefly in 1843), showed that phenol can be converted into aniline, thus providing support for Laurent's view that phenol is a hydrate of the phenyl radical. Hofmann continued to work on aniline and its derivatives in London; he and his associates also did research on aliphatic amines, as well as on organic phosphorus, sulfur and arsenic compounds. In 1856, his eighteen-year-old student, William Henry Perkin, accidentally prepared aniline purple (mauveine), and proceeded to found a

[90] For a biography of Schunck, see Farrar (1977).
[91] For biographies of Hofmann, see Volhard and Fischer (1902) and the series of articles by Playfair and others in the *Journal of the Chemical Society* 69 (1896):575–732; individual papers in this series are cited below. Hofmann's correspondence with Liebig has been edited by Brock (1984), who has prepared a valuable introduction.
[92] Liebig's first choices for the London appointment appear to have been, successively, Fresenius and Will, but both declined. For excellent accounts of the founding and later history of the Royal College of Chemistry, see Roberts (1976) and Bentley (1978).

commercial enterprise for its manufacture.[93] This celebrated discovery was soon followed by the work of other Hofmann students (notably Edward Chambers Nicholson) and of Hofmann himself, who developed the rosaniline series. The lead in the industrial manufacture of synthetic dyes provided by these contributions was short-lived, however. British research in the field slackened after 1870, but was intensively pursued in Germany and in France, and by the 1890s German companies controlled most of the market.

Hofmann's style of laboratory instruction in London was described by Frederick Augustus Abel [1827–1902], who was with Hofmann until 1851:

It was Hofmann's rule, to which, during the continuance of my stay at the College he strictly adhered, to visit each individual student twice during the day's work, and to devote himself as patiently to the drudg-, ery of instructing the beginner, or of helping the dull scholar, as he did, diligently, to the guidance of the advanced student, whom he would skilfully delude into the belief that the logical succession of steps, in making the first investigation which the master had selected for pursuit by the pupil, was the result of skill in research which he has already attained, instead of being simply or mainly the skilful promptings of the great master of original research.[94]

Among Hofmann's many students and assistants in London, in addition to Abel, were William Crookes [1832–1919] as well as several Liebig pupils: Hermann Bleibtreu, John Blyth, Reinhold Hoffmann, Georg Merck, James Sheridan Muspratt and Thomas Rowney.

Hofmann returned to Germany in 1865 to become professor of chemistry in Berlin, where he continued to do outstanding research largely along the lines he had developed in London. It should be noted, however, that his new laboratory did not become the leading center of chemical teaching and research in Germany; after 1874, it was the Munich institute of Adolf Baeyer, the pupil of Kekulé. A factor may have been that, although Hofmann came to accept Gerhardt's chemical theories, he was also a loyal disciple of Liebig and, in the words of Armstrong, "seems never to have thoroughly entered with full sym-

---

[93] See Perkin (1896), Caro (1892), Brightman (1956), Beer (1959) and Scholz (1980–1981).
[94] Abel (1896), p.588.

pathy into the spirit of the new chemistry, and rarely adopted the modern style in his writings."[95] Hofmann's departure was lamented by his English admirers. To cite only two excerpts from later biographical notices:

Hofmann's departure was not only a cause of regret to those who had worked under him and to all his friends; it was a heavy loss also to the country at large, as no one had ever done so much for the cause of chemical science in the kingdom as Hofmann did, nor had any one exercised to such an extent that wonderful power he had of stimulating the enthusiasm of his students and of inciting in them a love of chemistry and of scientific research.[96]

Armstrong added an acerbic note:

While our adopted countryman, Hofmann strove in every possible way to promote our interests. His pupils established the new industry, which he did his utmost to support and forward when its importance became manifest; but the circle within which favourable conditions prevailed was hopelessly narrow. We, in fact, denied him the opportunity he eagerly seized upon when it was offered to him by his own countrymen: and among them he found both *Boden und Nahrung,* for they had been properly tailored by their universities, and were prepared to accept this advice and *apply theory to practice.* Our universities, unfortunately, long thwarted us instead of aiding us, and are still too much engaged in the blind worship of the unpractical. True culture has been neglected.[97]

It should be added, however, that although British organic chemistry had suffered a severe loss, talented new leaders later emerged from the Royal College of Chemistry, headed by Frankland from 1865 to 1885. Among them were Raphael Meldola [1849–1915], William Henry Perkin, Jr. [1860–1929] and Arthur George Perkin [1861–1937].

Before concluding this sketch of Liebig's influence in Great Britain, some additional words may be apropriate regarding the impact of his ideas regarding agriculture and medicine. The acclaim he received during his 1837 visit prepared the ground for

[95] Armstrong (1896), p.704.
[96] Perkin (1896), p.620.
[97] Armstrong (1896), p.730.

the favorable reception accorded the English translation of his 1840 book on plant nutrition. At first there was wide-spread enthusiasm, although James Finlay Weir Johnston [1798–1855], a student of Berzelius, was among the few notable dissenters. In rejecting "humus" (decayed plant material) as a fertilizer, Liebig emphasized instead, without any experimental evidence of his own, the importance of ammonia as a source of nitrogen, and he also stressed the essential requirement for minerals such as phosphates and sulfates. The ammonia theory was soon disproved by Jean Baptiste Boussingault [1802–1887], the leading French agricultural chemist of his time; he showed that the nitrogen of all the plants he studied (except legumes) is derived from organic fertilizers. He then demonstrated the fixation of atmospheric nitrogen by legumes, and later studied the formation of soil nitrates. Liebig's mineral theory, based on the extensive work of Nicholas Théodore de Saussure [1767–1845], proved to be more durable and led to the hope that artificial fertilizers devised by Liebig would replace natural manures. At Liebig's request, in about 1843 his friend James Muspratt obtained, on Liebig's behalf, an English patent for such products, and initiated their manufacture in the Newton plant of his chemical firm.[98] These preparations were tested by Lawes and Gilbert, and were found to be unsatisfactory.[99] Moreover, in 1842, Lawes had taken out a patent for the manufacture of "superphosphate" (animal bones solubilized by treatment with sulfuric acid), which proved to be effective. Lawes set up a factory to produce this material and, soon afterward, Liebig's first Brit-

---

[98] This business arrangement led to the following letter from Liebig to James Muspratt:
I have just received through Dr. Petzhoer of Dresden all the recipes according to which our various manures are prepared and composed at Newton. These recipes have been communicated to Dr. Petzhoer by Dr. James Sheridan Muspratt. You will agree with me that he had no right to make such a communication, indeed that this right could neither be given him by me or by anyone of the partners. There lies in this an abuse of confidence placed in him which must have the effect of immediately removing him from the situation he occupies. Owing to this inconceivable proceeding we are left at the mercy of everyone, for our methods and our proportions will very soon find their way into the English journals. . . . Dr. James Sheridan Muspratt has by this criminal indiscretion cut off the possibility of introducing it in Germany. He was not permitted to act as he did without asking your permission and your advice. I cannot tell you how afflicted I am.
This quotation in Stephens and Roderick (1972), pp.295–296, is presumably a translation of a letter written in German.
[99] See Schreiber (1957).

ish student, Thomas Richardson, followed suit. It is not surprising, therefore, that despite the favor shown him by royalty, Liebig's attitude toward Britain cooled markedly.[100]

The English translation of Liebig's 1842 book on animal physiology and pathology, under the title *Animal Chemistry*, also aroused great interest in Britain. The main theme in the book was his explanation of metabolism (*Stoffwechsel*) through chemical speculations about the fate of foodstuffs in the animal economy; these ideas were pursued more intensively in Germany and in the United States than in Britain, and will be discussed later in this chapter. Another theme, derived from his speculations about fermentation and putrefaction, attracted the attention of some British physicians who found in Liebig's writings support for the theory of "morbid poisons," advanced in the eighteenth century by Joseph Adams, a pupil of the celebrated John Hunter. In explaining the decomposition of organic matter by fermentation or by putrefaction, Liebig countered Berzelius's concept of catalysis by invoking the idea of communicated molecular motion incited by oxygen in the presence of water. From this speculation Liebig drew the proposal that, among the products of putrefaction, there may be substances that can affect the blood and tissues in a manner to produce disease.[101]

United States. Much has been written by American chemists and historians of science about Liebig's influence on the development of chemistry in the United States during the nineteenth century, and particular attention has been paid to the response to his book on agriculture.[102] In a historical account of American chemical education at that time, Browne listed seven notable Liebig pupils from the United States: John Lawrence Smith, Eben Norton Horsford, Oliver Wolcott Gibbs, Charles Mayer Wetherill and John Addison Porter, as well as Frederick Augustus Genth (who came to the United States after his studies with Liebig and Bunsen) and Samuel William Johnson (who worked in Liebig's Munich laboratory in 1855). Browne also noted that

---

[100] See Grossmann (1917). Liebig appears to have refrained from writing polemical articles about British scientists until 1855, when he attempted to refute the criticisms of Lawes. At Liebig's request an English translation was prepared by Samuel W. Johnson, an American visitor (of whom more later in this chapter) in his Munich laboratory; the article appeared in the *Country Gentleman*.

[101] Pelling (1978), pp.113–145.

[102] The most important source is the book by Rossiter (1975). See also Browne (1932), Van Klooster (1944), Moulton (1942) and Resneck (1970).

these men "are only a few of the leading American teachers of
chemistry who were schooled under Liebig. The list could be
greatly extended."[103]

As was indicated in Table 2–1, there do not appear to have
been any American students of chemistry or pharmacy at Giessen before 1841, and the total during the succeeding decade was
16 (including one guest). The five men mentioned by Browne
were readily identified, as was Josiah Dwight Whitney. Also,
some biographical data were found for Franz Lennig (not listed
in the Appendices) and Samuel George Rosengarten, both of
whom joined family chemical firms in Philadelphia, and for
William Leonard Faber, who became a metallurgist and mining
engineer. Despite an intensive search, however, no information
was found about the teaching (or other) activities of the remaining seven Americans: Daniel Breed, Buckland W. Bull, Carl
(Charles?) Harris, Carl (Charles?) Johnson, Orlando B. Mayer,
William Pearce and Thomas Jefferson Summer, except for an
indication that Summer had died before 1851. It would seem,
therefore, that Browne's expectation of an enlargement of his
list was not justified. As teachers, Smith was professor of chemistry at the University of Virginia 1852–1866, Horsford at Harvard 1847–1863, Gibbs at the College of the City of New York
1848–1862 and Harvard 1863–1877, Wetherill at Lehigh 1864–
1871, and Porter at Yale 1852–1866. The lectures they presented and whatever laboratory instruction they could offer
were no doubt modeled on Liebig's program but, with the exception of Gibbs, they made few contributions in pure chemistry. Horsford's attempts to conduct sustained research at
Harvard failed, partly because of the demands of teaching and
the shortage of help, and also perhaps because of deprivation of
the kind of stimulation he found in Giessen.[104] Genth, who
came to the United States in 1848, was for a time (1872–1888)
professor of chemistry at the University of Pennsylvania.[105] Although he continued his pioneer research on cobalt-ammonia
compounds (later in collaboration with Gibbs), his main professional activity was as a commercial analyst. Indeed, it may be
said that the leading Liebig pupils in the United States brought
from Giessen mainly the influence of his personal research program of the 1840s, with its emphasis on routine chemical analy-

[103] Browne (1932), p.720.
[104] Rossiter (1975), pp.77–78.
[105] There were other Liebig pupils in the emigration from Germany to the United
     States at the time of the 1848 revolution; they included Wilhelm Keller, Schiel and
     Zoeppritz. Amend, another emigré, is listed in Appendix 2 as a Liebig pupil on the
     testimony of his son A. P. Amend: *The Chemist* 8 (1930):5–6.

sis as applied to the practical fields of agriculture (Smith, Wetherill, Porter), nutrition (Horsford), mineralogy (Genth, Smith, Wetherill) and geology (Whitney). Their regard for Liebig ranged from veneration (Horsford) to dislike (Gibbs), and they do not appear to have established friendships with fellow-students such as A. W. Hofmann, Erlenmeyer and Williamson who might have infected them with some of the ferment of the emerging new chemistry. No doubt for some of them, a stay in Giessen seemed to be essential for academic preferment in the United States; Whitney is said to have written: "I suppose I should be run after for a professorship, if I had studied in Giessen, as it seems to be a settled point that no young man can be expected to know anything of chemistry unless he has studied with Liebig."[106]

As in Britain, the English translation of Liebig's book on agriculture aroused wide interest in the United States, and stimulated a demand for experimental tests of his proposals for enriching soil fertility. All the young American chemists who went to Europe during the 1840s in order to learn modern techniques for the analysis of plants, soils and minerals did not go to Giessen; a notable exception was John Pitkin Norton [1822–1852], who spent two years with Johnston in Edinburgh and later also worked in Utrecht with another Berzelius pupil, Gerrit Jan Mulder [1802–1880], at a time when Liebig had excoriated Mulder unmercifully and had drawn a hysterical response from his victim.[107] Norton had studied at Yale with Benjamin Silliman [1779–1864], who regarded Johnston's 1841 book on agricultural chemistry more highly than Liebig's "ingenious conjecture." Upon his return to New Haven, Norton and Benjamin Silliman, Jr. [1816–1885] initiated at Yale a program of chemical studies similar to that of Horsford at Harvard and used the Giessen manuals of Will and Fresenius in the practical instruction. Norton's own laboratory activity at Yale was limited to soil analysis; he did no chemical research.[108] After his untimely death, Norton was succeeded by the Liebig pupil Porter, whose performance in teaching and research was undistinguished. Porter's principal contribution to science may be said to have been the arousal of the interest of his new father-in-law, the railroad tycoon Joseph Sheffield, in promoting the sciences at Yale. By

---

[106] Quoted from Miles (1976), p.508.

[107] See Mulder (1846). Another noted American chemist, James Curtis Booth [1810–1888], does not appear to have studied in Giessen during his stay in Europe; see Miles (1976), pp.42–43.

[108] An excellent account of Norton's career has been given by Rossiter (1975), pp.49–88. See also Resneck (1970).

1861 his gifts had become large enough to establish the Sheffield Scientific School, with Porter as its first dean.[109] A few years later, the School was appointed the land-grant school in Connecticut under the Morrill Land Act, and in 1874 the state legislature voted to establish the first agricultural experiment station in the United States; it was located in the Sheffield Scientific School from 1877 until 1882 when it was moved into its own quarters in New Haven. The director of the Station until 1900 was Samuel William Johnson [1830–1909], who had begun his chemical studies at Yale with Norton, and had then worked in the laboratories of Pettenkofer and Liebig in Munich and of Frankland in London. Apart from his many articles on topics in analytical chemistry, Johnson wrote a book, *How Plants Grow* (1868), that was received with acclaim and went through many editions. Of special relevance to the theme of this chapter is Johnson's expansion of the research program of the Station by bringing to its staff men who followed experimental lines in animal metabolism (Wilbur Olin Atwater [1844–1907]) and in protein chemistry (Thomas Burr Osborne [1859–1929]) derived indirectly from Liebig's speculations in his 1842 book on animal physiology.[110]

These speculations were, in part, elaborations of Lavoisier's analogy of animal respiration to combustion. They also depended heavily on the use of data for the elementary composition of food materials and excretory products to set up balanced chemical equations.[111] Because the elementary composition of plant proteins differed little from that of comparable animal proteins, it was claimed that dietary proteins are converted into those of the blood and tissues without appreciable chemical change. The oxidation of fats and carbohydrates was considered to be the principal source of animal heat, and urinary urea was considered to arise from the breakdown of proteins, without the production of heat, during muscular activity. Instead of the catalytic activity of "ferments" (later to be called enzymes), as suggested by Berzelius, Liebig invoked the operation of "molecular motions" or a "vital force." Nearly all of Liebig's physiological theories were disproved, but they also stimulated theoretical and experimental efforts that were sustained for many years afterward. Among them was the work of Hermann Helmholtz dur-

---

[109] See Chittenden (1928).
[110] For Johnson, see Osborne (1911); for Osborne, see Vickery (1931,1942); for Atwater, see Rosenberg (1970).
[111] For discussions of Liebig's theories of animal metabolism, see Holmes (1964), pp.vii–cxvi; Holmes (1963); Fruton (1972), pp.93–100,267–275,403–406,425–427; Mani (1976), pp.22–75.

ing the 1840s on chemical changes and heat production during muscular contraction, as well as his formulation of the principle of the conservation of energy. There were also the contributions of the *Liebig-Schüler* Moritz Traube, who in 1858 broke new ground in his studies on oxidative ferments, and disputed Liebig's assertion that it is the decomposition of such agents that causes chemical changes in other substances.[112]

The most active line of research stimulated by Liebig's physiological speculations involved metabolic balance studies and direct measurement or indirect estimation of heat production. Such investigations had been conducted before, but after 1860 the available methods were improved by several German physiologists, among whom were two former Liebig pupils, Max Pettenkofer and Carl Voit [1831–1908], as well as Voit's student Max Rubner [1854–1932]. In particular, Rubner's demonstration, in 1894, that the law of the conservation of energy is obeyed by animal organisms was widely hailed, and at the turn of the century, his approach to the study of metabolism became a prominent feature of physiological research in the United States. For this development, Wilbur Olin Atwater, who had begun as an agricultural chemist, must be given a large measure of credit; after a visit in 1887 to Voit's institute in Munich, he came away not only with admiration for the work done there but also with the conviction that calorimetry could provide better dietary standards. During the 1890s, with the help of the physicist Edward Bennett Rosa [1861–1921], Atwater designed a respiration calorimeter large enough to accommodate human subjects, and his techniques were used and further developed by later American leaders of nutrition research, notably Henry Prentiss Armsby, Francis Gano Benedict and Graham Lusk.

Osborne followed a different scientific track, also derived indirectly from Liebig.[113] However, although Liebig's name is usually assigned a high place in historical accounts of nineteenth-century protein chemistry, it was the agricultural chemist Heinrich Ritthausen [1826–1912] from whose work Osborne drew inspiration.[114] During the 1860s Ritthausen subjected various proteins to acid hydrolysis, and found that they yielded dif-

---

[112] For Moritz Traube, see Bodländer (1895) and Müller (1970).
[113] See Vickery (1931,1942).
[114] For Ritthausen, see Osborne (1913). After his studies with Erdmann in Leipzig, Ritthausen joined the world's first agricultural experiment station established in 1851 at nearby Möckern. According to Osborne, Ritthausen began as a student of Liebig, but I could find no evidence for this statement.

ferent amounts of the amino acids he isolated from the hydrolysates; from this result he concluded that plant and animal proteins given the same name (e.g., albumin) are different chemical
entities. Johnson had met Ritthausen and was familiar with his
work; in 1888 he suggested to Osborne, then his assistant (and
son-in-law), that he might undertake the study of plant proteins
along the lines initiated by Ritthausen. During the succeeding
two decades there flowed from Osborne's laboratory at the Connecticut Agricultural Station a series of important papers on the
crystallization of seed proteins, on the analysis of their amino
acid composition, and (with Harry Gideon Wells) on the specificity of their action as antigens. Moreover, after 1910, Osborne
collaborated with Lafayette Benedict Mendel [1872–1935] of
Yale in nutritional studies that established the need, in the animal diet, of certain amino acids, and also laid the groundwork
for later work on vitamins. Osborne's research was greatly aided
by the parallel work of Emil Fischer on the amino acid analysis
of proteins and of Frederick Gowland Hopkins on tryptophan as
an "essential" amino acid and on the need for what Hopkins
called "accessory factors" in animal nutrition.

It may be suggested, therefore, that Liebig's teachings and
writings influenced the development, in the United States, of
agricultural chemistry, analytical chemistry and some aspects of
biochemistry, both directly, through the efforts of pupils such as
Genth and Horsford, and indirectly, through the work of men
such as Atwater and Osborne. In contrast to Britain, however,
Liebig's influence on the development of organic chemistry in
the United States was negligible. This may have been due, in
large part, to the fact that his American students at Giessen had
only come there in the 1840s, when he had ceased to be an active participant in the field, and also perhaps to the limited opportunities available at American universities for sustained
research in pure chemistry. It was not until later in the nineteenth century, through the achievements of men such as
Arthur Michael [1853–1941] at Harvard and Moses Gomberg
[1866–1947] at the University of Michigan, both of whom had
studied in Germany, that teaching and investigation in organic
chemistry began to flourish at American universities.

Other Countries. As is indicated in Table 1, Liebig's renown
brought to Giessen students from many other nations. Of particular interest is the role played by several of his Russian pupils,
notably Zinin, Voskressensky, Khodnev and Sokolov, in the development of productive programs of teaching and research in
organic chemistry, especially at Kazan and at St. Petersburg.

Liebig and the Chemical Industry

By the time of Liebig's advent to Giessen in 1824, an embryonic chemical industry had arisen in some of the German states, largely through the influence of pharmaceutical chemists such as Johann Bartholomäus Trommsdorff [1770–1837], Johann Gottfried Dingler [1778–1855], Johann Andreas Buchner [1783–1852] and Heinrich Emanuel Merck [1794–1855]. These men not only practiced their profession and taught apprentices and students the practical chemistry of their time, but also manufactured pharmaceutical products that were sold to local pharmacists and wholesale dealers in chemicals.[115] In 1827, with Liebig's encouragement, Merck established the *Chemische Fabrik E. Merck* in Darmstadt for the large-scale manufacture of pharmaceutical chemicals. Liebig maintained close ties to the Merck family; Emanuel was for a time co-editor of the *Annalen,* and his son Georg received his Dr.phil. at Giessen after his studies with A. W. Hofmann in London. Many Liebig pupils, some of whom had started as pharmacists, set up chemical factories; among them was Ernst Sell, who supplied the coal-tar distillate for Hofmann's work in Giessen.[116] For some leading pharmacists of the older generation, such developments were not wholly welcome. Ernst Wilhelm Martius [1756–1849] wrote: "The multiplication of chemical factories alienates the apothecary from his laboratory, it leads him to give up the preparation of many materials within his knowledge and experience, and tempts him to depend on the chemical manufacturer rather than his own science."[117]

From his early days in Giessen, Liebig took an active interest in the development of German industry, especially in the Rhein-Main-Neckar region. On 6 May 1834, he wrote Wöhler: "You know that I have been traveling and have visited all the important factories in Berg [near Düsseldorf]; I have learned a great deal, much more than I expected, and will make such a journey every year; there is no better or more convenient way to remain *au niveau* of manufacturing without exertion."[118] Such visits provided a background against which Liebig viewed the state of the chemical industry in Britain; upon his return from the 1837 trip, he stated that

---

[115] See Schmauderer (1969), Gustin (1975) and Huhle-Kreutzer (1989).
[116] For Sell, see Flemming (1967).
[117] Martius (1847), p.273. See also Hickel (1978).
[118] Hofmann (1888), vol.1, pp.82–83.

I have seen the Glasgow region covered with every kind of chemical factories, it is the center, it is one of the most important wheels in the immense machinery of English industry. . . .Nowhere else in this so celebrated nation does one find anything like that in Glasgow. See how much English calico costs in our markets, their textiles are the same, their patterns are excellent, only their colors are not fast; in this respect the French have the decided advantage, because their dyers are chemists.[119]

Here Liebig was referring to the textile-dyeing industry of the Mulhouse region; as was noted above, the sons of several owners of these Alsatian factories came to Giessen to study chemistry.

In his propaganda for an enhanced status of chemistry in German and Austrian universities, Liebig emphasized the importance of practical chemical education for the development of a chemical industry to match that of France and especially that of Britain.[120] Although some of his own business ventures, to which he devoted considerable effort, were unsuccessful,[121] his message bore fruit in the industrial achievements of several of his pupils during 1850–1970, when the German chemical firms that later came to dominate the world market were founded.[122]

Among the manufactures stimulated by Liebig's ideas was the production of artificial fertilizers. Although his proposals for increasing soil fertility were greeted initially with wide, if not complete, approval by agriculturists in Britain and the United States, the response of their counterparts in the German and Austrian states was largely negative, even scornful. The intrusion of a man without experience into their domain was resented, and the later recognition of the shortcomings of Liebig's first artificial fertilizer were taken to justify their adherence to

[119] Liebig (1838), p.346. For a description of the chemical industry in the Glasgow region during the nineteenth century, see Clow and Clow (1952).

[120] See Borscheid (1976), pp.33–50.

[121] At the behest of the Hessian government, during 1824–1831, Liebig managed the Salzhausen factory for the production of inorganic chemicals; see Berl (1931). This venture failed, as did Liebig's first attempt during the 1840s to market his artificial fertilizer, obtained by fusing potash with lime (see note 98). Much of his correspondence with A. W. Hofmann during 1845–1846 relates to a business arrangement with the former Liebig pupil Bullock to manufacture quinine preparations; see Brock (1984), pp.15–17,27–61; they also corresponded in 1858 about Liebig's unsuccessful venture into the silvering of mirrors (ibid., pp.17–18,239–240). In 1868, Liebig devised a modified version of Horsford's baking powder, but this product did not sell well in Germany; see Volhard (1909), vol.2, pp.297–303. More successful was the commercial introduction, at about the same time, of the famous *Liebigs Fleischextrakt*, based on the work of the 1840s in Giessen; although the nutritional value of this preparation was disputed, especially by Carl Voit, it continued to be marketed for many decades after Liebig's death.

[122] See Haber (1958).

the agricultural practice he had rejected. As has been well described by Borscheid,[123] the famine that preceded the 1848 revolution forced German agriculturists to adopt new methods of
soil fertilization. One of the first enterprising chemists to set up
a plant for the large-scale production of artificial manures was
Ludwig Baist who, after his studies with Liebig, in 1856 established the *Chemische Fabrik Griesheim.*[124] By this time, Liebig had
adopted methods for the preparation of "superphosphates" similar to those proposed earlier by Lawes.[125] Other Liebig pupils
who became important industrialists were Karl Clemm (later
Clemm-Lennig) and ,his brother Gustav; both men made frequent trips to England to learn new techniques, and successively
enlarged the scope of their enterprise from the manufacture of
artificial fertilizers to soda and sulfuric acid (through the purchase of the Giulini form), as well as to dyes and dye intermediates; in 1865 it became the *Badische Anilin- und Soda-Fabrik
(BASF).* During 1860–1870, the seeds were planted for other future giants in German chemical manufacture: the Friedrich
Bayer factories in Elberfeld and Leverkusen, Kalle & Co. in
Mannheim, Meister, Lucius and Brüning in Hoechst, Cassella &
Co. in Frankfurt. They produced dyes for the textile industry,
and gradually took control of the world market from declining
British and French firms, they entered the production of heavy
chemicals (soda, sulfuric acid, nitrates, etc.) and later they
gained dominance in the manufacture of drugs.

Many other German chemical industries flourished during
1860–1914, and some of them owed their success to Liebig
pupils; Theodor Fleitmann introduced a new method of nickel
manufacture and Hermann Bleibtreu set up the first Portland
cement factory in Germany. Also, among Liebig's students at
Giessen were future partners in important family textile firms
(Christian Elbers, Georg Gail, James Gros), future dye manufacturers (Gustav Geiger, Carl Leverkus, Valentin Weidenbusch,
Julius August Wolff), as well as future producers of other industrial products, such as soap (Julius Unger). To these names
should be added those of some of the men who later became
chemical directors in various German factories (Carl Gustav
Guckelberger, Reinhold Hoffmann, Wilhelm Mayer, Carl Nöllner).

---

[123] See Borscheid (1976), pp.16–27 and Scharrer (1949).
[124] For Baist, see Flemming (1965).
[125] Welte (1968). See also Borscheid (1876), pp.83–93, who emphasizes the importance
  of the early production of artificial fertilizers in the rise of the German sulfuric acid
  industry.

This brief summary of Liebig's role in the upsurge of the German chemical industry after 1850 suggests that although the initial impetus came largely from men who had gained practical chemical experience through their association with pharmacy, the transition to large-scale manufacture required not only more scientific training, such as that provided by Liebig and his assistants at Giessen, but also study of the industrial methods developed in Britain. Moreover, in advising the sons of manufacturers and other students aspiring to industrial careers to concentrate exclusively on learning how to solve chemical problems through the application of the methods of analytical chemistry, Liebig did not prepare them to appreciate fully the practical significance of the new organic chemistry, based on the theoretical insights of Gerhardt and Kekulé. It was only after 1860 that these ideas were introduced into industrial practice, especially in the manufacture of synthetic dyes and drugs, through the close collaboration between university professors such as A. W. Hofmann, Adolf Baeyer and Emil Fischer with the next generation of industrial leaders, some of whom had been their students. One consequence of this development was pressure for a change from the Giessen system of chemical education by these leaders of the now powerful German chemical industry, which by the 1890s was employing more chemical Dr.phil. holders than the universities. In 1896, Carl Duisberg [1861–1935], a pupil of Baeyer and a friend of Fischer, described the hiring policy of the Bayer firm which he headed, and insisted that a more general scientific education, including physics, mathematics and biology, was required for achievement in industrial chemistry.[126] As was mentioned earlier in this chapter, the standards for the award of the Dr.phil. degree in chemistry at Giessen were rather lax, and this liberal attitude prevailed there and at other German universities for many years. By the 1890s, however, the pressure for reform led not only to the introduction of a certifying examination (*Verbandsexamen*) for chemists, but also to the upgrading of the doctoral degrees awarded by the technical universities.[127]

Conclusion

In this chapter I have attempted to examine Liebig's role as a leader, and presumed originator, of the kind of research group

---

[126] Duisberg (1896). See also Schütt (1973), Borscheid (1976), pp.51–71 and Burchardt (1980), pp.334–335.
[127] See Johnson (1985a, 1985b).

that has come to be a distinctive feature of the latter-day organization of chemical investigation. By offering a rather detailed
accounting of his pupils at Giessen during 1830–1850, in terms
of what they did there and what they did afterward, I have
sought to define more sharply the criteria for the evaluation of
their contributions to Liebig's research program and of his influence on their later professional activity, particularly in chemical research, education or industry. A necessary part of the
story has been, in the first place, the considerable role Liebig
played, especially through his collaboration with Wöhler, in the
development of organic chemistry. These achievements established his scientific reputation and, after 1840, certified the
public acclaim accorded his semi-popular writings on agriculture
and animal physiology.

Although the program of practical instruction that Liebig
developed in Giessen had its forerunners, by the mid-1830s it
was becoming widely regarded as a significant advance in chemical education. In achieving this aim, Liebig overcame many difficulties, not the least of which was the shortage of funds, and
he had to draw on his own stipend to supplement the allowance
for the expenses of the laboratory. In a long letter (12 August
1833) to the chancellor of the university, Liebig wrote:

The resources of the laboratory have been too small from the beginning. I was given four bare walls instead of a laboratory; despite my
solicitations, nobody thought of a definite sum for its outfitting [and]
for the purchase of supplies. I needed instruments and [chemical]
preparations and was obliged to use 3-400 fl. per year of my meager
stipend for the purchase of preparations; I have needed, in addition to
the attendant paid by the state, an assistant, who costs me 320 fl.; if
you subtract these two expenditures from my stipend, not much remains to clothe my children.[128]

Apart from student fees, during the 1830s Liebig began to derive additional income, some of which was used to support his
research and teaching, from his editorship of the *Annalen* and

---

[128] Volhard (1909), vol.1, pp.68–69. The salary for a laboratory attendant had been provided in 1827. It was not until 1835, however, that Liebig obtained state funds for his
assistant Ettling, and that the laboratory facilities were improved; a major renovation
and expansion occurred in 1839, and an annex (supervised by Will) was made available in 1843.

from the publication of the first of his many books.[129] These efforts, at a time when Liebig's productivity as an organic chemist was at its peak, increased the fame of his laboratory but also made severe demands on his health. There can be no question of Liebig's devotion or effectiveness during the 1830s as a leader of his research group, through his personal scientific contributions, the enthusiasm he imparted to his students, and his readiness to apply his financial resources and his seemingly inexhaustible energy to the furtherance of his educational enterprise.

After his 1837 visit to Britain, however, Liebig appears to have changed his professional goals. In ceasing his active participation, except as a critic, in the development of the organic chemistry of his time, he emerged as a propagandist for chemistry through his writings on the state of chemical education in Prussia and Austria and on agriculture and animal physiology. These activities transformed a talented chemist, highly esteemed by the leaders of his profession, into a public figure, thus enhancing further the fame of his laboratory so that, in the 1840s, it was widely believed that "no young man can be expected to know anything of chemistry unless he had studied with Liebig." As a consequence, the large increase in the number of chemical matriculants at Giessen after 1840 led to what Morrell has termed the beginning of "Big Science." It does not follow, however, that the scientist who has the largest research group in his field, and the greatest financial resources to support its activities, is the one most likely to be right in his scientific claims or to be most effective in the education of his students.

The account presented in this chapter and the data collected in the tables and appendices suggest that by the time Liebig was only forty years old, and both his public fame and the enrollment in chemistry at Giessen were nearing their peak, his role as the leader of a productive research group had changed markedly. The structure of the basic educational program, ably administered by assistants such as Will and Fresenius, was still in place, and Liebig still gave inspiring lectures, but with only a few exceptions (notably A. W. Hofmann and A. Strecker), the main service of his personal research assistants and advanced students was to provide routine analytical data Liebig wanted for his writ-

---

[129] Liebig's first book (1837) was *Anleitung zur Analyse organischer Körper* (a reprint of his contribution to Geiger's *Handbuch der Chemie*); it was soon translated into French, English and Italian. Liebig also collaborated with Poggendorff and Wöhler in the preparation of the *Handwörterbuch der reinen und angewandten Chemie;* the first volume appeared in 1837.

ings on agriculture and physiological chemistry or for his disputes with other chemists, especially Mulder. Moreover, most of
the *Liebig-Schüler* of that time who later achieved distinction as
organic chemists (for example, Williamson, Kekulé, Erlenmeyer,
Wurtz) followed the teachings of Gerhardt, who had presumed
to criticize the master, and upon whom Liebig had heaped
opprobrium.

Whatever questions may be raised about the designation, by
many of his biographers, of Liebig as the originator of the modern chemical research school, it cannot be denied that he was a
superb entrepreneur and a master of publicity, not without an
admixture of striving for self-advancement. His initiative in the
organization of the Giessen program provided additional impetus for the improvement, after 1840, of the laboratory facilities
for chemical instruction at German universities, although the influence of some of his less famous contemporaries, notably
Wöhler and Bunsen, with rather different styles of scientific
leadership, cannot be disregarded. As part of his propaganda
for the importance of chemistry, Liebig was a bold fashioner of
chemical speculations about biological problems. If his guesses,
more often than not, proved to be wrong, he stimulated, in their
disproof, lines of research that produced valuable scientific
knowledge. No German chemist before his time had attained
Liebig's public stature, but it is perhaps fair to say that this eminence had less to do with scientific or educational achievement
than with his extraordinary skill as an advocate for greater recognition of the role his discipline could play in the development
of German industry and commerce. Economic and political
factors may have been more significant in Germany's later rise to
pre-eminence as a supplier to the world of dyes, drugs and
heavy chemicals, but Liebig's stimulus, both directly and
through some of his pupils, must be counted among his most
important achievements.

An aspect of Liebig's rise to public fame, apart from the
influence of his semi-popular scientific writings, that perhaps
deserves more study by devotees of "Liebigiana," is the favorable
impression he made in his personal contacts, especially with
leaders of government and industry. The man who, in the pages
of his *Annalen* (which these worthies did not read), poured vit-
riol on his scientific adversaries, was the same charming person
whose presence they found to be impressive. His former assistant Fleitmann is reported to have said that "Never had he seen
more brilliant, tender and clever eyes than the large, wonder-
fully shining dark gray eyes of Liebig. He bore the stamp of ge-

nius on his brow, and he moved with the inborn elegance of an aristocrat of the intellect."[130] Liebig was warmly responsive to the adulation he received during his visits to England and Scotland, especially from wealthy manufacturers, political leaders and the royal entourage, and during his stay in Munich (1852–1873) he was an honored guest at the Wittelsbach court. Although sharply critical within his profession, Liebig appears to have been respectful toward established authority, and distrustful of radical political change; on 5 May 1848 he expressed to Wöhler the fear that "we approach a complete revolution, that under the flag of the republic communism will swallow us all," and also added later (21 October) in that eventful year that "instead of 36 to 40 laboratory workers I have only 10; Will, to whom I send all beginners, has 18. With fewer students, the assistants have more time to do research."[131]

In this chapter, I have attempted to dilute some of the hagiography that has characterized much of the writings about Liebig, and some readers may find my account too iconoclastic. I must add, therefore, that I share the view expressed by Brock that Liebig exerted a seminal influence on nineteenth-century chemistry.[132] If many of the seeds that he scattered brought disappointing or, at best, unexpected fruit, the ones he planted in the universities and in the embryonic chemical industry of Germany found fertile soil. I hope that he will continue to fascinate historians of science, not only because of his achievements and shortcomings as a research chemist and as a teacher, but also because of the complex and often contradictory qualities of his mind and character.

---

[130] Kohut (1904), p.222.
[131] Hofmann (1888), vol.1, pp.315, 321.
[132] In his article on "Liebigiana," Brock (1981) has rendered further valuable service to students of nineteenth-century chemistry in calling attention to the unpublished Liebig correspondence available in Giessen and in Munich.

Chapter Three

Felix Hoppe-Seyler and Willy Kühne

In his Huxley Memorial Lecture, delivered in 1924, Frederick
Gowland Hopkins stated:

The effective and continuous application of chemical methods and
chemical thought to the problems presented by living plants and animals began somewhat suddenly when the nineteenth century was already well advanced. Such effective applications were first due to
Justus von Liebig, who as a biochemist seems to me to have had no
predecessor in any sense that is real.[1]

Whatever questions this statement may bring to mind, especially
as regards the nature of the chemical methods available to Liebig and the mode of chemical thought permitted by the knowledge of his time, there can be no doubt that one important
aspect of his research program had been clearly defined by his
predecessors—namely, the isolation and characterization of
chemical substances derived from biological fluids or extracts of
plant and animal tissues.[2] This kind of separation chemistry continued a tradition established by eighteenth-century pharmacists
who were chemical craftsmen. During the first two decades of

---

[1] Hopkins (1924), p.1247.
[2] See Goodman (1972).

the nineteenth century, the most influential cluster of such craftsmen was in Paris, where Antoine François Fourcroy [1755–1809], Nicolas Louis Vauquelin [1763–1829], Louis Jacques Thenard [1777–1857] and Michel Eugène Chevreul [1786–1889] nurtured a tradition that carried into the twentieth century.[3] In particular, Chevreul's study, in the 1820s, of the constitution of the fats served as an exemplar to nineteenth-century chemists, and his book on organic analysis provided one of the clearest contemporary accounts of the strategy then used for the chemical study of the "immediate principles" of plants and animals.[4]

The novel aspect of Liebig's research program in physiological chemistry was his use of data for the elementary composition of isolated biochemical substances to set up balanced chemical equations for the changes undergone by foodstuffs in the animal body. He thus entered an area of science whose roots lay principally in Lavoisier's parallel studies on animal respiration and on chemical combustion.[5] This early connection of chemistry and physiology was emphasized by Berzelius, who in 1806 defined organic chemistry as "the part of physiology that describes the composition of living bodies, together with the chemical processes that occur in them."[6] Indeed, by the 1820s, there had been several notable achievements in the chemical study of physiological processes, among them the formation of urea and the digestion of food materials.[7] In the succeeding three decades, chemical methods were applied more frequently in physiological research, in large part because of the influence of François Magendie [1783–1855] and his pupil Claude Bernard [1813–1878] in Paris,[8] and of the remarkable scientific progeny of Johannes Müller [1801–1858], notably Theodor Schwann [1810–1882], Emil du Bois-Reymond [1818–1896], Ernst Wilhelm von Brücke [1819–1892], Hermann Helmholtz [1821–1894] and Rudolf Virchow [1821–1902].[9] Although not a

---

[3] For Fourcroy, see Smeaton (1962) and Kersaint (1966). For Vauquelin, see Kersaint (1958). For Thenard, see Thenard (1950). For Chevreul, see Malloizel (1886) and Costa (1962).

[4] See Chevreul (1823, 1824).

[5] For the most perceptive and detailed account of Lavoisier's research program linking chemistry and physiology, see Holmes (1985).

[6] Berzelius (1806), p.6. See also Simmer (1955).

[7] For urea formation, see Prévost and Dumas (1823). For their joint studies on digestion, see Tiedemann and Gmelin (1827); also see Mani (1956).

[8] For accounts of Bernard's attitude toward chemistry and chemists, see Schiller (1967), Grmek (1973) and Holmes (1974).

[9] For Schwann, see Florkin (1960) and Watermann (1960). For du Bois-Reymond, see Boruttau (1922), Marseille (1968) and Rothschuh (1971). For Brücke, see Brücke (1928). For Helmholtz, see Königsberger (1902–1903). For Virchow, see Ackerknecht (1953).

student of Müller, Carl Ludwig [1816–1895] became friendly
with several of these men during the 1840s, and also reflected
the recognition of the growing importance of chemistry in biological studies.[10] Some of these men played significant roles in
the development of physiological chemistry through their own
researches (especially Schwann, Bernard, Brücke). The others,
most of whom preferred to use physical apparatus in physiological research, nevertheless encouraged the efforts of younger associates who identified themselves as physiological chemists. For
most of these leading physiologists in France and in Germany,
Liebig's chemical equations for the processes in what was called
nutrition, metabolism or *Stoffwechsel* were not appropriate approaches to the problems of the dynamics of biochemical
change. One of the more kindly worded criticisms was offered in
1862 by the French biologist Henri Milne-Edwards [1800–1885]:

I am far from wishing to say that the speculative views of M. Liebig on
the transformations of organic matter in the interior of the animal
economy, and the use he has made of equations to show how it would
be possible to conceive the formation of the various products of
chemical-physiological work, have been useless for the progress of science. On the contrary, I believe that in giving a precise form to his
argument, he has rendered a real service and has accustomed physiologists to a mode of thought that is very useful for the study of nutrition. It is only necessary to take care in accepting these hypotheses as
the expression of what is actually occurring in the organism, where the
intermediate reactions are very complex and very important to know.[11]

This appraisal is somewhat different in tone from the retrospec-
tive one made by Hopkins in 1924, and hardly suggests the sud-
den beginning of the "effective and continuous application of
chemical methods and chemical thought to the problems presented by living plants and animals."
    Although by the 1860s the value of chemical methods in
biological research was widely acknowledged by physiologists,
they also emphasized the considerable limits of the available
chemical knowledge. Chemists (including Liebig) had not defined the properties of the "albuminoid substances," which

---

[10] For Ludwig, see Rosen (1936), Schröer (1967) and Zupan (1987). For a list of the
numerous persons who studied in Ludwig's laboratories, see Schröer (1967), pp.287–
312.
[11] Milne-Edwards (1862), vol.7, p.542.

many biologists then considered to represent the stuff of protoplasm. Also, the prevalent view that the chemical processes
whereby cellular constituents such as glycogen are made in the
"anabolic" phase of metabolism are linked to the life of tissues
and organisms placed a barrier to the idea that such processes
could ever be subjected to chemical dissection. To these limits
Louis Pasteur, a chemist turned microbiologist, added the insistence that some degradative processes, such as fermentations,
are linked to the life of microbial cells. The problems presented
by these apparent limits to the chemical study of biological phenomena did not figure largely in the research of the organic
chemists who followed Liebig, for the challenges presented by
the new ideas of structure, valence and stereochemistry, and the
fruit of their application to the synthesis of new industrial products, separated them from physiological chemistry, and they also
tended to scorn those chemists, whom they often termed
Schmierchemiker, who worked on non-crystalline materials derived
from biological sources. Consequently, very few of the German
physiological chemists who were productive during the second
half of the nineteenth century came from such laboratories as
those of Liebig, Wöhler or Bunsen. Instead, they were largely
from medical institutes of physiology or pathology headed by
former pupils of Müller, and brought to their research and to
the education of their junior associates attitudes that reflected
the scientific tensions within the physiological chemistry of their
time.

Two of these men, both of whom came to lead important
research groups, were Felix Hoppe-Seyler [1825–1895] and
Willy Kühne [1837–1900]. Both groups worked on a wide range
of biochemical problems, often the same problems, and both
made important discoveries. Moreover, many of the young men
associated with these groups later played significant roles in the
development of the biochemical sciences not only in Germany,
but also in other countries, especially England, Russia, Belgium
and the United States. Although their scientific beginnings were
similar, in their mature years Hoppe-Seyler and Kühne became
adversaries. Initially, what divided them were the usual matters
of priority and disputes about the validity of scientific claims,
but later their styles of scientific thought diverged markedly,
especially in regard to the importance of chemistry in the development of their common field of endeavor. These differences
found expression in the influence these two men exerted on
the later research and on the institutional roles of their junior
associates.

Felix Hoppe-Seyler[12]

Orphaned before he was ten years old, Hoppe-Seyler was adopted by the clergyman Dr. Seyler, husband of his eldest sister, and in 1864 he added the name of his benefactor to his original name of Hoppe. The spartan regimen of his schooling in Halle left a lasting mark on Hoppe-Seyler's character and habits. He worked hard, was attentive to duty, and enjoyed physical exercise. In 1846, he enrolled as a medical student at Halle, where he began chemical laboratory work. A year later, a chance meeting with the noted physiologist Ernst Heinrich Weber [1795–1878] led young Hoppe to transfer to Leipzig, where he worked in Weber's laboratory and studied chemistry with Otto Erdmann and with Carl Gotthelf Lehmann. Hoppe then completed his medical studies in Berlin, where he received his Dr.med. degree in 1850 for a chemical and histological study of cartilage. In 1852, after additional clinical training in Prague and Vienna, he entered private medical practice in Berlin, but found this profession to be unattractive, and two years later succeeded in obtaining an appointment as prosector in anatomy at Greifswald. He remained there only one year, because in 1856 Virchow invited Hoppe to join his new Institute of Pathology in Berlin as his assistant and also as head of the chemical laboratory that Virchow had established. This appointment marked a decisive turning point in Hoppe's career. Despite the demands of routine work, principally the preparation of histological sections, he established his reputation as a teacher and researcher in physiological chemistry.

In 1861, Hoppe became full professor of applied chemistry in the medical faculty at Tübingen. The faculty's first choice had been Carl Voit, but he declined the offer.[13] At that time there was much discussion at the University of Tübingen regarding the proposal, which was approved in 1863, to establish a separate faculty of science.[14] Hoppe accepted an invitation to join it and, as a consequence, he became closely associated with the professor of general chemistry Adolf Strecker, an outstanding Liebig pupil mentioned in the previous chapter. The two men

---

[12] The most extensive biography of Hoppe-Seyler is the obituary notice by Baumann and Kossel (1895). It includes a nearly complete list of his personal publications; those not bearing his name, by his research students and assistants, are not cited in the list. Another valuable biographical source is the lecture by Thierfelder (1926). See also Fischer (1895) and Gamgee (1895).

[13] Wankmüller (1980), pp.45–48.

[14] Engelhardt and Decker-Hauff (1963).

shared the lectures on organic and inorganic chemistry, and Hoppe-Seyler (as he now became known) adopted in his presentation the concepts that stemmed from the theories of Gerhardt and Kekulé. He also lectured on physiological chemistry and toxicology. His research laboratory at Tübingen was in the former kitchens and adjacent rooms of the old ducal castle, the famous *Schlosslaboratorium;* even by the standards of the time, the working conditions were primitive.

In 1872, Hoppe-Seyler left Tübingen to become full professor of physiological chemistry (and hygiene) in the medical faculty of the newly established German university in Strassburg, part of the booty of the victors in the Franco-Prussian war.[15] In this showcase university were assembled some of the most promising younger German scientists, among them the organic chemist Adolf Baeyer [1835–1917] and the pharmacologist Oswald Schmiedeberg [1838–1921]. At first, Hoppe-Seyler's laboratory facilities were not much better than those in Tübingen, but in 1884 he obtained a new building, the first to be constructed in Germany specifically for research and instruction in physiological chemistry.

In his personal research, Hoppe-Seyler made his most lasting contributions in the study of the chemistry of hemoglobin. He began this work in 1857, with an investigation of a phenomenon observed by the Silesian physician Wolff, who had written him that the blood of coal miners poisoned by carbon monoxide remained bright red in the venous portion of the cardiovascular system. This Dr. Wolff, about whom more deserves to be known, conducted experiments with rabbits, confirmed his clinical observation, and asked Hoppe to undertake a closer examination. From his initial studies, Hoppe concluded that CO is bound more tightly than O_2 by hematoglobulin (as the coloring matter of erythrocytes was then named). At that time, he did not know that the great Claude Bernard had already initiated experimental work on the toxicity of CO.[16] Hoppe-Seyler continued this work in Tübingen, where he carried out an impressive series of chemical and spectroscopic studies that indicated the relations among the chemical entities later known as hemoglobin, hematin, hemochromogen, hemin and hematoporphyrin. Although due credit must be given to George Gabriel Stokes [1819–1903], the Lucasian Professor of Mathematics at Cambridge, for his

---

[15] See Craig (1984).
[16] No historical account that I have read about this episode in the history of biochemistry can be compared in its completeness, accuracy or insight with that of Grmek (1973), pp.71–207.

discovery in 1864 of the spectroscopic changes undergone by oxy-hemoglobin upon the addition of reducing agents, it was the systematic work of Hoppe-Seyler that laid the groundwork for subsequent progress in this field, especially by Gustav Hüfner [1840–1908], his successor at Tübingen, and by Marceli Nencki [1847–1901], professor of physiological chemistry at Berne.

During the mid-nineteenth century, there was much speculation about the physiological mechanisms of metabolic oxidation, which was widely believed to occur in the blood, possibly with the intermediate conversion of respiratory oxygen to ozone. Hoppe-Seyler's studies on oxyhemoglobin led him into this arena, and his experimental results gave strong support to the view that oxyhemoglobin is not a physiological oxidant, but only a carrier of molecular oxygen, and the physiological oxidations are effected in the organ tissues, not in the blood.[17] Hoppe-Seyler then participated in the debates of his time on the role of "active" oxygen and "active" hydrogen in the intracellular conversion of metabolites. Although he argued with the chemist Moritz Traube [1826–1894] about this question, both men espoused the view that biological oxidations and fermentations are catalyzed by intracellular ferments whose activity is not indissolubly linked to the life of intact cells.[18] As Traube put it:

Even if all putrefactive processes depended on the presence of infusoria or molds, a healthy science would not close the road to further research; it would simply conclude that in the microscopic organisms there are present chemical substances which elicit the phenomena of decomposition. It would attempt to isolate these substances, and if they could not be isolated without changed properties, it would only conclude that all the separation methods which had been used had exerted a deleterious effect on these substances.[19]

Other aspects of Hoppe-Seyler's personal research dealt with the chemistry of milk and of bile, and some of many problems (to be mentioned later) that he suggested to his junior associates led to significant results. Also, an important part of the research done in his laboratories at Tübingen and Strassburg involved the development of new analytical and preparative methods, as well as the improvement of old ones. This emphasis found ex-

---

[17] Hoppe-Seyler (1867).
[18] Traube (1858, 1899). See also Müller (1970).
[19] Traube (1858), p.8.

pression in the successive editions of Hoppe-Seyler's *Handbuch der physiologisch- und pathologisch-chemischen Analyse,* which included not only chemical procedures, but also physical methods such as spectrophotometry and polarimetry.[20] He also wrote a textbook of physiological chemistry, published in 1877–1881.

Hoppe-Seyler figures importantly in the institutional history of biochemistry as the leading nineteenth-century advocate for the establishment of separate university departments devoted to research and instruction in this field.[21] In 1877, he launched the publication of the *Zeitschrift für physiologische Chemie.*[22] This new journal served not only in his campaign for the independent status of biochemistry, but also as the principal outlet for the research papers from his Strassburg laboratory. Before 1865, most of his papers appeared in Virchow's *Archiv für pathologische Anatomie und Physiologie.* Then, in 1866, he started a journal for the publication of papers from his Tübingen laboratory. Shortly before Hoppe-Seyler went to Strassburg, the separate issues of the journal were collected in a book[23] that contains the famous 1869 paper of Friedrich Miescher on nuclein. Between 1872 and 1877, most of the papers from the Strassburg laboratory were published either in Pflüger's *Archiv für die gesamte Physiologie des Menschen und der Tiere* or in the *Berichte der deutschen chemischen Gesellschaft.* I have dwelt on the publication of the papers from Hoppe-Seyler's successive laboratories because, as we shall see, Kühne's policy was different, and reflected the divergence in the attitudes of the two men toward biochemical problems and the institutional status of their common field.

---

[20] The first edition was entitled *Anleitung zur pathologisch-chemischen Analyse.* The sixth edition appeared in 1893, and was prepared jointly with Hans Thierfelder, under whose authorship the two subsequent editions were published in 1903 and 1909, respectively. In his obituary notice for Hoppe-Seyler, Gamgee (1895, p.624), a pupil of Kühne, wrote:

> In spite of a decided narrowness, amounting at times to unfairness, which asserts itself in nearly all Hoppe-Seyler's writings, and which caused him to attach undue importance to his own work and that of his pupils, and which explains some unfortunate omissions and deficiencies, the "Handbook" remains the recognized practical work consulted by the student of physiological chemistry.

[21] Hoppe-Seyler's most explicit appeal was in his speech at the opening of his new institute in Strassburg: *Ueber die Entwicklung der Physiologischen Chemie und ihre Bedeutung für die Medizin* (Strassburg 1884).

[22] After Hoppe-Seyler's death, his pupils Baumann and Kossel became the editors of the journal, and added his name to its masthead. See Karlson (1977) for the celebration of the centenary of the founding of the journal. For a valuable account of the history of biochemical journals, see Štrbáňová (1981).

[23] The full title was *Medicinisch-chemische Untersuchungen aus dem Laboratorium für angewandte Chemie zu Tübingen* (Berlin 1866–1871).

Willy Kühne[24]

Fredrich Wilhelm Kühne, who preferred to be known as Willy, was the son of a prosperous merchant in Hamburg. During his studies in Göttingen, where he came under the influence of Friedrich Wöhler and the physiologist Rudolph Wagner, Kühne decided to become a physiological chemist. His father wanted him to be an engineer, and sought the advice of the two elderly professors of chemistry in Berlin, Eilhard Mitscherlich and Heinrich Rose, who confirmed the elder Kühne's opinion that his son's choice was a hopeless one. Willy persisted, however, and won his father's consent. At Göttingen, Kühne did not enroll as a medical student, and he received his Dr.phil. in 1856 for a physiological study based on Claude Bernard's finding that glycosuria could be induced by puncture of the fourth ventricle of the brain. After a brief stay in the laboratory of Carl Gotthelf Lehmann at Jena, Kühne spent a year in Berlin, largely in the physiological institute of du Bois-Reymond. He met Hoppe-Seyler at that time, and although there does not appear to be any record of joint research, it is likely that Kühne learned something about hemoglobin. The high point in Kühne's education came during the next two years in Paris with Claude Bernard, whose approach to physiological problems was decisive in molding Kühne's later scientific style. Kühne then went to Brücke's laboratory in Vienna, to improve his histological technique, and in 1861 returned to Berlin as Hoppe-Seyler's successor in Virchow's institute of pathology. His lectures there on physiological chemistry were subsequently collected into a book.[25] Seven years later Kühne became a full professor of physiology in Amsterdam, and in 1871 he succeeded Helmholtz as head of the physiological institute at Heidelberg.

Kühne's research achievements were many and varied. His first important research papers, published between about 1859 and 1865, dealt with the physiological problems of muscular contraction and its innervation. During the course of this work, in which Kühne displayed exceptional skill as a histologist, he defined the structure of the myoneural junction, and also reported the preparation from minced frog muscle of a coagula-

---

[24] A biography of Kühne was prepared by Schalck (1940). See also the obituary notices by Hofmeister (1900) and by Voit (1900). Other valuable articles about Kühne are those by Leber (1903), Kronecker (1907) and Crescitelli (1977).

[25] Kühne (1866–1868).

ble protein fraction to which he gave the name *myosin*.[26] He considered this material to be the substance that solidifies upon the onset of rigor, and reported that the coagulation of myosin is hastened by the blood ferment that causes the formation of fibrin. Kühne also proposed that the red pigment in voluntary muscle is identical to blood hemoglobin. Toward the·end of his career, after about 1885, he returned to physiological and histological studies on muscle and nerve. In the intervening years (1867–1887), Kühne conducted intensive research on two topics. The first was the action of pancreatic juice on proteins, and led him to the preparation of a pancreatic ferment he named *trypsin*. During the course of this work he attempted to characterize the products formed by the action of pepsin and trypsin, and he also proposed the term *enzyme* to denote what were then called "unorganized ferments." I will return to this aspect of Kühne's research later in this chapter. The other important topic and, in my opinion, Kühne's most significant scientific contribution was his continuation of the work of Franz Boll [1849–1879] on the visual pigment of retinal rods. By the use of a solution of bile salts, Kühne succeeded in extracting the pigment (which he named *Sehpurpur* or *rhodopsin*) and showed that its bleaching upon illumination is reversed in the dark by contact with the retinal epithelium.[27]

Before he went to Heidelberg in 1871, most of Kühne's research papers appeared in Virchow's *Archiv*, but then reports of his group were published either in the *Verhandlungen des naturhistorisch-medicinischen Vereins in Heidelberg* or in his house organ *Untersuchungen aus dem physiologischen Institut der Universität Heidelberg*. In 1882, Kühne replaced Max von Pettenkofer as co-editor (with Carl Voit) of the *Zeitschrift für Biologie*. As Kühne wrote to his pupil Russell Henry Chittenden, "naturally, in the future, all my papers and those from my institute will appear in the *Zeitschrift für Biologie*, which will become essentially *my* organ."[28] Thus, between 1883 and 1900, many papers by Kühne and his associates were published there. From its inception in

---

[26] Kühne (1860, 1864). For a modern treatment of Kühne's work on muscle, see Needham (1971). For his last research on protoplasmic movement see Kühne (1888, 1898).

[27] A special issue of the journal *Vision Research*, vol.17 (1977), provides the best available historical account of the contributions of Boll and Kühne to the study of the photochemistry of the visual process, and includes English translations (by Ruth Hubbard and George Wald) of their key papers on the retinal pigments. An earlier translation of Kühne's writings on this subject was published by Michael Foster (see Kühne, 1878a).

[28] Letter from Kühne to Chittenden, 16 October 1882. Yale University Archives, Mss. Group 611, Box 1, Folder 5 (quoted by permission of the Yale University Archives). In the letter, the word *my* was underlined.

1877, the *Zeitschrift für physiologische Chemie* did not contain any papers from Kühne's institute, and it may be inferred that, during his lifetime, none were submitted for publication in that journal.

Hoppe-Seyler versus Kühne

In reading the published biographical writings about Hoppe-Seyler and Kühne, I found little mention of the adversarial relationship implied by Kühne's attitude toward Hoppe-Seyler's journal. It has been stated[29] that, during his initial stay in Berlin, Kühne became friendly with Hoppe-Seyler, but in view of later events it is possible that their mutual antagonism began at that time. In the absence of documentation on this point, I can only surmise that a contributing factor was the difference in their personalities. From all accounts, Kühne appears to have been a gregarious person, something of a gastronome, and a connoisseur of the arts. No such qualities have been attributed to Hoppe-Seyler, whose first steps in his career were fraught with more difficulties than those encountered by Kühne. It is likely that the critical sense Hoppe-Seyler displayed later, especially toward the chemical writings of physiologists, was a factor in his attitude toward Kühne, whom he may have considered to be a dilettante in physiological chemistry. Whatever the merits of my surmise may be, there is no doubt that by the 1860s their differences on scientific questions had appeared in print. Thus, in 1866 Kühne questioned the validity of Hoppe-Seyler's report that crystalline hemoglobin contains sulfur. In his textbook Kühne wrote: "Upon the combustion of several grams of recrystallized hemoglobin with sodium carbonate and potassium nitrate, I obtained no precipitate of barium sulfate by the addition of barium chloride"[30]; and he also suggested that Hoppe-Seyler's preparation was contaminated with protein material. Hoppe-Seyler's prompt reply was: "I regret that I cannot escape the thought that he [Kühne] did not wait long enough after dissolving the melt with barium chloride, since it is well known that in very dilute solutions the barium sulfate does not precipitate immediately." As for the idea that hemoglobin does not contain

---

[29] Crescitelli (1977), p.1317.
[30] Kühne (1866–1868), p.199.

sulfur, Hoppe-Seyler concluded his rebuttal with the suggestion that "Kühne does not appear to have considered sufficiently the various consequences of his assertion."[31]

This exchange was only a prelude to arguments about other matters, especially those related to Kühne's work on the digestion of proteins and his introduction of the words *trypsin* and *enzyme* in 1876. These two words, like myosin and rhodopsin, have come to occupy an important place in our present scientific vocabulary, and much has been written in recent years to celebrate the centenary of Kühne's contributions to enzymology.[32]

In 1867, Kühne published a long paper[33] on the action of pancreatic juice on proteins. He confirmed the results of Lucien Corvisart [1824–1882] who had shown in 1857 that the active agent (*pancreatin*) is effective in an alkaline medium, and is therefore different from pepsin. Kühne observed the presence of leucine and tyrosine among the products of the pancreatic digestion of proteins, but did not attach any physiological significance to these amino acids except to note that Virchow had previously found them in pathological tissues. Kühne also suggested that the metabolic fate of leucine and tyrosine falls within the domain of the *Luxusconsumption* in protein breakdown.[34] For Kühne, as for many of his contemporaries (notably Brücke), the important products were the *peptones,* which were taken to be the physiological intermediates between food proteins and blood proteins. Thus, by 1867, the gastrointestinal digestion of proteins was seen as a stepwise process wherein they are first converted by pepsin to albumoses, which are changed to the more readily diffusible peptones, whose uptake by the intestinal epithelium is immediately followed by their conversion into the proteins of the blood and lymph. This view came to be known among German physiologists as the *Peptonlehre.* During the 1880s, in addition to his other lines of research, Kühne and his pupils Chittenden and Neumeister attempted to separate the intermediates in this process by means of fractional precipitation. As others before and after them, they obtained a large number of fractions, which they characterized only in terms of solubility

---

[31] Hoppe-Seyler (1866), p.190.
[32] Gutfreund (1976). This reference provides photocopies of Kühne's two articles in the *Verhandlungen des Heidelberger Naturhistorisch-medicinischen Vereins* NF 1 (1876):190–198, in which the words *enzyme* and *trypsin* are introduced. See also Neurath and Zwilling (1986).
[33] Kühne (1867).
[34] The concept of "Luxusconsumption" had been advanced by Friedrich Bidder [1810–1894] and Carl Schmidt [1822–1894], *Die Verdauungssaefte und der Stoffwechsel* (Mitau and Leipzig 1852).

properties and elementary analysis. To denote these fractions, Kühne introduced a complex nomenclature that included names such as protoalbumose, deuteroalbumose, dysalbumose, hemialbumose and antialbumose, along with corresponding names for the various peptone fractions. For example, *antipeptone* denoted the peptone fraction which was resistant to further enzymatic attack.[35]

Franz Hofmeister, in his obituary notice for Kühne, suggested that Kühne's work on albumoses and peptones was directed to the problem of protein structure.[36] I believe that this suggestion reflected Hofmeister's aspirations in 1900, rather than Kühne's intentions, for Hofmeister had also worked on peptones during the 1800s and, in 1902, offered (independently of Emil Fischer) the peptide theory of protein structure.[37] In none of Kühne's papers, or in any of the letters of Kühne that I have read, have I found any evidence for Hofmeister's suggestion. Indeed, many noted physiologists of Kühne's time believed that the protein materials being studied by chemists had been derived from protoplasmic "living" giant albuminoid aggregates, whose structure was not accessible to study by the available chemical methods.[38]

A more recent eulogy of Kühne has also credited him with having proposed methods for the purification of trypsin that are similar to those later used in the crystallization of this enzyme, for discovering the inactive precursor of trypsin, and for recognizing the differences in the specificity of the action of pepsin and trypsin.[39] A more careful historical study would have shown these attributions of credit to be undeserved. Apart from neglecting the more definitive studies of Rudolf Heidenhain [1834–1897] in 1875 on the precursor, which he named a *zymogen*, this eulogy obscures the fact that the word *trypsin* has come to represent something rather different from the material that Kühne had prepared from pancreatic juice. He considered his "pure" trypsin to be "the most energetic of all albuminolytic enzymes"[40] because he attached great significance to the appear-

---

[35] For full accounts of Kühne's work on the digestion of proteins, see Gamgee (1893), vol.2, pp.216–230 and Moore (1898).
[36] Hofmeister (1900), pp.3877–3878.
[37] See Chapter Five of this book.
[38] Fruton (1979).
[39] Neurath and Zwilling (1986), pp.365–366.
[40] Kühne (1867), pp.158–159. Kühne's phrase calls to mind the statement by Claude Bernard (1856), p.129, that: "No other alkaline fluid of the economy brings about this kind of decomposition of albuminoid substances with such rapidity and so much intensity as the pancreatic juice. . . . This decomposing action of the pancreatic juice

ance of leucine and tyrosine among the products of its action. During the 1930s, after Moses Kunitz [1887–1978] had crystallized trypsin and chymotrypsin, and Mortimer Anson [1901–1968] had crystallized carboxypeptidase, studies on the specificity of the three enzymes[41] showed that the leucine and tyrosine observed by Kühne had arisen from the combined action of the latter two enzymes, and not from the enzyme that we now call *trypsin*, which, to use Kühne's phraseology, is one of the "less energetic" (as measured by the extent of the cleavage of most proteins) of the known proteolytic enzymes.

Research on the digestion of proteins had been proceeding in Hoppe-Seyler's laboratory in Tübingen before the appearance in 1867 of Kühne's long paper.[42] The difference in the approach from that of Kühne is suggested by a paper published in 1870 by Hoppe-Seyler's pupil Lubavin, who wrote about the decomposition of proteins by acids and pepsin as follows:

In summary, it appears that in all these decompositions of proteins there is only one principal reaction, and that with the various agents of decomposition it is only the rate and the side reactions that are different. This principal reaction is the uptake of water and cleavage, and one can distinguish two steps: the first is the formation of peptones, the second is the formation of leucine, tyrosine and probably other products.[43]

Lubavin's claim that leucine and tyrosine are among the products of the action of pepsin was another source of dispute between Kühne and Hoppe-Seyler, who defended his student's report. As we now know, the prolonged action of pepsin on proteins can give rise to the formation of these two amino acids.

Hoppe-Seyler's arguments with Kühne stemmed, in large part, from Kühne's insistence, in his 1867 paper, that the decomposition of proteins by pancreatin is not due to putrefaction (*Fäulniss*). Kühne followed Pasteur in stating: "Putrefaction is the decomposition of proteins caused by the metabolism of lower organisms, whereas any other decomposition, whether or

---

is all the more energetic when this fluid contains a larger proportion of its active organic matter." Bernard believed that he had shown that the presence of bile is required for the action of pancreatic juice on proteins.

[41] Fruton (1938).

[42] Bary (1866). This paper was based on de Bary's 1864 doctoral dissertation at Tübingen.

[43] Lubavin (1870), p.484.

not it is accompanied by the formation of smelly substances, is excluded from the definition."[44] Hoppe-Seyler, as an opponent of Pasteur's view that fermentation and putrefaction are processes linked to the activity of living organisms, did not accept Kühne's definition. In a paper published in 1870, he wrote:

I include among the putrefactive processes 1) the conversion of proteins into peptones, leucine, tyrosine, butyric acid, hydrogen sulfide, ammonia, carbon dioxide; 2) the hydration of urea to carbon dioxide and ammonia [and] of hippuric acid to glycine and benzoic acid; 3) the conversion of lactic acid to butyric acid, carbon dioxide and water [and] the similar fermentation of malic acid.[45]

Among the experiments reported in this paper was a study of the effect of phenol on the decomposition, by a mold, of urinary urea. As the concentration of phenol was increased, the organisms were killed, but the breakdown of urea to CO_2 and NH_3 continued; at the highest concentration of phenol tested, however, the cleavage of urea was inhibited. In a later paper,[46] Hoppe-Seyler offered additional evidence in support of his view that many of the fermentations considered to be linked to the life of microorganisms can proceed after the organisms had been killed by disinfectants.

By the 1870s, the agents which cause the conversion of sugar to alcohol or to lactic acid had come to be called "organized ferments," while agents such as pepsin or diastase, which cause the cleavage of proteins or starch, were termed "unorganized" (or "unformed") ferments. Hoppe-Seyler preferred to make the distinction between biological organisms and the catalytic ferments they elaborate, and insisted that, like the hydrolytic reactions, fermentations are chemical processes. Although in neither of the above two papers of 1870 and 1876 was there any direct criticism of Kühne's work, it seems likely that the appearance of the latter article stimulated Kühne to resume his studies on the digestion of proteins.

In 1876, Kühne wrote a short account of a lecture he had given in Heidelberg, and stated that "to obviate misunderstandings and to avoid cumbersome circumlocutions, the speaker proposes that the unformed or unorganized ferments, whose action

---

[44] Kühne (1867), pp.158–159.
[45] Hoppe-Seyler (1870), pp.564–565.
[46] Hoppe-Seyler (1876), pp.1–17.

can proceed without the presence of organisms or outside of them [organisms] be denoted enzymes."[47] In this account Kühne reported that he had confirmed the finding that salicylic acid is an excellent disinfecting agent, and that he had found that it does not inhibit the action of trypsin (his new name for pancreatin). Then, in 1878, Kühne published in his house organ a long polemical paper in which he stated his reasons for introducing the word *enzyme:*

The collection of the set of phenomena that we call fermentative in a single concept is, as is well known, a legacy that present-day biology received from chemistry; however, the value of the legacy became doubtful upon the acceptance of Mitscherlich's famous, though oft-fought and repeatedly newly discovered, proof that the norm of all fermentations [*Gährungsprocesse*] is the requirement for the presence of a living organism. Despite all the later discussion, there is no way around the simple fact that wort never ferments without yeast: the yeast may be able or unable to grow, it may be healthy or sick, it may require oxygen or carbon dioxide, all this does not change the fact that no one has yet found out how to make alcohol and carbon dioxide from sugar without yeast or without living organisms. The same fate of coming into the realm of biology befell in turn the fermentation of lactic acid, the formation of butyric acid, the formation of acetic acid from alcohol, as well as much else, and finally putrefaction and moldering [*Verwesung*], since all these were later recognized to be the work of organisms, although different from yeast. These vital fermentations [*Lebensgährungen*] remained joined to those elicited without organisms, such as the formation of sugar by diastase or ptyalin, the cleavage of glucosides by emulsin, the decomposition of fats by plant constituents or by pancreatic juice, as well as many other processes, and if one had considered these with some certainty to be chemical processes of hydrolytic cleavage, similar views about the others appeared to be justified, because they had been taken over into the fermentative processes [*Gährungsvorgänge*]. Nowadays, however, the necessity for the separation between the two classes of phenomena has been recognized and has already found expression in the terms "formed" ferments and "unformed" ferments, although the same person still approves the connection of the processes, and only separates them in relation to the effective agent.

The last-named designations have not, as is well known, gained general acceptance, since on the one hand it was explained that chemical bodies such as pepsin, ptyalin, etc. could not be called ferments,

---

[47] Kühne (1876). This is an abstract of Kühne's lecture to the Naturhistorisch-medicinisches Verein in Heidelberg on 4 February 1876. The translation of the quoted passage differs slightly from that of Boyde (1980), p.72.

because the name had already been assigned to yeast cells and other organisms (*Brücke*), whereas on the other hand it was said that yeast cells could not be, or be called, a ferment, because then all organisms, including man, would have to be so designated (*Hoppe-Seyler*). Without wishing to inquire further why the name has aroused such controversy [*Anstoss*], because of this contradiction I have only taken the occasion to take the liberty to propose a new one, and to denote some of the better-known substances, called by many "unformed" ferments, as enzymes. No particular hypothesis is attached, but it is merely stated that in zyme there is present something that has one or another activity that is counted as being fermentative. However, since I have not restricted the expression to the invertin of yeast, more complicated organisms from which the enzymes pepsin, trypsin and so forth may be obtained are not so fundamentally different from unicellular ones as *Hoppe-Seyler*, for example, appears to think. There was also a second reason why I decided to look for a new name (which I found to be much more attractive for our language). As is well recognized, the chemical processes thus far known to be activated by the so-called "unformed" ferments are, without exception, hydrolytic ones, whereas the "formed" ferments effect, for example, reductions and oxidations, and in many cases yield numerous products whose formation is not understandable chemically in terms of the simple cleavage and decomposition of the substance that is being fermented.[48]

Apart from warning against the acceptance of Hoppe-Seyler's distinction between organisms and ferments, Kühne's paper also included lengthy criticisms of experimental reports from Hoppe-Seyler's laboratory, in particular the claim that the cleavage of proteins by gastric juice leads to the production of some leucine and tyrosine, a phenomenon that Kühne considered to be specific for his trypsin. In his immediate reply, Hoppe-Seyler wrote:

The decisive question is whether it is too bold to hypothesize that in yeast there is a substance that decomposes sugar to alcohol and CO_2, and perhaps another that produces succinic acid.... I consider this hypothesis to be essential, because fermentations are chemical processes, and must have chemical causes.... Recently, Kühne has issued

---

[48] Kühne (1878b), p.293. The translation differs somewhat from that given by Bayliss (1908), p.6, or by Fruton (1972), pp.73–74. I cannot refrain from adding that I found the translation of this passage, as of other Kühne writings, into acceptable English, a formidable task. His convoluted sentences suggest hasty composition. This has been remarked upon by others. See Hubbard (1977). Also, Foster wrote in the preface (p.vii) to Kühne (1878a) that his wife "found the task of converting Prof. Kühne's somewhat idiomatic German into readable English not free from difficulty."

the demand that the distinction I made [between organisms and ferments] be opposed; however, since he offers no reason worthy of consideration, I do not find it necessary to say anything in reply. The new word enzyme may added to the numerous names Kühne has proposed, insofar as they are designations of substances that are still unknown.[49]

Although the last sentence was excessively harsh and unfair, it must be noted that, over the years, one of the ways to gain priority and to achieve fame in the biochemical sciences has been to propose new words to denote biochemical substances or processes. If sooner or later a word finds favor, even though its original definition has been changed, its author may be hailed as the discoverer of the substance or the process.

As it turned out, the term *enzyme* was generally adopted in England and the United States before it came into wide usage in German-speaking countries. The reason had little to do with Kühne's arguments, but rather with the close relationship between the English words *ferment* and *fermentation*.[50] For the Germans, the etymological problem did not exist, because their word for fermentation is *Gärung*, and the word *Ferment* continued to be used extensively in German writings well into the 1930s, long after some enzymes of oxidation-reduction and of fermentation had been separated from living cells.[51] For several decades, many of the French biochemists preferred the term *diastase*, introduced in 1833 by Payen and Persoz to denote what we now call amylase, and it was not until the 1920s that they began to adopt the English usage.[52] Some historians of biochemistry have sought to read into the gradual shift from *ferment* to *enzyme* a reflection of a new scientific attitude.[53] That there was a new attitude toward biochemical processes cannot be ques-

---

[49] Hoppe-Seyler (1878), pp.2–4.

[50] See, for example, Lea (1893), p.53. There, Lea (a Kühne pupil) wrote: "It appears desirable to use the word 'enzyme' to denote the soluble unorganized ferments generally, reserving the older name of 'ferment' for the organized agents such as yeast to which it was first applied." The word then appeared in the title of the book by Bayliss (1908), and in many English and American books, including the famous monograph by Haldane (1930).

[51] The word *Ferment* continued to be used by leading German biochemists, notably Otto Warburg, until the 1940s.

[52] Thus, Émile Duclaux [1840–1904] wrote in his *Traité de Microbiologie*, (1899), vol.2, p.9: "It is in recognition of this great service that I have proposed the generic name *diastases* for all substances belonging to this type. The term *enzymes*, used by some scientists, has no advantage, is newer, and does not remind one of any great discovery." See also Plantefol (1968).

[53] Kohler (1973); Teich (1981).

tioned, but one of its consequences was that by the 1930s the word *enzyme* had lost much of the meaning assigned to it by Kühne, and had acquired much of the definition of the German word *Ferment* given it by Traube and Hoppe-Seyler.

The rapid-fire exchange between Kühne and Hoppe-Seyler about enzymes lasted for only a few months, during the first part of 1878, and ended with Kühne's paper[54] entitled "Response to an Attack by Mr. Hoppe-Seyler." There had been a prelude to this exchange the year before, with the appearance of the first part of Hoppe-Seyler's textbook of physiological chemistry, and in December 1877 Kühne was moved to correct some statements about trypsin.[55] A few years later, there was another flurry after the publication of the section of the textbook dealing with Kühne's work on visual pigments. Hoppe-Seyler emphasized Boll's priority, gave him credit for some of Kühne's important findings, and also drew attention to the uncertainties in the chemical characterization of rhodopsin and of various eye pigments that Kühne had named chlorophan, xanthophan and rhodophan.[56] Kühne countered with another riposte, couched in rather uncomplimentary terms about Hoppe-Seyler's knowledge of physiology, and he asserted his priority with respect to the regeneration of rhodopsin.[57] It is not surprising, therefore, that Kühne later wrote to Chittenden that Erwin Herter (one of Hoppe-Seyler's students whom Chittenden had met) "probably is somewhat prejudiced against me by Hoppe-Seyler, who does not like me [*der mich nicht liebt*]."[58] This feeling may have been shared by others, notably Franz Boll, who angered Kühne by attempting to assert his own priority, but this dispute was terminated quickly, because Boll died in 1879, when he was only thirty years old.[59] Nor was Kühne generous in a later exchange with the able Dutch chemist Cornelis Adrianus Pekelharing [1848–1922] who criticized the work done in Kühne's laboratory on albumoses and peptones.[60]

---

[54] Kühne (1878c). In a later paper Kühne (1880, p.5) wrote "at present there is no factual basis for the explanation of bacterial putrefaction by means of enzymes present in lower organisms, because nobody has succeeded in isolating albuminolytic enzymes from bacteria." In the same paper, he also reported that extracts of liver and spleen do not contain trypsin-like enzymes because he did not observe the formation of leucine and tyrosine in the action of such extracts on proteins.

[55] Kühne (1877).

[56] Hoppe-Seyler (1881).

[57] Kühne (1882).

[58] Letter from Kühne to Chittenden, 21 November 1887. Yale University Archives Mss. Group 611, Box 1, Folder 7 (quoted by permission of the Yale University Archives).

[59] See Hubbard (1977) and Gamgee (1877).

[60] Kühne (1891).

If, in these confrontations, Kühne emerges as a somewhat less amiable person than the one described by his admirers, it must also be noted that, over the years, Hoppe-Seyler has acquired a rather unfavorable reputation as a scientific critic. His judgment has been found wanting in his rejection of the claim by Charles Alexander MacMunn [1852–1911] to have discovered distinctive intracellular respiratory pigments, and in his attitude toward the work of Johann Ludwig Wilhelm Thudichum [1829–1901] on the chemical constituents of the brain.[61] He also argued with Moritz Traube as to whether biological oxidations involve primarily the activation of oxygen (Traube's view) or the activation of hydrogen (Hoppe-Seyler's view) and also with Marceli Nencki about the chemistry of hemoglobin and about fermentation. In considering the adversarial relationship between Hoppe-Seyler and Kühne, however, it is not sufficient to attribute it to a clash of two argumentative personalities, or to a competition for public acclaim. Their disputes, I believe, reflected a difference of conviction about the proper place of modern chemical theory and methodology in the experimental study of biological problems. Despite his early chemical training, Kühne remained a physiologist who, like his mentor Claude Bernard, accorded an auxiliary, rather than a central, role to chemistry. Hoppe-Seyler, despite his early medical training, thought

---

[61] In 1884, MacMunn reported spectroscopic observations that indicated that presence in various organisms of distinctive intracellular pigments which he named *myohematin* and *histohematins*. He considered these pigments to have a respiratory function, because he found them to be capable of oxidation and reduction. A few years later, Ludwig Levy, an Alsatian medical student at Strassburg, while working on muscle pigments in Hoppe-Seyler's laboratory, came to the conclusion that MacMunn's myohematin was not a hitherto-unknown pigment, but a product of the decomposition of hemoglobin. After a brief rebuttal by MacMunn, Hoppe-Seyler entered the fray, and endorsed his student's conclusion. In a book published posthumously, MacMunn (1914, pp.72–73) wrote: "The Histohematins and Myo-hematin have not found their way into text-books because they do not belong to the ordinary pigments. . . . A good deal of discussion has taken place over this pigment, and the name of Hoppe-Seyler has prevented the acceptance of the writer's views. The chemical position is undoubtedly weak, but doubtless in time this pigment will find its way into the text-books."

  In his admirable account of the reception of MacMunn's report, David Keilin [1887–1963], who rediscovered the histohematins and named them cytochromes, indicated that not only were questions raised by Hoppe-Seyler about the validity of MacMunn's claims, but doubts were also expressed by British contemporaries who were competent in this field; see Keilin (1966), and also Margoliash and Schejter (1984). As for Hoppe-Seyler's treatment of Thudichum, the full, but overpassionate, biography of Thudichum by Drabkin (1958) makes much of Hoppe-Seyler's adverse influence, but casts Gamgee, a pupil of Kühne, as the principal villain. The importance of Thudichum's studies on the chemical constituents of the brain was not fully appreciated during his lifetime, and he engaged in acrimonious disputes with his critics, among whom was Hoppe-Seyler, who initially defended his pupil Oscar Liebreich against Thudichum's claim that Liebreich's "protagon" was a mixture. It must also be noted, however, that later work by Konstantin Diakonov in Hoppe-Seyler's laboratory demolished protagon as a biochemical entity.

TABLE 3–1

Research Groups of Hoppe-Seyler (H) and Kühne (K)

	Berlin		Tübingen	Amsterdam	Strassburg	Heidelberg
	H (56–61)	K (61–68)	H (61–72)	K (68–71)	H (72–95)	K (71–00)
Germany & Austria	6	5	22	2	51	29
Russia	4	3	12		25	6
United Kingdom	1		1	1	2	4
United States					6	6
Belgium					8	1
Other*			4		10	3
Total	11	8	39	3	102	49

*H (4 from Italy; 2 from Greece, Japan, Norway); 1 from Chile, Finland, France, Switzerland)
K (2 from The Netherlands; 1 from France)

otherwise and, in a sense, reflected Liebig's influence, as modified by the new theoretical and experimental advances in organic chemistry. Moreover, whereas Hoppe-Seyler sought to advance the institutional status of physiological chemistry, and thereby encountered the opposition of leading German physiologists,[62] Kühne does not appear to have considered these efforts worthy of support. One of the aims of this chapter is to ask whether these scientific attitudes were transmitted to Hoppe-Seyler's and Kühne's pupils, and to what extent such influences played a role in the later contributions of these junior associates to the development of the biochemical sciences.

Some numerical data for the size and national composition of the research groups led by Hoppe-Seyler and Kühne are given in Table 3–1. As best as I have been able to determine, the total for Hoppe-Seyler's junior associates at Berlin, Tübingen and Strassburg was 152, and that for Kühne's groups in Berlin, Amsterdam and Heidelberg was 60. No doubt, the difference in the totals was partly a consequence of the fact that, during the period from about 1870 to 1890, students who wished to do experimental work in physiological chemistry in Germany had a smaller choice of laboratories than those aspiring to a career in animal physiology. At that time, Kühne's institute was only one of many such laboratories, the most famous of which was that of Carl Ludwig in Leipzig. Consequently, Hoppe-Seyler's quarters in Tübingen and (until 1884) in Strassburg were crowded,

---

[62] See Pflüger (1877) and Hoppe-Seyler (1877).

TABLE 3–2

Principal Subsequent Activity of Members of Hoppe-Seyler's Research Group*

	Berlin (56–61)	Tübingen (61–72)	Strassburg (72–95)
Full (*ord*) prof. Physiol. or med. chem.		G1/2R	G3/R1/F2
Other acad. ranks physiol. or med. chem.		G2/R3	G2/R2
Full (*ord*) prof. physiol.	G1/3R	G2/F2	G4/R2/F4
Other acad. ranks physiol.	G1	G1/R1	G4/R5/F1
Full (*ord*) prof. pharmacol.		G2/R2	G2/F1
Full (*ord*) prof. pathology			R2/F1
Other acad. ranks pathol.			R2
Full (*ord*) prof. internal medicine	G2/R1/F1	R1	G2/R1/F1
Other acad. ranks int. med.			R1
Full (*ord*) prof. other clin. specialties	G2	G2	G4/F1
Other acad. ranks clin. spec.		R1	G2/R1
Full (*ord*) prof. other subjects (anatomy, botany, hygiene, pharmacy)		G1	G1/F4
Medical practice (private & hospital)		G2/R1/F2	G19/R2/F8
Pharmacy		G5	G1/F1
Other occupations			G1
Not determined		G3/R2/F1	G4/R5/F2

*G = German or Austrian; R = Russian; F = other foreign nationalities

whereas Kühne's laboratory in Heidelberg was not. In Appendices 3 and 4 are given the names, together with some biographical data, of the persons whom I have identified as having been junior members of the groups associated with Hoppe-Seyler and with Kühne.

The Hoppe-Seyler Research Group

As is indicated in Table 3–2, of the 135 (of 152) persons for whom I was able to find sufficient biographical information, 70 became full professors and 28 others attained lower academic positions. In some cases, a promising academic career was cut short by early death; for example, Konstantin Diakonov and Gustav Jüdell died before they reached the age of thirty. The total (98) far exceeds the number of Hoppe-Seyler's junior associates who later entered private or hospital medical practice (34), or who became pharmacists (7). It is likely that most of the

17 persons for whom I was unable to find biographical data became physicians.

A striking characteristic of the publications from Hoppe-Seyler's successive laboratories is that, except for his years in Berlin, when he did his initial studies on hemoglobin, most of the research papers do not bear his name, but include the statement that the investigation was conducted at Hoppe-Seyler's suggestion. His Dr.phil. and Dr.med. students, as well as his postdoctoral assistants and guests, were led to the study of a wide variety of biochemical problems drawn not only from Hoppe-Seyler's personal research interests (hemoglobin, biological oxidation and fermentation, chemistry of milk and bile) but also from other areas of the field. Moreover, Hoppe-Seyler welcomed proposals for research from his postdoctoral associates and, if he felt that an idea was worthwhile, he gave his encouragement and support. One item of testimony comes from the noted Russian physiologist Ivan Mikhailovich Sechenov [1829–1905], who went to Berlin primarily to study with Johannes Müller, but was instead drawn to du Bois-Reymond, from whom he learned electrophysiology, and to Hoppe in Virchow's institute. Sechenov later wrote:

. . . the principal place of studies at Berlin became for me the laboratory of medical chemistry just established at Virchow's institute, with its young director Hoppe-Seyler, a dear, able and lenient teacher who did not discriminate at all between the German and Russian students. . . . With Hoppe-Seyler the work consisted mainly in learning about the composition of animal fluids and we were so introduced to the methods that the research went easily and quickly. To us Russians, as mere beginners, he did not assign special topics, but he willingly heard out the proposals which came into our heads, and helped us in carrying them through with advice and action, if the plan seemed to be reasonable and feasible. Thus, he fully approved of the plan I had conceived of studying acute alcohol poisoning . . . and in his laboratory I made an analysis of expelled air for alcohol, the influence of alcoholic poisoning on body temperature (in the arteries, veins and rectum), and intoxication by inhalation of alcohol.[63]

---

[63] Sechenov (1952), pp. 115–116. My translation differs somewhat from that on p.68 of the American edition (Sechenov, 1965). During his studies abroad, Sechenov also worked in the laboratory of Otto Funke [1828–1879], head of the chemical section of Ernst Heinrich Weber's institute of physiology in Leipzig. Sechenov was not happy there, and in 1858 he returned for a few months to Hoppe's laboratory, where he studied the chemical composition of bile. He then continued his tour in Vienna at the institute of Carl Ludwig [1816–1895], with whom Sechenov established a close friend-

Sechenov's appreciation was later reflected in the relatively large number of Russian students who came to work with Hoppe-Seyler in Tübingen and in Strassburg.

Another item of evidence of Hoppe-Seyler's liberality comes from the encouragement he gave Friedrich Miescher [1844–1895], whose name in the list of Hoppe-Seyler's students is the one most likely to be recognized by present-day biochemists. Miescher, a young Swiss physician, was strongly influenced by his uncle Wilhelm His [1831–1904], professor of physiology at Basel, and went to Tübingen with the idea of studying the chemical composition of cell nuclei. This was not a line of research that Hoppe-Seyler was pursuing at that time and, as is well known, Miescher succeeded in obtaining from pus cells a phosphorus-rich preparation he named *nuclein*, which contained the substance now called DNA. The story of this discovery and of its aftermath has been told many times and in different ways,[64] and will not be repeated here, except to note that, with his usual skepticism, Hoppe-Seyler himself checked Miescher's finding.[65] In confirming Miescher's claim that nuclein is indeed a new biochemical constituent, and not an impure protein material, Hoppe-Seyler then also showed it to be present in yeast, and his pupil Pal Plósz found it in nucleated erythrocytes. After his return to Basel, where he succeeded his uncle as professor of physiology, Miescher did his outstanding work on the nuclein obtained from the sperm heads of the Rhine salmon and, with the help of the chemist Jules Piccard [1840–1933], showed it to be composed of what was later called a nucleic acid and a protein-like substance which Miescher named *protamine*. However, he did not create a research group in Basel, so that his pioneering contributions only served as the starting point for the work of others.

Although Miescher's nuclein, and its endorsement by Hoppe-Seyler, attracted the attention of leading cytologists, notably Eduard Zacharias [1852–1911] and Walther Flemming [1843–1905], it was not accepted by many physiological chemists, in part because of Miescher's idea that the phosphorus-

---

ship, and then in Helmholtz's institute of physiology at Heidelberg, where he attended Bunsen's lectures on chemistry and also met Mendeleev, who became a lifelong friend. Some years later (1862), Sechenov also worked in the laboratory of Claude Bernard.

64 Mirsky (1967); Fruton (1972), pp.183–204; Protugal and Cohen (1977), pp. 3–30. Miescher's scientific papers and some of his letters were collected in the volume *Die histochemischen und physiologischen Arbeiten von Friedrich Miescher* (Leipzig 1897), edited by his uncle. The principal biographical articles about Miescher are by Jaquet (1944), Meuron-Landot (1965) and Olby (1974).

65 Miescher (1869); Hoppe-Seyler (1869).

containing protein of egg yolk is also a nuclein. The confusion
was resolved by Albrecht Kossel [1853–1927], one of the three
most distinguished German students of Hoppe-Seyler at Strassburg, the other two being Eugen Baumann [1846–1896] and
Hans Thierfelder [1858–1930]. These men played important
roles in the later development of biochemistry in Germany.

The Influence of Hoppe-Seyler's Scientific Progeny

Kossel came to Hoppe-Seyler's laboratory in Strassburg in 1876,
after having completed his medical studies in Rostock. He remained with Hoppe-Seyler until 1883, when he moved to Berlin, where he succeeded Baumann as the head of the chemical
section of du Bois-Reymond's institute of physiology. After serving as professor of physiology (and hygiene) at Marburg from
1895 until 1901, Kossel became Kühne's successor as professor
of physiology in Heidelberg.[66]

At Hoppe-Seyler's suggestion, Kossel first worked in Strassburg on the chemistry of peptones and then on the nuclein of
yeast.[67] From this stimulus, there came, during the succeeding
twenty years, a series of important contributions to the elucidation of the nature of the cellular constituent that Miescher had
discovered. Kossel's initial achievement was to provide definitive
evidence for the separate identity of the nucleic acids (as they
were named by Richard Altmann [1852–1900]) by isolating adenine and thymine as cleavage products and by demonstrating
the presence of a carbohydrate component. Moreover, Kossel's
studies on the chemical composition of cell nuclei, with particular regard to the proteins that might be associated with nucleic
acids to form nucleins (nucleoproteins), led him to the discovery
of a peptone-like material which he named *histone.*[68] Indeed, after 1890 Kossel turned increasingly to protein chemistry, with
systematic studies on the protamines from fish sperm. The discovery that the protamines yield large amounts of the amino
acid arginine upon acid hydrolysis led Kossel to seek new methods for the quantitative determination of the amino acid composition of protein hydrolysates, an effort in which he was soon

[66] Among the biographical articles about Kossel, those by Edlbacher (1928), by Jones
(1953) and by Gerber and Sauer (1985) are especially noteworthy.
[67] Kossel (1876, 1878, 1881).
[68] Kossel (1884, 1928). See also Doenecke and Karlson (1984).

overtaken by Emil Fischer.[69] Nevertheless, the methods introduced by Kossel for the analysis of the so-called hexone bases
(arginine, lysine, histidine) continued to be widely used in the
years before the Second World War. A less enduring outcome of
Kossel's studies was his hypothesis that in the protamine clupein,
arginine recurs regularly in the repeating triplet Arg-Arg-X.
Such temptation to infer periodicity in amino acid sequences
from data on the amino acid composition of protein hydrolysates was a feature of subsequent speculations, offered during
the 1930s.[70]

At his successive laboratories in Berlin, Marburg and
Heidelberg, Kossel led sizable research groups, totalling about 85
persons. His junior associates included some of the later leaders
in research on the chemistry of nucleic acids, protamines and
histones: Kurt Felix [1888–1960], Walter Jones [1865–1935],
Phoebus Aaron Levene [1869–1940] and Hermann Steudel
[1871–1967]. Others who achieved a measure of distinction were
Henry Drysdale Dakin [1880–1952], Andrew Hunter [1876–
1969], Ernest Kennaway [1881–1958] and Albert Prescott
Mathews [1871–1957]. The research activities of Kossel's group
were channeled along the particular lines of his own scientific
interest and, in further contrast to Hoppe-Seyler's practice,
nearly all the publications bore Kossel's name as either the sole
author or as the first-named co-author.

From the available accounts, Kossel appears to have been a
rather reserved, and occasionally secretive, person in his interactions with some of his junior associates. Dakin, who was at
Heidelberg in 1904, later recalled that "Kossel gave me a copy of
Kahlbaum's catalog and told me to mark everything likely to be
of use in getting crystalline derivatives of hexone bases . . . one
thing that sticks in my mind is that except for the one in use I
was to keep the rest out of sight."[71] Although Mathews reported
that around 1900 Kossel came every afternoon to speak to the
members of his research group, Kennaway, who was in Heidelberg ten years later, wrote:

Once or twice a week he would appear, very neatly dressed, in the laboratory where the assistants and the visitors worked, and would then
have a short talk with each one in turn. His approach and manner

[69] See Kossel (1901).
[70] See Fruton (1979).
[71] Quoted from a letter, dated 10 January 1950, from Henry Dakin to Mary Ellen Jones,
with the kind permission of Professor Jones.

were abrupt; he would click his heels together and say 'Nun, wie geht es mit Ihren Untersuchungen?' One would then ask his advice; any proposed course of action was pretty sure to elicit one of his favorite phrases—'Es hat Vorteile und es hat Nachteile.' One got the impression that he was not interested in any compound, or for that matter in any research worker, which, or who, did not produce 'sehr interessante Salze' or at any rate something crystalline.[72]

In assessing these descriptions of Kossel's style of leadership, it is appropriate to note that he was largely self-taught as a chemist, except for his association with the better-trained members of Hoppe-Seyler's group in Strassburg. The skill he acquired through persistent effort, and which he imparted to his students, nevertheless produced knowledge of great importance for further advances in biochemistry, and his contributions were recognized in the award of the 1910 Nobel Prize for physiology and medicine.

One fact of Kossel's personality deserves special mention. In 1914, although he was a noted German scientist, and a Nobel Prize winner, his name was missing from the list of signatories to the notorious *Aufruf an die Kulturwelt*.[73] Moreover, according to Edlbacher, during the war years Kossel refused the demand "of high officials to appease the public by asserting that the allotted food rations were sufficient."[74]

When Kossel came to Strassburg, Eugen Baumann had already been associated with Hoppe-Seyler for several years. The son of a pharmacist, Baumann first aspired to a career in his father's profession, and prepared himself through chemical training with Herman Fehling in Stuttgart, followed by apprenticeships in Rostock and Gothenburg. In 1870, Baumann completed his studies at Tübingen, largely in the organic chemistry laboratory of Rudolf Fittig. One of his examiners was Hoppe-Seyler, who discerned Baumann's talent, at once offered him an assistantship, and then brought him to Strassburg, where Bau-

---

[72] Mathews (1927); Kennaway (1952), p.394.
[73] Upon the outbreak of the First World War, 93 leading members of the German cultural establishment signed an appeal, dated 4 October 1914, disputing German responsibility for the outbreak of the war, defending the invasion of Belgium, and denying the reports of atrocities there. The chemists on the list were Adolf von Baeyer, Emil Fischer, Karl Engler, Fritz Haber, Wilhelm Ostwald and Richard Willstätter. The medical sciences were represented by Emil von Behring, Paul Ehrlich, Albert Neisser, Wilhelm Röntgen, Max Rubner, Wilhelm Waldeyer and August von Wassermann. For the full text of the *Aufruf* and a complete list of the signers, see Kellermann (1915), pp.64–68. See also Schwabe (1969), pp.22–23.
[74] Edlbacher (1928), p.4.

mann served as Hoppe-Seyler's chief junior associate until he went to Berlin in 1877. There, Baumann became head of the chemical section of du Bois-Reymond's new institute of physiology, and in 1883 he was appointed professor of chemistry in the medical faculty at Freiburg im Breisgau. His untimely death at the age of fifty cut short a productive career marked by important biochemical discoveries.[75]

Baumann's initial research in Strassburg dealt with the organic chemistry of compounds related to creatine, but in 1875, at Hoppe-Seyler's suggestion, he shifted his attention to the problem of the metabolic origin of substances, present in urine, whose hydrolysis yields products such as phenol, catechol or indole. Baumann's isolation of a crystalline "phenol-forming" compound, and his demonstration that it is a salt of *ortho*-phenylsulfonic acid, was a major step in the study of synthetic processes in the metabolism of animals.[76] He showed that the phenol is formed from the amino acid tyrosine by bacterial action in the gut, and that a similar "detoxification" of a product of intestinal decomposition occurs in the metabolic conversion of indole to indoxylsulfonic acid. These findings led Baumann to embark on extensive chemical and metabolic studies on sulfur compounds, including the amino acid cystine; a by-product of this research was the development of *sulfonal*, for a time widely used as a dormitive agent. His chemical skill was also evident in his later isolation, characterization and synthesis of homogentisic acid (2,5-dihydroxy phenylacetic acid), found in the urine of patients with alcaptonuria. Baumann considered this substance to be derived from tyrosine, and subsequent work (notably by Archibald Garrod [1857–1936]) showed alcaptonuria to be an "inborn error of metabolism" in which an important step in the normal metabolic degradation of tyrosine is blocked. Three years before his death, Baumann began work on the chemistry of the thyroid gland. Although the favorable effect of the ingestion of iodine in the treatment of goiter had long been known, it was Baumann's success in the isolation from thyroid tissue of an organic iodine compound which exerted a curative action, and which he named *Thyrojodin*, that laid the groundwork for the subsequent advances in this field. In addition to his experimental contributions, Baumann was a severe critic of ill-founded speculations about the chemical nature of "living proteins."[77]

---

[75] For biographies of Baumann, see Tiemann (1896), Kossel (1897), Orten (1956) and Spaude (1973).

[76] Baumann (1878); see also Roy (1976).

[77] Baumann (1882); see also Fruton (1979).

Although Baumann was not the only physiological chemist of his time to bring modern organic chemistry into the study of metabolic processes (Max Jaffé [1841–1911] was another), his work set a standard that was reflected in the successes of bio-chemists such as Franz Knoop [1875–1946] and Gustav Embden [1874–1933] during the first decade of the twentieth century; their research will be discussed in a later chapter. Baumann's in-fluence is also evident in the oft-quoted passage from a lecture given by Frederick Gowland Hopkins in 1913:

My main thesis will be that in the study of the intermediate processes of metabolism we have to deal, not with complex substances which elude ordinary chemical methods, but with simple substances undergo-ing comprehensible reactions. By simple substances I mean such as are of easily ascertainable structure and of a molecular weight within a range to which the organic chemist is well accustomed. I intend also to emphasize the fact that it is not alone with the separation and identi-fication of products from the animal that our present studies deal; but with their reactions in the body; with the dynamic side of biochemical phenomena.[78]

In his elaboration of this theme, Hopkins drew special attention to Baumann's contributions.

During his relatively short scientific career, Baumann had approximately 50 junior associates in his laboratories in Berlin and Freiburg. Many of them later gained recognition for work that derived from their association with Baumann; Ludwig Brieger [1849–1919], Sigmund Fränkel [1868–1939] and Franz Röhmann [1856–1919] are examples. Baumann's pupils also in-cluded Hoppe-Seyler's son Georg [1860–1940]. Many papers from his two laboratories did not bear his name as an author and, from the available biographical accounts, it appears that Baumann was a kindly and helpful leader of his research group.

After Baumann and Kossel had left Strassburg, Hoppe-Seyler's chief junior associate was Hans Thierfelder, who studied the chemistry and metabolism of glucuronic acid, and contrib-uted to the establishment of its structure. Later, he led an active research group in Tübingen, where in 1909 he succeeded Hüfner as professor of physiological chemistry. Thierfelder's most important work was in the field opened by Hoppe-Seyler

---

[78] Hopkins (1914), p.653. This lecture was reprinted in Needham and Baldwin (1949), pp.136–159.

and Thudichum in the previous century—the chemistry of phosphatides and cerebrosides—and among his students were Karl Thomas [1883–1969] and Ernst Klenk [1891–1971], both of whom played a significant role in the development of biochemistry in Germany.[79]

Many of Hoppe-Seyler's other German students also continued to do biochemical research, despite the fact that before the First World War there were only two independent institutes of physiological chemistry at German universities (Tübingen and Strassburg). As in the case of Kossel, whose research was entirely chemical in nature, several of the former Hoppe-Seyler pupils who became professors of physiology, pharmacology, pathology or medicine in Germany later directed biochemical work at their institutes. For others, who had to content themselves with remaining heads of the chemical sections of such institutes, the opportunities for sustained research were more limited but, in some cases, not unfavorable. An example is the productive biochemist Ernst Leopold Salkowski [1844–1923] who, after working with Hoppe-Seyler in Tübingen and (briefly) with Kühne in Heidelberg, became head of the chemical division of the institute of pathology in Berlin.[80] It may be said, therefore, that although Hoppe-Seyler's hopes for the recognition, in German universities, of physiological chemistry as an independent discipline were not fully realized until the 1930s, his influence made itself felt, through the activity of his pupils, in the continued growth of biochemical research in Germany.

Hoppe-Seyler's influence was also evident abroad, especially in Russia and in Belgium. It may be seen from Table 3–1 that of the 74 junior members of his research groups who had come from countries other than the German or Austrian states, by far the largest number (42) were from Russia. In part, the influx of Russian students was a consequence of Sechenov's experience in Berlin.[81] Among the many Russians who worked with Hoppe-Seyler in Tübingen were former students of Sechenov, and who later became professors of medical chemistry, physiology or pathology at leading Russian universities. They, in turn sent some of their best pupils to Hoppe-Seyler's laboratory in Strassburg. Indeed, 16 of the 42 Russians who had been with Hoppe-Seyler in his three laboratories became full professors and another 17

---

[79] See Klenk (1931). It should be noted that Thierfelder's work on conjugated glucuronic acids owed its stimulus to prior research in Schmiedeberg's laboratory.
[80] See Neuberg (1923).
[81] Although Sechenov is best known for his research on the physiology of the brain, he also studied the respiratory exchange of carbon dioxide, and had a keen interest in chemistry (see note 63 in this chapter).

held lower ranks at Russian universities.[82] It may fairly be said, therefore, that the achievements of Russian biochemists both before the October Revolution and afterward stem to a considerable degree from the association of their teachers with Hoppe-Seyler.

It is also noteworthy that, of the eight Belgians who studied with Hoppe-Seyler, six became full professors in their home country. They included Léon Fredericq [1851–1935], who founded the distinguished school of physiology in Liège, and the great plant physiologist Léo Errera [1858–1905], who became professor of botany at Brussels.[83] On the other hand, so far as I have been able to determine, none of the future leaders in the biochemical or medical sciences in Britain came to Hoppe-Seyler for postdoctoral study, and only one of those who came from the United States gained distinction. He was William Henry Welch [1850–1934], who had gone to Strassburg to study pathology with Recklinghausen, and had also done practical work in physiological chemistry. In a letter dated 3 July 1876, Welch wrote his father that

I shall never repent the course in physiological chemistry which I am taking here. . . . The teacher, Hoppe-Seyler, may almost be said to have founded the science. He is certainly the leading authority upon the subject at the present day. I work about three hours or four a day and during that he comes to me twice and looks over my work. The German professors as a rule give the students very little personal attention, but Hoppe-Seyler is an exception in that respect.[84]

As will be mentioned later in this chapter, another American, Russell Henry Chittenden, arrived in Hoppe-Seyler's laboratory two years later, found it to be uninviting, and went to Kühne instead.

The Kühne Research Group

The fact that Kühne's group was smaller than that of Hoppe-Seyler is less significant for our purpose in this study than the large proportion of Kühne's junior associates who continued to

---

[82] For the many Russian students of Hoppe-Seyler, see Appendix 3.
[83] For Fredericq, see Florkin (1943). For Errera, see Fredericq and Massart (1908).
[84] Quoted from Bonner (1963), p.138. See also Flexner and Flexner (1941), pp.78–79. It is uncertain whether Welch belongs in the list of Hoppe-Seyler's research associates, as I found no record of a publication based on work in Hoppe-Seyler's laboratory.

TABLE 3–3

Principal Subsequent Activity of Members of Kühne's Research Group*

	Berlin (61–68)	Amsterdam (68–71)	Heidelberg (71–00)
Full (ord) prof. physiol. or med. chem.	R1		R1/F2
Other acad. ranks physiol. or med. chem.	R1		G1/R1
Full (ord) prof. physiol.	G1		G5/F5
Other acad. ranks physiol.			G6/R2/F1
Full (ord) prof. anatomy		G2	
Other acad. ranks anatomy			G2
Full (ord) prof. pharmacol.			G2
Full (ord) prof. pathology	G1/R1		R1
Full (ord) prof. internal medicine			G1
Full (ord) prof. other clinical specialties	G1		G3/F1
Full (ord) prof. other subjects (biology, plant physiology)			R1/F1
Medical practice (private & hospital)		F1	G6/F2
Other occupations			G3
Not determined	G2		F2

*G = German or Austrian; R = Russian; F = other foreign nationalities

conduct research, and the quality of their later scientific contributions. As best as I have been able to determine, of the 60 persons whom I have identified as having been associated with Kühne, 30 attained full professorships and another 14 held lower academic ranks (see Table 3–3). Apart from the four men whose later activity I have not ascertained, most of the others (at least 9) became physicians in hospitals or in private practice.

The list of Kühne's junior associates (see Appendix 4) includes many prominent names in the history of the medical sciences. Thus, at Berlin, while serving as Hoppe-Seyler's successor in Virchow's institute, Kühne taught physiological chemistry to Julius Cohnheim [1839–1884], Hugo Kronecker [1839–1914] and Theodor Leber [1840–1917]. Cohnheim, whose later studies on inflammation stand among the great nineteenth-century achievements in pathology, worked in Kühne's laboratory on the cleavage of starch by saliva.[85] Kronecker became famous not only for his notable contributions to the study of cardiac physi-

---

[85] Cohnheim (1863). This paper is reprinted on pp.10–20 in the *Gesammelte Abhandlungen von Julius Cohnheim*, E. Wagner, ed. (Berlin: Hirschwald 1885), which also includes a biographical memoir by Kühne (pp.vi–xlvi) and a list of the publications by Cohnheim and his associates (pp.xlvii–li). Also see Maulitz (1978).

ology, but also for his role in furthering the efforts of many younger scientists.[86] Leber, who became professor of ophthalmology at Heidelberg, later reported that in Berlin "with heartwarming kindness Kühne had also accepted me, as a young beginner, into his laboratory and drew me into personal relations."[87] Among the foreign guests in Kühne's laboratory in Berlin was Aleksandr Danilevski [1838–1923], later professor of medical chemistry at Kharkov, where his research program included extensive studies on myosin and on peptones.

Kühne's stay in Amsterdam was relatively brief and, from all accounts, rather unhappy. He appears to have had only three junior associates, but all of them later made important scientific contributions. One of them was Walther Flemming [1843–1905], a brilliant microscopist, who became professor of anatomy at Kiel, where he performed his pioneering studies on cell division. He introduced the word *chromatin* to denote the stainable material of the nucleus, and set the stage for the subsequent development of cytogenetics.[88] The second associate was Gustav Schwalbe [1844–1916], who later became an anthropologist.[89] Kühne's choice of Flemming and Schwalbe appears to reflect his concentration at that time on histological, rather than biochemical, problems. Kühne's third associate in Amsterdam was the English physician Thomas Lauder Brunton [1844–1916]. As a member of the staff of St. Bartholomew's Hospital in London, Brunton worked on problems in digestive and cardiovascular physiology, and also promoted the development of pharmacology in Britain.[90]

It is clear from the data in Table 3–1 that Kühne's role as a leader of a significant research group only became evident after his move to Heidelberg in 1871. He had many German pupils there, some of whom later achieved distinction, notably Leon Asher [1865–1943], who became professor of physiology at Berne, and Rudolf Magnus [1873–1927], who became professor of pharmacology at Utrecht.[91] The German students at Heidelberg who may be considered to represent Kühne's biochemical progeny were few in number; apart from Salkowski, mentioned

[86] For Kronecker, see Meltzer (1914).
[87] For the quotation, see Leber (1903), p.214. Also see Wagemann (1917).
[88] For Fleming, see Spee (1906) and Doenecke (1983). Although Flemming's work indicated the importance of nuclein in the function of the cell nucleus, Miescher dismissed the evidence obtained by means of cytological stains as having nothing to do with heredity; see Fruton (1972), pp.198–200.
[89] For Schwalbe, see Keibel (1916).
[90] For Brunton, see Anon. (1917). A glimpse of Kühne's influence on Brunton is provided by Brunton (1896).
[91] For Asher, see Muralt (1950). For Magnus, see Swazey (1974).

above, only Carl Friedrich Wilhelm Krukenberg [1852–1889], Richard Neumeister [1854–1905] and Otto Cohnheim [1873–1953] later did noteworthy biochemical research that derived its stimulus from their association with Kühne. Salkowski's studies in pathological chemistry reflect primarily the influence of the Hoppe-Seyler school, but Krukenberg's contributions to comparative biochemistry and Neumeister's work on peptones were continuations of the research they had begun in Kühne's laboratory. However, neither of the latter two men attained full professorships. According to his biographer, Krukenberg committed suicide in desperation at his unfavorable academic prospects.[92]

Special mention must be made of Otto Cohnheim, the son of one of Kühne's first pupils and one of the last of his junior associates. After receiving his medical degree at Heidelberg in 1896, Cohnheim worked in Kühne's institute on the absorption of sugar in the intestine and in the peritoneal cavity.[93] He then undertook a closer investigation of the observation made earlier by Franz Hofmeister and others that the intestinal mucosa causes the disappearance of peptones. As was mentioned above, most physiologists then believed that these products of the action of pepsin and trypsin on dietary proteins are converted to blood proteins in the intestinal wall. However, when Cohnheim examined the action of intestinal mucosa on peptones, instead of finding protein-like material, the products turned out to be amino acids, indicating the presence of an enzyme (he named it *erepsin*) which completed the gastrointestinal breakdown of food proteins to amino acids.[94] His discovery, reported in the *Zeitschrift für physiologische Chemie* a year after Kühne's death, was soon confirmed, and later work showed that Cohnheim's erepsin is a mixture of peptide-hydrolyzing enzymes (peptidases). A year after the appearance of Cohnheim's first paper on this subject, Otto Loewi [1873–1961] demonstrated that an extensively digested autolysate of pancreatic protein can replace intact protein in the animal diet. It was not long before the physiological rationale which had motivated Kühne's laborious efforts to characterize albumoses and peptones was recognized to have been incorrect; as the biochemist Otto von Fürth wrote in 1912:

---

[92] For Krukenberg, see Völker (1984). For Neumeister, see Weinland (1906).
[93] Cohnheim (1898).
[94] Cohnheim (1901).

There are few chapters in physiological chemistry that show more
clearly the rapidity with which scientific ideas change than the doctrine of the albumoses and peptones. . . . How little of what I learned
with great pains has importance today, and how many of the problems
that the previous generation of physiologists fought about so passion-
ately have lost for us any sense or significance.[95]

According to Cohnheim's biographer, Professor D. M. Matthews,
it was not Kühne's idea, but his own, that led Cohnheim to the
discovery of erepsin.[96]

Although Kühne's junior associates included many men
who made sterling contributions in the medical and biological
sciences, the role of his scientific progeny in the development of
biochemistry in Germany may be judged to have been much less
significant in this regard than that of the pupils of Hoppe-
Seyler.

Kühne's Influence Abroad

Hoppe-Seyler's considerable role in the education of the future
leaders of Russian biochemistry has been noted above, but mention must again be made of Kühne's influence in the Russian
states through the work of Aleksandr Danilevski. Also Kühne's
guests in Heidelberg included the great plant physiologist
Valdimir Palladin [1859–1922] and the notable Polish pathologist Edmund Biernacki [1866–1911]. There can be little doubt,
however, that in the Russian territories Hoppe-Seyler (along
with Marceli Nencki) outshone Kühne as a physiological chemist. During the latter half on the nineteenth century, the opposite appears to have been true in England and in the United
States, largely as a consequence of the influence of a few British
and American scientists who had been students in Kühne's institute in Heidelberg.

---

[95] Fürth (1912), vol.1, p.77.
[96] Matthews (1977, 1978). Otto Cohnheim changed his name to Kestner. He became
professor of physiology at Hamburg, but was obliged by the Nazis to leave Germany
in 1934. After a period of exile in England, he lived out his last years in Hamburg.

The British contingent included Arthur Gamgee [1841–1909], John Newport Langley [1842–1925] and Arthur Sheridan Lea [1853–1915].[97] Like Brunton, all three were elected Fellows of the Royal Society. Gamgee's scientific contributions dealt largely with the properties of hemoglobin, but in spite (or because) of this interest in a field associated with Hoppe-Seyler's name, Gamgee's attitude toward him was not entirely favorable. Gamgee wrote extensively on many biochemical subjects and, as noted above, he hailed Kühne's first report on rhodopsin. His academic career was checkered; for a time (1873–1885) he was professor of physiology at Owens College, Manchester, but he then turned to the practice of medicine. According to Sharpey-Shafer,

Gamgee may certainly be described as a genius, but erratic. If he had kept to Physiology he might have been the first physiologist of the day—at any rate the first in biochemistry, which was the part of the subject he favoured. But he divided his energies between science and clinical medicine, and thus failed that high distinction in either, to which his intellectual powers seemed to entitle him.[98]

Lea and Langley were pupils of Michael Foster [1836–1907] at Cambridge.[99] Although Foster's own research achievements were modest, he inspired a succession of men who, during the twentieth century, made Britain a leading center of physiological research. He also appears to have had an appreciation of the importance of chemical methods in biological investigation, and his admiration of Kühne's work on rhodopsin is indicated by his publication, in 1878, of a translation of Kühne's writings on this subject. At Foster's suggestion, Lea went to Kühne's laboratory in Heidelberg, where he performed a histological study of pancreatic secretion. Later, at Cambridge, Lea turned to an extension of Kühne's studies on the enzymatic digestion of proteins,[100] and wrote the book *The Chemical Basis of the Animal*

---

[97] For Gamgee, see Anon. (1909) and D'Arcy Thompson (1974). For Langley, see Fletcher (1926) and Geison (1973). In 1903, Langley succeeded Foster as professor of physiology at Cambridge; for an appreciation of Langley's talents in that capacity, see Hill (1965), pp.3–5. For Lea, see Langley (1917).

[98] Sharpey-Schafer (1927), p.30.

[99] Foster's role in the efflorescence of British physiology has been admirably treated by Geison (1978). Another of Foster's pupils, William Horsecroft Waters [1855–1887] was stated to have been in Kühne's laboratory, as well as that of Ludwig. Although Waters published a paper based on his work in Leipzig, I could not find a report of any research done in Heidelberg, and his name is not included in Appendix 4. His early death cut short what appears to have been a promising career. See O'Connor (1988), p.235.

[100] Lea (1890).

Body, which appeared as an appendix to the sixth edition of Foster's *Text-book of Physiology.* However, Foster's hope that Lea would develop an active program in physiological chemistry was not realized, in large part because of Lea's progressive spinal disease. Moreover, as Langley wrote in his obituary notice for Lea, physiological chemistry

as it was carried in England in the '70's and '80's did not involve any profound chemical knowledge; it was mainly concerned with the experimental determination of simple but fundamental reactions of the fluids and tissues of the body. Lea's knowledge of chemical theory and methods was more than adequate for the conditions of the time.[101]

In 1898, Foster replaced Lea by Frederick Gowland Hopkins.

The British student of Kühne who achieved the greatest distinction was Langley. He worked in Heidelberg on salivary secretion and, upon his return to Cambridge, made important contributions to the study of pepsinogen and of the physiological mechanisms of glandular secretion. Langley then performed brilliant research on what he called the *autonomic* nervous system and, through the use of drugs and histological methods, he clarified many previously obscure aspects of the transmission of nerve impulses to effector tissues.

The high regard in which Kühne was held in England is evident not only from the acclaim that greeted his publications on rhodopsin, but also from the invitation to deliver the Croonian Lecture to the Royal Society in 1888. It would seem, however, that his immediate influence, through his pupils, on the development of physiological chemistry in Britain was not great. Indeed, in a lecture he gave in 1924 on the status of biochemistry, Hopkins said:

It is unfortunately the case that until the beginning of the present century, English work did painfully little to advance the subject. There was less interest in it here than in any other country of importance, and this even in the case of professed physiologists and biologists, and quite certainly in the case of pure chemists. In Germany and Scandinavia much work was being done even 50 years ago, and Slavonic

_____
[101] Langley (1917), p.xxv.

names have always been prominent in the literature. The schools of Hoppe-Seyler, Hofmeister, Nencki and Hammarsten were centres of highly productive activity.[102]

When Hopkins spoke these words, he had been professor of biochemistry at Cambridge for ten years, and had gained fame for his work on tryptophan and glutathione, for his studies on what he called "accessory food factors" and (with Walter Fletcher) on the formation of lactic acid in muscle. It was not until 1925, however, when adequate quarters were provided for Hopkins's department of biochemistry, that it could become a "centre of highly productive activity."[103]

Clearly, Hopkins was deploring the delay in the establishment, at British universities, of sizable research groups in his field, and was well aware of the important contributions to biochemical knowledge made in Britain during the latter half of the nineteenth century. Most of these achievements, however, came from the efforts of individual investigators, such as Thudichum or the so-called chromatologists, notably Henry Clifton Sorby [1826–1908] and MacMunn (mentioned above), as an adjunct to their principal professional activity, whether in medicine or geology. Also, several British chemists and botanists, some of whom were associated with the thriving brewing industry made significant contributions to enzyme chemistry; among these men were Cornelius O'Sullivan [1847–1907], Joseph Reynolds Green [1848–1914] and Sidney Howard Vines [1849–1934].[104] A notable exception to Hopkins's generalization appears to have been the professor of physiology at King's College London, William Dobinson Halliburton [1860–1931] who, before the end of the nineteenth century, had initiated a program of research in chemical physiology that attracted many students.[105] Halliburton's principal scientific contributions lay in the fractionation of muscle proteins. He obtained, in addition to Kühne's myosin, a protein fraction whose properties were later noted to accord with those of actin.[106] Also, Halliburton reported that he could

---

[102] Hopkins (1924), p.1247. This important address was not included in the commemorative volume edited by Needham and Baldwin (1949), and merits closer study by historians of the biochemical sciences.

[103] Among the numerous writings about Hopkins, the articles by Needham (1962) and by Pirie (1983) are especially noteworthy.

[104] For O'Sullivan, see O'Sullivan (1934). For Green, see Vines (1915). For Vines, see Randle (1934). See also Morgan (1980).

[105] See Morgan (1983).

[106] See Needham (1971), p.133.

not confirm Kühne's claim that the coagulation of myosin is caused by the fibrin-forming agent in blood.

As regards the United States, there can be little doubt that Kühne exerted a considerable influence on the development of physiological chemistry primarily through the significant role that his loyal disciple Russell Henry Chittenden [1856–1943] played in the education of many future leaders of American biochemistry. Of the five other American pupils of Kühne, three also became university professors. Henry Sewall [1855–1936], as professor of physiology at the University of Michigan, stands out among them.[107] The other two were Sigmund Pollitzer [1859–1937], who became professor of dermatology at the New York Postgraduate Medical College and the other was Herbert Eugene Smith [1857–1933], who received an M.D. degree after his undergraduate studies with Chittenden at Yale, and then followed him to Kühne's laboratory. Upon his return to Yale, Smith served from 1885 until 1910 as professor of chemistry and dean of the medical school. He does not appear to have had a flair for research and, like Sewall and Pollitzer (though for different reasons), did not contribute significantly to the advance of physiological chemistry.[108] What follows, therefore, will deal largely with Chittenden's association with Kühne, and his place in the history of the development of the biochemical sciences in the United States.[109]

Before he was twenty years old, as an undergraduate in the Yale Sheffield Scientific School, Chittenden had achieved a modest success in the study of the chemical composition of scallops. He found glycine, as well as glycogen, thus satisfying the curiosity of Professor Samuel W. Johnson, who had suggested the problem because he had noticed that when cooked scallops were kept for a time, and then reheated, they tasted sweeter. Before he received his bachelor's degree in 1875, Chittenden also served as a laboratory assistant in the chemistry course for pre-medical students and, according to his account, was placed

---

[107] Webb and Powell (1946); see also Davenport (1982).
[108] For Pollitzer, see Anon. (1938). For Smith, see Anon. (1935). The other two Americans in Kühne's laboratory were William C. Ayres and Benjamin Miller Van Syckel, both of whom entered medical practice upon their return from Heidelberg. In a postal card to Chittenden (dated 13 October 1883), Kühne recommended Van Syckel, who had applied (unsuccessfully) for a post at Yale, as an able histologist. Ayres worked with Kühne on rhodopsin and then established an ophthalmological practice, first in New York City, and later in New Orleans. In paper entitled "Postgraduate study in Europe," New York Medical Journal 37 (1883):204–205, Ayres offered useful advice to American physicians who were planning to study abroad, and recommended Kühne's laboratory highly.
[109] Vickery (1945). Much of this memoir is based on Chittenden's unpublished autobiography, written during the 1930s, and deposited in the Yale University Archives.

in charge of instruction in physiological chemistry. In 1878, he went to Germany with letters of introduction to Hoppe-Seyler. As Chittenden reported many years later:

On reaching Strassburg, however, and presenting the letters of introduction, and after viewing the facilities provided in the laboratories, there arose a grave suspicion that perhaps Strassburg after all was not best adapted for my needs. There were so many people working in the laboratory, everything was so crowded and lacking in orderly arrangement, there seemed so little opportunity for much personal attention, that there was a natural hesitation to take the decisive step. . . . There was nothing attractive about either the city or the university in those early years after the Franco-German war of 1870. To be sure, Hoppe-Seyler was a great man and the laboratory was a beehive of activity, yet it seemed to lack something—atmosphere if you choose—essential for the proper development of a youthful mind that needed guidance and encouragement.

Intuition is not wholly to be ignored, and I went on to Heidelberg. . . . The situation was somewhat awkward, but assuming a confidence that was not wholly felt, Kühne was sought and the hope expressed that a place might be found in his laboratory. . . . [M]uch to the writer's surprise, he said, 'Are you the Chittenden who published in Liebig's *Annalen* a year or two ago an article on glycogen and glycocoll?'. . . . The atmosphere was completely changed, and my spirits rose accordingly, reaching a higher level when Kühne remarked that he would find a place for me in the laboratory at once.

Going to Heidelberg was a wise choice, as the experiences of the next few weeks showed, and indeed, the experiences of later years likewise proved. The Heidelberg laboratory was characterized by the great diversity of interests represented, many sides of physiology being studied, by the German, English, Russian and American workers there. At the same time, numbers were small, hence there was ample opportunity for Kühne to exercise his influence over us all. He himself carried on his own experimental work nearby, from which we all derived some profit. . . .

More than twenty years of close friendly relations with Kühne, up to the time of his death in 1900 only strengthened the first, early impressions of the man. He was as large and broad mentally as he was physically, kind-hearted to a high degree, an ideal teacher and an experimental worker of great skill and daring, whose every thought was given to the advancement of physiology.[110]

Two of the three publications that came from Chittenden's work in Heidelberg were based on histological studies. The third re-

---

[110] Chittenden (1930), pp.29–30.

ported the formation of xanthine from albumin; as noted above, this was soon disproved by Kossel. According to Chittenden, Kühne also asked him to serve as an assistant in his lecture demonstrations.

After his return to Yale in 1879, Chittenden received his Ph.D. degree, and in 1882 was appointed professor of physiological chemistry, thus becoming the first person to hold this title in the United States. In that year, as Chittenden later wrote:

... at the request of Professor Kühne, I went back to Heidelberg for the summer months to undertake with him an investigation of the primary cleavage products of albuminous bodies, which Kühne had recently discovered. This was the beginning of a period of cooperation in research between the Heidelberg laboratory and the Sheffield laboratory of physiological chemistry which lasted for eight years, with results decidedly stimulating to the scientific atmosphere of the New Haven laboratory. It also resulted quite naturally in focussing the attention of the workers in the laboratory on problems connected with digestion, so that in the Sheffield laboratory many experimental studies were carried on dealing with the chemico-physiological processes of the gastrointestinal tract.[111]

Indeed, between 1883 and 1900, many of Chittenden's students at Yale worked on the digestion of proteins, but his laboratory also concerned itself with problems of toxicology, especially as related to the effects of arsenic and alcohol. Chittenden's more general biochemical writings at that time echo some of the chemical speculations of leading physiologists, notably those of Pflüger.[112] In 1898, Chittenden was appointed director of the Sheffield Scientific School; soon after that date he ceased to direct biochemical research and his scientific writings dealt largely with problems of human nutrition. His standing in the American academic community was high (he had been elected to the National Academy of Sciences in 1890) and, when new professors of physiological chemistry were sought, Chittenden was asked for recommendations.

[111] Ibid., p.38. This joint research on albumoses and peptones was conducted entirely by correspondence. The Yale University Archives hold, in their Chittenden collection, over 100 letters and postal cards from Kühne to Chittenden. Regrettably, the correspondence in the other direction does not appear to have been preserved.
[112] See, for example, Chittenden (1894), pp.112–117, where he discusses the possible transformations of the purines without the use of a single structural formula or any reference to the work of Emil Fischer during the 1880s.

Chittenden's account of his decision to work with Kühne, rather than with Hoppe-Seyler, has been taken by Kohler to indicate that

wittingly or unwittingly, Chittenden's abrupt change of plans was a strategic decision. Kühne's program was closer to what Chittenden had known at the Sheffield Scientific School and was more appropriate to his future plans there. Hoppe-Seyler's institute combined physiological and pathological chemistry and had strong links to clinical medicine. The Sheffield School had no connection with clinical medicine and offered physiology as a basic biomedical course, as Kühne did.[113]

The accuracy of this description may be questioned, because Hoppe-Seyler and Kühne were both members of the medical faculties of their respective universities, and I know of no evidence to suggest that Kühne's ties to his clinical colleagues were any less close than those of Hoppe-Seyler at Strassburg. Also, the fact that in 1885 Chittenden's Yale student Smith became dean of the Yale medical school suggests some "connection" between Chittenden and that school. It seems more likely that what Chittenden wrote in 1930 about the relation of physiological chemistry to medicine may have been strongly influenced by his bitterness at the decision, in 1921, of the Yale administration, to abolish the Sheffield Scientific School as a separate educational division of the university, and to transfer the department he had fathered to the medical school. As regards what Kohler termed Chittenden's "strategic decision," it may be suggested that Kohler imputes more calculated thought than seems warranted to a 22-year-old American who, though "impressed by the pace of research" at Strassburg, was reportedly so repelled by the crowded conditions of Hoppe-Seyler's laboratory and the unattractiveness of the city that, after a brief inspection, he chose to go elsewhere. Since Chittenden's autobiographical writings are the only available source of information about this episode in his career, the truth may never be known, but it seems at least equally likely that this proud young man may have sensed that he would be outclassed in Hoppe-Seyler's institute, and that he may have considered Hoppe-Seyler's reception insufficiently hospitable. According to Kohler, Chittenden recalled his "surprise and delight that he was a better chemist than any

[113] Kohler (1982), pp.98–99.

of the European students" in Kühne's laboratory. If true, this claim says more about the level of chemical knowledge and skill in that laboratory than about Chittenden's ability as a chemist, as measured by the German standards of the 1870s.

In their chemical content, the papers published by Chittenden from the Sheffield Laboratory before he ceased his research compare unfavorably with those then coming from the laboratories of Baumann and Kossel. In calling attention to this contrast between the biochemical work of a famous pupil of Kühne and that of two of Hoppe-Seyler's scientific progeny, I only wish to suggest that Kühne, a great physiologist who used the methods of histology and chemistry to approach important physiological problems, placed less emphasis on the thorough chemical training of his pupils than did Hoppe-Seyler. If, as Chittenden claimed, his modest chemical experience was greater than that of most of Kühne's other junior associates, Kühne may have taken advantage of this circumstance and thus encouraged an ambitious young man, later to emerge as a successful academic entrepreneur, to believe that he was following in the path of his teacher.

Several of Chittenden's students at Yale rose to professorships. In the development of physiological chemistry in American universities, the most important one was Lafayette Benedict Mendel [1872–1935]; after his post-doctoral studies with Rudolf Heidenhain in Breslau and with Eugen Baumann in Freiburg, Mendel succeeded Chittenden as head of the Department of Physiological Chemistry at Yale. Yandell Henderson [1873–1944] also remained at Yale, as professor of applied physiology and head of a section devoted to alcohol studies. William John Gies [1872–1956] became chairman of the Department of Biological Chemistry at the Columbia University College of Physicians and Surgeons; his tenure of this office was undistinguished, as were the research achievements of his junior colleagues. The most eminent among Chittenden's students was Alfred Newton Richards [1876–1966] who, as professor of pharmacology at the University of Pennsylvania, made outstanding contributions to the study of the physiology of the kidney.[114]

Under Mendel's leadership, the Department of Physiological Chemistry at Yale became a leading center of research on problems of nutrition, largely as a consequence of his collaboration with the outstanding protein chemist Thomas Burr Osborne [1859–1929] of the nearby Connecticut Agricultural

---

[114] For Mendel, see Chittenden (1937). For Henderson, see Haggard (1944). For Gies see Clarke (1956). For Richards, see Schmidt (1971).

Experiment Station. Their studies on the nutritive qualities of highly purified proteins showed the need, in the animal diet, for certain amino acids, and also laid the groundwork for later research on vitamins. Among Mendel's students who later exerted significant influence on the development of biochemistry in the United States were Howard Bishop Lewis [1887–1954], who became professor of physiological chemistry at the University of Michigan medical school, and William Cumming Rose [1887–1985], who became professor of biochemistry in the Department of Chemistry at the University of Illinois.[115]

By 1930, Chittenden's original conception of physiological chemistry, which he owed to Kühne, as "simply a part of physiology," had given way to the wider recognition by other American educational leaders that this field is also a part of chemistry. One sign of this change in attitude was the appointment, in the late 1920s, of the physical chemist William Mansfield Clark [1884–1964] and the organic chemist Hans Thacher Clarke [1887–1972] to head the departments of biochemistry in the medical schools at Johns Hopkins and Columbia respectively.[116] Another sign was the recognition of the contributions of men such as Donald Dexter Van Slyke [1883–1971], a well-trained organic chemist at the Rockefeller Institute for Medical Research, to the development of modern clinical chemistry.[117] Indeed, by 1930, the Rockefeller Institute had emerged as a leading center of research into a variety of fundamental biochemical problems, such as the constitution of nucleic acids and the purification of enzymes.[118] However, in one important area of investigation, namely the elucidation of the chemical pathways of intermediary metabolism, research efforts in the United States lagged behind those of German and Austrian biochemists, and the study of animal metabolism was largely dominated at American medical schools by the physiologists who drew inspiration from the calorimetric approach of the Pettenkofer-Voit-Rubner schools. The situation changed dramatically when Hitler came to power in Germany; among the younger exiles was Rudolf Schoenheimer [1898–1941], a physician well trained in organic chemistry, who came to Columbia University, and introduced the isotope technique into the study of intermediary metabolism. It should be added that Hans Clarke's welcome, not only to Schoenheimer, but also to several other emigrés, transformed the De-

[115] For Lewis, see Rose and Coon (1974). For Rose, see Roe (1981).
[116] For Clark, see Vickery (1967). For Clarke, see Vickery (1975).
[117] See Hastings (1976).
[118] See Corner (1964).

partment of Biological Chemistry into an outstanding center of biochemical research.[119] As the United States approached the Second World War, what remained of Chittenden's view of biochemistry had largely been supplanted by a chemically oriented approach, more akin to that derived from the aspirations of Hoppe-Seyler than from those of Kühne.

Conclusion

In this chapter, I have attempted to indicate some of the tensions in the emergent physiological chemistry of the latter half of the nineteenth century by focusing attention on the careers and research groups of two great scientists who became adversaries. I have suggested that their arguments arose from a deeply felt difference in the views about the role of chemistry in the study of biological phenomena, and that this difference was reflected in the later activities of their scientific progeny. For Kühne, as for other physiologists of his time, chemistry was a valuable tool, together with histology, in biological research, but he does not appear to have considered a thorough training in the modern chemistry of the day to be essential for the education of experimental biologists. Indeed, an examination of Appendix 4 indicates that the work done by his junior associates at Heidelberg involved rather more histology than chemistry. It was otherwise in Hoppe-Seyler's laboratories in Tübingen and Strassburg; in holding to the conviction that the understanding of biological phenomena lies in the elucidation of the structure and interactions of the chemical substances of which living things are composed, Hoppe-Seyler carried forward a modernized version of Liebig's teachings.[120] For most of the leading organic chemists of Hoppe-Seyler's time, however, the main tasks were the promotion of their subject for its own sake and for the benefit of chemical industry, and they tended to regard the physiological chemists who studied such materials as proteins, nucleins or ferments as *Schmierchemiker*. This attitude began to change at the turn of the century, when some of the biochemical

---

[119] For Schoenheimer, see Peyer (1972) and Kohler (1977). For an account of Clarke's reception of the German and Austrian exiles, see Chargaff (1978), pp.66–76.

[120] It is worthy of note that Hoppe-Seyler's predecessor at Tübingen was Julius Eugen Schlossberger [1819–1860], a student of Liebig. For valuable accounts of Schlossberger's role in establishing physiological chemistry at Tübingen, see Hesse (1976) and Wankmüller (1980).

problems tackled in Hoppe-Seyler's laboratories came into the mainstream of organic-chemical research. Among the leaders in this development were men who had been junior associates of Adolf von Baeyer, the pupil of Kekulé.

Chapter Four

Adolf Von Baeyer and His Scientific Progeny

Adolf Baeyer [1835–1917] led research groups at three institutions: the *Gewerbeinstitut* in Berlin (1860–72), the University of Strassburg (1872–75) and the University of Munich (1875–1915).[1] Initially a pupil of Bunsen, he came under the spell of Kekulé's ideas; Baeyer's Dr.phil. dissertation (Berlin 1858) was based on work he had done in Kekulé's private laboratory in Heidelberg, and he then joined Kekulé at Ghent. In 1905, Baeyer wrote:

Although upon my entry into his [Kekulé's] laboratory I had only learned how to analyze, the knowledge I had acquired during my boyhood allowed me to undertake independent experimental work. On the other hand, my theoretical knowledge was very inadequate. . . . In this respect, his lectures and my association with him opened a new

---

[1] Baeyer was ennobled by Ludwig II of Bavaria in 1885; the *von* will be omitted in the text and notes of this chapter. For an extensive list of biographical writings about him, see Poggendorff (1971), pp.45–48. Among them the book by Schmorl (1952) and the tributes by Perkin (1923) and by Rupe (1923) are especially noteworthy. On the occasion of Baeyer's seventieth birthday, there appeared a two-volume set containing his autobiographical notes; see Baeyer (1905), pp. vii–xx,xxviii–lv. It also included a list of research publications from 1857 until 1905. (ibid.,pp.lvi–cxviii) and the text of about 300 papers that had appeared during this period. For his eightieth birthday, a series of articles in *Naturwissenschaften* by Willstätter and other noted German chemists described various aspects of Baeyer's scientific work; these and other articles will be cited individually below. A valuable recent account, written from the vantage point of more modern organic chemistry, was prepared by Huisgen (1986).

world for me. Younger chemists cannot gain from the literature an adequate idea of the influence that the young Kekulé exerted on his contemporaries. . . .Carried away by the logical coherence of the new doctrine, later named *Strukturchemie,* he constructed for his enthralled listeners the edifice of theoretical chemistry in which we still reside.[2]

Baeyer's personal contributions to chemical knowledge were considerable, and bear the stamp of Kekulé's influence, but they are outweighed in their importance for the development of modern organic chemistry by his role as the teacher of an outstanding group of younger chemists. For this reason, in this chapter I consider not only the research group associated with Baeyer in his own investigations, but also the junior associates (and their students) who worked independently in the laboratories he headed. Baeyer's style of leadership of this larger community of scientific effort made his successive laboratories, especially the one in Munich, major breeding grounds of chemical talent during the last quarter of the nineteenth century, when Germany emerged as the world leader in university chemical research and in the industrial production of chemicals.

The Berlin and Strassburg Periods

Baeyer's entry into the chemical profession appears to have been inauspicious. According to his account, a cool reception was accorded in Berlin to his dissertation "of which I had right to be proud, as the work was done on my own initiative and performed by me entirely independently."[3] After his stay in Ghent, with modest success in the study of the constitution of uric acid, Baeyer used this work to become *Privatdozent* in Berlin.[4] Although his *Habilitation* lecture (in Latin) was received more favorably, no space was available to him at the university, and he did not have the financial means to set up a private laboratory. Fortunately for Baeyer, his family was well connected; his father

---

[2] Baeyer (1905), pp.xiv–xv. The warmth of the relationship between Baeyer and Kekulé is evident from their personal correspondence during the period 1860–1868. I am indebted to Prof. Klaus Hafner, Darmstadt Technische Hochschule, for sending me photocopies of these letters.

[3] Ibid, p.xiii.

[4] Baeyer recalled in 1905 (ibid., p.xiii) that, on his way to Ghent in 1858, he met Adolf Schlieper (see Appendix 1), who gave him a small box containing the remainder of the uric acid derivatives Schlieper had prepared during the 1840s in Liebig's laboratory, and asked Baeyer to undertake their further investigation.

(General Johann Jacob Baeyer) had many influential friends, and from one of them he learned of an opening at the *Gewerbeinstitut* (a technical vocational academy) in Berlin. Baeyer's application was initially declined, but it seems that the Crown Prince, a friend of Adolf's brother, "interceded with the *Kultusminister* on his behalf."[5]

The personal qualities and scientific acumen of the young Baeyer became evident during twelve years at the *Gewerbeinstitut*. By 1862, adequate laboratory facilities for about 15 people had been provided; with the help of several assistants (Damm, Deichsel, Heintzel, Herzog), Baeyer continued his work on the constitution of uric acid, and established its relationship to malonylurea, which he named barbituric acid.[6] After 1864, he moved to the study of the condensation reactions of aldehydes and ketones, the reduction of isatin and other organic compounds by means of sodium amalgam or by distillation in the presence of zinc dust, the constitution of mellitic acid, and the reactions of acetylenic compounds. Many of the results of these investigations provided the background for the later research of Baeyer and his junior associates. Moreover, by 1865, he had begun to attract able research students, among them Max Berend and Carl Liebermann, and, as Baeyer later wrote: "A great event for the laboratory was the arrival of C. Graebe, who had already gained extensive training as a chemist."[7]

Carl Graebe's role in the emergence of Baeyer's laboratory from relative obscurity cannot, in my opinion, be overestimated. A pupil of Bunsen, Kolbe and Erlenmeyer, Graebe had already made important contributions to the study of quinones, and he brought to Berlin the idea that the commercially important dye alizarin (from madder) is a quinone. He wrote in a letter (27 February 1868) to his parents:

---

[5] Willstätter (1956), p.111; in the English translation, see p.118.

[6] For a description of Baeyer's laboratory in Berlin, and the terms of his appointment at the *Gewerbeinstitut*, see his letter to Jean Servais Stas [1813–1891], who had befriended Baeyer in Ghent; Gillis (1960), pp.25–26. For accounts of Baeyer's work on the constitution of uric acid, see Dieckmann (1915) and Huisgen (1986).

[7] For the quotation, see Baeyer (1905), p.xviii. It may be noted that in 1865 there was an interregnum in organic chemistry at the University of Berlin, as the successor of Eilhard Mitscherlich [1794–1863], August Wilhelm Hofmann, had just been appointed and his new institute was under construction. Baeyer later wrote: "For me the appointment of Hofmann was actually a happy development, since I no longer stood alone and limited to contacts with my own research associates. Hofmann also showed his good will in 1866, by meeting my request for an appointment as an unpaid *Extraordinarius*" (ibid., p.xvi).

In order to advance more rapidly, I joined with Dr. Liebermann, who works in our laboratory. Last Friday we went to work energetically, and were fortunate to get at once on the right track. Sunday was spent in the laboratory and on Monday evening at about 7 o'clock we had come so far that we were able to report a very important result at the meeting of the chemical society that began at 7:30.[8]

The "very important result" was the demonstration that the reduction of natural alizarin with zinc dust produced anthracene, and this discovery was followed a few months later by the synthesis of alizarin (dihydroxyanthraquinone) from anthracene, readily available from coal tar. This achievement attracted wide public attention, for it was comparable in importance with Perkin's synthesis of mauveine, and brought the growing German dye industry into active competition with its British and French rivals.[9] Also, it brought luster to Baeyer's laboratory in Berlin.

For the theme of this chapter, Baeyer's role in the success of Graebe and Liebermann is of special significance. It is not usual for a chemist at the beginning of his scientific career to be self-effacing when junior associates in his laboratory make a striking new discovery. It would seem, however, that the style of leadership that Baeyer displayed repeatedly in later years, when he had attained great fame, was evident in the encouragement he gave to his research associates in Berlin. According to Willstätter,

It was Baeyer's idea and wish that Graebe should apply the zinc dust distillation method to alizarin in order to determine the nature of its parent substance and to elucidate its constitution. Graebe, however, was not inclined to follow this advice. He replied, "Do it yourself!"; zinc dust distillation was Baeyer's method and it was up to Baeyer to apply

[8] The quotation is from Duden and Decker (1928), p.31. In this valuable obituary notice, Decker notes (p.27) that, in his later research, Baeyer owed more to Graebe than had previously been acknowledged. Although the personal relationship between the two men appears to have been cordial, it is noteworthy that in 1916 Baeyer begged Willstätter not to write his biography and is reported to have said: "As long as Graebe is alive, you should not write about the history of alizarin"; Willstätter (1956), p.103; in the English translation, see p.110. For a biography of Liebermann, see Jacobson (1918).
[9] For accounts of the commercial rivalry in the patenting of the synthesis of alizarin, see Duden and Decker (1928) and Schmauderer (1971).

it. Finally, Baeyer phrased his suggestion in more definite terms: "Graebe, you are my assistant and I order you to distill alizarin with zinc dust." Baeyer reported this course of events to me in these words.[10]

Although this account is somewhat at variance with the picture drawn by some of Graebe's biographers,[8] and its accuracy may be questioned, there is no doubt that Baeyer's justifiable pride in Graebe's success, which he hailed on repeated occasions in succeeding years, was not tainted by any public claim of his own for credit in that success.[11]

Graebe left Baeyer's laboratory in 1869, and shortly afterward became full professor of chemistry at Königsberg; despite unfavorable conditions, he continued to do brilliant work in the synthesis of dyes and in the study of aromatic compounds. He was unable to get a professorship at a better German university, and in 1878 found a haven in Geneva where he was happier and highly productive. Upon his retirement in 1906, Graebe lived out his years in Frankfurt a.M. (his birthplace), and devoted his energies to the preparation of an important treatise on the history of organic chemistry. As for Liebermann, he chose to remain in his native Berlin, and succeeded Baeyer as head of the chemical laboratory in the *Gewerbeinstitut,* which became part of the Charlottenburg Technische Hochschule, established in 1879. There he led a sizable research group in research on synthetic dyes and on the structure of natural products.

Shortly before Graebe's departure in 1869, Baeyer was joined in Berlin by the twenty-year-old Victor Meyer who, like Graebe, had received his Dr.phil. for work done with Bunsen in Heidelberg.[12] A man of verve (he had initially aspired to be an actor), intellectual brilliance and phenomenal knowledge of the chemical literature, he captivated the members of Baeyer's laboratory. In Berlin, Meyer worked independently on several topics before undertaking one of the fundamental problems raised by Kekulé's proposed structure of benzene, namely the ring position of substituents in isomeric aromatic compounds; his contributions were decisive in the solution of this problem. Meyer's

[10] Willstätter (1956), pp.114–115; in the English translation, see p.122.
[11] For example, see Baeyer (1878), pp.20–24. This lecture was reprinted by Kolbe (1878) in his own journal, with sarcastic comments; see also Rocke (1987). For Graebe's attitude toward Baeyer in their declining years, see Graebe (1915), who wrote an appreciative eulogy on the occasion of Baeyer's eightieth birthday.
[12] The biography of Victor Meyer by his brother Richard (Meyer, 1917) is the primary source; see also Liebermann (1897), Thorpe (1900) and Costa (1974).

later achievements, as professor at the Stuttgart Technische Hochschule and at the universities in Zurich and Heidelberg were even more striking, for he opened new vistas to the understanding of the structure of organic nitrogen compounds and invented the vapor density method for the determination of the molecular weight of organic substances. Afflicted by illness, he took his life in 1897.

In calling attention to the achievements of Graebe, Liebermann and Meyer in increasing the fame of Baeyer's laboratory in Berlin, it should also be emphasized that they did not overshadow his own successes as a research chemist. As was noted above, after leaving the problem of the constitution of uric acid, Baeyer embarked on work that opened new lines of investigation, and during his final years in Berlin (1866–1872) he took his first steps toward the eventual synthesis of indigo, and discovered a new class of dyes—the phthaleins.[13] For the development of organic chemical thought, perhaps his most important success was the demonstration that mellitic acid is benzene hexacarboxylic acid, as this discovery broadened the scope of experimental work based on Kekulé's formula for benzene.

Kekulé's picture of the benzene ring was first presented by his friend Adolf Wurtz on 27 January 1865 before the chemical society in Paris. Much has been written about this memorable theory, partly because of Kekulé's later references to the way in which it came to his mind, but more importantly because it channeled the thought of his disciples into new directions.[14] This influence is clearly evident in the publications from Baeyer's laboratory in Berlin after 1865, especially in those of Graebe and Meyer, and of Baeyer himself. In particular, Kekulé's idea of the ring structure of benzene led them to conceive of the ring structure of many other organic compounds, whether they be barbituric acid, anthracene, pyrrole or quinoline. In view of these and later discoveries made by Baeyer and his students, it may perhaps be appropriate to suggest that his

[13] Baeyer's researches on indigo and the phthaleins have been discussed by Friedländer (1915) and Meyer (1915) respectively.
[14] For Baeyer's work on the structure of benzene, see Dimroth (1915), Holleman (1915), Anschütz (1929) and Huisgen (1986). The twenty-fifth anniversary of Kekulé's paper was held in Berlin on 11 March 1890, and he reluctantly agreed to attend; as he wrote to Baeyer on 2 December 1889, "I therefore prefer to declare at once that I am ready to travel to Berlin on the date to be determined in order to allow myself to be slaughtered, unless unforeseen obstacles make the journey impossible" (letter in the Kekulé collection at the Darmstadt Technische Hochschule). See also Wotiz and Rudofsky (1987).

research group played a major role in providing a sounder basis for Kekulé's theories, and in demonstrating the practical value of these theories in the synthesis of commercially valuable organic chemicals.[15]

In 1905, Baeyer received a Nobel Prize "in recognition of his service to the development of organic chemistry and of the chemical industry through his researches on organic dyes and hydroaromatic compounds." Baeyer's connection with the emerging German chemical industry began in Berlin, where he established a cordial relationship with Heinrich Caro, a well-trained organic chemist, who in 1868 had joined the newly established Badische Anilin- und Sodafabrik (BASF) in Ludwigshafen.[16] Caro was also a friend of Liebermann's father, a manufacturer of chemicals. The patents issued to Graebe and Liebermann for the synthesis of alizarin were assigned to BASF.[17]

From the accounts of his contemporaries in Berlin, the thirty-year-old Baeyer appears to have been a man of fine presence, and some of the qualities he displayed in later years—"a steadfastness of purpose and determination to allow nothing to interfere with his duties as a teacher or with the course and development of his investigations"[18]—were already evident. His style of chemical research became proverbial, in that he depended largely on his own assiduous test-tube experimentation, and even when mechanical devices (for example, shaking machines) became available, he preferred stirring rods. Baeyer's relationship to Graebe—encouragement to do independent research, readiness to help and on occasion to guide—was to be mirrored with respect to Baeyer's attitude toward his later junior associates. To these qualities must be added his extreme reticence, which often made him appear to be cold, unapproachable and haughty. "In large gatherings or toward strangers he could be formal, stiff, monosyllabic, somewhat uncivil . . . [but] if he found a conversation interesting he entered into it with the same animation that he displayed in the lecture hall."[19]

During Bayer's twelve-year stay at the *Gewerbeinstitut*, at least 38 men worked in his laboratory. Apart from Graebe, Liebermann and Meyer, this group included Emile Ador (from Switzerland) and Willem Anne van Dorp (from The Netherlands), who later conducted valuable organic-chemical research in the pri-

[15] See Russell (1987).
[16] For a biography of Caro, see Bernthsen (1912).
[17] See Duden and Decker (1928), p.14 and Caro (1892), pp.1041–1047.
[18] Perkin (1923), p.1541.
[19] Penzoldt (1917), p.1332; for Penzoldt, see Grote (1923), pp.169–186.

vate laboratories they had established, as well as Adolf Emmerling, later to become head of the agricultural experiment station at Kiel. Baeyer's assistant Conrad Alexander Knop became a pharmacist, and Otto Haussknecht a teacher at a technical school. Heinrich Ludwig Buff, a pupil of Liebig, died shortly after his stay with Baeyer, as did the young Russian nobleman Joseph von Korff. A sizable proportion of the group (at least 17) entered industrial chemistry (Ascher, Berend, Jaffé and Marasse set up new factories); this number may be larger, as I was unable to determine the later activity of six men (Damm, Deichsel, Kretschmer, Mizerski, Mohs, Ulrich). Also, it has been difficult to establish the number of students who received a Dr.phil. for the work they did in Baeyer's laboratory. According to his account, Baeyer's first research student, Max Berend, obtained this degree at the University of Berlin; however, Berend's name is not included in the published list of doctoral dissertations at Berlin before 1885 and of the 38 people whom I have identified as members of his group at the *Gewerbeinstutut,* only four (Jaffé, Korff, Liebermann, Marasse) are listed there.[20] It is likely that Berend and perhaps seven others took their degrees at another university after completing their work in Berlin.

Of special interest in relation to Baeyer's role in the development of nineteenth-century biochemistry is the fact that five of his junior associates in Berlin made significant contributions in that field. The first to arrive in Baeyer's laboratory was the young physician Otto Schultzen, who had received his Dr.med. degree in Berlin and then worked with Graebe in 1867; shortly afterward, Schultzen and his pupil Marceli Nencki published an important paper on the physiological formation of urea from amino acids. At Schultzen's suggestion, Nencki spent two years in Baeyer's laboratory, where he worked on the chemistry of uric acid. Schultzen became professor of medicine at Dorpat, but his promising scientific career was cut short by his untimely death in 1875. Nencki, as the leader of a productive research group in Berne, actively applied the modern organic chemistry of his time to the study of many biochemical problems.[21] Two other junior members of the Baeyer laboratory in Berlin, Oskar Liebreich and Nikolai Lubavin, were also students of Hoppe-Seyler, and Ernst Ludwig later became a professor of medical chemistry in Vienna.

The presence, in Bayer's laboratory, of these chemically minded biologists suggests that, in his early years, he had an ac-

---

[20] See Berlin (1899).
[21] For Nencki, see Nencki (1904), Szwejcerowa and Grosyńska (1956) and Bickel (1972).

tive interest in the problems of physiological chemistry. This surmise finds support in the semi-popular lecture he presented in
1866 on the carbon cycle in nature and, more importantly, in his
paper in 1866 (with Knop) on indigo and the one in 1870 on
condensation reactions in plant metabolism and in fermentation. In the latter article, Baeyer offered speculations based on
reactions observed in the chemical laboratory, in the tradition of
the ideas offered by Liebig and Dumas thirty years earlier, and
brought up to date by Liebig in 1870.[22] However, Baeyer did not
publish any further papers of this kind. It may only be a coinci-
dence that by about 1870 Baeyer had become dissatisfied with
his situation at the *Gewerbeinstitut,* and was discussing with Theodor Frerichs, the professor of pathology at the University of
Berlin, the possibility of an appointment as a professor of medical chemistry in the medical faculty, but that this plan failed owing to the opposition of Rudolf Virchow.[23] Moreover, in 1866,
Baeyer had begun to give evening lectures on organic chemistry
to the physicians at the University of Berlin.[24]

Baeyer must have counted himself fortunate, therefore, in
his appointment as professor of chemistry at the university the
victors in the Franco-Prussian war had established in Strassburg.
With the help of Friedrich Rose [1839–1925], whom Baeyer
brought from Heidelberg to head the inorganic division of the
chemical institute, adequate teaching and research facilities were
made available by the fall of 1872, and this space was expanded
in 1874. Among the first students were Emil Fischer and his
cousin Otto Fischer, who entered Baeyer's organic division in
1873. In describing his initial experiences there, Emil Fischer
later wrote:

Instruction in organic chemistry was different then from what it is today. It was not customary to make preparations [of organic sub-
stances]. There were no booklets or other introductory guides to the
conduct of organic-chemical preparations. When I began my doctoral
work I was therefore unable to distill ether or to perform an elementary analysis. All these things were learned by chance from a neighbor
or from a good-natured or friendly assistant. . . . [Baeyer] did not concern himself with trifles such as mechanical operations, analytical de-

---

[22] See Baeyer (1866, 1870), Baeyer and Knop (1866), p.1 and Liebig (1870).
[23] See Baeyer (1905), p.xvii and Pennzoldt (1917), p.1332.
[24] Duden and Decker (1928), p.27.

terminations and the like, but we had a superb adviser in his very skilled and experienced private research assistant, Julian Grabowski, whose friendship and help we all sought.[25]

From this account it would seem that Baeyer did not supervise closely the initial efforts of his doctoral students in Strassburg. No doubt his teaching obligations were considerable, as there was a large influx of Alsatian students of pharmacy, and Baeyer's insistence on pursuing his personal research probably contributed to his attitude toward the beginners in his laboratory.

During the three years he stayed in Strassburg, Baeyer continued his work on the condensation of phenols with aldehydes and with aromatic compounds, especially phthalic anhydride. From these studies emerged new phthaleins, such as phenolphtalein and fluorescein.[26] This line of research brought Baeyer into closer connection with Caro, because some members of the phthalein group showed commercial promise as textile dyes. Also, Caro's earlier discovery of nitrosodimethylamine led to their collaboration, as well as extensive work by Baeyer's group, in the study of aromatic nitroso compounds.

As regards the composition of Baeyer's research group in Strassburg, two postdoctoral guests (Goldschmiedt, Hemilian) and 17 research students worked in his laboratory. Ten of the students (E. Fischer, O. Fischer, Fuchs, Grabowski, Hepp, E. Jaeger, ter Meer, Schraube, Weiler, Zeidler) received their Dr.phil. in Strassburg[25]; some of them stayed on as Baeyer's assistants. The other seven (Fitz, Gerber, Grimm, C. Jaeger, Kimich, Kopp, Schiff) took their doctoral degrees elsewhere. Of these 19 men, six later became full professors of chemistry: E. Fischer (Erlangen, Würzburg, Berlin), O. Fischer (Erlangen), Goldschmiedt (Prague, Vienna), Grabowski (Cracow), Hemilian (Kharkov), Schiff (Modena, Pisa). Nine men (Fuchs, Gerber, Hepp, C. Jaeger, E. Jaeger, Kimich, ter Meer, Schraube, Weiler) entered the chemical industry, Kopp and Zeidler became pharmacists, Grimm a physician, and Fitz established a private research laboratory in Strassburg (for further data, see Appendix 5).

---

[25] Fischer (1905), pp.xxii–xxiii.

[26] The further study of the chemistry of fluorescein (and the phthalein of orcinol) was assigned to Emil Fischer, whose dissertation on this subject marked his first appearance in the chemical literature. This effort proved to be more successful than Fischer's unhappy beginning as a doctoral student. His first assignment involved the reduction of 500 grams of mellitic acid with sodium amalgam (made with 25 kilograms of mercury) but, as he was carrying the heavy flask, the laboratory floor gave way, with the loss of this valuable material; ibid., p.xxiii.

Baeyer's Institute in Munich

The professorship vacated by the death of Liebig in 1873 remained unfilled until 26 May 1875, when a royal decree named Baeyer to be full professor of chemistry at the Ludwig-Maximilian University and the head of the chemical laboratory of the Bavarian Academy of Sciences.[27] In fact, however, there was no laboratory to speak of because, during his twenty-one years in Munich, Liebig had largely basked in the adulation accorded his lectures but had not done sustained experimental work or guided the research of junior associates or doctoral students. Consequently, upon his arrival later in 1875, Baeyer faced not only the challenge of building a research school, but also the necessity of adding to Liebig's splendid lecture hall adequate laboratory facilities for practical instruction and chemical investigation. Moreover, he encountered the traditional antipathy of the more easy-going Bavarians toward Prussians.[28]

It seems that Baeyer did not see fit to inform his associates in Strassburg of his possible move to Munich, for Emil Fischer later wrote: "In the beginning of the summer semester 1875 there came into this peaceful and productive laboratory a serious disturbance with the news that our esteemed teacher had received and accepted a call to Munich as the successor of Liebig."[29] Emil Fischer, Otto Fischer, Hepp and Schraube decided to follow Baeyer to Munich. This was fortunate for Baeyer, for the only scientific people he found in Liebig's former institute were Jacob Volhard, who had manfully held the fort during the interregnum, and his small research group. Volhard remained in Munich until 1879, when he attained a full professorship in Erlangen, where he remained for three years before moving to Halle.

Baeyer's first requirement, the design and construction of adequate laboratory facilities, was aided by a generous allocation of state funds. The building, completed in 1878, embodied his concept of the organization of instruction in a university insti-

---

[27] For the best available historical summary of the development of the Munich laboratory, see Prandtl (1949, 1952).

[28] Among the many anecdotes that embellish accounts of Baeyer's career is the report that Liebig's former *Diener*, a retired Bavarian non-commissioned officer, had protested to the *Staatsminister* (von Lutz) that he could not serve a Prussian professor. To which von Lutz is said to have replied: "Dear Diegele, now we must all do so," for by 1875 Bavaria had been brought into the German *Reich* created by Bismarck; see Schmorl (1952), pp.93–94.

[29] Fischer (1905), p.xxvii.

tute of chemistry, with two sections, one for organic chemistry
and another for inorganic (and analytical) chemistry. The design
of the ground floor, for the organic section, was essentially du-
plicated on the floor above for the inorganic section. On each
floor, there were two large laboratory halls with working space
for about 120 people, separate laboratory space for the head of
the section, and numerous rooms for special purposes. Of particular interest was the stipulation that "smaller laboratories for
the more experienced [persons] will not be provided in order to
promote the interaction of the practical workers. Corridors are
to be avoided as much as possible."[30]

The cornerstone was laid in July 1876 and by the fall of
1877 the building was partially occupied; during the construction, makeshift laboratories were provided in Liebig's former
dwelling. By the winter semester 1881–82, there were 92 *Praktikanten* in the inorganic section and 181 in the organic section;
these figures rose rapidly during the 1880s and by 1890 had lev-
eled off at about 230 and 270 per semester respectively. It
should be emphasized that only about 15 per cent of these laboratory workers had matriculated in chemistry; the others were
students of other sciences, medicine or pharmacy.[31] Moreover,
the group of those specializing in chemistry included predoctoral students who went elsewhere for their dissertation research
as well as postdoctoral guests whose work in the Munich laboratory did not lead to a research publication. As best as I have
been able to determine, the total number of persons who may
be considered to have constituted Baeyer's research school from
1875 until his retirement in 1915 is approximately 560, of whom
395 received the Dr.phil. degree for work done in his institute.
The names of these people, and some biographical data for
most of them, are given in Appendix 5, together with those for
Baeyer's junior associates in Berlin and Strassburg.[32]

---

[30] Baeyer and Geul (1880), p.4; Prandtl (1952), p.54. Most of the *Privatdozenten* and research students in organic chemistry worked in one of the two large laboratories on
the ground floor (Saal IV).
[31] See Prandtl (1952), pp.124–133.
[32] In compiling the list of names in Appendix 5, the first source was the bibliography in
Baeyer (1905), pp.lvi–cxviii. Apart from a very few omissions and some typographical
errors, this bibliography up to 1905 is most valuable. The junior associates and
Dr.phil. research students at the Munich institute during the period 1905–1915 were
identified by means of Prandtl (1952), pp.51–98, and the appropriate volumes of the
Jahres-Verzeichnis der an den deutschen Universitäten erschienen Schriften. The resulting list
was checked against entries in *Chemisches Zentralblatt* and *Chemical Abstracts,* as well as
bibliographies of Baeyer's associates in Munich. In the collection of biographical data
about many of the Dr.phil. students at Munich, I was greatly aided by Frau Professor
Dr. L. Boehm and her associate Frau Spin of the University Archive, as well as by
Archivoberrat Dr. Hecker and his associate Frau Knüttel of the Munich *Stadtarchiv.*

As at other major German universities of Wilhelmine Germany, Baeyer was the only *ordentlicher Professor* of chemistry, and instruction in that field involved the services of many teaching assistants.[33] A distinctive feature of Baeyer's institute, however, was the presence of a succession of talented *Privatdozenten* or *ausserordentliche Professoren* who led independent research programs of their own. Thus, after Baeyer turned the direction of the organic-chemical section over to assistants, who were promoted to the higher rank, the section was headed successively by Bamberger (1885–1893), Thiele (1893–1902), Willstätter (1902–1905), Dimroth (1905–1913) and Wieland (1913–1915), all of whom later gained distinction for their contributions to organic chemistry. Although the successive heads of the inorganic section were mostly specialists in that branch of chemistry, there were also two noted organic chemists: Emil Fischer (1879–1882) and Pechmann (1887–1895). Among the many other outstanding organic chemists who were *Privatdozenten* (in the order of their *Habilitation*) were Otto Fischer (1878–1884), Koenigs (1881–1915), Einhorn (1891–1915), Friedländer (1883–1884), William Henry Perkin, Jr. (1884–1886), Claisen (1887–1890), Dieckmann (1898–1915) and Pummerer (1911–1915). If one adds the sizable number of Dr.phil. recipients and postdoctoral guests (some of whom will be mentioned below) who worked productively in Baeyer's institute on organic-chemical problems, and later achieved distinction, it is evident that this laboratory was the principal seedbed for the next generation of leaders in the development of organic chemistry in Germany. To this should be added the equally compelling conclusion that Baeyer was less concerned with the promotion of branches of chemistry other than his own. Although a corps of able inorganic chemists (Zimmerman, Krüss, Muthmann, Hofmann, Prandtl) served as heads of the inorganic division, their research formed a relatively small part of the scientific achievements of Baeyer's institute. Of the approximately 1600 research publications during 1875 to 1915, only about 350 dealt with work on problems in inorganic chemistry.[34] Moreover, even by the turn on the century, when physical chemistry came to be recognized at other major German universities to be an important specialty, the Munich institute was conspicuously deficient in this regard. One reason for this neglect, apart from considerations of space,

[33] A separate doctoral program in applied chemistry was administered at the University of Munich by Albert Hilger [1839–1905], who was succeeded by Theodor Paul [1862–1928].
[34] See Baeyer (1905), pp.lvi–cxiii.

seems to have been related to the state of development of the German chemical industry, which at that time drew its major profits from the production of organic chemicals and gave generous financial support to university research in organic chemistry.[35]

The large student enrollment placed a heavy burden on Bayer's teaching assistants, and made it necessary for them to depend for the conduct of their research on the efforts of private assistants (if personal or other funds were available for this purpose) and of as many Dr.phil. candidates and postdoctoral guests they could attract to their lines of investigation. In this respect, Baeyer appears to have been more generous than his counterparts at other university chemical institutes in Germany, where most of the research done by predoctoral students and postdoctoral guests was done as part of the chief's research program. During his forty years in Munich, Baeyer had a succession of postdoctoral *Privatassistenten,* usually drawn from his former Dr.phil. students (for example, Villiger), but only 45 doctoral dissertations (out of 395) were based on work within the areas of Baeyer's personal research interest. The work for the other 350 dissertations was supervised by 35 junior associates during their stay at the Munich institute, and formed an essential part of their scientific achievements. Eleven of them had supervised more than 10 dissertations: Bamberger (33), Claisen (12), Dimroth (14), Einhorn (45), Hofmann (32), Koenigs (16), Pechmann (16), Piloty (12), Thiele (30), Wieland (18), Willstätter (16). In addition, the individual research groups included most of the postdoctoral guests from abroad.

William Henry Perkin, Jr., who was with Baeyer during the 1880s, described Baeyer's attitude as follows:

Perhaps for the reason that he was accustomed to do all the experimental work himself, Baeyer had little inclination to work with others, and the titles of his papers show that, whilst he frequently published with one or another of his assistants . . . the actual number of his co-workers was relatively small. This is all the more surprising when it is remembered that the laboratory was always overcrowded with the most promising material from all parts of the world, and that every newcomer would have considered it a great honor to have been allowed to work with the head of the laboratory. Neither did Baeyer often suggest subjects for the researches which so many of the younger men were carrying out for their dissertations for the Ph.D. degree. When the

[35] See Bock (1972), p.144.

time came to do original work, these young researchers were usually handed over to one of the many Privatdozenten attached to the laboratory, and it was the duty of these senior men, who were often men of great experience, to suggest the theme for investigation and to superintend the work and help to bring it to a satisfactory conclusion. This plan worked well, and gradually there rose up a great school of research which has rarely if ever been equalled. . . . There was a feeling in the laboratory that no one was of any account who did not research, and, moreover, the position of each researcher and the esteem in which he was held depended solely on the quality of the work he was engaged in. This was the atmosphere which produced the greatest chemists of the day and weeded out those who were of no account. It was only necessary that the commanding figure of Baeyer should stroll through the research laboratories each day and for him to chat with the various workers, criticise their results, and admire their preparations, to make it out of the question for anyone to forget for a moment that research was the only thing that really mattered.[36]

A more intimate glimpse of Baeyer's attitude toward some of the predoctoral students has been provided by Willstätter in his recollections:

The rule was that one had to be accepted by a *Dozent* who would assign a topic and supervise the work. Baeyer, whose pupil I wanted so much to be, had no more need for doctoral candidates, Bamberger could no longer be reached, no one knew when his successor would arrive. Nothing remained but to ask Professor Baeyer to help me in my disappointment. Of course I did not dare to ask the great man [to accept me], and he made me no offer, but took me to Prof. Alfred Einhorn, who accepted me gladly. . . . The problem assigned to me was to examine a nitrogen-free cleavage product of cocaine. . . . During the next few months it so happened that Baeyer took over the direction of my work for the dissertation and an insertion (on the dehydrogenation of hydrogenated benzene carboxylic acids), while Prof. Einhorn put a good face on the not entirely unwished development of the matter, and withdrew more and more. When I wrote up the results for papers in the *Annalen*, Baeyer took it for granted that that not he but Einhorn should publish with me, even on the inserted work with which only Baeyer was more familiar; and so it was.[37]

---

[36] Perkin (1932), pp.1545–1546.
[37] Willstätter (1956), pp.49–53. This translation differs from the one published in 1965, pp.51–53. One reason is that Willstätter's writings, full of circumlocutions and incomplete sentences, present problems similar to those encountered in the translation of Liebig texts. Willstätter may have learned much from Baeyer but, unlike Emil Fischer, he did not acquire Baeyer's skill in writing clear and well-formed German sentences.

In reciting this passage, and in reading the entire section of Willstätter's book in which it appears, I cannot escape the impression that he was adept in eliciting Baeyer's favor. Despite this favor, even when Willstätter had become a *Privatdozent*, like the others of that rank he merely retained his workbench in the large hall; also, he had to register each semester for Baeyer's organic-chemical course and to pay the regular student's fee.

Another glimpse of student life in the Munich laboratory has been provided in an obituary notice for Kipping, who worked there for his Dr.phil. with Perkin:

Work in the laboratory began at eight and continued until six or after, with a rather long interval for Mittagessen—and chess. Kipping saw very little of von Baeyer. He refers to one occasion, however, on which he showed him with some pride a product he had isolated. The Geheimrat looked at it under a lens, snapped out "Ach, Harz" and stalked away. This may have been the origin of Kipping's critical and, to say the least, realistic attitude to his students' laboratory efforts.[38]

Other Baeyer anecdotes, some no doubt apocryphal, abound in the writings about him. Rupe's account of Baeyer's visits to the student laboratory is perhaps the most revealing:

During my time in Munich the *Praktikanten* in the organic section did not come face to face with the "old man" often, and the ones in the inorganic section on the floor above even less frequently. When he occasionally strode through the halls, erect, every inch a king, in his indigo blue long cutaway, the hat on his head, with a stern countenance, there swept through the room a slight wave of uneasiness, everyone crouched behind his workbench, and sociable conversations were abruptly interrupted. He was considered to be the "stiff Prussian," unapproachable, strict, pitiless. Whoever had ears to hear, however, knew that at bottom Baeyer was a very kindly soft-hearted person who quietly did much good and eased many difficulties. He belonged to the few scholars and chemists who acknowledged wholly and ungrudgingly the merits of others. The colleagues in the laboratory thought "he can afford it"; yes, but many others could also afford it but did not do it.[39]

---

[38] Challenger (1950), p.185. The term *Harz* (resin) was one of opprobrium, as organic chemists of the time considered only nicely crystalline substances to be worthy of attention.

[39] Rupe (1932), p.24.

According to Rupe, Baeyer became quite disagreeable on days when he had to administer oral examinations, particularly those for students of medicine or pharmacy. Also, he was inexorable in his rule that the laboratory be closed to all, including the *Dozenten*, for the summer holiday.

The teaching load was especially heavy for the assistants in charge of instruction in inorganic and analytical chemistry. This course of studies had to be completed before a student was allowed to take a preliminary examination in these subjects as well as in elementary organic chemistry. Admission to the organic section was conditional on passage of this examination and the availability of working space in the laboratory. In the case of Clemens Zimmermann, who served as head of the inorganic section from 1882 until 1885, his death in the latter year (at the age of 29) was attributed to overwork.[40] Zimmerman had been an assistant in that section from 1879, when Emil Fischer was its head. As Fischer later wrote:

Upon taking on the new position, which of course I owed primarily to Baeyer's good will, my activity at the institute suffered a radical change. Until then I had been a free researcher, without any responsibility for instruction. I had been able to do all my scientific work myself, except for the joint research on rosaniline with my cousin Otto, and this collaboration went off very smoothly. From then on I could not operate in this manner any longer, because most of my time was devoted to practical instruction in chemical analysis. The help of assistants in the personal research became absolutely necessary.[41]

To this quotation should be added the note that, in order to employ private research assistants, a university chemist in Wilhelmine Germany had to be independently wealthy or to receive funds from industrial firms in the form of grants, consulting fees or royalties on patents. Indeed, even an institute director such as Baeyer received state funds for only one *Privatassistent* and had to pay the salaries of any additional ones from his own purse. Also, the academic prospects of these research assistants, especially of those who served a professor for more than a year or two, were not bright.[42] As may be seen in

---

[40] For Zimmermann, see Baeyer (1885).

[41] Fischer (1922), p.80. Shortly before leaving Strassburg, Fischer had discovered phenylhydrazine, and in Munich he began to develop this line of research in addition to his initial studies on triphenylmethane dyes and on purines.

[42] See Chapter One, note 10.

Appendix 5, many of the men who worked in that capacity for Baeyer disappeared from the scientific literature after they left his laboratory because they entered industry or for other reasons that I was not able to determine. The same be said of the many more recipients of the Dr.phil. degree for work they did as pupils of junior members of Baeyer's institute. If, as I have suggested above, during the last quarter of the nineteenth century Baeyer's style of leadership made his institute the principal breeding ground of the next generation of leading German organic chemists, it must also be recognized that, as at other German university institutes of chemistry, one of its major social functions was to provide manpower for the laboratories of a burgeoning and research-minded German chemical industry.[43]

Baeyer's Personal Research in Munich

Soon after the new laboratory was available, Baeyer turned from the continuation of his work on phthaleins to resume his research on the constitution and synthesis of indigo. The problem had been undertaken in 1869 by Kekulé, whose proposed structure of the indigo derivative isatin differed form the one that Baeyer had suggested but in 1878, as Baeyer wrote later: "Since eight years had elapsed after the first publication of Kekulé, without him and his pupils having succeeded in achieving a result, I considered myself justified in testing the validity of his views."[44] Between 1878 and 1885 Baeyer published a series of papers on this problem, the high point of the investigation coming in 1880–1882, when he described the synthesis of indigo in three ways: from o-nitrocinnamic acid, from o-nitrobenzaldehyde and from di(o-nitrophenylacetylene). This achievement brought Baeyer considerable public fame, and attracted the attention of the German chemical industry. None of Baeyer's procedures were found to be commercially feasible, and large sums of money were expended by the leading manufacturers to develop a feasible method based on his work. A practical solution came in 1890 from the contribution of Carl Heumann [1850–1894], professor of chemical technology at the Zurich Polytechnicum, who devised a method (starting with anthranilic acid)

---

[43] For valuable discussions of the education of chemists in Wilhelmine Germany, see Schütt (1973), Burchardt (1978, 1980) and Johnson (1985a, 1985b).

[44] Baeyer (1900), p.lx. This article is reprinted in Baeyer (1905), pp.xxxviii–lv.

that was successfully exploited by the *Badische Anilin- und Soda-
fabrik* (BASF) to produce synthetic indigo at prices that drove
the natural product off the world market. The extent of Baeyer's involvement, if any, in these later developments, is unclear,
as the director of BASF was his friend and former collaborator
Heinrich Caro. In this connection, Baeyer's pupil Liebermann
later wrote:

Among the general public it is widely believed that Baeyer participated
in the most active manner in these technical efforts, even at the cost of
his scientific work. The opposite is the case. Although especially the
dye industry owes its greatest successes primarily to Baeyer and his
school, not even one indication can be found for the view that his scientific activity was influenced in any way by industrial considerations.
He constantly gave industry only stimuli, indeed set them technical
problems, but without exception left their solution to others.[45]

Baeyer abandoned the indigo problem in 1885; he later wrote
that "I was exhausted by the strenuous activity in the indigo
field and had such an aversion to it that I had to find another
problem."[46]

    He chose to return to the study of acetylenic compounds, a
line of investigation which, like the indigo problem, had been
one of his research interests at the Berlin *Gewerbeinstitut*. In Munich Baeyer attempted to synthesize polyacetylene dicarboxylic
acids, but when he reached the tetra-acetylene compound he
stopped, because in anything but very small quantities it was
spontaneously explosive. In his Baeyer Memorial Lecture, Perkin reported: "It was the author's good fortune to be allowed to
take part and watch these experiments, and the experience of
the experimental skill with which Baeyer carried out these very
ticklish operations, with no other apparatus than test-tubes and

---

[45] Liebermann (1915), p.576. According to Schmorl (1952), p.109, Baeyer was granted
the first patent for the synthesis of indigo on 19 March 1880. On the other hand,
Prandtl (1952), p.61, states that "Baeyer never patented any of his discoveries." In this
connection, it should be noted that the final paragraph of a paper by Baeyer (1880),
p.2262, reads as follows: "Experiments are underway at the Badische Anilin- und
Soda-Fabrik under the direction of Dr. Caro to make available to industry the preparation of indigo blue described above. For this reason I could only now publish in this
paper facts which I had already discovered at the beginning of this year, but I hope
that before long I will be able to present to my colleagues a more detailed treatment
of the field." According to Weinberg (1924), p.15, Paul Friedländer, Baeyer's *Privatassistent* at that time, "was sent frequently to Ludwigshafen to follow the progress of
the work on o-nitrophenylpropiolic acid at the Badische Anilin- und Soda-Fabrik."
[46] Baeyer (1905), p.xxxv.

glass rods, is to this day a very lively recollection."[47] Apart from demonstrating Baeyer's chemical skill, this work led him to propose a theory (*Spannungstheorie,* strain theory) to account for the explosive instability of compounds containing triple carbon-carbon bonds, and then to use this theory in an impressive series of experiments designed to determine the bond structure of benzene derivatives and of their reduction (hydroaromatic) products.[48]

These latter studies, between 1885 and 1893, also represented a resumption of a line of work that Baeyer had undertaken in Berlin, where he had examined the properties of the products of the reduction of mellitic acid and of phthalic acid. As in Berlin, Baeyer's later work in this field was based on Kekulé's proposed structure of benzene as a ring of six CH groups with alternating single and double bonds between the carbon atoms. During the twenty years after Kekulé's enunciation of this theory, the question of its validity became the subject of lively speculation and dispute.[49] However, although no one before him had tackled the problem experimentally in so masterly a fashion, Baeyer was forced to conclude that "It is not possible to draw any conclusion about the constitution of benzene from the course of the reduction of a benzene derivative."[50]

The efforts of Baeyer and his contemporaries to elucidate the bond structure of the benzene ring were superseded during the twentieth century by those based on the electronic theory of valency. There was, however, one theoretical consequence of Baeyer's experiments that was recognized in 1890 by a young chemist, Hermann Sachse [1862–1893], but whose importance was not appreciated until much later. Among Baeyer's findings was the phenomenon of *cis-trans* isomerism in the products of the reduction of aromatic polycarboxylic acids; in discussing this property Baeyer considered the parent substance of these products (cyclohexane) to be, like benzene, a planar molecule, but having a strained structure. Sachse questioned this assumption, and suggested that the interconversion of the isomers observed

---

[47] Perkin (1932), p.1533.
[48] In addition to the article by Dimroth (1915), see the excellent account by Ramsay (1987) of Baeyer's strain theory and of the early history of conformational analysis. For the later development of Baeyer's original concept of strain, see Wiberg (1986); for conformational analysis, see Eliel (1975).
[49] See note 14, this chapter.
[50] Baeyer (1892), p.178.

by Baeyer could be explained by the theory that the cyclohexane ring can exist in two strain-free "configurations" (the later term was "conformations").[51] In 1905, Baeyer wrote:

I had calculated that with cyclohexane there arises a strain that increases with the number of carbon atoms. Sachsse (*sic*) has objected to this, on the ground that it is not necessary to accept my assumption that the larger rings lie in one plane. If one considers the surface formed by the carbon atoms to be bent, one always creates shapes in which there is no strain. This is certainly correct, but oddly enough my theory appears to be correct. It is not clear why this should be so, and I think that it is not yet the time to formulate hypotheses to explain this phenomenon.[52]

Baeyer's concept of a planar cyclohexane ring was accepted by most organic chemists and (later) X-ray crystallographers well into the 1920s, although Sachse's view was echoed during the 1890s by Carl Bischoff [1855–1908] in Riga. Indeed it was not until Leopold Ruzicka [1887–1976] demonstrated in the 1930s the applicability of Sachse's theory to macrocyclic compounds that conformational analysis became an important activity of leading organic chemists.

Before proceeding to the next stage in the sequence of Baeyer's research activities, it may be worth adding to the above sketch of Baeyer's work on hydroaromatic compounds another episode that suggests something of his attitude toward the post-doctoral guests in his Munich laboratory. According to a biographical account of the career of the noted Finnish chemist Ossian Aschan,

Baeyer had about 50 students from many countries working in big laboratory rooms. Naturally, he had no time to speak or discuss with all of them, not even with Aschan. Aschan therefore worked by himself with his naphthenic acids, the smell of which was abominable. One day Baeyer was conscious of a terrible smell in the laboratory, and angrily asked for the cause of it. Thus Aschan found an opportunity to speak with the great master, who finally asked him: "what are you working on?" Aschan was then able to explain the analogy between the naph-

---

[51] Sachse (1890). Sachse had received his Dr.phil. degree in 1889 for work with Carl Liebermann in Berlin.
[52] Baeyer (1905), p.xxxv.

thenic acids and hydrogenated benzoic acids, which were then the main interest of v.Baeyer. The latter became interested, and after that he talked to Aschan every day.[53]

The logic of Baeyer's work on hydroaromatic compounds led him next into the study of the terpenes, about which he published an extensive series of papers during 1893–1901; this field was being explored by others, notably Otto Wallach [1847–1931], and Baeyer's contributions have been deemed "less brilliant and convincing than much that had gone before."[54]

During the course of the research on the terpenes, Baeyer had occasion to use as an oxidant "Caro's acid," formed by the treatment of a persulfate with concentrated sulfuric acid, whose properties he studied, and from this work he turned to an investigation of peroxides. Stimulated by the discovery in England of oxonium salts, during the final years of his scientific activity Baeyer undertook research on compounds of this type, in relation to the basic properties of oxygen, and also on triphenylmethyl compounds in which carbon can exhibit basic properties.[55] In spite of his advanced age, Baeyer displayed in all these studies the same virtuosity of chemical technique as before.

Baeyer published somewhat over 300 papers bearing his name from his laboratories in Berlin, Strassburg and Munich. In reporting the results of his personal research, it appears to have been Baeyer's policy to add the name of a co-worker (a research assistant or a Dr.phil. student) at the head of a paper, or of a section of a set of papers published together with a lengthy introductory section under his own authorship. However, some of his important papers, especially those containing theoretical discussions, appeared under Baeyer's name alone, with an acknowledgment at the end to co-worker (usually a *Privatassistent*) who contributed to the work. Thus, in addition to the 28 papers bearing Villiger's name as a co-author, there were 15 others in which only an acknowledgment indicates his participation in the research. In general, it may be said that Baeyer was rather more generous than most other leading German professors of chemistry in allowing his Dr.phil. students and the postdoctoral guests who worked with him to publish their results alone.

---

[53] Enkvist (1972), p.84.
[54] See Harries (1915); for the quotation, see Perkin (1923), p.1536.
[55] See Wieland (1915) and Schlenk (1915).

TABLE 4–1
The Baeyer Research School

	Berlin (60–70)		Strassburg (72–75)		Munich (75–15)	
	Predoct.	Postdoct.	Predoct.	Postdoct.	Predoct.	Postdoct.
Germany & Austria	21	11	12	2	357	83
United Kingdom					32	1
United States					20	8
Switzerland	1		2		16	2
Russia & Poland	2	2		2	8	5
France			1		3	3
Belgium					1	1
Netherlands		1			3	1
Sweden					1	3
Other countries					8*	4[+]

* 3 from Greece, 1 from Bulgaria, Canada, Denmark, Japan, Norway
[+] 2 from Finland, 1 from Italy, Japan

The Composition of Baeyer's Research School

In Table 4–1, some data are given for the national origin of the 617 persons listed in Appendix 5. Those coming from countries other than the German and Austrian states (including Hungary) represented 21 per cent of the total. It is noteworthy that of the 11 foreign students in Berlin and Strassburg six came from Russia and Poland (Chojnacki, Grabowski, Hemilian, Korff, Lubavin, Nencki), three from Switzerland (Ador, Gerber, Jaeger) and none from the United Kingdom or the United States. One reason for the absence of British and American students must have been the fact that the Berlin *Gewerbeinstitut* did not award doctoral degrees. Also, as was noted in Chapter Two, after Liebig's departure from Giessen in 1852, the most attractive laboratory for aspiring young British and American chemists was Wöhler's institute in Göttingen.[56]

The data in Table 4–2 summarize the information available to me regarding the later professional activity of 522 of the per-

---

[56] According to the data collected by Jones (1983), between 1860 and 1875 about 15 British and 25 American students received Dr.phil. degrees in chemistry at German universities. Most of these men (8 British, 17 American) worked in Wöhler's institute in Göttingen. The others took the degree at Heidelberg (7), Tübingen (5), Freiburg i. B. (1), Marburg (1) and Rostock (1).

TABLE 4–2
Principal Subsequent Activity of Members of the Baeyer School*

	Berlin (60–72)	Strassburg (72–75)	Munich (75–15)
Full (*ord*) prof. chem. University	G2	G3	G24/F30
Higher techn. school	G2		G15/F3
Full (*ord*) prof. applied chem.	G2	F1	G3/F2
Other acad. ranks in pure or applied chem.	G2		G39/F8
Full (*ord*) prof. other subjects	G2/F2		G7/F2
Research institute			G2/F2
Private research lab.	F2	G1	G3/F1
Medical practice		G4	G2
Pharmacy	G1	G1/F1	G16/F1
Hospital chemistry			G3
Industrial manufacture	G10/F1	G5/F1	G162/F27
Commercial or consulting chemists	G6	G2	G59/F12
Govt. techn. depts.			G18/F8
Secondary school teaching	G1		G5/F3
Other occupations (politics, philosophy, economics, publishing, etc)	F1		G9/F3
Not determined	G4		G68/F17

*G = German or Austrian; F = other nationalities

sons listed in Appendix 6. I was unable to find such information for four members of the Berlin group and for 85 (68 German or Austrian, 17 from other countries) of the Munich group; also, six members of the Munich group died within a year or two of the award of the doctoral degree (see Appendix 5). In what follows, the foreign students will be considered first, and separate sections of this chapter will be devoted to the much larger proportion (79 per cent) who came from Germany or Austria.

Most of the sizable number of British students (33) in Baeyer's laboratory in Munich did not pursue a career in university research, and few of these men had the opportunity to contribute significantly to a British chemical industry that was being eclipsed by the growing German firms. Of the nine men who entered academic life, three (Perkin, Kipping, Cohen) merit special mention because the experience they had gained in Munich was reflected in the later development of organic chemistry in England.

Of these men, the one most influenced by Baeyer's personality, style of leadership and chemical orientation was William

Henry Perkin, Jr., who had come to Baeyer's laboratory after having received his Dr.phil. in Würzburg. His most brilliant scientific achievement was the preparation, while in Munich, of the first derivatives of cyclopropane and cyclobutane. A part of the account by Greenaway of the background of this work, which Baeyer later used for his strain theory, deserves quotation:

That rings containing less than six atoms of carbon might exist had early suggested itself to Perkin, and he had not been long in Munich before he felt strongly tempted to try to effect the synthesis of such rings. This view was entirely opposed to the idea held at that time, that carbon rings containing three, four or five atoms of carbon were not capable of existence, and older chemists to whom he submitted his proposal were far from encouraging. . . . Baeyer, however, seems to have had some doubts as to the opinion he had at first expressed on the question, because shortly afterwards he asked Perkin on more than one occasion whether he was still interested in the possible existence of the smaller rings, and made some encouraging remarks when Perkin told him that he was hoping to start work on them.[57]

After a professorship at Manchester (1892–1902), Perkin went to Oxford, where he established an active department of chemistry. His extensive researches on the structure and synthesis of natural products such as camphor, terpenes, berberine, brazilin and harmine were in the tradition of Baeyer's laboratory, and involved the efforts of many able younger co-workers. Among them were Jocelyn Field Thorpe [1872–1940], later professor of organic chemistry at the Imperial College of Science and Technology in London, and Robert Robinson [1886–1975], Perkin's successor at Oxford.

As *Privatdozent* in Munich Perkin had, as a research student, his countryman Frederick Stanley Kipping, later professor of chemistry at Nottingham, where he conducted pioneering research on organosilicon compounds. Another British student in the Baeyer institute at that time was Julius Berend Cohen, who worked with Pechmann, and who became professor of chemistry at Leeds. Cohen was an inspiring teacher, and had an interest in biochemical problems; among his students were the noted biochemists Henry Drysdale Dakin and Henry Stanley Raper.[58]

---

[57] Greenaway (1932), pp.17–18. See also Finley (1965).
[58] For Kipping and Cohen, see Challenger (1950) and Raper (1935) respectively.

There can be little doubt that Perkin, Kipping and Cohen brought to the education of British chemists something of the scientific spirit which they had imbibed in Munich, as well as the chemical skill and discipline they had acquired there. It should be added, however, that in the years before World War I important new currents were flowing in British organic chemistry, notably through the influence of Perkin's successor at Manchester, Arthur Lapworth [1872–1941], who had not studied in Germany. In introducing electrochemical ideas into organic chemistry, Lapworth played a leading role in preparing the ground for the interpretation of organic-chemical reactions in terms of the electronic theory of valency.[59]

The American contingent in Baeyer's Munich institute was somewhat smaller (28) than the British one, but as many as 11 of them later held academic posts, no doubt as a consequence of the growth of universities in the United States during the early years of the twentieth century. Moreover, in contrast to the situation during the middle years of the nineteenth century, chemical teaching and research at some American universities began to take account of the revolution in organic chemical thought after 1860.[60] Although the technical and agricultural aspects of chemistry still retained their primacy, and the importance of theoretical achievements such as those of Josiah Willard Gibbs were not yet appreciated, modern organic chemistry was introduced in the United States by a few outstanding men, foremost of whom was Arthur Michael [1853–1942], who had studied with A. W. Hofmann, Bunsen and Wurtz.[61] Three others, who had worked in Baeyer's Munich institute, were Nef, Noyes and Gomberg.

John Ulrich Nef received his Dr.phil. in 1886 for research with Baeyer, and also worked in the Munich institute with Koenigs. Upon his return to the United States, Nef held brief academic appointments at Purdue and Clark before becoming professor of chemistry at the University of Chicago in 1892. During his relatively short scientific career, he displayed originality and skill in his research, especially in his study of organic compounds of divalent carbon, and engaged in a lengthy scientific polemic with Arthur Michael. Nef also worked on the chemistry of sugars, in the hope of elucidating the mechanism

---

[59] For Lapworth, see Robinson (1947). Lapworth, Kipping and W. H. Perkin, Jr., were brothers-in-law.
[60] See Browne (1926).
[61] For Michael, see Costa (1971) and Fieser (1975). There are many brief notices about Michael, but a full biography of this important figure in the development of organic chemistry in the United States appears to be lacking.

of their fermentation by yeast. According to one of Nef's biographers, "Professor Baeyer was very fond of this intense young man and frequently entertained him as a guest in his home. Professor R. Willstätter has stated that Professor Baeyer had remarked to him that Nef was the most brilliant of all the students he had had during his tenure of the chair of chemistry at Munich."[62] Nef brought to the education of his students at Chicago some of the qualities he had acquired in Munich:

Nef was a strict disciplinarian in his department. He would not permit the slightest deviation from his own meticulous cleanliness and rigorous accuracy in every step of his laboratory procedures. . . . Some students resented his manner, but most of those who worked for their Ph.D. degrees under him held him in high regard. . . .He always worked alone and, with the exception of three papers at the beginning of his career, his was the only name on his publications.[63]

Moreover, at Chicago, Nef attempted to introduce the German system of a single professor with a corps of *Dozenten* to help with the formal instruction, and Julius Stieglitz [1867–1937] was his first appointee. This plan failed, and in 1905 Stieglitz was made a full professor and in 1915 (after Nef's death) he became head of the department. Stieglitz conducted a productive program of research on the mechanisms of organic-chemical reactions from the point of view of the electronic theory of valency. Nef also brought to Chicago the Scotsman Alexander Smith who had received his Munich Dr.phil. in 1889 for work with Claisen. Although Smith published valuable research papers (especially on forms of sulfur), his fame rested primarily on his distinction as a lecturer to beginners in chemistry; he remained with Nef from 1894 until 1911, when he became head of the chemistry department at Columbia University.[64]

William Albert Noyes, a Ph.D. student of Ira Remsen at Johns Hopkins, worked with Baeyer in 1888–1889, but it was not until 1907, when he was appointed head of the chemistry department at the University of Illinois in Urbana, that he had an opportunity to bring to a major American university some of what he had learned in Munich. Although, like other former workers in Baeyer's institute, Noyes worked on natural products

---

[62] Wolfrom (1960), p.207.
[63] Miles (1976), p.362.
[64] For Stieglitz, see Noyes (1941); for Smith, see Kendall (1923) and McKee (1932).

(especially camphor), he also wrote on the electronic theory of valence. When he retired in 1926, the Urbana laboratory had become one of the leading centers of chemical research and instruction in the United States. Among the many able people Noyes brought to Urbana was Roger Adams [1889–1971], who succeeded Noyes as head of the department. Before World War II, few chemical research groups in the world matched that of Adams in size or in the production of skilled organic chemists; his wide-ranging research program involved the efforts of 184 Ph.D. recipients, many of whom played important roles in the growth of industrial chemistry in the United States.[65] It should be added that at Urbana (where there was no medical school), biochemistry was recognized to be a part of chemistry, and Adams's department included an important biochemical research group led by William C. Rose [1887–1985].

The influence of Baeyer's institute was felt in the American Middle West not only in Chicago and Urbana, but also in Ann Arbor, Michigan, where Moses Gomberg created another of the great schools of organic chemistry in the United States. After he received his Ph.D. at Michigan in 1894, he joined the faculty there, and during 1896–1897 worked successively in Munich and in Victor Meyer's institute at Heidelberg. In Munich Gomberg worked with Thiele on organic nitroso compounds, but he also conceived a plan to synthesize the hitherto elusive tetraphenylmethane. Like the younger Perkin, Gomberg faced the skepticism of older chemists, including Meyer, in this case on the ground that this compound could not be made for steric reasons. In Heidelberg, Gomberg succeeded in isolating a small amount of tetraphenylmethane, and upon his return to Michigan undertook the synthesis of hexaphenylethane, the next member of the series. Instead of the colorless and rather unreactive compound that was to be expected, Gomberg obtained (upon the exclusion of air) a colored material which he recognized to be triphenylmethyl, a highly reactive free radical. This great discovery opened a new area of organic chemistry, and many chemists (including Wilhelm Schlenk in Baeyer's institute) rushed into the field. In the years before his retirement in 1936 (he served as head of the department in Michigan during 1927–1936), Gomberg pursued research on other topics in organic chemistry, and of his many doctoral students several men (most notably Donald Dexter Van Slyke and Werner Bach-

---

[65] For Noyes, see Hopkins (1944) and Adams (1952); for Adams, see Tarbell and Tarbell (1981, 1982).

mann) later achieved great distinction. Gomberg was a man of exceptional human qualities—modest, courteous, generous—with a genuine concern for the welfare of his research students.[66]

If the influence of the Baeyer school on the development of organic chemistry in Great Britain and the United States is unmistakable, the extent of that influence in other non-German-speaking nations is less clear, with the notable exception of Finland. There, Edvard Hjelt and Ossian Aschan successively held the professorship of chemistry at Helsingfors (Helsinki). Although there was a sizable contingent in the Baeyer laboratories from the Russian territories (including Poland and Latvia), the distinguished Polish biochemist Nencki made his most significant contributions in Berne and the brilliant Polish organic chemist Grabowski died shortly after his professorial appointment in Cracow. Of the others, Vladimir Ipatieff was professor of chemistry at the Military Artillery Academy in St. Petersburg (Leningrad) from 1899 until 1930, where he conducted his important studies on catalysis, but he then moved to the United States, where his efforts were largely devoted to the application of catalysis to petroleum technology. Paul Walden, as professor in Riga, did his important work on stereochemistry there, but left in 1918 to become professor of chemistry at Rostock.[67] To the above names may be added those of others who had worked in the Baeyer institute and later became noted professors of chemistry in their homeland: Lovén and Widman in Sweden, Holleman in The Netherlands and Henry in Belgium. Two outstanding French chemists, Ernest Fourneau and René Locquin, also were in the Munich laboratory.

Baeyer's Scientific Progeny in German-Speaking Countries

As was noted above, Baeyer attracted promising young German chemists to his laboratories in Berlin and Strassburg. In their scientific achievements, the one who stands out most prominently is Emil Fischer, who indeed overshadowed all other Baeyer pupils in later public prestige. Since Fischer's career forms part of the story in the next chapter, all that will be said

---

[66] For Gomberg, see Bailar (1970), McBride (1974) and Walling (1977). Willstätter (1956), p.64, wrote that in Munich Gomberg was "very reserved and modest, kept entirely to himself, and never engaged in small talk in or out of the laboratory."
[67] See Ipatieff (1946); for Walden, see Hückel (1958).

here is that although many of Baeyer's personal qualities were reflected in Fischer's demeanor, Fischer's scientific outlook and style of leadership differed markedly from those of his teacher. Also, perhaps nothing need be added to the sketches of the careers of Graebe, Liebermann and Victor Meyer, the outstanding members of Baeyer's group in Berlin, except to recall that they were the first in the long line of Baeyer pupils and assistants who later made significant contributions and held important professorships in chemistry (Graebe in Königsberg and Geneva, Liebermann at the Charlottenburg TH, Meyer in Heidelberg).

Among the hundreds of Germans and Austrians who worked in Baeyer's institute in Munich between 1875 and 1915 are a dozen men who, as teachers and researchers, left an indelible imprint on the later development of organic chemistry and, in a few cases, of biochemistry. Some of them had been doctoral students at the institute, some were brought by Baeyer to be heads of the teaching sections, and there were also others for whom his laboratory provided the starting place for what turned out to be distinguished scientific careers. In the following brief sketches of the research and personal qualities of these men (in alphabetical order), I attempt to indicate the extent of Baeyer's influence on their chemical thought and work as teachers and investigators.

After receiving his Dr.phil. degree in Berlin, Eugen Bamberger worked in the Munich laboratory for eleven years (1882–93), seven of them as head of the organic-chemical section. He then became professor of chemistry at the Zurich ETH, but in 1905 was obliged to resign this post because of a serious neurological disorder. Bamberger was one of the most productive investigators in the Baeyer institute; among his most notable studies were those on polycyclic hydrocarbons (especially naphthalene) and on organic nitrogen compounds. Like Baeyer, he was a highly skilled experimenter who preferred to work with the simplest chemical apparatus and, in addition, he exhibited remarkable ingenuity in work involving very small amounts of material.[68] Among his many predoctoral and postdoctoral students was the Austrian physician Karl Landsteiner who, after receiving his Dr.med. degree in Vienna, spent two years doing organic-chemical research; he also worked with Hantzsch in Zurich and with Emil Fischer in Würzburg before embarking on his memorable research in immunology.

---

[68] For Bamberger, see Blangey (1933).

Eduard Buchner received his Dr.phil. degree in 1888 at Munich for work with Theodor Curtius (of whom more shortly). Nine years later Buchner discovered a method to prepare, from yeast, a cell-free extract that ferments sugar.[69] In the intervening years, he had been a *Privatdozent* in Baeyer's institute, where he continued to work on diazoacetic acid ester (the subject of his dissertation) and also did research on pyrazoles. According to Willstätter, Buchner was passed over for promotion in the organic-chemical section of the institute and "never forgave this snub and bitterly blamed Baeyer personally and his circle. . . . Upon the appearance of the first communication [on cell-free fermentation] in the *Berichte der chemischen Gesellschaft*, Baeyer said to me: 'This will make him famous, even though he has no talent for chemistry.' "[70] Indeed, although Buchner was awarded the 1907 Nobel Prize in Chemistry, his contributions to the subsequent study of the entity he had named *zymase* were quickly overshadowed by the work of others, notably Arthur Harden [1865–1940].[71]

Because of the practice of naming organic-chemical reactions after their discoverers, Ludwig Claisen's name is particularly well known to students of organic chemistry through such terms as the Claisen condensation or the Claisen rearrangement.[72] He came to Baeyer's institute in 1886 after having spent ten years in Bonn as a student (Dr.phil. 1874) and *Privatdozent* with Kekulé, and three years in Manchester with Roscoe and Schorlemmer. Claisen began his famous studies on the condensation of aldehydes with ethyl acetoacetate and ethyl malonate in Bonn, and greatly extended his exploration of this field in Munich. Baeyer gave him much encouragement, and Claisen was extremely productive; during his four years in Munich, he had 13 research students, and their combined work led to 41 papers. Claisen left in 1890 to become professor of chemistry at the Aachen TH, and in 1897 he went to Kiel, where he remained only until 1904, when he was obliged to resign owing to a cardiac disorder. After three years in Berlin, Claisen spent the rest of his life in Godesberg, where he had a private laboratory, and

---

[69] For accounts of Buchner's discovery and its historical background, see Kohler (1971, 1972) and Fruton (1972), pp.22–86.
[70] Willstätter (1956), p.65; in the English translation, see p.66. As much else in Willstätter's autobiography, this recollection should be viewed in connection with his own research. In one of his last papers he denigrated the significance of Buchner's discovery by stating that "Buchner's press juice and macerated yeast act upon sugar in a manner which differs from that of living yeast"; see Willstätter and Rohdewald (1940), p.52.
[71] For Buchner, see Harries (1917) and Buchner (1963).
[72] Surrey (1954), pp.30–36.

his last papers appeared in 1926. One of his associates at Kiel described him as follows:

Claisen was a high-minded person, extremely considerate of others, even-tempered, steady and reserved. A typical old bachelor who cared for his health. . . . [His] scientific strength was closely linked with a limitation. He was a synthetic organic chemist. . . . With a wonderful sensitivity and a wealth of ideas he found the conditions to overcome [difficulties presented by] recalcitrant reactions; perhaps by the addition of a small amount of a catalyst no one else would have thought of.[73]

Claisen's pioneering achievements in the art of organic synthesis give evidence of the importance of craftsmanship, as well as theoretical insight, in the development of modern organic chemistry.

Among Baeyer's junior associates in Munich an especially interesting person is Theodor Curtius, whose life was not constricted by a highly productive (though narrowly channeled) scientific career, but included a passion for music and mountaineering.[74] A student of Kolbe in Leipzig, Curtius's Dr.phil. dissertation (1882) dealt with derivatives of glycine. Despite his implacable hostility to the chemical ideas of Kekulé and his pupil Baeyer, Kolbe revealed something of his inner nature in advising Curtius to go to Munich for postdoctoral work. During his four years (1882–1886) with Baeyer, who gave him encouragement and advice, Curtius embarked on a line of chemical investigation that was to occupy him the rest of his scientific life. Possibly at Baeyer's suggestion, Curtius examined the action of nitrous acid on glycine ethyl ester, and thus prepared the first known aliphatic diazo compound (diazoacetic acid ester), which proved to be a remarkably versatile organic reagent.[75] Curtius then spent four years at Erlangen (with Otto Fischer), where his research led him to the first preparation of hydrazine, and in 1890 he became professor of chemistry at

---

[73] The quotation is from Anschütz (1936), p.139. This biographical article is the best available source of information about Claisen's life and work.

[74] For Curtius, see Darapsky (1930) and Freudenberg (1963). Curtius was an accomplished pianist and had a good baritone voice; he wrote several musical compositions. His achievements as a mountaineer in the Swiss Alps have been described by the famous guide Christian Klucker (1930).

[75] In his *Habilitationsschrift* at Erlangen, Curtius (1886) wrote: "The auspicious interest of Ad. v.Baeyer, under whose eyes the first of these researches came to life and were developed, was of incalculable value and remains one of my happiest recollections."

Kiel, where his intensive studies revealed the utility of hydrazine and its organic derivatives (notably the azides[76]) in a great variety of synthetic reactions. In 1897 Curtius was appointed to be Kekulé's successor in Bonn and, in the following year, he succeeded Victor Meyer at Heidelberg. During the next 28 years, Curtius's research continued to be almost entirely devoted to the further study of the reactions of organic hydrazides and azides. Stimulated by Emil Fischer's entry into the protein field, in 1902 Curtius resumed his studies of the 1880s on glycine derivatives and used azides to synthesize what Fischer called *peptides*; this aspect of Curtius's work is considered in the next chapter, in connection with Fischer's research on proteins. At Erlangen, Kiel and Heidelberg, Curtius had relatively large research groups and over 150 Dr.phil. dissertations emerged from the work he parcelled out to his students.

Otto Dimroth, a research student of Thiele in Munich (Dr.phil. 1895), was brought back by Baeyer ten years later to head the organic-chemical section. A skilled experimenter, Dimroth made notable contributions to the study of heterocyclic compounds and was one of the first members of the Baeyer school to use physical-chemical ideas and methods in organic chemistry. He attained full professorships of chemistry at Greifswald (1913–1918) and Würzburg (1918–1940). From the accounts of his biographers, Dimroth appears to have been completely devoted to chemistry, an able administrator, and to have embodied most of the qualities of pre-World War I upper-middle-class Bavarian morality. The principal source of information about Dimroth is the report of a memorial meeting held on 4 July 1940.[77] In the introductory statement by Harms, there occurs the following passage: "What those close to him treasured most highly was his complete trustworthiness, his constant helpfulness and his cheerful optimism, also in political matters. How happy he would have been if he had lived to witness the events of the past weeks!" The "events" were of course those of the Nazi Blitzkrieg through The Netherlands, Belgium and France in June 1940.

Alfred Einhorn was less successful in fulfilling his academic aspirations than the others in my list of notables in the Baeyer laboratory, and relatively little has been written about him.[78] I include him because he appears to have played a significant role

---

[76] For a modern account of the development of the chemistry of azides, see Scriven and Turnbull (1988).
[77] See Harms et al. (1941); the quotation that follows is on p.3 of this report.
[78] See Uhlfeder (1917).

in the Munich institute, through his supervision of 45 research students (including Willstätter) who received the Dr.phil. degree, and because of his achievements in pharmaceutical chemistry. After his doctoral degree in Tübingen, Einhorn came to Baeyer's institute for two brief periods (1882–1885, 1885–1886) before returning in 1891 to live out his years in a small room in the Munich laboratory. It is a mark of Baeyer's generosity that he gave Einhorn a haven to pursue his researches and to send him many co-workers. Einhorn brought to the laboratory his interest in the chemistry of cocaine and introduced into medical and dental practice the anaesthetic Novocaine. This practical achievement, along with others, was the result of Einhorn's systematic and skillful approach to organic-chemical problems.

Wilhelm Koenigs came to Baeyer's institute in 1876, soon after his doctoral work with Kekulé in Bonn, and remained in Munich until his death thirty years later.[79] A man of wealth, Koenigs chose to devote his time (and money) to chemical research and good living; he declined the offer of a full professorship in Aachen, as the successor of Claisen. Although only a *Dozent* (he was promoted to *ausserordentlicher Professor* in 1892), Baeyer gave him a private laboratory, and consulted him frequently on scientific and other matters. Willstätter described Koenigs as follows: "A good-natured, light-hearted and humorous Rhinelander, who always looked grumpy, he was a thoroughly kind and high-minded person. He was reserved and always spoke softly. His style of living, his domicile, kitchen and cellar were of a higher standard; agreeable good fellowship was his element."[80] Along with the good spirit that his easy-going manner and generous hospitality brought to Baeyer's laboratory, Koenigs's scientific contributions were considerable. He was an outstanding experimenter and his sustained research on the chemistry of alkaloids (especially quinine) served as a valuable background for later work in this field.

Kurt Heinrich Meyer came to the Munich institute in the final years of Baeyer's professorship.[81] The son of the noted pharmacologist Hans Horst Meyer [1853–1939], young Meyer had been exposed in his parental home to the relation of chemistry to medicine. As a student, he was attracted to the applica-

---

[79] See Curtius and Bredt (1912).
[80] Willstätter (1956), p.63; in the English translation, see p.66.
[81] There is some confusion in the biographical literature about the first names of K. H. Meyer. In most articles (including the one in the *Dictionary of Scientific Biography*), he is listed as Kurt Heinrich Meyer. However, in the most extensive notice about him, by Hopff (1959), it is Kurt Otto Hans Meyer, and in Prandtl (1952), p.90, it is Kurt Horst Meyer. His full name appears to have been Kurt (Otto) Heinrich Meyer.

tion of physical chemistry to the study of problems in organic chemistry, and worked for his Dr.phil. (Leipzig 1907) with Arthur Hantzsch [1857–1835]. After a stay in England, Meyer secured a place in Baeyer's laboratory, and soon showed his chemical talent in an independent study of the keto-enol equilibrium of ethyl acetoacetate. By that time, Baeyer may have recognized that his earlier neglect of physical chemistry required correction, and Meyer, having proved his mettle as an organic chemist, was encouraged to give instruction in physical chemistry.[82] After his war service, Meyer returned to the Munich institute (then headed by Willstätter), but in 1921 he left academic life to become head of the research laboratory of BASF in Ludwigshafen. In addition to the supervision of work on dyes, he developed an important program of research on natural high polymers, and was aided in this effort by the young physical chemist Herman Francis Mark [b.1895]. In 1932, Meyer left BASF (by that time part of I. G. Farben) and Germany for Geneva, where he lived out his years as professor of chemistry. There, he conducted outstanding research on polysaccharides (cellulose, glycogen) and on the enzymes that catalyze their cleavage. There can be little doubt that, in versatility of chemical talent, few of Baeyer's junior associates can be compared to Meyer who, in my opinion, is a prime examplar of the expansion of the horizon of organic chemistry beyond that envisioned by Baeyer.

Hans von Pechmann, a scion of Bavarian nobility, received his Dr.phil. (1874) in Greifswald with a burst of activity during a nine-year period of intermittent university studies and gadding about. He came to Baeyer's laboratory in 1878; Koenigs described the early stages of his activity as follows:

At the start, the new colleague was seldom seen in laboratory and therefore seemed to belong to the so-called "rare elements." When he came, he wandered up and down the hall, reserved and taciturn, hands behind his back, top hat on his head, monocle in his eye, and he seemed to be considering what he would do next in the laboratory. After he had then worked zealously for a few hours, he usually disappeared again for several days or weeks. . . . Emil Fischer told us that . . . Baeyer expressed regret at the fact that Pechmann did not come to the laboratory more often, because he possessed a quite outstanding chemical talent.[83]

---

[82] See Prandtl (1952), p.111.
[83] The quotation is from Koenigs (1903), pp.4421–4422. This obituary notice is the principal published source of information about the life and work of Pechmann.

Baeyer's discernment was soon validated, for after Pechmann was made a teaching assistant in the organic-chemical section in 1882, he took full advantage of Baeyer's liberal policy of allowing his junior associates to select the subjects for the advanced practical work done by the students. During the succeeding thirteen years, Pechmann became one of the most productive members of the Munich institute and supervised the Dr.phil. research of many students, several of whom (notably Cohen and Wedekind) later made valuable contributions in organic-chemical research. Among his postdoctoral students was Carl Duisberg, who became head of I. G. Farben. Pechmann was, as Baeyer had sensed, a remarkably gifted organic chemist. He introduced into the art of synthesis important reagents, such as acetone dicarboxylic acid, diacetyl and diazomethane, and also explored fruitfully the synthesis of natural products such as the coumarins. In 1895, Pechmann became full professor of chemistry at Tübingen, where he brought together an effective research group and attracted many students. His stay there was relatively brief; in 1901 Pechmann fell into a serious mental depression that led to his suicide in the following year.

Johannes Thiele was Baeyer's choice in 1893 to succeed Bamberger as the head of the organic-chemical section.[84] A doctoral student of Volhard in Halle, Thiele had attracted notice through his work on guanidine compounds, and in Munich he developed a productive line of research on these and other organic nitrogen compounds. His best known studies, however, were those on aliphatic compounds with alternating single and double bonds (he named them conjugated double bonds) which led to his concept of partial valences. This theory was an important contribution to the continuing discussion about the constitution of benzene. Thiele was an effective teacher, but also a strict disciplinarian. Willstätter recalled that after Thiele's arrival,

Combustion analysis was improved, the requirements for training in making organic preparations were increased, the student was encouraged and forced to do experiments with clean test tubes, dozens of which had to be placed in the drawer in a precisely prescribed manner. Doctoral students in great number were trained by means of well-

---

[84] For Thiele, see Straus (1927).

chosen and well-supervised graduate research topics. Military discipline prevailed. And the students readily submitted themselves to this discipline, to which they were not accustomed, because they respected him.[85]

Indeed, among Baeyer's associates, Thiele was outstanding in the number of his students who later gained distinction as research chemists. Among the 31 men whose research Thiele supervised during his nine-year stay in Munich were Dimroth, Gomberg, Manchot, Meisenheimer and Wieland. In 1903, Thiele succeeded Rudolf Fittig [1835–1910] as professor of chemistry at Strassburg, where he had many Dr.phil. students who worked on a variety of organic-chemical topics. According to one of them, Thiele once declared "enlightened despotism to be the only correct form of organisation," an attitude consistent with his political opinions, which Willstätter described as "resembling those of an East Prussian Junker."[86] Thiele died in April 1918, and thus was spared the indignity of removal from his professorship after the French re-occupation of Strasbourg several months later.

After his Dr.phil. (1901) with Thiele, Heinrich Wieland remained in the Munich institute until 1917, and headed the organic-chemical division during 1913 to 1917. He then became full professor of chemistry at the Munich TH (1917–1921) and at Freiburg i.B. (1921–1925), before returning to Baeyer's former institute as Willstätter's successor.[87] During the first years of his independent research, Wieland confined his efforts to the study of organic nitrogen compounds, especially such topics as fulminic acid and derivatives of hydroxylamine and hydrazine; a high point in this work was the preparation of tetraphenylhydrazine and, from it, the free radical diphenylnitrogen. He continued to explore intensively the chemistry of nitrogen compounds until the 1930s, but by 1914 he had entered three additional areas of investigation: the chemistry of bile acids, the alkaloid field and the mechanism of enzyme-catalyzed oxidation processes. Wieland's major contributions in all these areas gave evidence of his remarkable chemical insight, his ca-

[85] Willstätter (1956), p.59; in the English translation, see p.61.
[86] Straus (1927), p.112; Willstätter (1959), p.59. Thiele's xenophobia, especially toward Frenchmen, was evident in his recommendation to Baeyer that Ernest Fourneau not be admitted in 1901 to the Munich laboratory as a research student of Willstätter; ibid., p.91.
[87] The best brief account of Wieland's scientific work is the one by Huisgen (1958). For more detailed summaries of Wieland's various researches, see Dane et al. (1942).

pacity for sustained intensive laboratory work and his ability to direct effectively the efforts of a large group along disparate lines of research. He received the 1927 Nobel Prize in chemistry for work on the ring structure of the bile acids, but his achievements in the study of other natural products (especially the pterins) were no less impressive. Also, his extensive studies on dehydrogenases and their role in biological oxidations greatly influenced biochemical thought. Huisgen has described Wieland as follows:

One could only marvel at Wieland's singular working capacity and self-discipline. He spent hours and hours daily at the benches of his co-workers whom he guided to careful observation, critical judgment, and enthusiasm for the experiment. His mastery of test-tube technique often helped to overcome many an obstacle. Moreover, Wieland never stopped experimenting himself and laying the ground work for problems to be taken over by his collaborators at an appropriate stage. . . . What attracted the considerable number of young chemists to work under Wieland? It was not only the scholar, wholly dedicated to his research, but also the simplicity and ingenuousness of the great man. Wieland regarded clarity of scientific thinking as an exercise in logic as well as training of character. . . . His uprightness and his personal courage remain unforgotten. He made no secret of his hatred for a totalitarian regime. In defiance of official orders, he protected in his Institute a group of persons persecuted for so-called racial reasons.[88]

There can be little doubt that Wieland was a leading figure in the transformation of what has been called the "classical" organic chemistry of Baeyer and his nineteenth-century contemporaries into a broader branch of science, not only with new and more refined methods, but also deriving renewed vigor from a developing biochemistry. Moreover, Wieland preserved in German chemistry some of the humane qualities of Baeyer's style of scientific leadership.

Richard Willstätter's career provides another outstanding example of the power of twentieth-century organic chemistry, when used by a gifted scientist, to elucidate the chemical structure of constituents of biological systems, and also reveals some of the limitations of organic chemists in their approach to bio-

---

[88] Huisgen (1958), p.219.

chemical problems.[89] As noted earlier in this chapter, Willstätter received his Dr.phil. (1894) largely for work done with Einhorn on cocaine. Willstätter then undertook the study of the closely related alkaloid tropine, whose structure he established, and this success led to the correction of the formula of cocaine proposed by Einhorn. In 1905, Willstätter succeeded Bamberger at the Zurich ETH, where he achieved further notable success in the synthesis of cyclo-octatetraene (a higher homologue of benzene) and in the study of the constitution of chlorophyll. During the period 1912–1915, Willstätter was head of a division of the newly established Kaiser-Wilhelm Institute of Chemistry in Berlin, and he then moved back to Munich as Baeyer's successor. The work on chlorophyll led Willstätter further into plant physiology, with extensive work on photosynthesis, notable success in the elucidation of the structure of various plant pigments, and the investigation of the enzyme peroxidase. His final sustained research effort (1918–1925) dealt with the purification and characterization of several enzymes; although valuable contributions were made to some aspects of enzyme chemistry, Willstätter's conviction that enzymes are relatively small unstable organic molecules and not proteins, as many biochemists had thought, proved to be erroneous.[90] During the course of his organic-chemical and biochemical researches, Wilstätter had many outstanding co-workers; among them were Yasuhiko Asahina, Michael Heidelberger, Richard Kuhn, Kaj Linderstrom-Lang, Jacob Parnas, Rudolf Pummerer, Arthur Stoll and Laszlo Zechmeister.

In 1924, at the height of his public fame, Willstätter resigned from his professorship in Munich because of the decision of the faculty (led by the physicist Wilhelm Wien) not to approve the appointment of Victor Goldschmidt as the successor of Paul von Groth in the chair of mineralogy. According to Willstätter, this blatant example of anti-Semitism had been preceded by inci-

---

[89] Willstätter's discursive autobiographical notes, completed in December 1940 when he lived as an exile in Switzerland, were prepared for publication by his former student Arthur Stoll, and the first edition appeared in 1949. A second edition (referred to in this chapter) and an English translation (by Lilli S. Hornig) were published in 1956 and 1965 respectively; in the latter version, the valuable postscript by Stoll has been greatly shortened. This book is an important source of information about Willstätter and many of the people (especially Baeyer) with whom he had been associated during his brilliant scientific career. His account of his personal triumphs and tragedies is frequently moving, and reveals much about his personality. As in all autobiographical writings, Willstätter's recollections are tinged with the memory of past controversy. Although later articles about him have drawn extensively from the book for details of his life, the scientific work has been described somewhat more objectively by others, notably Kuhn (1949), Robinson (1953) and Huisgen (1961).

[90] See Fruton (1977).

dents involving two other noted scientists named Goldschmidt: the organic chemist Stefan Goldschmidt, whom Willstätter wanted to appoint as Pummerer's successor in his institute, and the biologist Richard Benedict Goldschmidt, who was the leading candidate for the chair being vacated by Richard Hertwig.[91] Despite the relaxation, during the mid-nineteenth century, of rules regarding the appointment of Jews to professorships in German universities, anti-Semitism remained endemic in most faculties, as it was in other countries (Russia, France, United States). However, before the barbarism that followed the advent of Hitler, middle-class German Jews had an opportunity to make their way, despite odds, in science and medicine. The encouragement that Baeyer gave to Victor Meyer, Eugen Bamberger and Richard Willstätter, his friendship with Heinrich Caro, and the high industrial position attained by Baeyer's student Arthur Weinberg all attest to Baeyer's immunity to the anti-Semitic virus. Baeyer's attitude was no doubt influenced by the fact that his mother (née Hitzig) had Jewish forebears; this did not go unnoticed by the Nazis; his sons Hans von Baeyer [1875–1941], professor of medicine at Heidelberg, and Otto von Baeyer [1877–1946], professor of physics at Berlin, were both dismissed "auf Grund rassischer Abstammung" from their university posts during the 1930s.

Willstätter's resignation, which was officially deplored and which many of his scientific colleagues thought foolhardy, essentially marked the end of his productive career. A man of considerable wealth, he could continue to live comfortably and travel widely, and he directed (by telephone) the work of his associate Margarete Rohdewald, to whom Wieland had given laboratory space in the Munich institute. That research, largely on aspects of carbohydrate metabolism, was completely overshadowed by the advances then being made by others, notably Carl and Gerty Cori, Gustav Embden and Otto Meyerhof. Any thought Willstätter may have had that his eminence would protect him in Nazi Germany was rudely shattered in 1938, and he managed to escape to Switzerland in the following year.

The twelve men whose scientific careers I have sketched were of course not the only ones who transmitted, as university teachers and investigators, the traditions of the Baeyer institute to German students of chemistry. Among the names that were omitted, apart from Emil Fischer, were Walter Dieckmann, Otto Fischer, Ferdinand Henrich, Karl Andreas Hofmann, Jakob

[91] Willstätter (1956), p.341; in the English translation, see pp.361–362.

Meisenheimer, Rudolf Pummerer and Wilhelm Schlenk (see Appendix 5). With some exceptions (Giessen, Göttingen, Leipzig, Marburg), all the major pre-1914 German universities had, at one time or another, full professors of chemistry who had been members of the Baeyer research school. In all, about 50 Germans or Austrians attained full professorships at German, Austrian or Swiss universities and higher technical schools, and about 40 more remained in academic life at lower ranks (see Table 4–2). It may be estimated, therefore, that approximately 18 per cent continued to contribute to the chemical literature and to participate in the chemical education of students.

The Baeyer Research School and the German Chemical Industry

Although a considerable proportion of the members of the Baeyer research school sought academic careers, more than half of the 486 German and Austrian students listed in Appendix 5 entered chemical industry or became commercial chemists.[92] This influx of young Dr.phil.'s in chemistry from Munich and other German universities (especially Heidelberg, Rostock, Berlin, Erlangen, Göttingen, Marburg, Kiel and Freiburg) coincided with the expansion of the chemical industry during 1880–1914. The five-fold increase (from 1800 to 9000) in the number of industrial chemists in Germany paralleled a roughly comparable increase in the number of chemical companies and in the share capital of these firms.[93] Some companies, notably those in the dye industry, were greatly strengthened by the patent law of 1877, applicable to the whole Reich, and began to set up their own laboratories for research and development.[94] By 1900, about 400 graduate chemists were employed by the three leading German companies—Bayer, Hoechst (formerly Meister, Lucius and Brüning) and Badische Anilin- und Soda-Fabrik (BASF). In a lecture delivered in 1901, a noted English dye chemist, Arthur George Green [1864–1941], called attention to

---

[92] In addition to the 243 persons indicated in Table 4–2, it is likely that, apart from those who may have died soon after their doctorate, most of the 74 persons whose later professional activity I did not determine also entered industry or became commercial chemists.

[93] See Borscheid (1976), p.235, and Haber (1958), p.170.

[94] See Beer (1959) and Schmauderer (1971a, 1971b).

the fact that, during 1886–1900, 948 English patents on coal-tar products had been issued to the six largest German firms, and only 86 to the six leading English firms.[95]

By the 1890s the growing German chemical industry was employing graduate chemists with widely different educational backgrounds. As was noted in Chapter Two, the requirements for the Dr.phil. degree varied considerably among the 20 German universities, and all too often the work for the degree was narrowly focused on an assigned laboratory task that formed part of the research effort of a *Dozent,* with insufficient education in mathematics, physics, biology or, for that matter, other areas of chemistry. Moreover, the nine German higher technical schools were also turning out inadequately educated chemists who were required to study such subjects as mechanical engineering; these institutions were only authorized to issue a diploma. As a consequence, calls came from spokesmen of the chemical industry for the establishment of a state examination in chemistry, which was to replace the Dr.phil. *Promotion* as basis for entry into the profession. The response of the university professors was predictable; Rudolf Fittig bluntly stated: "Our institutes are establishments for teaching and research, independently of any application of their findings."[96] Wilhelm Ostwald was strongly opposed, as was Baeyer, who felt that the kind of qualifying examination he had introduced in Munich met the concerns of the industrialists, among whom was Carl Duisberg, one of his former pupils. The active debate continued in print, in the Reichstag, and at meetings of chemical societies until 1897, when it was ended by the intervention of Baeyer in the discussion of the question at a meeting of the Electrochemical Society in Munich. His biographer Schmorl describes this event as follows:

Some participants vented long speeches and rejoinders from which it soon became evident that one could not rightly tell what had been decided. During this wordplay Adolf Baeyer entered. The scientists knew his views, but they were not certain whether he would engage in the debate. As it surged back and forth, Adolf Baeyer asked to be allowed to speak. With blunt frankness he declared his complete opposition to the introduction of the state examination because it was not the responsibility of the state to educate chemists in accordance with the views of the industry. Scientific work, which is far more significant

[95] Green (1902), p.11; for a biography of Green, see Baddiley (1943).
[96] Fittig (1895), p.6.

than an examination, should rather be kept at its present high level. As for the basic education in inorganic as well as organic chemistry, this must be considered differently. It is the business of laboratories of universities and higher technical schools, not of the state.[97]

Later in 1897, Baeyer convened a meeting of the laboratory directors at German universities (*Verband der Laboratorium-vorstände an deutschen Hochschulen*) at which it was agreed to institute what came to called the *Verbandsexamen*, to be required of all chemistry students before they were permitted to do research for the Dr.phil. degree.

This incident appears to be one of the few occasions when Baeyer entered the arena of scientific politics, and the manner of its resolution says much about the esteem in which he was held by leaders of the German chemical industry. Baeyer's association with Caro, his role (indirectly) in the achievements of Graebe and Liebermann, and his direct role in the indigo synthesis were important in forging personal links to the chemical companies. Also, as it became an increasingly widespread practice in the relationship between the industry and leading academic chemists, consultant fees were paid and royalty payments were made, and generous samples of chemical intermediates or natural products were provided to Baeyer and his associates to aid the work of their assistants and Dr.phil. students. Moreover, several of the men who went from the Baeyer laboratory into industry became leaders of important chemical firms. As head of Bayer & Co. (and member of the Reichstag), Carl Duisberg, who had worked with Pechmann in Munich, was influential in the industrial and political world of pre-Hitler Germany. At the turn of the century, Duisberg visited the United States where he saw the rise of industrial trusts, and upon his return he began to advocate the formation of a cartel that would include the major German dye and pharmaceutical firms. In 1925, with the formation of the mammoth I. G. Farben (*Interessengemeinschaft Farbenindustrie Aktiengesellschaft*), he became its director. Another influential figure was Friedländer's research student Arthur von Weinberg, who entered the Cassella firm, owned by members of his family. After many years of productive research on dyes,

---

[97] For the quotation, see Schmorl (1952), p.122; an account of the controversy is given on pp.119–125. For later accounts of the establishment of the *Verbandsexamen*, see Schütt (1973), Burchardt (1978) and Johnson (1985a). In addition to the introduction of the *Verbandsexamen*, the higher technical schools were authorized to award the degree of *Diplom Ingenieur*, which was considered to be equivalent to the Dr.phil. given by universities.

Weinberg became an executive of the company and later of I. G. Farben, after the entry of Cassella into the consortium.[98] Others of the Baeyer school who became prominent in the German chemical industry were Gustav von Brüning, Edmund ter Meer and Julius Weiler. At least 81 of the 179 German pupils of the Baeyer school whom I have identified as having entered the chemical industry were employed by the three major firms that entered the I. G. Farben combine—BASF, Bayer and Hoechst (formerly Meister, Lucius and Brüning).

Conclusion

In assessing the role of Baeyer's research school in the development of organic chemistry, one must agree with Graebe that "Baeyer undertook [in Munich] to establish a teaching and research laboratory of a magnitude and purpose that had not existed anywhere before on such a scale. Thanks to his organizational talent and the joy he took in his work, he succeeded brilliantly in this enterprise."[99] In particular, Baeyer's generous style of leadership, with encouragement to promising young people to develop their individual talent, was a decisive factor in his success. It must also be recognized, however, that Baeyer remained a disciple of Kekulé, and that during his forty years in Munich Baeyer's scientific outlook was largely derived from the structural concepts advanced by Kekulé, Couper and Butlerov and the stereochemical ideas offered by van't Hoff and LeBel. Baeyer's experimental achievements, and those of his associates in Munich, amply demonstrated the fruitfulness of this theoretical groundwork for the advancement of chemical knowledge and its application in industrial practice.[100] With some notable exceptions, Baeyer's scientific progeny reflected his indifference to the role that the emerging physical chemistry might play in offering a more secure theoretical basis for organic chemistry.[101]

In one important respect, several of Baeyer's outstanding pupils went beyond his vision of the scope of organic-chemical

---

[98] For Duisberg, see Duisberg (1933) and Stock (1935); for Weinberg, see Ritter and Zerweck (1956).

[99] Graebe (1915), p.435.

[100] This orientation is particularly evident in the books by Hjelt (1916) and Walden (1941) on the history of organic chemistry.

[101] See, for example, Henrich (1908); the fifth edition of this book appeared in 1924 under the title *Theorien der organischen Chemie.*

research. Rupe, a research assistant of Baeyer in 1890, later reported that one day Baeyer said: "I have just received a letter from E. Fischer, in which he writes me that he has completed the total synthesis of glucose. Now the field of organic chemistry is exhausted, now we have to finish up the terpenes, and then all that remains is the chemistry of the *Schmiere*."[102] This statement, perhaps made partly in jest, suggests that, in the twenty years after he left Berlin, he had come to consider the study of the ill-defined substances studied by physiological chemists— proteins, nucleic acids, enzymes, lecithin, cholesterol, among others—to be outside what for him was the field of organic chemistry. At the risk of oversimplification, this attitude may be said to imply that after the nineteenth-century organic chemists had succeeded in displacing the natural dyes used in the textile industry by synthetic ones, the remaining challenges to their ingenuity came from natural substances such as the sugars, the medically interesting alkaloids and the terpenes (important in the perfume industry). Clearly, the chemical achievements of Emil Fischer, who brought the proteins into the orbit of respectable organic chemistry, as well as those of Willstätter and Wieland, demonstrated the narrowness of Baeyer's vision of the future of his discipline. Moreover, in addition to his seeming indifference to the place of physical-chemical concepts and methods in organic-chemical research, Baeyer's attitude encouraged a disdain, among his disciples, of the efforts of physiological chemists to tackle the problems presented by the chemical structure of complex cellular constituents and by the transformation of these substances in metabolic processes.

---

[102] Rupe (1932), p.435.

Chapter Five

Emil Fischer and Franz Hofmeister

During the first decade of this century, Emil Fischer [1852–1919] headed, at the University of Berlin, a chemical institute which rivalled the one that Baeyer had set up in Munich. Among the leading organic chemists of his time, Fischer was exceptional in his sustained interest in the constitution of biologically important substances, and for about ten years (1899–1908) the principal activity of his large research group was in the field of protein chemistry. After 1860, this field had been relegated by organic chemists to the physiological chemists associated with university medical faculties or agricultural experiment stations. One of these physiological chemists was Franz Hofmeister [1850–1922], who succeeded Hoppe-Seyler at Strassburg, where Hofmeister led a productive research school. Fischer and Hofmeister, each in his own way, exerted a significant influence on the later development of the biochemical sciences, and both men were interested in the same scientific problem, albeit from different points of view.

A Meeting in Karlsbad

The scientific link between Fischer and Hofmeister lies in the problem of protein structure. By an accident of history, that link was forged on a single day, 22 September 1902, at the 74th annual meeting of the prestigious *Gesellschaft der deutscher Naturfor-*

scher und Ärzte in Karlsbad.[1] The first of the plenary lectures was presented during the morning by Hofmeister, under the title *Ueber den Bau des Eiweissmoleküls.* On the afternoon of the same day, in the third paper at the first session of the Section of Chemistry, Fischer gave a report entitled *Über die Hydrolyse der Proteinstoffe.* In both talks, the theory was advanced that in proteins the constituent amino acids are joined to each other by the condensation of the amino group of one amino acid with the carboxyl group of another amino acid to form amide (CO-NH) bonds in a linear structure to which Fischer gave the name *peptide.* In subsequent accounts of the history of protein chemistry, these two lectures mark the appearance of the so-called Fischer-Hofmeister peptide theory of protein structure.[2]

The organizers of the Karlsbad meeting clearly attached importance to Hofmeister's lecture, and the complete text appeared shortly afterward in successive issues of the widely circulated weekly *Naturwissenschaftliche Rundschau.*[3] A major part of the lecture dealt with the mode of linkage of amino acids in proteins. After considering various earlier proposals, Hofmeister presented several arguments in favor of the view that the amino acids are joined largely by amide bonds. He attached special significance to the biuret reaction—the purple color given with alkaline copper sulfate by proteins and by intermediate products of their enzymatic digestion (the so-called albumoses and peptones).[4] Among the synthetic materials (apart from biuret) that give this reaction, Hofmeister cited the products obtained by Curtius during the early 1880s either by the self-condensation of glycine ethyl ester to form a *biuret-base* or by the reaction of benzoyl chloride with silver glycinate. In the latter process, one of the isolated products was benzoyl-glycyl-glycine, which in retrospect may be considered to represent the first synthetic peptide derivative.[5] In favor of his theory, Hofmeister also offered evidence from physiological studies on the enzymatic cleavage of proteins and of hippuric acid (benzoyl-glycine). It should also be noted that he took occasion to hail Fischer's entry into the protein field.

---

[1] For a brief history of the Society of German Scientists and Physicians, see Bochalli (1948). Unaccountably, he refers to the 1902 Karlsbad session as the 75th meeting.

[2] For the history of theories about protein structure, see Plimmer (1908), Vickery and Osborne (1928), Lieben (1935), pp.359–391, Fruton (1972), pp.148–179, (1979) and Shamin (1977).

[3] Hofmeister (1902a). The text of Hofmeister's lecture was an adaptation for a general scientific audience of his review article; see Hofmeister (1902b).

[4] For a comprehensive account of the history of the biuret reaction, see Kurzer (1956).

[5] For Curtius, see Chapter Four, note 74.

The paper read by Fischer on the afternoon of 22 September was not published in full, but an abstract prepared by him was printed a few weeks later in the *Chemiker-Zeitung*. In this *Autoreferat* Fischer summarized his recent studies on the isolation of amino acids and peptides from protein hydrolysates and proposed that "in analogy to the known designation of carbohydrates as disaccharides, trisaccharides, etc., the substances of the type glycyl-glycine be named dipeptides and that anhydride-like combinations of a greater number of amino acids be denoted tripeptides, etc."[6] This statement appears to mark the introduction of the word *peptide* into the language of chemistry. Later, Fischer also began to use the term *polypeptide*. In the abstract, Fischer also wrote:

Finally the speaker discussed the coupling of the amino acids in proteins. The idea that acid-amide-like groups play the principal role comes to mind most readily [*liegt am nächsten*], as Hofmeister also assumed in his general lecture this morning. The same conviction led him [Fischer] more than 1 ½ years ago to initiate experiments to effect the synthetic linkage of amino acids.[6]

The last sentence was clearly intended to draw attention to Fischer's paper on the synthesis of glycyl-glycine and its derivatives.[7] It should be noted that the idea of amide linkages in proteins had appeared in the chemical literature before 1902. During the period 1860–1900, several chemists attempted to prepare peptone-like materials by subjecting amino acids to dehydration at high temperatures. One of these investigators was Édouard Grimaux [1835–1900] who in 1882 depicted alanyl-alanine as an amide, and also wrote: "If one succeeds in producing anhydrides containing the residues of aspartic acid, of tyrosine etc., one may thereby be able to transform them into amides that resemble more and more the nitrogenous colloids made by living organisms."[8] The Karlsbad meeting that links Hofmeister and Fischer thus provides still another apparent in-

---

[6] Fischer (1902).
[7] Fischer and Fourneau (1901). In a letter dated 21 October 1901 (Bancroft Library), Fourneau acknowledged the receipt of reprints "du petit travail sur l'anhydride du glycocolle."
[8] Grimaux (1882); the quotation is on p.69.

stance of simultaneous scientific discovery.[9] For the purposes of this chapter, however, their encounter is of greater interest in offering a contrast in the style of two scientists with different backgrounds and personalities, who worked in different institutional settings, and who had met in the consideration of the same scientific problem.[10]

The biologist Hofmeister had already done extensive and important work on proteins, and held a large view of the role of chemistry in the study of physiological problems. The chemist Fischer had just entered the arena of protein research, and was emboldened by his notable success in the synthesis of sugars and purines to believe that synthetic organic chemistry could also solve the problem of protein structure. In scientific stature, as measured by public esteem, Fischer vastly overshadowed Hofmeister. At the time of the Karlsbad meeting Fischer already held the title *Geheimer Regierungsrat*,[11] and a few months later he was awarded a Nobel Prize in Chemistry, whereas Hofmeister received no honors of this sort. Also, in the hierarchy of the university disciplines in Wilhelmine Germany (as indeed elsewhere) organic chemistry had a far higher status than did physiological chemistry, for reasons discussed in the previous two chapters. Nevertheless, despite the difference in their public prestige, Hofmeister's impact on the later development of the biochemical sciences was as significant as that of Fischer. The influence exerted by the two men came not only from the research achievements in their respective laboratories but also from the experience gained there by predoctoral students, research assistants and postdoctoral investigators from many countries. In succeeding parts of this chapter, I will attempt to sketch the personal and scientific qualities of Fischer and Hofmeister, to examine more closely the development during ca. 1899–1914 of their common area of interest and to compare the effectiveness of

---

[9] See Merton (1961) and Baumann (1972). In the latter source, the Fischer-Hofmeister peptide theory of protein structure is listed on p.66. One view of the priority in this case was offered hy Hofmeister's former assistant Otto von Fürth (1919): "In 1902, on the occasion of the Karlsbad meeting, Franz Hofmeister first formulated scientifically the assumption of an amide-like linkage of the individual amino acids in proteins. E. Fischer, however, was the one who attacked the problem systematically."

[10] Whether Fischer and Hofmeister met face-to-face at the Karlsbad meeting, or anywhere else, is uncertain. I have found no reference to such a personal meeting, nor have I found any evidence for an exchange of letters between them. None are included in the extensive collection of Fischer's correspondence at the Bancroft Library in Berkeley, California. The only indication of a contact is through Hofmeister's associate Karl Spiro, who had received his Dr.phil. in chemistry for work with Fischer. The Bancroft Library collection contains seven letters from Spiro to Fischer, over the period 1901 to 1913; the one of 27 January 1901 refers to Hofmeister.

[11] In 1910, Fischer was elevated to the rank of *wirklicher Geheimer Regierungsrat,* which required that he be addressed as *Exzellenz.*

their research groups in the education of the next generation of productive investigators in the chemical and biochemical sciences.

Emil Fischer

Fischer's scientific career spans almost exactly the years of the first German Reich. His first published paper (1874) was based on the work he did with Baeyer in Strassburg. Fischer's death in July 1919 came eight months after the capitulation of Germany in the First World War. In the bitterness of defeat, one of his eulogists wrote: "Emil Fischer represents a symbol of Germany's greatness."[12]

The number of biographical writings about Fischer is legion,[13] and for the purposes of this chapter I will only refer to some aspects of his life, work and personality that suggest something of his style as a scientist and as an educator. That Fischer possessed exceptional chemical talent was evident from his first publications from Baeyer's laboratories in Strassburg and Munich. At first, Fischer continued to work on dyes, the subject of his outstanding Strassburg dissertation,[14] but later he published relatively few papers in this field. Instead, he chose to exploit his discovery of a compound which he identified as C_6H_5-NH-NH_2, the phenyl derivative of the substance he named *hydrazine*. Fischer's masterly elaboration of this work was described his *Habilitationsschrift*, and he became a *Privatdozent* in 1878. In the following year came a promotion to *ausserordentlicher Professor* at Munich, and he then held full professorships at Erlangen (1882–1885) and Würzburg (1885–1892) before he went to Berlin in 1892 as the successor of August Wilhelm Hofmann.

Fischer's finding in 1884 that phenylhydrazine gives crystalline derivatives of natural sugars (glucose, fructose, galactose, etc.) was the starting point for his entry into biochemical research. Although he did not neglect other opportunties pre-

---

[12] Harries (1919a); the quotation is on p.843.

[13] An extensive bibliography of writings about Fischer may be found in Poggendorff (1971), pp.204–207. Among the many excellent accounts of his scientific work those by Harries (1919a), Forster (1920), Hoesch (1921) and Hilgetag and Paul (1970) are especially valuable. In 1918, Fischer wrote an autobiography (up to about 1900) which was published posthumously; see Fischer (1922, 1987). It contains brief and occasionally uncomplimentary remarks about some of the many people who had been his pupils or assistants.

[14] Fischer (1874). Most of the subsequent work on triphenylmethane dyes was done in collaboration with his cousin Otto Fischer (see Appendix 5).

sented by his systematic investigation of the reactions of phenylhydrazine derivatives (for example, in the synthesis of indole derivatives), the remarkable work performed by his research group on the degradation and synthesis of carbohydrates stands out, in my opinion, as the high point of his scientific career.[15] The manner in which Fischer applied and developed the van't Hoff-LeBel concept of the asymmetric carbon atom in elucidating the stereochemistry of the sugars bespeaks the theoretical insight he brought to the problem. The methods he developed for the synthesis of sugars and their derivatives were elegant and lasting. Although it remained for others, notably the great British school of carbohydrate chemistry (Purdie, Irvine, Haworth), to define more precisely the structure of the sugars, Fischer's achievements before 1900 provided the basis for further investigation. And, a by-product of his organic-chemical work was the study of the fermentation of sugars and of the cleavage of glycosides by enzymes, leading to Fischer's famous lock-and-key analogy of the specificity of enzyme action.[16]

Before initiating his work on sugars, Fischer undertook the systematic study of caffeine because he questioned the validity of the proposal made in 1875 by Ludwig Medicus that its structure is closely related to that of uric acid, whose constitution Baeyer had not succeeded in determining during the 1860s. In a series of elegant synthetic experiments between 1881 and 1898, Fischer provided definitive evidence for the correctness of Medicus's views, and showed that caffeine and uric acid, as well as xanthine and guanine, are derivatives of a parent substance which Fischer named *purine*.[17] During the 1880s, largely through the work of Albrecht Kossel, it became known that the purines guanine and adenine are constituents of nucleic acids.[18] The Nobel Prize in Chemistry awarded to Fischer in 1902 was in recognition of his synthetic work on sugars and purines.

As was noted above, Fischer entered the protein field in 1899, and during the succeeding ten years the efforts of his research group were directed to the study of proteins, peptides and amino acids. This aspect of Fischer's research activity will be considered later in this chapter. His success fell short of his hopes, and after 1908 Fischer returned to carbohydrate and purine chemistry, attempted to enter the nucleic acid field through

---

[15] For Fischer's work on indoles, see Roussel (1953) and Robinson (1983). The most valuable accounts of Fischer's work on sugars are those by Hudson (1941, 1948), Freudenberg (1966) and Mason (1987).

[16] Fischer (1894).

[17] Fischer (1899). For Ludwig Medicus [1847–1915], see Reitzenstein (1915).

[18] For Kossel, see Chapter Three, note 66.

the synthesis of nucleotides, and initiated a sustained and fruitful research program on compounds which he termed *depsides* (Greek *depsein,* to tan), present in lichen substances and tannins.[19] In the final years of his life, Fischer also turned to the synthesis of glycerides, thus adding the fats to the other classes of chemical constituents of biological systems—carbohydrates, nucleic acids and proteins—whose structure he had studied during the course of his remarkable career.

Fischer's scientific papers and lectures were collected in eight volumes, which include approximately 600 experimental articles and about 20 lectures.[20] Of the articles, 185 appeared under his sole authorship and 295 as joint publications, with Fischer's name first in all but two of them. Nearly all the remaining 120 experimental papers were published under the sole authorship of his Dr.phil. students, and represent extracts from their dissertations. In the case of work done by his postdoctoral assistants, Fischer often published important papers alone, with an acknowledgment to the co-worker at the end of the text. For example, the final sentences in a long paper of which he was the sole author (on the hydrolysis of casein) reads as follows: "The above experiments were in part exceptionally difficult and tedious. I therefore feel all the more obliged to express my great thanks to my assistant Dr. Otto Wolfes for the help he has given."[21] The collective volumes do not contain all the published reports of the experimental work done in Fischer's laboratory; in particular, some of the papers by his associate Emil Abderhalden are not included or are given in abbreviated form.

The testimony of Fischer's colleagues and junior associates suggests that before he went to Berlin in 1892, the style of his leadership resembled, in some respects, that of his teacher Baeyer. At Erlangen and Würzburg, dictatorship appears to

---

[19] See Ratman (1961).

[20] The format of all these volumes was established by Fischer himself, and he edited the first four: *Untersuchungen über Aminosaüren, Polypeptide und Proteine 1899–1906* (Berlin 1906); *Untersuchungen in der Purinreihe 1882–1906* (Berlin 1907); *Untersuchungen über Kohlenhydrate und Fermente 1884–1908* (Berlin 1908); *Untersuchungen über Depside und Gerbstoffe 1908–1919* (Berlin 1919). The fourth one was completed shortly before Fischer's death, and the final stages of its publication were supervised by Max Bergmann, who also edited the later volumes: *Untersuchungen über Kohlenhydrate II 1908–1919* (Berlin 1922), *Untersuchungen über Aminosäuren, Polypeptide und Proteine II* (Berlin 1923), *Untersuchungen über Triphenylmethanfarbstoffe, Hydrazine und Indole* (Berlin 1924); *Untersuchungen aus verschiedenen Gebieten* (Berlin 1924). The last contains scientific articles that did not appear in the earlier volumes, as well as various lectures and obituary notices written by Fischer. All the above volumes were published by Springer.

[21] Fischer (1901), p.176. Wolfes's help is similarly acknowledged in five other papers. Likewise, Georg Pinkus, a *Privatassistent* of Fischer during 1893–1898, did not appear as an author or co-author of any publication from Fischer's laboratory, but only through acknowledgments in thirteen papers.

have been blended with encouragement of young talent, as in the case of Ludwig Knorr. One of Fischer's many English students in Würzburg later wrote:

Physically commanding, his authority rested on the solid foundation of natural dignity unmarred by self-assertion. The brisk, upright carriage marked the man of action; the glowing eyes revealed his attitude of constant, keen enquiry; the impatience with trivialities was one aspect of his dominating, steadfast control of essentials. With ordinary human perception, it was impossible for anyone to escape his contagious enthusiasm, and yet all the time the master did not obscure the man, for although his daily demeanour was tinged with severity, his heart when revealed was deeply kind, and, in circumstances of relaxation, joyous.[22]

The recollections of those who knew Fischer in Berlin convey a somewhat more somber picture. According to Willstätter,

Emil Fischer ruled his laboratory with absolute authority. The princely man who gathered about him most of the doctoral students and other young investigators overshadowed everyone in greatness, spirit and scientific insight. His lively and penetrating participation spurred the independent scientists of the institute to the greatest efforts. In the Chemical Society, his criticism was feared, no contradiction was tolerated. On occasion, in the laboratory, a vigorous young scientist suffered under Fischer's autocratic nature.[23]

Willstätter then refers to Carl Harries, whom Fischer had taken on from the Hofmann school as an assistant and whom he had promoted to be head of a section of his institute. In his obituary notice for Fischer, Harries wrote that

He demanded that each assistant should choose a research topic, and took great interest in this work, but hounded them from one experiment to the next and egged them on to premature publication. . . . If a chemist did not publish anything for a while, he at once concluded: "The fellow has become lazy.". . . . In later years there developed be-

---

[22] Forster (1920); the quotation is on p.1158.
[23] Willstätter (1956), p.211; in the English translation, see p.224.

tween Fischer and me an ever-stronger antagonism which I sincerely regretted. However, I could not change our relationship without relinquishing some of my rights.[24]

It would appear that Fischer considered the organic-chemical part of his Berlin institute to be largely a laboratory for the investigation of problems of immediate interest to him, and not for research along other lines by junior associates of independent spirit. Harries had been kept on at the institute on the recommendation of Friedrich Althoff of the *Kultusministerium* and in 1899 had married Hertha von Siemens, then a student in Fischer's laboratory, and the daughter of Werner von Siemens, the noted industrialist.[25] Other independent-minded junior members of Fischer's laboratory may have had aspirations similar to those of Harries, but were obliged to accept Fischer's rule.

For a beginner in Fischer's laboratory, his appearance must have been awesome. A few days after Otto Diels began work there in 1896, "a tall, very erect man with a pince-nez and black beard strode through the hall, followed by a laboratory servant. Diels was struck by the fact that he wore a formal black hat and a blue tunic. Diels asked his neighbor who the man was, and received the amazed reply: 'Don't you know? Why, that was Fischer!' "[26] Another report that suggests something of the atmosphere of Fischer's research laboratory states:

With a stern eye he inspected the laboratory workers, who reported to him the progress of their experiments. Fearsome was his *Flügelschlagen* [flapping of wings], without further comment, for the poor wretch if something had gone thoroughly wrong. Only rarely did the chief sit on a stool and conduct a brief private conversation. Then it was even permissible to laugh. However, the slightest attempt at intimacy would terminate the conversation immediately.[27]

It may be that Fischer reserved his most severe treatment for the German students and treated the foreign visitors more kindly.

---

[24] Harries (1919b); the quotation is on pp.611–612. For biographies of Harries, see Nagel (1924) and Willstätter (1926).
[25] See Fischer (1922), pp.191–192, and Willstätter (1956), pp.131–132.
[26] See Olsen (1962), p.xiv; for Diels's recollections of his association with Fischer, see pp.xix–xxii.
[27] Herneck (1970), p.45.

The American physician James Bryan Herrick, who worked in Fischer's laboratory in 1905, described him as follows:

He was modest, kindly, always the gentleman. Twice a day he made the rounds, moving quietly from desk to desk inspecting the work, always seeming interested, criticizing, helpfully suggesting. He had the faculty of seeing quickly where one's trouble lay. So gentle in manner was he that one scarcely realized that he was a good executive commanding officer.[28]

Another aspect of Fischer's style of leadership has been noted by his last chief research assistant, Max Bergmann, who wrote that Fischer was reticent to his co-workers

when he gave them instructions for the conduct of experiments or himself did laboratory work in their presence. Then, an indication of the purpose and goal and expected outcome of the experiment was either not given or stated very incompletely. The explanation of this behavior may be found in a printed guide for the conduct of scientific experiments, which Fischer regularly presented to the older students of his institute and to his own assistants. One sentence was: "You are urgently warned against allowing yourself to be influenced in any way by theories or by other preconceived notions in the observation of phenomena, the performance of analyses and other determinations."[29]

This injunction comes to mind on reading the doctoral dissertations from Fischer's laboratory, and one cannot but be struck by their limited scope and by the absence of a significant intellectual input on the part of the student. To be sure, the completion of moderately difficult tasks may have been preceded by failures, and the mere performance of elementary analyses was time consuming, but the skimpiness of content raises questions about the educational benefits the student may have derived from his participation in a significant research program.

To these facets of life in Fischer's institute must be added the matter of secrecy, not an uncommon phenomenon in organic-chemical laboratories. Hans Thacher Clarke, an English-

---

[28] Herrick (1949), p.130–131.
[29] Bergmann (1930), p.415.

man who worked at the Berlin institute from 1911 until 1913, later wrote that upon his arrival there he was advised

not to ask the other members of the laboratory what they were doing. This was so contrary to British tradition that I was interested to find out the reason; it appeared that most of the chemists who were working on topics of their own were retained as consultants by one or another of the German manufacturing firms, which had priority on any patentable discoveries made by the individuals concerned. This system appeared to me, as it still does, as being at variance with the prime function of an academic laboratory.[30]

Heinrich Wieland also deplored the fact that "Among chemists there was at that time the custom to conceal ongoing research in deep mystery in order to protect the field from foreign intrusion; many *Dozents* solemnly swore their doctoral students to secrecy."[31]

Any assessment of Fischer's role as the leader of a research group must take into account the fact that his personal qualities were affected by continued ill-health and, in later life, successive family tragedies. Throughout his professional career he was afflicted by respiratory and intestinal problems, arising from and exacerbated by the poor ventilation of the laboratories in which he had worked. His discovery of phenylhydrazine in 1875, and its frequent use thereafter, exposed him to an agent whose toxicity he recognized only about fifteen years later.[32] In 1881, Fischer had an attack of mercury poisoning, as a consquence of the generation of the volatile mercury diethyl during experiments on the reaction of mercuric oxide with alkyl phenylhydrazines. Indeed, in 1885 some members of the Würzburg faculty opposed Fischer's appointment because of his respiratory difficulties.[33] His correspondence of later years is replete with letters from various spas reporting on the state of his health to his chief laboratory assistants.

To his medical problems must be added the tragic death in 1895 of his wife Agnes (née Gerlach) whom he married in 1887; according to Harries, Fischer "treated her in the same high-handed manner as were the assistants and co-workers."[34] From this marriage, about which Fischer wrote with extreme reserva-

[30] Clarke (1958), p.2.
[31] Wieland (1950), p.1.
[32] Lewin (1919).
[33] Günther (1980).
[34] Harries (1919b), p.611.

tion, came three sons: Hermann Otto Laurenz [1888–1960] who
laters became a distinguished organic chemist; Walter Max
[1891–1915] who interrupted his medical training because of an
acute depression and committed suicide; and Alfred Leonhard
Joseph [1894–1917], a physician who died of typhus while on
military service in Rumania.

Although personal illness and family tragedy must have
contributed to molding Fischer's temperament, there can be
little doubt that the dominant factor was his uncompromising
singleness of purpose as a research chemist. Although Fischer is
not reported to have had exceptional manual skill in the laboratory, such as that ascribed to Bunsen or Baeyer, he possessed a
remarkable memory and, despite his constant ailments, a seemingly superhuman capacity for sustained work. If he drove his
assistants and students, he demanded no less of himself. The
polished and often inspiring lectures he delivered were prepared
with painstaking care. And, above all, he was (as Herrick termed
it) "a good executive commanding officer" of his research
group.

At least until he became a public figure in 1902, with the
award of the Nobel Prize, Fischer eschewed involvement in academic or governmental business that did not affect directly his
research interests, although he continued to maintain a close relationship to several important chemical firms. After 1902,
Fischer participated actively in public affairs, initially (1905) in
promoting the idea of a Reich Institute of Chemistry as a counterpart of the *Physikalische Reichsanstalt* then under consideration, and later (1909) in the successful collaboration with Adolf
von Harnack to persuade government and industrial leaders to
establish the Kaiser-Wilhelm Society for the Promotion of the
Sciences. Fischer played a decisive role in the organization of the
various research institutes founded by the Society, in particular
those for chemistry in Berlin-Dahlem and for coal research in
Mühlheim-Ruhr.[35]

Fischer's political outlook during the First World War was
described by Georg Klemperer, his personal physician, as being
that of a "liberal." In particular, mention has been made of his
attitude toward the appointment of able Jewish chemists. Willstätter wrote that

In the [Kaiser-Wilhelm] institutes that were being established in
Berlin-Dahlem, it was obvious that a disproportionately large number

---

[35] See Trendelenburg (1919), Burchardt (1975) and Wendel (1975).

of Jewish or non-Aryan scientists were appointed, too large, in my opinion. They were readily available, since the universities certainly did not make their departure difficult. No wonder that Emil Fischer was asked: "How is it possible that Your Excellency is not antisemitic?" Fischer replied: "We Rhinelanders are not so stupid tht we must be afraid of the Jews."[36]

Likewise, Klemperer wrote soon after Fischer's death: "Very indicative [of his liberalism] was his attitude to antisemitism at a time many faculties openly paid homage to this trend. He said: 'If I had the aptitude for it, I would only need to consider my teacher Bayer [sic] to recognize the absurdity of antisem- itism.' "[37] Both Willstätter and Klemperer were obliged to leave Germany during the 1930s.

As was noted in an earlier chapter, upon the advent of the First World War, Fischer joined 92 other leading members of the German cultural establishment (including Willstätter) in signing the notorious *Aufruf an die Kulturwelt*.[38] Also, because of his sci- entific eminence and his close connections to the German chem- ical industry, Fischer became involved in the work of numerous war committees, and headed several of them.[39] As the war pro- gressed, Fischer, along with other signers of the *Aufruf*, became increasingly despondent about the prospects of victory. His last letters include expressions of the hope that a firm and respect- able government might be established in a defeated Germany, and that a socialist revolution might be averted.[40]

Franz Hofmeister

Hofmeister's scientific work began in Prague, where he was born the son of a well-respected prosperous physician, and

---

[36] Willstätter (1956), p.210; in the English translation, see p.222.
[37] Klemperer (1919), p.2. According to Harries (1919b), during the 1890s Fischer was more conservative in his political views.
[38] See Chapter Three, note 73.
[39] For valuable summaries of Fischer's close relationship to the German chemical indus- try, see Duisberg (1919) and Wohl (1919). His pharmaceutical patents, notably those for the barbiturate Veronal and for Sajodin (an iodine compound designed to release iodine slowly), were especially profitable. In addition to his close association with the BASF and Bayer firms, Fischer had ties to other chemical companies (for example, Boehringer-Mannheim) where several of his former students or assistants had risen to executive positions. These connections were important in Fischer's role in mobilizing the industry for war production during 1914–1918; see Weinberg (1919).
[40] For a perceptive account of Fischer's attitude to the First World War, based largely on his letters at the Bancroft Library, see Feldman (1973). The letter from Fischer to Duisberg, dated 14 November 1918, of which I have a copy, bears out Feldman's con- clusions.

where he attended its ancient university at a time when the German-Austrian influence was predominant.[41] During the course of his medical studies, which he completed in 1872, Hofmeister showed aptitude in chemistry and upon the advice of the professor of physiology, Ewald Hering [1834–1918], he became an assistant in the laboratory of Hugo Huppert [1834–1904], the newly appointed professor of medical chemistry, who specialized in clinical chemical analysis.[42] Huppert had been a student of Carl Gotthelf Lehmann [1812–1863], whose treatise on physiological chemistry had considerable influence, and who had introduced the term *peptone*, which figures largely in the story of the peptide theory of protein structure. Indeed, Hofmeister's unpublished *Habilitationsschrift* in 1879 dealt with the analysis of peptones; this work preceded that of Kühne, mentioned in an earlier chapter. In Huppert's laboratory, Hofmeister examined various analytical methods for the identification of amino acids and for the detection of proteins in biological fluids, and also showed that the sugar excreted in the urine during pregnancy is not glucose, as had been thought, but lactose. Although Hofmeister learned analytical methods from Huppert, he found that association unsatisfying, in part because Hofmeister had been influenced by Hering's broader view of biological problems.

In 1881, the Prague medical faculty decided to establish a new institute of experimental pharmacology, with Hofmeister as its chief, so he had to obtain a *Habilitation* in that subject. In 1883, after six months in Strassburg, where Oswald Schmiedeberg [1838–1921] headed the leading German institute of pharmacology, Hofmeister was appointed *ausserordentlicher Professor* of pharmacology at Prague, and two years later he was made a full professor. There can be little doubt that Hofmeister's visit to Strassburg, apart from introducing him to the teaching of pharmacology, greatly broadened his scientific outlook. Schmiedeberg's research program was not limited to pharmacological studies, but also included work on such biochemical topics as the crystallization of seed globulins, the nature of glycoproteins, the metabolic oxidation of foreign organic substances and the bio-

---

[41] The published biographical material about Hofmeister is much less extensive than that for Emil Fischer. The most important source is the obituary notice by Pohl and Spiro (1923), which also includes a nearly complete list of the papers published from Hofmeister's successive laboratories in Prague, Strassburg and Würzburg.

[42] For a history of the medical faculty at Prague, see Koerting (1968). Ewald Hering is best known for his theories of vision and the fact that in 1870 he succeeded Jan Purkyně at Prague and, in 1895, Carl Ludwig at Leipzig; see Kruta (1972). For Hugo Huppert, see Zeynek (1904).

synthesis of urea and of hippuric acid.[43] In addition to his association with Schmiedeberg, Hofmeister had the opportunity to make contact with Hoppe-Seyler's research group.

Before going to Schmiedeberg's laboratory, Hofmeister had continued to work on the chemistry and metabolism of peptones and in 1882 had shown that they largely disappear in the intestinal tract, but his evidence regarding their subsequent fate was inconclusive. After his return to Prague, his research program changed markedly. Beginning in 1887, he and his students (Lewith, Limbeck, Münzer) published a series of papers in which they reported that different inorganic salts could be placed in a regular order with respect to their ability to effect the precipitation of proteins, the order remaining essentially the same for different proteins. This relationship later came to be known as the *Hofmeister series* or the *lyotropic series*.[44] Hofmeister recognized that the differences among inorganic salts in the "salting-out" effect is a general function of their hydration, but he did not write of ions; Svante Arrhenius's theory of electrolytic dissociation was not accepted widely until the 1890s. An important consequence of Hofmeister's systematic study was his use of ammonium sulfate (introduced into protein chemistry by Camille Méhu in 1878) to effect the crystallization of egg albumin.[45] Hofmeister also examined the swelling and adsorption phenomena exhibited by proteins, of interest to many nineteenth-century biologists in relation to the properties of protoplasm. In addition to these pioneering physical-chemical experiments, Hofmeister and his associates in Prague made contributions to the study of some aspects of intermediate metabolism. For example, Julius Pohl examined the oxidation of alcohols and fatty acids in the animal body, and Hofmeister himself investigated the newly discovered process of biological methylation and the long-fashionable problem of urea formation.[46]

Until his departure from Prague in 1896, Hofmeister and his junior colleagues published about 75 papers on protein chemistry, intermediary metabolism, and assorted other topics in biochemistry, physiology and pharmacology. Only 30 of these publications bear Hofmeister's name, and most of them were based on work he had done at the beginning of his scientific career. It is also noteworthy that much of the output of

---

[43] For Schmiedeberg, see Meyer (1922) and Koch-Weser and Schachter (1978).
[44] A useful discussion of the Hofmeister series was prepared by Abernethy (1967). For a modern treatment, see Collins and Washabaugh (1985).
[45] For the crystallization of egg albumin, see Hofmeister (1889). A brief biographical note about Camille Méhu [1835–1887] was written by McCollum (1956).
[46] Pohl (1893, 1896), Hofmeister (1894, 1896).

Hofmeister's Prague laboratory served as a background for the efforts of the students and junior associates who joined him after his move to Strassburg in 1896.

In 1895, the chair of physiological chemistry at Strassburg became vacant, owing to the death of Hoppe-Seyler. It would hardly be appropriate to say that this full professorship was the most prestigious one in its field in Germany, because there was only one other, in Tübingen, whence Hoppe-Seyler had come to Strassburg in 1872, when it was made a showcase German university. In contrast to Tübingen, the Strassburg institute was in the medical faculty, and its senior members proceeded to seek a replacement. According to the archival records in Strasbourg,[47] Schmiedeberg was placed in temporary charge, and the post was offered to Gustav Hüfner, who was highly regarded for his work on hemoglobin. He declined on 12 December 1895, and gave his advanced age as the reason, but a later biographer stated that Hüfner did not wish to leave Tübingen to join a medical faculty.[48] The post was then offered to Eugen Baumann at Freiburg i.B.; he declined because he did not wish to leave his chemical laboratory there.[49]

The Strassburg medical faculty then proposed that physiological chemistry only have an *ausserordentlicher Professor* within an institute of hygiene and bacteriology (which Hoppe-Seyler had also represented), whose head would be the Amsterdam professor Josef Förster [1844–1910], a member of the Voit-Pettenkofer school. Indeed, the Strasbourg archives contain a copy of a memorandum (25 February 1896) to this effect from the curator of the university to Förster. At this point some of the original 1872 faculty, notably Schmiedeberg and the physiologist Friedrich Goltz, intervened. According to Hofmeister's biographers, these two men "immediately asked an audience with the Governor [of Alsace-Lorraine], Count Hohenlohe, and explained to him that because of the extensive space that had been allotted to Hoppe-Seyler, the establishment of two separate full professorships with separate institutes was possible without any new construction and without additional cost."[50]

Their argument proved to be persuasive and, owing to the influence of Schmiedeberg, Hofmeister was offered the chair of physiological chemistry. He accepted it on 23 June 1896, and took up his duties later that year. Schmiedeberg's sponsorship of

---

[47] Archives du Bas-Rhin, Strasbourg, AL103, Paquet 194 no.1028, Paquet 258 no.1208.
[48] For Hüfner, see Zeynek (1908) and Wankmüller (1980).
[49] For Baumann, see Chapter Three, note 75.
[50] Pohl and Spiro (1923), p.14.

Hofmeister bespeaks his generosity, since at that time the two men had similar interests in the study of metabolic problems.

I have described the circumstances of Hofmeister's appointment because they suggest something of the attitude of the Strassburg medical professors toward the discipline Hoppe-Seyler had represented, and also toward Hofmeister as a colleague. Although Hofmeister later served twice as dean of the medical faculty, the relatively meager archival record of his relationship to the administration of the university during 1896–1918 gives hints, but no more, of difficulties with regard to the appointment of teaching assistants and the like.[51] Whatever future historical study may uncover about Hofmeister's institutional problems, the circumstances of his appointment alone suggest a situation repeated later in many medical schools in many countries. With few exceptions, for the clinical professors and the professors of physiology who controlled the medical faculties of German universities, biochemistry was considered to be useful in medical education and research only as a part of physiology or in relation to its application to practical clinical problems.

During the 22 years of Hofmeister's stay at Strassburg, approximately 300 papers were published from his institute. A noteworthy feature of the list is that only fourteen bear Hofmeister's name. Among this small number there were several general articles that attracted wide interest. I have already mentioned the two on the structure of proteins that appeared in 1902, and there was another on this subject in 1908.[52] Also, in 1901, Hofmeister published a small brochure which dealt with the integration of enzyme action in metabolic processes. Some historians of biochemistry have considered this publication to have been influential in the emergence of a modern biochemistry based on the enzyme theory of life.[53] Thirteen years later, Hofmeister wrote another article in which he modified his views on this subject in line with the then-current ideas about the colloidal nature of living matter.[54] In 1912 and 1913, he published lectures on the chemical regulation in the animal body and on carbohydrate metabolism in the liver, and nearly all of his suc-

---

[51] Archives du Bas-Rhin, Strasbourg, AL103, Paquet 259 no.1211.
[52] Hofmeister (1908).
[53] Hofmeister (1901); see Kohler (1973) and Florkin (1975), p.26.
[54] Hofmeister (1914).

ceeding papers dealt with problems of nutrition, with particular attention to what he called "accessory nutrients" (Hopkins had termed them "accessory food factors" and Funk "vitamines").[55]

Thus, after 1900, Hofmeister's name appeared in the scientific literature principally as the author of general surveys of various important biochemical problems. The significant experimental papers from his Strassburg institute were published under the authorship of Hofmeister's assistants, students (chiefly candidates for the Dr.med. degree), and postdoctoral guests. These papers represented a wide range of interests, encompassing protein chemistry, enzyme action, hormones and intermediary metabolism. In all these fields, Hofmeister's research group at Strassburg made sterling contributions and, as was noted above, some of the research was an elaboration of work done in his Prague laboratory.

At the start, Hofmeister's chief junior associates were Otto von Fürth, who had accompanied Hofmeister from Prague, and Karl Spiro. Fürth remained in Strassburg until 1905, when he moved to Vienna, where he later became a noted professor of medical chemistry. At Strassburg, he worked on muscle proteins, tyrosinase and adrenalin.[56] Spiro joined Hofmeister in 1896 as his principal assistant, and remained with him until they both left Strassburg in 1919. After his Dr.phil. work with Emil Fischer, Spiro studied medicine and turned to pharmacology. This brought him to Schmiedeberg's laboratory in Strassburg, and he then moved to Hofmeister's institute, where he embarked on a sustained study of the physical-chemical properties of proteins. There can be little doubt that Hofmeister's earlier work in this field stimulated Spiro's productive efforts. Of special historical interest is his 1914 paper, with Max Koppel, on buffers.[57] The importance of Spiro's role in the chemical education of the students and guests who worked in Hofmeister's laboratory cannot be overestimated.

Other notable contributions of the Strassburg group to protein chemistry included the application of the salting-out

---

[55] Hofmeister (1912, 1918).

[56] Fürth's work on the purification of adrenalin (he named it *Suprarenin*) was published under his sole authorship, but two German patents were issued to Hofmeister and Fürth, as joint inventors, for Fürth's process. Shortly afterward, this work was superseded by the crystallization of the hormone by Aldrich and by Takamine in the United States, and its synthesis by Stolz at the Hoechst works; see Weisser (1984), p.139.

[57] For Spiro, see Asher (1932) and Leuthardt (1932). Max Koppel was a promising medical student who died in action during the First World War; see Spiro (1915). Their paper (Koppel and Spiro [1914]) has been translated into English by Roos and Boron (1980).

method, mentioned above, to the crystallization of human serum albumin (Hans Theodor Krieger, 1899) and of Bence-Jones protein (Adolf Magnus-Levy, 1900), as well as the isolation of thyroglobulin (Adolf Oswald, 1899). Moreover, a study by Spiro and Otto Porges on the fractionation of serum proteins led to their discovery that at least three types of globulins are present, and foreshadowed the demonstration many years later of the identity of the so-called gamma-globulin fraction with the immunoglobulins. Among other contributions in the protein field, that of Ernst Friedmann in establishing the structure of the amino acid cystine and Walther Hausmann's method for the analysis of proteins are especially noteworthy.

Highlights of the extensive research in Hofmeister's institute on enzymes included the discovery by Jacob Parnas of what he called *aldehyde mutase*, Fürth's studies on tyrosinases, and the work of Julius Schütz on the kinetics of pepsin action. It is perhaps fair to say that the greatest single success was made in the field of intermediary metabolism, with Franz Knoop's demonstration of the beta-oxidation pathway for the metabolic breakdown of fatty acids,[58] but it also must be noted that Gustav Embden and Parnas both began their fruitful research on the metabolism of carbohydrates at Strassburg. If to these achievements one adds such important studies as those of Friedrich Bauer on inosinic acid and of Wilhelm Stepp on what was later to be called vitamin A, it is evident that, during the first decade of this century, the productivity of Hofmeister's institute in the biochemical sciences was not matched by any other laboratory.

This incomplete summary of the research problems attacked by Hofmeister's assistants, students and postdoctoral guests indicates clearly that the research pursued in his institute was not a concerted attack along narrowly prescribed lines dictated by the leader of the group. From the published recollections of several of Hofmeister's associates, and from obituary notices about them, there can be little doubt that instead of assigning problems, he suggested them, and that he also helped significantly both in the conduct of the work and in the preparation of papers for publication. Indeed, according to his biographers, Hofmeister himself "wrote many of the papers of younger and foreign co-workers, and also in many others one frequently recognizes, especially in the opening and closing sentences, the master's characteristic style."[59] Between 1901 and

---

[58] Knoop (1904).
[59] Pohl and Spiro (1923), p.28.

1908 many of these papers appeared in a journal that Hofmeister founded, *Beiträge zur chemischen Physiologie und Pathologie: Zeitschrift für die gesamte Biochemie.* In the absence of adequate documentation, his reasons for establishing this journal are unclear; one possibility is that his associates had encountered difficulties in gaining access to the *Zeitschrift für physiologische Chemie,* although many of the Strassburg papers appeared there between 1897 and 1901. Hofmeister's journal ceased publication because he decided to merge it with the *Biochemische Zeitschrift,* founded by Carl Neuberg in 1906.

The scope of the work done by Hofmeister's group, and the manner of its publication, suggest a strategy of research that placed breadth of scientific education and encouragement of individual talent above his own ambitions to solve particular experimental problems. The diverse lines of investigation, with relatively few participants in each of them, are in sharp contrast to Fischer's concentration on one or two problems at a time, and Hofmeister differed strikingly from Fischer in the assignment of credit. These differences are consistent with the available accounts of Hofmeister's personal qualities. In his obituary notice about Knoop, Karl Thomas wrote of the lifelong reverence and gratitude Knoop bore toward Hofmeister, a man "of wide-ranging spirit, who commanded a view of the whole of biology and who was full of new ideas which he generously handed out to his co-workers."[60] Similar sentiments may be found in articles about other junior associates of Hofmeister. For the pharmacologist Alexander Ellinger, "[t]he years he spent in Hofmeister's institute under the guidance of this brilliant teacher and his pupil Karl Spiro determined the course of his life in research and his scientific goals. With pride, he [Ellinger] always called himself a *Hofmeisterschüler.*"[61] Otto Loewi, who worked in the Strassburg institute in 1897, later described Hofmeister as "an ingenious biologist and an unforgettable personality," and credited him with guiding Loewi on a path that led him to the discovery that completely digested protein can replace intact protein in the animal diet.[62] Other appreciations of Hofmeister's influence on the careers of his students may be found in obituary notices for the biochemists Gustav Embden, Ernst Friedmann and Henry Stanley Raper, the clinician Gustav von

---

[60] Thomas (1948), p.2.
[61] Ellinger (1924), p.146.
[62] Loewi (1960), p. 7.

Bergmann and the botanist Friedrich Czapek.[63] Moreover, in the face of the tension between the German rulers and the French population of Alsace, Hofmeister won the affection of his Alsatian students, and made special efforts to promote their interests.

A glimpse of the atmosphere in the Strassburg laboratory has been provided by Lawrence Joseph Henderson, who worked there during 1902–1904. Many years later he wrote[64]:

The experimental problem on which I went to work within a few weeks of the opening of the semester, after I had satisfied Hofmeister that I possessed a very rudimentary skill in chemical work, was badly chosen. I remember saying to him at the outset that I should like to work on some problem about blood and I also remember that this was due to a clear perception that I was going to work on the acid-base equilibrium in blood at some time when I returned to America. But although Hofmeister had himself done important physico-chemical work I willingly accepted his proposal to study the products of partial hydrolysis of hemoglobin and on this problem I worked steadily throughout my first year in Strasbourg and intermittently throughout my second, without ever making any progress, so that nothing ever came of my experimental work during these two years. And yet, strange to say, during all this time I was happy, contented and, I think, never seriously disturbed at the failure of the investigation. . . . I was for the first time surrounded, at least during the working hours in the laboratory, by a group of active, young biological chemists, the assistants Spiro and Fürth, and in addition, Embden, Knoop, Friedmann, Blum and others. Then, Hofmeister's lectures were excellent and, considering his age and lack of training, remarkably modern and adequate to the needs of the immediate future. Moreover, he was not merely a chemist, for he possessed a real understanding of at least some of the complexities of biological phenomena and therefore of the characteristic difficulties in the application of physical and chemical methods to their study.

In 1908, Henderson spent a month in Strassburg:

---

[63] For Embden, see note 138, this chapter; for Friedmann, see Mitchell (1956); for Raper, see Hartley (1953); for Bergmann, see Katsch (1955); for Czapek, see Boresch (1922).

[64] The following two quotations are from pp.76–77 and 161–162 of the typescript of Henderson's unpublished *Memories*, written about 1936–1939, and deposited in the Harvard University Archives (HUG 4450.7.2). I am greatly indebted to Henderson's son, Lawrence J. Henderson, for permission to quote selected passages from this autobiographical memoir, and to Clark A. Elliott, Associate Director of the Archives, for making it available to me for perusal. A description of the Henderson papers has been prepared by Parascandola (1971).

Spiro and I did two little pieces of experimental work in cooperation. This work arose from his remark that it was a pity that I had never published anything from the Strasbourg laboratory and that we ought each to propose a problem that could be finished in a month and work both out, and so we did. The problem I proposed was an uninteresting routine measurement, but Spiro suggested the physico-chemical explanation of the long-known phenomenon of the movement of water and of chloride ion between cells and plasma when the pressure of carbon dioxide is varied. For some reason that I can't explain, I had up to that time overlooked these facts. I was the more impressed by them because I felt guilty at my ignorance. There was no difficulty in formulating the theoretical explanation of the facts through the application of the theory of acid-base equilibria and of the theory of osmotic pressure. The problem having been thus clarified, it was easy to construct simple systems in which the same processes should be observable, and, having performed the experiments, we found what we were looking for.

Although in his later career as a professor at Harvard, Henderson made valuable contributions to the study of acid-base equilibria in blood, he was not an enthusiastic experimenter, and the fact that his work at Strassburg during 1902–1904 was unsuccessful may perhaps be taken as an early manifestation of a preference for theoretical approaches to biological problems. According to his own account, Henderson spent much of his time in Strassburg in doing thermochemical calculations.

Hofmeister appears to have been a man of good humor and great energy, and "in his personal relations he exhibited a delightful kindness and was a sparkling conversationalist, full of the grace of the Austrian." Also, he had a happy family life and, until shortly before his death, enjoyed good health. There were leisure hours doing watercolors or listening to music. But at root, according to Spiro, Hofmeister "was a solitary person, who desired solitude, . . . and he regarded the world around him with complete skepticism and often with scorn."[65] This attitude is consistent with Hofmeister's avoidance of scientific meetings, except when he was invited to present a lecture; Emil Abderhalden, who attended many meetings at which physiological chemists were present, wrote that he had never met Hofmeister.[66] It would seem, therefore, that although Hofmeister was energetic within his institute, he was not an active participant in the aca-

---

[65] The two quotations in this paragraph are from Spiro (1922), p.vii.
[66] Abderhalden (1922), p.1168.

demic politics of Wilhelmine Germany and hence received few public honors. He appears to have been content to remain in Strassburg, for he declined invitations to move to Heidelberg, Berlin and Vienna.

Hofmeister's last years were sad ones. After the return of Strasbourg to French rule, he was officially classified as a citizen of Czechoslovakia, but he chose to leave with the German professors, and found a haven in Würzburg, where he was given an honorary professorship in physiological chemistry.[67] In a laboratory provided by Martin Benno Schmidt, the professor of pathology, Hofmeister worked alone on the isolation of the antineuritic vitamin until shortly before his death on 26 July 1922.[68]

In assessssing Hofmeister's scientific contributions, it is noteworthy that a recurrent theme in the writings about him is that he saw biochemical problems with the eyes of a biologist.[69] I cannot dispute this judgment, but I find a curious paradox. Whereas Hofmeister considered the proteins that he and his coworkers had crystallized to be discrete substances of high molecular weight, the chemist Emil Fischer did not. For this reason, I now return to the problem both scientists discussed at the 1902 meeting in Karlsbad, with special reference to Fischer's role in the historical development of protein chemistry and his attitude toward the physiological chemists of his time.

Peptides, Peptones and Proteins

In a letter to Baeyer dated 5 December 1905, Emil Fischer wrote to his former teacher as follows:

On January 6th I will present a lecture at the Chemical Society summarizing my work on amino acids, polypeptides and proteins, and

---

[67] Archiv des Rektorats und Senats der Universität Würzburg, Nr.556, Schrank Ia, Fach 9, Nr.10.

[68] This sketch of the life and personality of Hofmeister is based almost entirely on published material, and may require revision if a major collection of his personal papers is found. Despite an intensive search in Strassbourg, Würzburg and elsewhere, I have located only a dozen of his letters. Perhaps his papers were destroyed during the World War II bombardment of Würzburg, but there appears to be no record of them in the university archives. Of special interest would be Hofmeister's *Tagebuch*, mentioned in Pohl and Spiro (1923), p.30. This document, if it has survived, would provide valuable further insight into Hofmeister's career and personal qualities.

[69] See Embden (1922).

then early next year I will publish the collected papers in the form of a book. The material has grown splendidly and there is much detail in it. Recently I have also prepared the first crystalline hexapeptide and hope to obtain a matching octapeptide before Christmas. Then we should be close to the albumoses. Unfortunately, then begin the principal difficulties, namely the identification of the artificial products with the cleavage products of proteins.... My entire yearning is directed toward the synthesis of the first synthetic enzyme. If its preparation falls into my lap with the synthesis of a natural protein material, I will consider my mission fulfilled.[70]

In the lecture, Fischer summarized the work of his research group on the isolation of optically active amino acids by the resolution of the available racemic forms. This work represented his entry into the protein field in 1899, and successive students and assistants were assigned the task of applying his method to the resolution of many synthetic racemic amino acids, some of which were first made in his laboratory. After a description of the synthesis and properties of various amino acid derivatives, Fischer turned to the polypeptides, and stated that he had expected from the start that such artificial substances would resemble the peptones.

As was noted above, Fischer's work on peptides began in 1901, with the preparation of the first free peptide—glycyl-glycine—by partial hydrolysis of glycine anhydride (2,5-diketopiperazine), a compound made by Curtius in the 1880s. This is not a general method for coupling amino acids to form polypeptides, and Fischer's first attempt to develop such a procedure failed. The later development of peptide chemistry showed that his proposed strategy of blocking the amino group with a substituent that could be removed selectively at the end of the peptide synthesis was sound, but the protecting group he chose (the carboethoxy group) proved to be unsuitable. It was not until 1903 (after the Karlsbad meeting) that he described a procedure that circumvented the need for an amino-protecting group. In the new method, the synthesis began with the reaction of a halogenacyl halide (e.g., chloroacetyl chloride) with an amino acid derivative to yield a product which was converted into a dipeptide derivative by means of ammonia. Treatment of the latter derivative with the halogenacyl chloride derived from another amino acid gave, after amination, the expected tripep-

---

[70] Letter from Fischer to Baeyer, 5 December 1905, Bancroft Library. For the lecture, see Fischer (1906); for the book, see note 20, this chapter.

tide derivative, and so on. By 1906, Fischer's associates had used this method to prepare about 65 peptides of different chain length and amino acid composition; three years later the total exceeded 100, the longest of which an octadecapeptide composed of 15 glycyl and 3 leucyl units. This formidable synthesis was completed by Fischer's assistant, Walter Axhausen.[71]

At each stage of this work, the synthetic peptides were subjected to the tests devised earlier by physiological chemists to study proteins and the products of their cleavage by enzymes— the biuret reaction, precipitation by inorganic salts, cleavage by proteolytic enzymes. From such studies Fischer concluded that the lower members of the series resembled the peptones and that, as the chain length was increased, the properties of the synthetic materials began to approach those of natural proteins. Indeed, he claimed that "*1*-leucyl-triglycyl-tyrosine prepared artificially has all the properties of the albumoses. These observations are of importance in casting doubt on the view which formerly prevailed that, being intermediate products between proteins and peptones, the albumoses are substances of considerable molecular complexity."[72] Nor was Fischer reticent in his claims for the power of his synthetic procedure: "The synthesis of the higher terms [peptides] has been restricted hitherto to the combinations of glycine, alanine and leucine; there is not a shadow of doubt, however, that all the remaining amino acids could be associated in complicated systems with the aid of our present methods."[73] Fischer's confidence in the power of his synthetic method proved to be shortlived, for it was not only cumbersome and costly, but also was unsuitable for the preparation of polypeptides containing more complex amino acid units, for example lysine or glutamic acid. The enormous effort of his assistants and students produced less than he had hoped for, and Fischer's disappointment may be inferred from the fact that after 1910 there were no further experimental publications on peptide synthesis from his institute. Fischer's halogenacyl halide method became obsolete when he ceased to use it, and only survived in the laboratory of Emil Abderhalden in Halle, where many peptides were produced, albeit of uncertain identity or purity. Among them was a nonadecapeptide, containing one more leucine than Fischer's longest polypeptide.

[71] Fischer (1907a), pp.387–388. The description of the synthesis and properties of the octadecapeptide is written out in one of Axhausen's notebooks (No.2, pp.83–86) in the Fischer Collection at the Bancroft Library. For later use of the Fischer method of peptide synthesis, see Fruton (1949).
[72] Fischer (1907b), p.1761.
[73] Ibid., p.1762.

Whatever may be said about the many practical limitations
of Fischer's method, and its rapid obsolescence, they do not di-
minish his decisive role in providing an impetus to later advances in the art of peptide synthesis. The high point came in
1932, with the invention of the so-called carbobenzoxy method
by Max Bergmann and Leonidas Zervas.[74] Their procedure is an
ingenious modification of Fischer's unsuccessful use of the carboethoxy group, and made possible the facile synthesis of polypeptides containing all protein amino acids. Also, it should be
noted that during 1900–1910 Theodor Curtius was stimulated
by Fischer's entry into the protein field, and resumed his work
of the 1880s on peptides. In 1902, Curtius developed a coupling
method involving the azides of acylamino acids; some harsh
words were exchanged, and each of these great chemists felt that
the other had not cited his work properly. The public accolade
went to Fischer, but today peptide chemists have little occasion
to use halogenacyl halides, while the azide coupling method is
still valuable.[75]

Special mention must also be made here of a by-product of
Fischer's work on peptide synthesis. Since proteins are composed of amino acids which, with the exception of glycine, have
a center of asymmetry and are optically active, it was necessary
to prepare the halogenacyl halides from optically active amino
acids. The reaction leads to an inversion of configuration about
the center of asymmetry, and the inversion is reversed upon
treatment of the halogenacyl peptide with ammonia at the final
stage of the synthesis. This type of double inversion had been
discovered in 1893 by Paul Walden, but Fischer's work between
1907 and 1911 shed new light on the process and gave renewed
evidence of his remarkable stereochemical insight.

The final section of the 1906 lecture dealt with proteins,
and began with a detailed description of the method Fischer had
devised in 1901 for the amino acid analysis of the mixture produced upon the hydrolysis of proteins with acids. In this
method, the mixture was esterified with ethanol and, after further operations, the amino esters were subjected to fractional
distillation under reduced pressure. The esters in the individual
fractions were then converted to free amino acids, which were
crystallized, weighed and characterized. Although it was clear
from the start that this complicated procedure was not likely to
give reliable quantitative data for the amino acid composition of

[74] Bergmann and Zervas (1932); for a brief account of the historical background of the
carbobenzoxy method, see Fruton (1982c).
[75] See Curtius (1904), p.64.

proteins, the method was valuable in demonstrating the general occurrence of amino acids such as alanine and phenylalanine, and in revealing the presence of proline and hydroxyproline. A third new protein amino acid, diaminotrioxydodecanoic acid, was reported, but Fischer was later obliged to withdraw this claim. Most of the work in Fischer's laboratory on the application of his method to the study of the amino acid composition of many protein preparations was done under the direction of Emil Abderhalden by a series of postdoctoral medical men. The method was subjected to a critical study by the American protein chemist Thomas Burr Osborne, who improved it somewhat but also called attention to its shortcomings.[76] It is difficult to evade the surmise that Osborne's critique may have incited the last two sentences in the following excerpt from a letter that Fischer sent to Abderhalden:

I consider that because of their greater wealth the Americans will beat us in several fields, and I have expressed this opinion at every opportunity. However, we can withstand this competition for a time because of our greater inventiveness and more distinguished individual achievements. That the gentlemen in America are also rather presumptuous is nothing new to me, but one can defend oneself against this at a suitable opportunity. As soon as I find the time, I will discuss this question in a retrospect on chemical research on proteins during the past ten years.[77]

To my knowledge, the promised retrospect was never written.

Fischer concluded his 1906 lecture with a discussion of the structure of proteins and stated that

So far as I can judge, the earlier view is that, in protein molecules, the amino acids are joined by amide bonds. This idea was most fully treated by Hofmeister, but he surely will not claim to be its originator, for all synthetic efforts to couple amino acids, among them the discovery of glycyl-glycine, which antedate his publication, are based on this assumption. . . . I wish however to note that simple amide formation is not the only possible mode of linkage in the protein molecule. On the contrary, I consider it quite probably that on the one hand it contains piperazine rings, whose facile cleavage by alkali and reformation from

[76] See Osborne and Jones (1910) and Osborne and Guest (1911). For Osborne see Chapter Two, note 110.
[77] Letter from Fischer to Abderhalden, 14 March 1912, Bancroft Library.

the dipeptides or their esters I have observed so frequently with the artificial products, and on the other hand the numerous hydroxyls of the oxyamino acids are by no means inert in the protein molecule. The latter could be transformed by anhydride formation to ester or ether groups, and the variety would increase further if poly-oxyamino acids are assumed to be probable protein constituents.[78]

Apart from his ambivalent commitment to the peptide theory of protein structure, Fischer also took a stand against the evidence that proteins have molecular weights greater than about 5,000. For example, in a letter to Lewellys F. Barker, Fischer wrote: "The synthesis of polypeptides is advancing briskly. I have recently made the first decapeptide and will now try to reach an eikosapeptide, whereupon one should be midway in the protein group."[79] In 1913, Fischer reported the synthesis of a depside derivative having a molecular weight of 4,021. This led him to state that the value is three times greater than that of the octadecapeptide he had described in 1907, and he added: "I believe that it also exceeds that of most natural proteins."[80] Three years later, in commenting on the value of 15,000–16,000 cited by Hofmeister for the molecular weight of hemoglobin, Fischer stated:

In my opinion, the methods applied to the determination of the molecular size of the hemoglobins are less certain than had been assumed previously. Although they crystallize beautifully, no guarantee of homogeneity is given, and even if one concedes this and accepts the validity of a molecular weight of 15,000–17,000 for several hemoglobins, it should always be remembered that the hematin, from all that we know of its structure, can bind several globin units. . . . On the other hand, I gladly concur in the view of Hofmeister and other physiologists that proteins of molecular weight 4000–5000 are not rare. If one assumes an average molecular weight of 142 for the amino acids, this would correspond to a content of 30–40 amino acids.[81]

In questioning the homogeneity of crystalline hemoglobin, Fischer implicitly also raised doubts about the crystalline pro-

---

[78] Fischer (1906), pp.607–608. The influence of Fischer's suggestion about piperazine rings on protein research during the the 1920s has been discussed by Fruton (1979).
[79] Letter from Fischer to Barker, 18 April 1906, Bancroft Library.
[80] Fischer (1913), p.3288.
[81] Fischer (1916), p.1006.

teins isolated from plant seeds by Osborne or the egg albumin crystallized by Hofmeister. For Fischer, " . . . the existence of crystals does not in itself guarantee chemical individuality, since isomorphous mixtures may be involved, as is frequently evident in mineralogy for the silicates."[82] When Fischer expressed these opinions about the molecular weight and homogeneity of the crystalline proteins, he had already given up his hope of synthesizing a natural protein, such as silk fibroin. It would seem that he recognized that his acceptance of the claims of the physiological chemists would imply his concession that, despite his previous statements to the contrary, his method of peptide synthesis was not adequate to solve the problem of protein structure. Indeed, this conclusion was stated by Hofmeister in an article published in 1908: "The rapid progress made in the synthesis of polypeptides under the aegis of Fischer could lead to the view that one may expect the elucidation of protein structure to come entirely from this approach. This hope at once comes to naught if one considers the enormous number of synthetic possibilities."[83] In this paper, Hofmeister reiterated his conviction that the isolation of well-defined cleavage products was the most important requirement for the determination of the structure of proteins. He noted, however, that "their enzymatic breakdown always leads to mixtures of many substances of unequal size . . . and the isolation of chemically characterized albumoses or peptones is a thankless task . . . as a consequence of the size of the protein molecule."[84]

Fischer's views regarding the molecular weight of proteins suggest his indifference to data from osmotic-pressure measurements or from the determination of the freezing-point depression of protein solutions. These methods represented some of the important contributions of the new physical chemistry, and before 1910 gave values (later shown to be too low) of 14,000 for egg albumin and 48,000 for hemoglobin.[85] Wilhelm Ostwald, the chief propagandist for this branch of chemistry, later wrote as follows about his encounter with Fischer at the 1889 meeting of the *Gesellschaft der Naturforscher und Ärzte*:

I found myself in a swarm of organic chemists gathered about Emil Fischer, already regarded as the future leader of our science, since

---

[82] Fischer (1913), p.3288.
[83] Hofmeister (1908), p.277.
[84] Ibid., p.276. For a more extensive contemporary summary of efforts to isolate peptones, see Siegfried (1916).
[85] See Schulz (1903), pp.94–104.

what was not organic chemistry was not recognized as chemistry. To
his disparaging remark about our new direction I answered that the
organic chemists owe us thanks for the possibility of determining the
molecular weights of non-volatile substances. Fischer replied: "That
was entirely unnecessary; I see directly the molecular weight of every
new substance, and do not need your methods." I took the matter as it
was intended: a figure of speech, in order make a good exit, and
hoped that time would bring about a reconciliation. This did not happen, however, as I have since then had repeated evidence of that un-
conditionally negative attitude.[86]

It should be added, however, that Fischer promoted the development of physical chemistry in Berlin, and held van't Hoff and
Nernst in high regard, but he did not like Ostwald.[87]

I have dwelt on Fischer's work and writings about proteins
for two reasons. The first is that during 1900–1910 they dominated discussion in this field. The accounts of his 1906 lecture
in newspapers and in popular science journals encouraged the
belief that the preparation of synthetic proteins was around the
corner.[88] Fischer attempted to disclaim the exaggerations, but
his embellishment of the biological significance of his work on
peptides could not fail to attract wide attention and to increase
his already considerable public prestige.[89]

The second reason for the focus on Fischer's work in the
protein field is that, in his papers and lectures, he repeatedly
disparaged the efforts in that field of physiological chemists,
whom he often only denoted as physiologists, presumably to indicate that, in his opinion, they were not proper chemists. I have
found no better evidence of his prejudice, no doubt bolstered by
his own aspirations, than in a letter Fischer wrote in 1904 to his
Berlin colleague, the physiologist Theodor Wilhelm Engelmann,
about Emil Abderhalden:

---

[86] Ostwald (1927), vol.3, pp.111–112.

[87] Fischer's attitude toward physical chemists, especially Ostwald and members of his
school, was well known to his colleagues. In the discussion after 1919 about Fischer's
possible successor at Berlin, Duisberg is reported to have said: "Fischer would turn in
his grave ... if he could know that a physical chemist might succeed him"; see
Kröhnke (1961), p.256.

[88] For an example of the popular accounts that appeared after Fischer's 1906 lecture,
see Kautsch (1906).

[89] For Fischer's reaction to the newspaper reports, especially the one by Abderhalden,
see Fischer's letter to Carl Oppenheimer, 27 January 1906, Bancroft Library. The
public acclaim of Fischer's work on proteins is in sharp contrast to the fact that his
more elegant and significant work in the sugar field largely went unnoticed outside
the chemical community.

Because of his unusual capacity for work, in a short time Abderhalden
has become so adept in the difficult methods of organic chemistry that
I was able to accept him last fall as a collaborator in my private laboratory. I note that that I had not dared to do this before with a medical
man. He is a good observer, and is an enemy of all superfluous hypotheses. Regrettably, biological chemistry is that part of our science in
which imprecise and incomplete experiments are often heavily padded
with the dazzling ornamentation of so-called ingenious reflections to
produce pretentious treatises. For this reason, people like Abderhalden
are needed.[90]

Abderhalden had joined Fischer's research group in 1902, after
receiving his Dr.med. in Basel, and working there with Gustav
von Bunge, the professor of physiology.[91] Fischer's early opinion
of Abderhalden's merit as a chemist led him to turn over to Abderhalden the succession of post-M.D. students who flocked to
Fischer's laboratory at that time, and most of them worked under Abderhalden's direction on the application of the ester
method to the amino acid analysis of a great variety of protein
preparations. However, the soundness of Fischer's assessment of
Abderhalden's chemical talent proved to be incorrect, for a good
part of Abderhalden's work both as a member of Fischer's group
and as an independent investigator proved to be irreproducible.
In particular, Fischer was later obliged to withdraw diaminotrioxydodecanoic acid from his list of protein amino acids, and
Abderhalden's claim for the existence of protective enzymes
(Abwehrfermente) elicited upon the parenteral administration of
foreign proteins, and his advocacy for at least 35 years of this
supposed phenomenon as a test for pregnancy or cancer, can be
described most generously as an example of sustained
self-delusion.[92]

Fischer's encouragement of Abderhalden, taken together
with his denigration of the efforts of physiological chemists such
as Hofmeister, may be taken simply as an expression of his assertion of superiority in the study of proteins, but his public
stature also gave his disapproval the stamp of sound scientific

---

[90] Letter from Fischer to Engelmann, 24 February 1904, Archiv der Humboldt-
Universität, Akte 1450, Bl. 172–173. This letter is cited by L. Läsker in Mikulinskij et
al. (1977), p.111.

[91] In his letter (25 December 1901, Bancroft Library) thanking Fischer for having accepted Abderhalden into his laboratory, Bunge felt it necessary to note that "I myself
am not originally a medical man, but a chemist."

[92] For Abderhalden, see Heyns (1951), Hanson (1970), Picco (1981), pp.24–30 and Karlson (1986), pp.385–386. A list of Abderhalden's publications is given in the *Jahrbuch
der Akademie der Wissenschaften und Literatur 1951*, pp.163–204.

judgment. What Fischer failed to see, or did not wish to acknowledge, was that in order to explore such difficult problems as the structure of proteins, the nature of enzyme action or the pathways of intermediate metabolism, physiological chemists had to resort to many experimental methods, including not only the procedures of organic chemistry, but also physical-chemical and biological methods. This diversity of approach, evident in the work of the research groups associated with Hoppe-Seyler and with Hofmeister, stood in sharp contrast to the highly disciplined methodology of the organic-chemical laboratory. In this connection, it is perhaps worth noting that in his autobiographical memoir, prepared during the 1930s, Lawrence J. Henderson wrote as follows of his time with Hofmeister:

No doubt Hofmeister and others in Germany, and an occasional more or less isolated investigator like Hopkins in England were at that time masters of a body of knowledge differing materially from what passed as either chemistry or physiology and which might fairly be called biological chemistry. But unlike organic chemistry and physical chemistry, biological chemistry did not possess and indeed hardly possesses today a sufficient equipment of methods of its own to make possible the sort of disciplined training that has so long been characteristic of organic chemistry and in a measure of physical chemistry. . . . Further, I have always had the feeling that organic chemistry is a very peculiar science, that organic chemists are unlike other men, and that there are few occupations that give more satisfactions than masterly experimentation along the old lines of this highly specialized science.[93]

Although questions may be raised about Henderson's views regarding "old line" organic chemistry or the state of biochemistry either at the turn of the century or during the 1930s, there is no doubt that, in part, his opinions echo those of Fischer. Thus, in 1893, before he entered the protein field, Fischer stated:

So long as we know little more about the chemical bearers of life, the proteins, than their percentage composition, so long as we cannot explain the most fundamental process of organic nature, the conversion of carbon dioxide into sugar by green plants, we must admit that physiological chemistry still remains in baby shoes. Will it ever be able to follow in detail the complicated processes in plants and animals and to

---

[93] See note 64, this chapter; the quotation is on pp. 75,85–86.

determine their effect on morphogenesis [*Formbildung*]? Will it be possible to control, on the basis of clear chemical principles, the metabolic disturbances in our own bodies by disease, and thus to partially realize the alchemists' dream of a vital elixir? I do not doubt it. But the means to attain this knowledge must be provided to physiology by organic chemistry, and this seems to me so necessary a mission of the latter that I wish to participate in its solution to the extent of my strength. Indeed, one cannot escape the fact that the first step toward the attainment of this goal will be the complete analysis and synthesis of proteins and that, despite the advanced methods of our time, the effort of several generations will be required.[94]

This oratorical flourish, with its seeming disregard of the work on proteins after Liebig, and announcing Fischer's intention to enter the protein field, was not muted 14 years later, when it was widely believed that he was about to achieve the synthesis of a protein. At a meeting of the *Verein deutscher Chemiker* (a society of industrial chemists), Fischer noted that the artificial proteins were not likely to supplant the cheaper natural ones in the diet, but that

The utility of protein synthesis must therefore be sought in the scientific, and especially in the biological, fields. One may expect that from the combined synthetic and analytical development of the whole group, there will emerge methods which will enable physiologists to elucidate the metabolism of animals and plants. Chemistry, however, should not only serve as a handmaiden, but will undoubtedly find large new areas for independent research and probably even in industrial work. I think chiefly of the study of ferments and of fermentative processes which occur everywhere in the organism, and which are certainly closely related to the metamorphosis of proteins. When we master them, then in a manner somewhat similar to that for the transformations of benzene and its derivatives, there will surely develop new branches of chemical manufacture which will rival in significance the splendid fermentation industry of today.[95]

---

[94] Fischer (1893), p.636. In an obituary notice for Kossel, Mathews (1927) stated that Fischer's work on proteins was undertaken "at Professor Kossel's suggestion and request." I have not found any documentary evidence for the validity of this claim. The Fischer-Kossel correspondence in the collection at the Bancroft Library deals mostly with the papers Fischer had submitted to the *Zeitschrift für physiologische Chemie* edited by Kossel, and no mention is made of their respective research on proteins.

[95] Fischer (1907c), p.917. Those who rejoice in the discovery of seemingly prescient statements in the writings of noted scientists may find in this quotation an anticipation of the rise of biotechnology during the 1980s. Although organic chemists played a role in the developments leading to this new industry, that history was rather different from the one dealing with the manufacture of dyes and drugs.

Despite his public words about the importance of biochemical
studies, through his disdain for the chemical research done by
most of the physiological chemists of his time Fischer aligned
himself with the physiologists and clinicians in the German medical faculties who opposed the aspirations of Hoppe-Seyler and
his students for the establishment of separate institutes for their
discipline at German universities.

Emil Fischer's Research Group

In Appendix 6 are listed 354 persons, nearly all of whom had
been predoctoral students, research assistants or postdoctoral
guests in Fischer's successive laboratories in Munich, Erlangen,
Würzburg and Berlin.[96] The starting point for the compilation
of the list was the set of eight volumes of Fischer's collected papers; as noted above, it was necessary to check each paper for
acknowledgments made to assistants whose names did not appear on the title page. Moreover, from other sources I learned
of research students, assistants and guests whose names did not
appear in these volumes. Thus, in the list of Abderhalden's publications from Fischer's laboratory I found numerous papers
that were not included in the two collective volumes dealing with
Fischer's research on proteins, and the co-authors of these papers are included in Appendix 6. I have also included the names
of several men who, as *ausserordentliche Professoren*, had a nominally independent status in Fischer's institute in Berlin. Several
of them worked on problems directly related to Fischer's research program; their Dr.phil. students or assistants are not included, however, unless they had joined Fischer's personal
research group. Also not included in Appendix 6 are several
people with whom Fischer collaborated on the basis of equality,
for example his cousin Otto Fischer (triphenylmethane dyes),
the microbiologist Paul Lindner (fermentation of sugars) and
the physician Joseph von Mering (Veronal). The numerical data
in Table 5-1 are derived from the biographical information collected in Appendix 6; it is important to note that the numbers

---

[96] The list in Appendix 6 is a corrected and expanded version of the one published
previously (Fruton, 1985b) and reproduced in Fischer (1987). It should be noted that
in Fischer's autobiography (note 13, this chapter), the names of some of his co-
workers are written differently from their appearance in his scientific papers. Examples are O. Bülow for Carl Bülow (p.93) and Roese for Ludwig Reese (p.94).

TABLE 5-1
The Fischer Research Group

	Munich (78–82)		Erlangen (82–85)		Würzburg (85–92)		Berlin (92–19)	
	Predc.	Postdc.	Predc.	Postdc.	Predc.	Postdc.	Predc.	Postdc.
Germany & Austria	4	2	15	4	49	2	118	62
United Kingdom					15	1	4	5
United States	1		1		5		11	18
Switzerland							4	3
Russia & Poland							2	5
France							2	2
Japan								6
Other countries							6*	7+
Total	5	2	16	4	69	3	147	108

*2 from Finland, Turkey; 1 from China, Rumania
+2 from Italy, Rumania; 1 from Belgium, Denmark, Greece

given for the postdoctoral people do not include those who had received their Dr.phil. for prior work with Fischer.

Before 1892, when he became A. W. Hofmann's successor in Berlin, Fischer's research output depended largely on the efforts of Dr.phil. candidates and of his personal research assistants, many of whom were drawn from the ranks of his own students. During the four years between his *Habilitation* in Munich and his move to Erlangen, Fischer had five research students (one of them was the American Edward Renouf) and two postdoctoral assistants, one of whom (Troschke) proved to be unsatisfactory.[97] In his succeeding three years at Erlangen, Fischer's research group included 16 Dr.phil. students and four postdoctoral assistants; upon his arrival there, Fischer inherited two *Privatdozenten*, Oscar Hinsberg [1857–1939] and Carl Paal [1860–1935], who were permitted to work on problems not set by Fischer but, as he put it, "gladly accepted my advice and help on theoretical and experimental matters."[98]

It was in Würzburg that Fischer's research programs, especially the one on sugars, began to flower, and the size of his group increased greatly. At any one time, Fischer supervised the work of about twelve Dr.phil. candidates and four research assistants. As is indicated in Table 5-1, a striking feature of the

[97] Fischer (1922, 1987), p.80.
[98] Ibid., p.94.

Würzburg group is the relatively large number of British predoctoral students. In moving to Würzburg, Fischer brought with
him Ludwig Knorr, who took over the inorganic and analytical
section, and Julius Tafel, who was to remain there after Fischer's
departure in 1892. When Knorr became professor of chemistry
at Jena in 1889, Fischer appointed Wilhelm Wislicenus to be
head of the inorganic section.

Some features of Baeyer's elevation to the professorship in
Munich, described in the preceding chapter, are evident in
Fischer's move to Berlin. Just as, in 1875, Liebig's institute was
hopelessly inadequate to accommodate the kind of program of
chemical research and instruction that Baeyer wished to establish in Munich, Fischer found the building occupied by Hofmann's group to be totally unsuited to his purposes and
ambitions.[99] That building, opened in 1869, was at that time the
most elegant set of chemical laboratories anywhere, with space
for about 70 people, two lecture halls, two private laboratories,
and rooms for special operations. According to Fischer, however,
although the architecture and facade were attractive its interior
design was impractical for chemical work:

Everywhere there was a shortage of air and light and a large portion
of the space consisted of dark and unused corridors. . . . The ventilation was entirely unsatisfactory, and my principal concern, as in the
Würzburg institute, came from the installation of a whole series of
chapels. . . . Even the heating arrangements were in a sorry state.[100]

After vigorous arguments with Friedrich Althoff [1839–1908],
the enlightened but dictatorial *Hochschulreferent* of the *Kultusministerium*,[101] Fischer succeeded, with the help of a generous
donation from chemical industrial firms, in securing the funds
(over 1.5 million marks) needed for the construction and equipment of the building he wanted for what was then called the *I.
Chemisches Institut*.[102] He expended much of his energy on the

[99] Heinig (1960), pp.342–343.
[100] Fischer (1922, 1987), p.146.
[101] For Althoff, see Sachse (1928), Schnabel (1953) and Brocke (1988). A glimpse of Althoff's manner of handling professorial appointments at Prussian universities is provided by Reinke (1925), pp.176–177.
[102] See Lenz (1910), vol.3, p.301. The *II. Chemisches Institut* of the University of Berlin
was headed until 1905 by the physical chemist Hans Heinrich Landolt [1831–1910].
He was succeeded, with Fischer's support, by Walther Nernst [1864–1941], and the
institute was renamed *Physikalisch-chemisches Institut*. After 1905, Fischer's institute was
the *Chemisches Institut der Universität*.

interior design of the building, which was not opened until 1900. The design was similar in some respects to that of Baeyer's institute, in that it provided for four large laboratories rather than the "chapels" in Hofmann's building, and could accommodate about 250 workers, but it was larger than the one in Munich, with more space for about 50 additional persons (including 16 *Privatdozenten* and assistants) and for rooms in which special operations were to be conducted.[103] As a consequence of his improved facilities in Berlin, Fischer had more research assistants and doctoral students, and was able to accept more postdoctoral guests. During 1900–1914, at any one time his group was composed of about 25–30 persons. The usual length of stay of the assistants was two to three years, of the students about 18 months, and of the guests 6–12 months. After the outbreak of the First World War, the work of his laboratory proceeded on a greatly reduced scale owing to the shortage of personnel, and he was obliged to accept women as research students. In 1915, only two people (Max Bergmann, Charlotte Rund) occupied Fischer's private laboratory, and in the large student laboratory, where he previously had 12–15 research students, there were only six *Praktikanten*, three of whom were women.[104]

Three research assistants (Lorenz Ach, Johan Fogh, Oskar Piloty) and one Dr.phil. student (Arthur Crossley) at Würzburg accompanied Fischer to Berlin. From the Hofmann regime, there remained two older men, Ferdinand Tiemann [1848–1899], who withdrew from active instruction, and Siegmund Gabriel [1851–1924], as well as the assistants Harries, Pulvermacher and Richter. In his autobiography, Fischer acknowledged his debt to Gabriel, who remained at the Berlin institute throughout Fischer's professorship: "He has always been a dear colleague and friend. I therefore gladly take this opportunity to express to him my heartfelt thanks for the friendly help he gave me so frequently, because of my illness or for other reasons, in lecturing in my place or in the administration of the institute."[105] Gabriel, an outstanding experimenter and teacher, was a man of exceptional kindness, and a Jew. So far as I have been able to learn, he did not receive any offers of a full professorship at a leading university, and he remained an *ausserordentlicher Professor* at Berlin until 1920. In the planning of the new

[103] See Fischer and Guth (1901).
[104] For Fischer's negative attitude on the place of women in chemistry, see Fischer (1922, 1987), p.192.
[105] Ibid., p.187. For Gabriel, see Colman and Albert (1926).

institute, Gabriel helped Fischer greatly, as did Piloty, who re-
acted to Fischer's rule differently. Later, Harries wrote:

From 1897 on, Emil Fischer was busy with the preparatory work for
the construction of the First Chemical Institute, and drew us assistants,
especially Piloty, into this activity very intensively. At that time, the
master was not in a good frame of mind. Sickly and nervous, he could
be extremely insulting if we often in unison proposed something he
found to be unacceptable. . . . But on the whole, as is widely recognized today, a model institute was created, in whose success Piloty
played a large part.[106]

In 1900, however, Piloty returned to his native Munich because,
according to Harries, Fischer had not fulfilled what Piloty considered the promise of a promotion to be head of a section of
the institute.

   In the years after 1900, apart from Gabriel and Harries,
there was a succession of section heads, among them Otto Diels,
Franz Fischer, Josef Houben, Hermann Leuchs, Robert Pschorr,
Alfred Stock and Wilhelm Traube, all of whom are listed in Appendix 6. Although Gabriel, Leuchs and Traube achieved independent status in Fischer's institute, and most of their students
are not listed as members of Fischer's research group, it is noteworthy that the lines of investigation these three men pursued
in Berlin were in areas derived from Fischer's previous research,
such as the synthesis of amino acids, purines and pyrimidines.
Moreover, of the three other nominally independent investigators at the institute, Otto Ruff, Alfred Wohl and Franz Sachs,
the first two worked on the chemistry of sugars.

   From the data in Table 5-1 it is evident that of the 255 people whom I have listed as members of Fischer's group in Berlin,
at least 108 were postdoctoral assistants or guests. Of the latter
number, 44 had received medical degrees before coming to
Fischer's laboratory, and nearly all of them were routinely
turned over to Abderhalden for work on the amino acid analysis
of proteins by the ester method. Among the twenty men from
Germany or Austria were such future notables as Fritz Pregl and
Alfred Schittenhelm, and the ten from the United States included Lewellys Franklin Barker, James Bryan Herrick, Phoebus

---

[106] Harries (1920), p.160.

Aaron Levene and Harry Gideon Wells. This relatively high proportion of post-M.D. guests reflects both the important place
that chemical training had come to occupy in the education of
those who aspired to a career in medical research and the prestige of Fischer's laboratory as a place in which to gain advanced
chemical experience. It may be questioned, however, whether
the association of these men with Abderhalden fulfilled their
hopes, although in the obituary notices for many of the noted
clinicians, especially the German ones, there is proud mention
of the fact that the deceased had been a *Fischer-Schüler.* Fischer's
attitude to the medical men who came to his Berlin institute
may be contrasted to that he evinced toward Karl Landsteiner,
who worked in his Würzburg laboratory on a straightforward
problem in organic chemistry.

The foreign contingent (75) in Fischer's Berlin institute represented about 30 per cent of the total, with the largest number
(29) coming from the United States. The decrease in the proportion of those from Britain, as compared with that in
Würzburg, is striking. I shall not speculate about the possible
reasons for this drop, except to note that none of the men who
had worked with Fischer before 1892 attained a professorship in
chemistry at a British university.

In Table 5-2 I have attempted to translate into numbers the
principal subsequent activity of Fischer's junior associates for
whom I was able to find biographical data. Of the 354 people
listed in Appendix 6, information was not available for 25 Germans or Austrians and for nine from other countries. Also, at
least nine men died soon after receiving their Dr.phil. degree or
during the course of their doctoral research.[107] Although it is
likely that some of the others whose professional acitivity I did
not determine also perished during the First World War, the
more probable explanation for my failure to find many names
in the later chemical literature is that these people had entered
industrial or commercial chemistry.

The data in Table 5-2 for the remaining 311 persons indicate that at least 136 of the Germans and Austrians, and 36 of
the foreign students, became commercial chemical consultants
or joined industrial firms or government bureaus such as the

---

[107] At least 15 of Fischer's German students died while in military service during the
First World War. One of them was Richard von Grävenitz, who presented his Dr.phil.
dissertation on 5 August 1914, at once entered the army, as befitted the scion of a
noble family with many generals, and who was killed in action a month before the
Armistice.

TABLE 5-2
Principal Subsequent Activity of Members of the Fischer Research Group[*]

	Munich (78–82)	Erlangen (82–85)	Würzburg (85–92)	Berlin (92–19)
Full (*ord*) prof. chem.	F1	G3	G1/F1	G21/F12
Full (*ord*) prof. applied chem.			G1/F1	G7/F3
Other acad. ranks pure & applied chem.		G1	F1	G18/F7
Full (*ord*) prof. other subjects (med., pathol., pharmacol., mineralogy)			G4/F1	G4/F11
Research institute			F1	G9/F7
Private research lab.				F1
Hospital chemistry				G4/F1
Medical practice				G9/F2
Industrial manufacture	G2	G6	G32/F6	G57/F11
Commercial & consulting chem.	G1	G3	G5/F8	G11/F6
Govt. techn. depts.	G2	G4	G3/F1	G10/F4
Other occupations			F1	G4/F2
Not determined	G1	G2/F1	G5	G17/F8

[*]G = German or Austrian; F = foreign nationalities

Patent Office. Consequently, except for one or two minor publications, or notices of patents, about 55 per cent of the Fischer pupils disappeared from the chemical literature. As in the case of the Baeyer school, some of the men who entered industry rose to high office in chemical firms such as Hoechst or Boehringer-Mannheim.

The Fischer students who remained in academic life or joined scientific research institutes include some of the distinguished chemists or biochemists of the twentieth century. More will be said about them later in this chapter, in relation to the influence of Fischer's scientific progeny on the development of the chemical and biochemical sciences. It may noted here that, except for Ludwig Knorr, the men who achieved distinction in chemistry all were associated with Fischer after about 1900. Also, it is of interest that a sizable number of the men who later gained fame worked in research institutes not attached to universities. Among the Germans or Austrians were Max Bergmann, Franz Fischer, Karl Landsteiner and Otto Warburg, and the foreigners included Ernest Fourneau, Walter

TABLE 5-3
The Hofmeister Research Group

	Prague (79–96)		Strassburg (96–19)			
	Predoc.	Postdoc.	PreMD	PrePhD	PostMD	PostPhD
Germany & Austria	13	6	41	3	49	9
Japan			3		7	
Italy			2		8	
Russia & Poland			2		4	1
United States					7	2
France			3		1	
United Kingdom			2		1	
Belgium			1		2	
Switzerland				2		
Other countries					3*	1 +
Total	13	6	54	3	84	13

*One each from Norway, Spain, Sweden
+ From Australia

Abraham Jacobs, Phoebus Aaron Levene and Donald Dexter Van Slyke.[108]

Franz Hofmeister's Research Group

At Prague, Hofmeister's institute of pharmacology consisted of five small rooms, in which a total of about 17 people worked between 1879 and 1896.[109] As is indicated in Table 5-3, most of his junior associates there were medical students, and there were also a few postdoctoral medical men. This group included such later notables as the botanist Friedrich Czapek, the physicians Adalbert Czerny and Friedrich Kraus and the pharmacologist Julius Pohl.

---

[108] All these men later worked at one of the Kaiser-Wilhelm Institutes (Bergmann, Franz Fischer, Warburg), the Rockefeller Institute for Medical Research (Bergmann, Jacobs, Landsteiner, Levene, Van Slyke) or the Pasteur Institute (Fourneau).

[109] Pohl and Spiro (1923), p.2. From the available records in the university archives of the medical faculty in Prague (Carton No.42) it seems that the financial support given Hofmeister's institute was rather modest.

In view of the circumstances of Hofmeister's appointment at Strassburg, recounted earlier in this chapter, it is not surprising that in the division of the space in Hoppe-Seyler's former building, the portion assigned to the institute of physiological chemistry was smaller than that given to the one for hygiene. At the start, Hofmeister's laboratory facilities were adequate; upon his arrival he found none of Hoppe-Seyler's former associates and apart from Hofmeister's two assistants (Spiro and Fürth), there were only five students (Ellinger, Franz Goldschmidt, Krieger, Reye, Schneider). During the first year, however, the population grew rapidly and by 1900 the laboratory became rather crowded, but Hofmeister does not appear to have made any demands for an enlargement of his facilities. According to Spiro, "if he had not been so modest, simple and thrifty, he could have improved the institute considerably; despite all the crowding, he was essentially satisfied with the situation."[110] Nor did Hofmeister take advantage of the invitation from Heidelberg to succeed Kühne as professor of physiology, for he declined the offer, reportedly on the ground that he was not willing to undertake instruction in a field in which he was not conducting research and "because he saw that . . . chemical physiology was being taught by pure electrophysiologists, he did not wish to commit the same mistake in reverse."[111]

Although, unlike Hoppe-Seyler and Hüfner, Hofmeister did not offer public lectures or write articles on the need for the enhancement of the institutional status of physiological chemistry at German universities, he appears to have felt keenly an obligation to promote that cause by remaining in Strassburg. Another factor in his decision may have been the friendship extended to Hofmeister by Friedrich Goltz, the professor of physiology, who had agreed that what was then termed *vegetative Physiologie* would be the sole province of the institute of physiological chemistry.[112]

The data in Table 5-3 indicate that about 90 per cent of the 155 persons associated with Hofmeister from 1896 until 1919 were either candidates for the Dr.med. degree or postdoctoral workers who had previously received that degree. It is significant that one-third of them (52) had come from countries other

---

[110] Pohl and Spiro (1923), p.15.

[111] Ibid., p.16. It should be noted that after Hofmeister declined the offer from Heidelberg, Albrecht Kossel accepted it and continued his line of chemical research on proteins; see Chapter Three, note 66.

[112] At that time, the term *vegetative Physiologie* denoted the area encompassed by physiological chemistry, not in the sense of plant physiology, but in relation to the study of metabolic processes.

than Germany or Austria, with the largest contingents from Italy, Japan and the United States. Another feature of the group that deserves mention is the appearance in biochemical research of Germans who had first acquired a Dr.phil. degree for work in organic chemistry, and who had then completed the requirements for the Dr.med. degree. In Hofmeister's group at Strassburg, Karl Spiro, Alexander Ellinger and Ernst Friedmann represent outstanding early examples of such preparation for research in the biochemical and medical sciences.[113]

From the biographical information I was able to gather about the persons listed in Appendix 7, I have attempted, as before, to determine the principal later professional activity of Hofmeister's junior associates at Prague and at Strassburg. The people who had worked with him in Prague presented special difficulty, and I succeeded in finding the needed personal data for only twelve of the 17 men listed in Appendix 7. For the Germans in the Strassburg group, I learned that two (Gross, Koppel) died soon after their doctoral work, and it would appear that three others (Franz Goldschmidt, Krieger, Schneider) all had died before 1920.[114] Also, I was unable to determine the later activity of six of the people from foreign countries.

The data in Table 5-4 thus refer to 156 of the 172 persons listed in Appendix 7. At least 62 of them (40 percent) attained full professorships, eleven in physiological or biological chemistry, but it should be noted that only two of them (Knoop, Schulz) gained this rank at German universities. The other nine were appointed in Austria (Fürth, Pauli), Switzerland (Oswald, Spiro), the United States (Henderson, Swain), Poland (Parnas), Russia (Kuraev) and Japan (Ishimori). The other 51 men held full professorships in medical subjects, such as internal medicine (14), physiology (10), pharmacology (10), pediatrics (8) and other clinical specialties (7), or in botany (2). More will be said later in this chapter about some of these Hofmeister pupils in relation to their contributions to the development of the biochemical sciences. It will be seen in Table 5-4 that a relatively large number of the Germans or Austrians remained in academic life, but did not attain full professorships, or were later associated with hospitals. Most of these 40 people were individuals who continued to publish in scientific or medical journals,

---

[113] Although Friedmann, Ellinger and Spiro may have aspired at the start of their studies to engage in chemical research on biological problems, the fact that all three men were Jews suggests the possibility that they saw better prospects for a career in medicine than in the academic chemistry of Wilhelmine Germany.

[114] Pohl and Spiro (1923), p.15.

TABLE 5-4

**Principal Subsequent Activity of Members
of the Hofmeister Research Group***

	Prague (79–95)	Strassburg (96–19)
Full (*ord*) prof. physiol. or med. chem.	G2	G3/F8
Other acad. ranks physiol. or med. chem.		G5/F1
Full (*ord*) prof. pharmacol.	G2	G5/F4
Full (*ord*) prof. physiol.		G2/F4
Full (*ord*) prof. internal medicine	G1	G7/F4
Full (*ord*) prof. pediatrics	G2	G5
Full (*ord*) prof. other clinical specialties	G2	G5/F3
Other acad. ranks med. subjects	G5	G10
Full (*ord*) prof. botany	G1	G1
Hospital research & administration		G27/F11
Private medical practice		G24/F10
Industrial research & manufacture		G2/F1
Pharmacy		G1/F1
Not determined	G4	G5/F5

*G = German or Austrian; F = foreign nationalities

---

and because of the limited academic opportunities for biochemists, many of them did biochemical research as members of medical school departments of physiology or pathology, or as heads of chemical laboratories in hospitals. Clearly, the number and quality of the publications per individual varied widely, but they included numerous significant biochemical contributions. If one adds the 13 foreign students who remained in academic life or conducted research in hospital laboratories, it would seem that nearly 75 per cent of the former members of the Hofmeister group continued to appear in the literature of the biochemical and medical sciences. The data in Table 5-4 indicate that most of those whose names disappeared from that literature had entered private medical practice.[115]

To this numerical summary of some of the features of Hofmeister's research group, I add a somber note. After Hitler's

---

[115] My claim for the disappearance of some names listed in Appendices 6 and 7 from the chemical, medical or technical literature is based on an examination of the following sources: *Chemisches Zentralblatt, Chemical Abstracts, Royal Society Catalogue of Scientific Papers, International Catalogue of Scientific Literature, Index Medicus, Quarterly Cumulative Index to Current Medical Literature* and *Quarterly Cumulative Index Medicus.* The search covered a 30-year period after the individual's association with Fischer or Hofmeister.

rise to power, of the approximately 80 former Hofmeister pupils whom I know to have been living in Germany or Austria in 1933, at least 19 emigrated, most of them to the United States and England, and a few to Turkey and Palestine. Three others perished in Nazi concentration camps. Of the comparable members of the Fischer group (approximately 110), at least ten emigrated and five died at the hands of the Nazis.[116]

Emil Fischer's Scientific Progeny

Since Nobel Prizes are widely considered to be the principal twentieth-century marks of scientific distinction, the Fischer group, together with the Baeyer school, stand out among their contemporaries as seedbeds of talent in the chemical and medical sciences. Four *Fischer-Schüler* (Otto Diels, Hans Fischer, Fritz Pregl, Adolf Windaus) received the prize in chemistry, and two (Karl Landsteiner, Otto Warburg) in physiology or medicine. As will be seen in what follows, however, the work that was honored in the award of these six prizes bore little relation to Fischer's own scientific achievements. Consequently, for an assessment of the influence of his scientific progeny on the later development of the areas of organic and biological chemistry that Fischer had explored, it is also necessary to consider the work and style of other noted members of his research group, particularly those who followed in their teacher's footsteps.

Diels was associated with Fischer for an extended period (1896–1916), as a Dr.phil. student, teaching assistant and section head. During that time Diels discovered carbon suboxide and worked on a variety of problems, including the chemistry of cholesterol. After he became professor of chemistry at Kiel, he made a decisive contribution to the elucidation of the structure of the sterol nucleus and, with his student Kurt Alder [1902–1958] developed the important method of diene synthesis, for which the two men were awarded the Nobel Prize.[117]

The three other prize winners in chemistry stayed in Fischer's institute only briefly. Hans Fischer had obtained both a

---

[116] See Pross (1955), Widmann (1973), Strauss and Roeder (1983), Kröner (1983) and Gedenkbuch (1986). Some names were found in the confidential *List of Displaced German Scholars*, prepared in 1936 by the Academic Assistance Council in London. For a valuable review of the literature on the emigration of German scholars during the 1930s, see Möller (1984).

[117] For Diels, see note 26, this chapter.

Dr.phil. in chemistry and a Dr.med. degree before his short association with Emil Fischer. On the advice of the noted Munich clinician Friedrich von Müller, who had concluded that bile pigments arise from hemoglobin, Hans Fischer entered a field opened in the previous century by the physiological chemists Hoppe-Seyler and Nencki. Hans Fischer's remarkable achievements over a 20-year period reached their peak in the synthesis of hemin, bilirubin and many porphyrins, thus establishing their chemical structure and paving the way for the later study of their metabolic formation and interconversion.[118]

Another prize winner, Fritz Pregl, was an Austrian physician who entered physiological chemistry at the turn of the century, and in 1904 spent a few months in Fischer's institute, where he was turned over to Abderhalden for training in the ester method of protein analysis. Upon returning to his native Graz, Pregl made his lifework the improvement of the available methods for the quantitative elemental analysis of organic compounds, and by 1917 he had devised new apparatus and techniques for the analysis of 2-3 milligram samples. The significance of this technical advance cannot be overestimated, for at a time when elemental analysis was still indispensable for the determination of organic-chemical structure, Pregl's microanalytical methods made it possible to analyze natural materials that were available to chemists in only small amounts.[119]

Adolf Windaus received his Dr.phil. in 1899 for work on digitalis alkaloids with Heinrich Kiliani [1855–1945], professor of medical chemistry at Freiburg im Breisgau. After a short time with Fischer, Windaus returned to Kiliani who advised him to enter the sterol field. Among Windaus's many distinguished contributions, of special importance was his demonstration (with Alfred Fabian Hess [1875–1933]) that a vitamin D can be produced by irradiation of yeast ergosterol with ultraviolet light. At Göttingen, where Windaus was professor of chemistry for almost 30 years, he led an outstanding research group.[120]

The two winners of Nobel Prizes in physiology or medicine (Landsteiner, Warburg) were very different men. After receiving his Dr.med. degree in Vienna, Karl Landsteiner spent the following three years (1891–1894) in several organic-chemical laboratories. In addition to a year with Fischer in Würzburg, he also

---

[118] For Hans Fischer, see Wieland (1950), Treibs (1971) and Watson (1965); for Friedrich von Müller [1858–1941], see Gruber (1958).
[119] For Pregl, see Philippi (1952) and Szabadváry (1966), pp.302–303.
[120] For Windaus, see Josephson (1933), Hückel (1947) and Butenandt (1961); for Kiliani, see Hückel (1949).

worked with Bamberger in Munich and with Hanztsch in Zurich. Landsteiner's remarkable achievements from 1897 onward began with his discovery of human blood groups; his subsequent systematic work on the specificity of serological reactions laid the foundations of modern immunology. At the Rockefeller Institute for Medical Research, where he worked after 1922, he led a small but highly productive research group. Landsteiner was an unassuming person of broad culture and considerable chemical insight.[121]

Otto Warburg completed his chemical studies in 1906 with Fischer; his Dr.phil. dissertation dealt with the synthesis of several peptides by the halogenacyl halide method. An ambitious young man, with his sights set on more rewarding fields of endeavor (including the cancer problem), Warburg at once left organic chemistry and turned to cell physiology, apparently influenced by the writings of Jacques Loeb [1859–1924]. Warburg's work on the respiration of fertilized sea urchin eggs was conducted largely at the Naples biological experiment station during the years when he was studying medicine in Heidelberg. Three years after receiving his Dr.med. degree in 1911, Warburg became head of a laboratory of the Kaiser-Wilhelm Society; as vice-president of the Society, Fischer was responsible for this appointment. Upon the outbreak of the First World War, Warburg at once joined a cavalry regiment in which he served until the Armistice. He later spoke enthusiastically about his war service: "I learned to handle people; I learned to obey and to command."[122] During the 1920s Warburg resumed his studies on cellular respiration, with special reference to oxidation and glycolysis in tumor cells. In connection with this work he devised new or improved experimental methods in manometry and in the use of tissue slices. He concluded that anaerobic glycolysis plays a primary role in the causation of cancer, a view he continued to reiterate into the 1960s, despite extensive evidence to the contrary. On the basis of experiments with model systems, Warburg insisted that the primary process in cellular respiration is the activation of protoplasmic iron, and he scornfully dismissed the work of Heinrich Wieland, Thorsten Thunberg and David Keilin showing that dehydrogenases and cytochromes play important roles in biological oxidations. Warburg received his Nobel Prize for a brilliant spectrophotometric study of what he called the *Atmungsferment*. His greatest achievements came dur-

[121] For Landsteiner, see Speiser (1961), Simms (1963), Heidelberger (1969) and Mazumdar (1975).
[122] Krebs (1972), p.635.

ing the 1930s when he discovered the first of the flavoproteins, when he showed that the action of dehydrogenases involves the participation of a nicotinamide-containing substance related to cozymase, and when his assistants Negelein and Wulff reported the first clear-cut demonstration of a coupling between oxidation and phosphorylation. In his polemics about the cancer problem and cellular respiration, as well as about photosynthesis, Warburg exhibited exceptional arrogance. His laboratory was run in semi-military fashion, and discipline was enforced by his personal servant Jacob Heiss. Warburg preferred to employ as his research assistants men who had been trained in technical schools rather than ambitious young scientists because "they never troubled him with requests for testimonials and for support in obtaining better posts."[123] Although more generous than Fischer in allowing the names of these men (Walter Christian, Fritz Kubowitz, Erwin Negelein) to appear as authors of important papers from his institute, at the end of the Second World War Warburg abruptly dismissed them.

I have dwelt at somewhat greater length on the career and personality of Otto Warburg not only because his work during the 1930s represents a peak of biochemical achievement in this century but also because in later life he repeatedly stressed the importance of his association with Emil Fischer. Although Warburg's claims for the special significance of the work he did in Fischer's laboratory either for protein chemistry or in his own later success can be described most charitably as exaggerations, there can be little doubt that Warburg adopted much of Fischer's style in his conduct toward his collaborators.

Among Fischer's German associates who also made important contributions in fields unrelated to his main research interests, but did not win Nobel prizes, were the following: Franz Fischer, who became head of the Kaiser-Wilhelm Institute for Coal Research (organized by Emil Fischer) and who is best known for his work with Hans Tropsch on the artificial synthesis of benzene by high-pressure hydrogenation; Hermann O. L. Fischer, who did his doctoral work with Knorr and, after serving as one of his father's research assistants, did important research on the structure of quinic and shikimic acid and on the synthesis of biologically important derivatives of glyceraldehyde; Carl Harries, whose studies on the ozonization of organic compounds and on rubber won him wide acclaim; Hermann Leuchs, who after his work with Fischer on amino acids and peptides,

---

[123] Ibid., p.636. A revised and illustrated version of this article appeared in book form (Stuttgart 1979; Oxford 1981).

discovered the N-carboxyanhydrides, but whose main subsequent activity was in other fields, notably the strychnine alkaloids; Otto Ruff, who began as a pharmacist, then worked with Fischer on sugars, and after 1903 played a leading role (comparable to that of the Nobel Prize winner Henri Moissan) in the study of fluorine compounds; and Helmuth Scheibler who, after his doctoral work with Fischer on peptides, turned to other organic-chemical problems, especially compounds containing divalent carbon, a subject which later became the important field of carbene chemistry.[124] To this group may be added Karl Spiro who, as was noted above, began as an organic chemist with Fischer but, under the influence of Hofmeister, then worked on the physical chemistry of proteins and on other biochemical problems.

The first of the notable Fischer pupils whose later research was clearly derived from his teacher's line of investigation was Ludwig Knorr, whose association with Fischer began in Munich, and continued in Erlangen and Würzburg. Knorr's dissertation dealt with a derivative of phenylhydrazine, and shortly afterward the study of the reaction of phenylhydrazine with ethyl acetoacetate led him to discover a pyrazole derivative that proved to be pharmacologically effective as a febrifuge. Fischer allowed Knorr to obtain a patent, and the manufacture of this product (*Antipyrine*) marked the entry of the Hoechst dyeworks into pharmaceutical chemistry.[125] In 1889 Knorr became professor of chemistry at Jena, where he directed the work of a sizable research group (approximately 150 students and assistants) that included such later notables as Fritz Haber [1868–1934]; several of Knorr's doctoral students joined Emil Fischer as research assistants. Knorr's outstanding studies on pyrazoles and on the keto-enol tautomerism in substances related to ethyl acetoacetate stemmed directly from his association with Fischer. In addition, until he ceased research in 1912, Knorr conducted a sustained program of work on alkaloids related to morphine. As was noted above, Fischer's liberality in encouraging Knorr to continue research along lines that Fischer had initiated was not evident in his attitude toward most of his junior associates in Berlin.

---

[124] For Franz Fischer, see Pichler (1967); for Hermann O. L. Fischer, see Sowden (1962), Stanley and Hassid (1969) and Lichtenthaler (1987); for Harries, see note 24, this chapter; for Leuchs, see Kröhnke (1967); for Ruff, see Hückel (1940); for Scheibler, see Wanzlik (1969).

[125] For Knorr, see Duden and Kaufmann (1927), Anon. (1965) and Flemming (1968). For the early history of the Hoechst firm, see Fischer (1958).

Among those who worked with Fischer after 1900, five men (Abderhalden, Bergmann, Freudenberg, Helferich, Zemplen) continued research on problems to which they had been introduced by Fischer. Of these, Emil Abderhalden achieved the greatest public fame in his lifetime. As professor of physiology at the veterinary school in Berlin (1908–1911) and then at Halle until 1945, Abderhalden followed Fischer along the tracks that proved to be dead ends, notably in the use of the halogenacyl halide method of peptide synthesis and in his espousal of Fischer's suggestion that diketopiperazines may be structural elements of proteins. In addition to his many papers on these subjects, Abderhalden wrote extensively on other biochemical topics, among them protein metabolism, enzymes, vitamins and hormones, as well as on the *Abwehrfermente*, mentioned earlier in this chapter. His seemingly inexhaustible energy, his skill as a scientific entrepreneur, and his enormous output of scientific papers and books, as well as of writings on other subjects, gained him the kind of public applause provided in 1937 by an American historian of chemistry:

No contemporary biochemist is better known than the editor of the "Handbuch der biologischen Arbeitsmethoden." However, Emil Abderhalden's enviable reputation is not primarily derived from his efficient leadership in this monumental production; rather, his recognized eminence has been attained through the constant stream of high-class experimental studies that has flowed from his laboratory. His researches have covered or touched so many areas of his chosen field that an abstract of hundreds of published papers might almost serve as a syllabus of a comprehensive course in physiological chemistry.[126]

This opinion was not shared by the leading investigators in the areas of Abderhalden's scientific efforts. Indeed, from Fischer's correspondence it would seem that his enthusiasm in 1904 about Abderhalden's chemical talent had cooled considerably in succeeding years.[127]

The scientific stature of Max Bergmann was rather different. As Fischer's chief research assistant during the First World

---

[126] Oesper (1937).

[127] Fischer's changed attitude toward Abderhalden is evident from their correspondence after 1908; for example, on 25 September 1909, Fischer chided Abderhalden for encroaching on his territory in studies on silk fibroin, and on 14 March 1912, Fischer offers only sympathy in Abderhalden's search for funds to support his research at Halle (Bancroft Library).

War, Bergmann worked on amino acids, depsides, carbohydrates and glycerides. Two years after Fischer's death he was appointed director of the newly established Kaiser-Wilhelm Institute for Leather Research in Dresden. There, in addition to his service to the leather industry, he and his associates did distinguished work, along the lines of his research in Berlin, on carbohydrates, with special reference to unsaturated sugars such as glucal and to the polysaccharides cellulose and chitin, as well as on amino acids and peptides. For several years, Bergmann, like Abderhalden, accepted the diketopiperazine theory of protein structure, and published a series of papers on unsaturated diketopiperazines. As was noted above, in 1932 Bergmann and Zervas reported the invention of the carbobenzoxy method of peptide synthesis; this advance marked a new era in this field. In the following year Bergmann was forced by the Nazis to leave Germany and he found a haven at the Rockefeller Institute for Medical Research, where he worked until his untimely death in 1944.[128] In his new laboratory, of which I was a member, there were two main lines of research. One involved the use of the use of the newly available peptides for the study of the specificity of proteolytic enzymes; my discovery of the first synthetic peptide substrates for pepsin and other well-defined proteinases provided strong evidence for the peptide theory of protein structure at a time when the theory was under serious challenge. The other main line, actively pursued by Bergmann himself, was the development of new methods for the amino acid analysis of proteins. With Carl Niemann [1908–1964] he advanced the hypothesis that each kind of amino acid unit recurs at a regular periodic interval in the polypeptide chain of a protein; this excursion into numerology was soon shown to be invalid.[129] Except for the interruption by work done in connection with the Second World War, the research on the amino acid composition of proteins was continued at the Rockefeller Institute by Bergmann's associates William Howard Stein [1911–1980] and Stanford Moore [1913–1982], who shared the 1972 Nobel Prize in Chemistry for their achievements in this field.

Karl Freudenberg worked with Fischer between 1909 and 1914 on depsides and tannins. After the First World War he became professor of chemistry at Heidelberg, where over a period

---

[128] For Bergmann, see Harington (1945); the more extensive notice by Helferich (1969) is inaccurate. For Zervas, see Katsoyannis (1973), pp.1–20.

[129] See Witkowski (1985). Bergmann's interest in chemical numerology finds early expression in the article by Fischer (1916), in which he acknowledges Bergmann's help in the calculations.

of thirty years he conducted research on the chemistry of carbohydrates and lignins as well as in the field of stereochemistry,
including the mechanism of the Walden inversion. Burckhart
Helferich, also a doctoral student and assistant with Fischer,
worked in Berlin on the synthesis of glycosides. As professor of
chemistry, first in Greifswald, then in Leipzig, his main achievements were in the synthesis of oligosaccharides, through the introduction of the triphenylmethyl (trityl) group for this purpose,
and in the systematic study of the specificity of enzymes that
hydrolyze glycosides. Geza Zemplén came to Fischer's laboratory
after his doctoral work in Budapest and initially worked on
amino acids but toward the end of his stay in Berlin he was directed into the field of carbohydrate chemistry. He continued in
this field afterward and made valuable contributions to the synthesis of oligosaccharides and of naturally occurring glycosides.
As professor of organic chemistry at the Technical University in
Budapest, Zemplén played a leading role in the development of
organic chemistry in post-war Hungary.[130]

   The influence of Fischer's scientific progeny in other nations was particularly evident in the twentieth-century development of biochemistry in the United States. The first of the
notable Americans to work in Fischer's Berlin institute was
Phoebus Aaron Theodore Levene. An emigré physician from
Czarist Russia, Levene entered biochemical research after several years of private medical practice in New York. During a
visit to Europe in 1901 he spent a summer with Fischer, and was
assigned the task of analyzing gelatin by the ester method. In
1905 Levene was appointed to the staff of the newly created
Rockefeller Institute for Medical Research, where he worked until the end of his life on an extraordinary variety of biochemical
problems. Chief among them was the constitution of the nucleic
acids, a field that Fischer attempted to enter when he gave up
the hope of making a synthetic protein. Among Levene's 700-
odd publications there were many papers dealing with proteins,
and even before 1910 he did not hesitate to invade Fischer's territory by isolating peptides from partial hydrolysates of proteins.
There were also experimental papers on the chemistry of phosphatides, cerebrosides and sugars, as well as on stereochemistry.
It was in the nucleic acid field, however, that Levene had the
greatest impact, especially through the efforts around 1910 of
his associate Walter Abraham Jacobs, who demonstrated that
yeast nucleic acid is composed of ribonucleotides. Jacobs had re-

[130] For Freudenberg, see Freudenberg (1967) and Lichtenthaler (1987); for Helferich,
see Bredereck (1957) and Lichtenthaler (1987); for Zemplén, see Schmidt (1959).

ceived his Dr.phil. degree in 1907 for work with Fischer on the resolution of several amino acids. Five years later he became an independent investigator at the Rockefeller Institute, where he remained the rest of his life. During the First World War Jacobs's main activity was in the field of chemotherapy, and he made a major contribution through his discovery of Tryparsamide, effective in the treatment of African sleeping sickness. After the war he turned to the study of alkaloids, showed that many of those used in the treatment of heart disease were sterol derivatives, and later elucidated the complex structure of the ergot alkaloids. A modest person of exceptional chemical insight, Jacobs had an outstanding group of junior associates, among them Michael Heidelberger [b. 1888], Robert Cooley Elderfield [1904–1979] and Lyman Creighton Craig [1906–1974].[131]

Another distinguished member of the Rockefeller Institute who had worked with Fischer was Donald Dexter Van Slyke. After his Ph.D. with Gomberg at Michigan, Van Slyke joined Levene, with whom he worked on the amino acid composition of casein hydrolysates. During his brief stay in 1910 at Fischer's institute, Van Slyke studied the properties of pyrrole carboxylic acid. After returning to the Rockefeller Institute, Van Slyke invented a gasometric method for the determination of amino-nitrogen, and used it to demonstrate (with Gustave Meyer) that proteins are cleaved completely to amino acids in the mammalian digestive tract and that amino acids disappear from the blood as it passes through the tissues; this work marked the final burial of the peptone theory of protein synthesis. In 1914 Van Slyke became the chemist of the newly opened hospital of the Rockefeller Institute; he worked there until 1949, and ended his days at the Brookhaven National Laboratory. Through his many contributions to quantitative clinical chemistry, and the postdoctoral training he provided to many future leaders of American medicine, Van Slyke profoundly influenced medical practice in the United States. During the 1920s, Van Slyke and his associates conducted fruitful research on acid-base equilibria and CO_2 transport in blood, as well as on the distribution of chloride and bicarbonate between erythrocytes and plasma. He remained a chemist to the end; for many years he worked on the structure of a new amino acid he had isolated from gelatin, and showed it to be delta-hydroxylysine.[132]

---

[131] For Levene, see Van Slyke and Jacobs (1944); for Jacobs, see Elderfield (1980).
[132] For Van Slyke, see Hastings (1976) and Edsall (1985).

A fourth American visitor to Fischer's Berlin institute was the pathologist Harry Gideon Wells, who was turned over to Abderhalden for training in the ester method, and was assigned keratin for analysis. In addition to his many later contributions to pathology, of special biochemical significance was Wells's demonstration of the immunological differences among the crystalline seed proteins provided by Thomas Burr Osborne.[133]

To these important participants in the development of the biochemical sciences in the United States must be added the Englishman Hans Thacher Clarke who spent two years in Fischer's laboratory before coming to the United States in 1914, when he joined the staff of the Eastman Kodak Company in Rochester. Fourteen years later Clarke became head of the department of biological chemistry at the College of Physicians and Surgeons of Columbia University, and made it one of the leading centers of biochemical research and education in the United States. This eminence was not attained through the efforts of a closely directed research group, and Clarke's output of scientific papers was modest, but through a liberal policy of encouraging the independent work of his departmental associates. Moreover, in selecting students for admission to graduate study, Clarke's decision

was based entirely on his judgment of the student's chemical knowledge and his ability to recall and coordinate his chemical experiences. This in itself cannot be considered unique. However, considering the times and social climate of the universities in the thirties, it was rather extraordinary, for no other extraneous factors entered into his decision. The students were mainly New Yorkers who received their undergraduate training in the east, from Harvard to the College of the City of New York. This attitude of Clarke's also opened the laboratory to students who reached New York as refugees from the racial policy of the Germans under the Third Reich and to scholars such as E. Brand, E. Chargaff, Z. Dische, K. Meyer, D. Nachmansohn, R. Schoenheimer, and H. Waelsch, who arrived in the United States for similar reasons. Clarke's humanism which thus opened the department to talent, and his introduction of organic chemistry and its techniques and approaches, soon made it one of the leading biochemical departments in the United States. It was not an accident that Schoenheimer, with H. Urey's student D. Rittenberg, introduced the use of stable isotopes for studies of metabolic reactions, nor was it an accident that R. R.

---

[133] For Wells, see Long (1949).

Williams isolated and determined the structure of vitamin B [thiamine] in the laboratory space of the large student laboratory which was assigned to him by Clarke.[134]

It should added that, as a student in London, Clarke was introduced to biochemistry by Robert Henry Aders Plimmer, who had worked in 1902 with Fischer on the amino acid composition of gelatin. After his return to England, Plimmer continued to work on the chemistry of proteins, and also played a leading role in the founding of the Biochemical Society.[135]

As was indicated in Table 5-1, relatively few members of Fischer's research group came from France. One of them was Ernest Fourneau, who worked in the Berlin institute after having been with Curtius in Heidelberg and before going to Baeyer's institute in Munich, where he worked with Willstätter. During his brief association with Fischer, Fourneau performed the conversion of glycine anhydride to glycyl-glycine, mentioned earlier in this chapter.[136] After several years in the research laboratory of the Poulenc chemical firm, in 1911 Fourneau joined the Pasteur Institute in Paris, where he conducted a sustained and fruitful search for effective new pharmaceutical agents. Among his many contributions were the preparation of the local anesthetic Stovaine and the demonstration that the active component of Gerhard Domagk's Prontosil is sulfanilamide.

These biographical notes about the more notable members of the Fischer group indicate the complexity of the problem of assessing the role that direct association with a famous leader has played in the later work of his successful progeny. In the case of the few who followed in Fischer's scientific footsteps after an extended stay in his laboratory, a positive influence is unmistakable. It is more difficult, however, to estimate the extent to which membership in his research group, especially when relatively brief, was reflected in the subsequent scientific output of the others. Indeed, it is conceivable that, whatever respect they may have had for Fischer's chemical genius, some of those who later achieved distinction found the tasks he assigned to them uninteresting. Also, whereas men like Otto Warburg may have found Fischer's dictatorial style of leadership worthy of emulation, others were repelled by it and, as in the case of Hans

---

[134] Shemin (1974), pp.135–136; see also Vickery (1975).
[135] For Plimmer, see Lowndes (1956).
[136] See note 7; for biographies of Fourneau, see Delépine (1950), Henry (1952) and Fourneau (1987).

Clarke, when given the opportunity to lead a research school, chose to adopt a more liberal attitude toward their students and postdoctoral associates. There can be little doubt that, as for the *Liebig-Schüler* during the nineteenth century, the mere designation as a *Fischer-Schüler* carried great weight, and a favorable recommendation from Fischer could be decisive in a person's professional advancement, but this consequence of Fischer's public prestige says little about that person's choice of problems for independent research, the manner in which he tackled them or the way he treated the junior members of his research group.

Franz Hofmeister's Scientific Progeny

In considering the later contributions of Hofmeister's pupils to the advancement of the biochemical sciences, it should be remembered that, in the German-speaking countries (Germany, Austria, Switzerland), the biochemical professorships were almost exclusively in the medical faculties of their universities, and that the appointees to such professorships were required to hold a Dr.med. degree. Moreover, as was noted in earlier chapters, the number of such professorships was small as compared with that for chemistry. Thus, before 1914, apart from the two in Germany (Tübingen, Strassburg), there were three in Austria (Graz, Prague, Vienna) and three in Switzerland (Basle, Berne, Zurich). Indeed, the opportunities to become a *Privatdozent* in physiological chemistry were far more limited than those in chemistry.[137]

It is understandable, therefore, why nearly all the Germans and Austrians who worked in Hofmeister's laboratories in Prague and Strassburg were either candidates for the Dr.med. degree or had received that degree before joining Hofmeister's group. Nor is it surprising that many of these men made outstanding biochemical contributions while holding university appointments in one of the better-established medical disciplines closely allied to biochemistry, such as pharmacology or physiology, and some of the clinical specialties, such as pediatrics, pathology or internal medicine. For this reason I will consider the later achievements of some of the more notable members of

---

[137] In 1919, according to Ferber (1956), pp.87–88, at the 21 German universities and nine higher technical schools there were 92 *Privatdozenten* in chemistry, 35 of whom were assistants, whereas for physiological chemistry the total was four, two of whom were assistants.

Hofmeister's research group by placing them in three categories: those whose principal scientific contributions were in the biochemical sciences, those who specialized in closely related disciplines (animal physiology, plant physiology, pharmacology) and those who became clinical investigators.

Among the biochemists, Gustav Embden is especially outstanding. He received his Dr.med. degree in Strassburg in 1899 and during the succeeding four years largely worked in Hofmeister's institute. There he investigated several biochemical problems before choosing for his lifework the study of intermediary metabolism. In 1904 he moved to a hospital laboratory in Frankfurt am Main; ten years later he was appointed professor of vegetative physiology at the newly created university in that city, and remained there until his untimely death in 1933, when he was subjected to harassment by Nazi students. Embden was a pioneer in the use of perfusion techniques for the study of chemical changes in mammalian organs, and applied these methods fruitfully to the metabolic transformation of amino acids, fatty acids and sugars. During the 1920s, Embden's studies on carbohydrate metabolism in muscle led him to conclude that the formation of lactic acid occurs largely after contraction. This view challenged the prevailing opinion, based on the work of Archibald Vivian Hill and of Otto Meyerhof (for which they received the 1922 Nobel Prize in Physiology or Medicine), that the formation of lactic acid is the energy-yielding process in muscular contraction. It was not until after the "revolution in muscle physiology" of the late 1920s that the validity of Embden's claim was widely recognized. The high point of Embden's efforts in the field of carbohydrate metabolism came in the final year of his life. He concluded that, in anaerobic glycolysis, fructose-1,6-diphosphate is cleaved into glyceraldehyde-3-phosphate and dihydroxyacetone phosphate, and he proposed a scheme for the conversion of glucose to lactic acid in muscle; this scheme was confirmed by Meyerhof in succeeding years. The establishment of the so-called Embden-Meyerhof pathway of anaerobic glycolysis represents one of the great successes of modern biochemistry. In Frankfurt Embden led a sizable research group (about 110 persons), among whom was the noted biochemist Gerhard Schmidt [1901–1981].[138]

Franz Knoop, one of Embden's contemporaries at Strassburg, obtained his Dr.med. degree in 1900 at Freiburg i.B.

---

[138] For Embden, see Deuticke (1933), Lipmann (1975), Nachmansohn (1979), pp.327–336 and Cori (1983); for Schmidt, see Kalckar (1987). For an account of the revolution in muscle physiology, see Hill (1932) and Fruton (1972), pp.364–368.

There, he came under the influence of the outstanding organic chemist Heinrich Kiliani, who encouraged youthful talent; another of his protégés was Adolf Windaus. Knoop then spent three years at Hofmeister's institute, where he performed the experiments that led him to propose that the metabolic breakdown of fatty acids involves the removal of two-carbon units through oxidation at the beta-carbon. In this work, Knoop fed dogs the phenyl derivatives of fatty acids and isolated from their urine the phenyl compounds formed upon the metabolic degradation of the artificial fatty acids. This application of synthetic organic chemistry to a problem that Pohl had explored in Hofmeister's laboratory in Prague provided one of the early examples of the use of the labeling technique to the study of intermediary metabolism. The dual influence of Kiliani and Hofmeister thus brought into physiological chemistry a well-trained organic chemist who also had acquired an appreciation of the physiological methods of his time. In 1904 Knoop returned to Freiburg where he later became professor of physiological chemistry, and after 1927 he held the same appointment at Tübingen. At both these universities he led a relatively small research group in important work on the metabolism of amino acids and of pyruvate; the aerobic oxidation of the latter metabolite was the subject of especially lively interest during 1920–1935. In particular, serious attention was given to a cyclic pathway via four-carbon dicarboxylic acids (succinate, fumarate, malate, oxaloacetate), based largely on the results of Knoop, Thorsten Thunberg and Albert Szent-Györgyi, but this scheme had to be modified when Knoop and his associate Carl Martius showed in 1936 that a six-carbon tricarboxylic acid (citrate) is also a possible intermediate in aerobic oxidation of pyruvate. The solution of the problem came in the following year through the decisive experiments of Hans Adolf Krebs [1900–1981] to demonstrate the operation of what has come to be known as the Krebs citric acid cycle.[139]

Another junior member of the Hofmeister group who made outstanding contributions to the study of intermediary metabolism was Jacob Karol Parnas, who was born in Poland (then under Russian rule) and in 1908 obtained his Dr.phil. in organic chemistry for the work he did with Richard Willstätter. Parnas then went to Hofmeister's institute, where he worked on cephalin and aldehyde mutase before embarking on a lifelong investigation of carbohydrate metabolism in muscle. Parnas remained

[139] For Knoop, see Ohlmeyer (1948) and Thomas (1948). For Krebs, see Holmes (1984) and Kornberg and Williamson (1984).

in Strassburg until 1916, when he returned to Poland; from 1920 to 1941 he was professor of physiological chemistry in Lwów. There he established an important biochemical research school, whose achievements included the elucidation of the phenomenon of ammonia formation during muscular contraction and the discovery of the enzymatic phosphorolysis of muscle glycogen. The process of anaerobic glycolysis in muscle is often termed the Embden-Meyerhof-Parnas pathway.[140]

To the three important biochemists mentioned above must be added Ernst Josef Friedmann, who received both a Dr.phil. (1902) and a Dr.med. (1905) for work he did at Strassburg. Apart from his chemical contributions in the study of the constitution of cystine and of adrenaline, Friedmann also did valuable experimental work in the field of intermediary metabolism. In 1907 he returned to his native Berlin, where he became head of the chemical laboratory of the Charité hospital; there, his research included a distinguished series of studies on the metabolic formation of ketone bodies. Friedmann left Germany in 1930 and, after brief stays in Moscow and Basle, settled in Cambridge, England, where he worked on several biochemical topics, including the development of methods for labeling organic compounds with radioactive halogens.[141]

Hofmeister's personal research in Prague dealt largely with proteins, and several of his more notable associates later made important contributions in this field. As was noted earlier in this chapter, in Strassburg Karl Spiro took up the study of the physical chemistry of proteins and was Hofmeister's most valuable colleague both in the formal chemical instruction of medical students and in the supervision of their doctoral research on proteins. In his later professorship of physiological chemistry in Basle, Spiro continued work in this field, but also published papers on a variety of other biochemical topics. A man of liberal spirit, Spiro followed Hofmeister's example in encouraging the independent efforts of his junior associates.[142]

Hofmeister's interest in the chemistry of proteins is also reflected in the later research of two men who worked with him in Prague. One of them was Wolfgang Pauli, who worked for his Dr.med. dissertation (1892) in Hofmeister's laboratory and, after a few years of clinical service, turned to the study of the physical chemistry of proteins. Pauli was among the first to follow up the

[140] For Parnas, see Mochnacka (1956), Heller and Mozolowski (1958), Ostrowska (1980) and Lutwak-Mann and Mann (1981).
[141] For Friedmann, see note 63, this chapter.
[142] For Spiro, see note 57, this chapter.

observations of William Bate Hardy [1864–1934] on the movement of proteins in an electric field. During the course of his
extensive studies, Pauli devised some of the early apparatus for
the measurement of the electrophoretic mobility of macromolecules. Through his many writings he became one of the leading
advocates of the colloid-chemical approach to the study of proteins. In 1922 Pauli was appointed professor of medical colloid
chemistry at Vienna, but was obliged to leave after the Nazis occupied Austria. The other Hofmeister pupil, also an Austrian,
who entered the protein field was Otto von Fürth, who came to
Prague after his medical studies in Vienna, and accompanied
Hofmeister to Strassburg. During his association with Hofmeister until 1905, when he returned to Vienna, Fürth began work
on muscle proteins, a subject he continued to pursue intensively
in later years, and clarified some of the uncertainties arising
from the earlier studies of Kühne, discussed in Chapter Three.
As was noted above, Fürth also worked at Strassburg on adrenaline; his interest in this substance stemmed from his previous
work with Hofmeister on the chemical reactions of the tyrosine
units of proteins. Later he studied the enzyme tyrosinase and its
role in formation of melanins.[143]

The problem of the mode of tyrosinase action also attracted
the Englishman Henry Stanley Raper, who began scientific work
with Julius Berend Cohen in Leeds and then joined John Beresford Leathes [1864–1956] in studies on fat metabolism at the
Lister Institute in London. Leathes encouraged Raper to spend
a year (1905–1906) in Hofmeister's laboratory, and he worked
there on the separation of the enzymatic cleavage products of
proteins. In 1910 Raper obtained a D.Sc. at Leeds and also qual-
ified for medical practice and, after several junior appointments
in Canada and in England, he became professor of physiology at
Manchester, where he conducted a sustained program of research on tyrosinase.[144]

Mention has been before of Lawrence J. Henderson's association with Hofmeister and Spiro. As professor of biological
chemistry at Harvard, Henderson performed outstanding studies on acid-base balance in biological fluids and he wrote several
books (*The Fitness of the Environment, The Order of Nature* and
Blood: A Study in General Physiology) that influenced contemporary scientific thought. Moreover, after the First World War,

---

[143] For Pauli, see Valko (1957) and Schwarzacher (1958); for Fürth, see Lieben (1948)
and note 56, this chapter.
[144] For Raper, see Kendal (1952), Hartley (1953) and Folley (1955).

Henderson headed the newly established Department of Physical Chemistry at the Harvard Medical School, but soon turned over the leadership of this group to Edwin Joseph Cohn [1892–1953].[145]

In addition to the above Hofmeister pupils who may be considered to have become biochemists, several others, as professors of pharmacology, conducted research programs that included work on biochemical problems. The most famous of these men was Otto Loewi, whose initial work after leaving Hofmeister's laboratory dealt with problems of metabolism. As was noted in a previous chapter, in 1902 Loewi showed that an extensively digested (biuret-free) autolysate of pancreatic protein can replace intact protein in the animal diet. This finding, together with Otto Cohnheim's discovery of erepsin, provided decisive evidence against the peptone theory of protein synthesis. After ten years with the noted pharmacologist Hans Horst Meyer [1853–1939] at Marburg and Vienna, in 1908 Loewi became professor of pharmacology at Graz, where he discovered the role of the *Vagusstoff* (later identified as acetyl choline) in the chemical mediation of nerve impulses. Henry Dale [1875–1968], with whom Loewi shared the 1936 Nobel Prize in Physiology or Medicine, later wrote of Loewi's association during 1896–1897 with Hofmeister,

whose Department was already one of the centres of the effective emergence of the then new discipline of biochemistry. And if one looks at the list of all the scientific papers which Loewi was to publish, during the four decades of his full activity in pharmacological teaching and research, and notes how often, and how recurrently, they were concerned with problems of nutrition and metabolism, it is easy to form the impression that he might well have chosen biochemistry as the subject of his life's work, if he had earlier had the opportunity, and the impulse, to obtain a stronger educational background in chemistry, and in organic chemistry especially.[146]

Loewi remained in Graz until 1938, when the Nazi occupation obliged him to leave Austria, and he moved to New York City.

[145] Of the many biographical articles about Henderson, the one by Cannon (1943) is particularly valuable. For a history of the Harvard Medical School laboratory of physical chemistry, see Edsall (1950).

[146] Dale (1963), p.2. For Loewi, see also Dale (1962), Lembeck and Giere (1968) and note 62, this chapter.

One of Loewi's contemporaries in Strassburg was Alexander Ellinger, an organic chemist who had received his Dr.phil. degree in Berlin for work with Willy Marckwald [1864–1942]. He then pursued medical studies in Munich and Strassburg, and worked with Spiro in Hofmeister's laboratory on peptones. In 1897 Ellinger went to Max Jaffé [1841–1911], the noted biochemist who was professor of pharmacology at Königsberg, where Ellinger obtained a Dr.med. degree. After Jaffé's death, Ellinger was appointed his successor, but in 1914 he moved to the newly created university in Frankfurt am Main. Before his untimely death in 1923, Ellinger conducted outstanding biochemical research on the chemistry and metabolism of several amino acids, especially tryptophan, and on various pharmacological problems. Like other pupils of Hofmeister, Ellinger encouraged the independent efforts of his junior associates; among them was the noted pharmacologist Werner Lipschitz [1892–1948].[147]

Another Hofmeister pupil who later achieved distinction in pharmacology was Ernst Peter Pick, who came to Strassburg after obtaining his Dr.med. degree (1896) in Prague. During the succeeding three years he used the salting-out method in efforts to separate the products of the cleavage of proteins by pepsin. Pick then went to the Austrian Serum Institute in Vienna where, in 1906, he and Friedrich Obermayer [1861–1925] reported the important discovery that chemical modification of a protein can produce antigens which elicit the formation of antibodies specific for the modified protein. In 1911 Pick joined Hans Horst Meyer, and succeeded him as professor of pharmacology in 1924. As in the case of many other Austrian students of Hofmeister, Pick had to leave Vienna in 1938; he came to the United States and continued pharmacological work at the Merck Institute for Therapeutic Research.[148]

To the above three pharmacologists must be added Walter Siegfried Loewe who worked with Hofmeister for his Dr.med. degree in Strassburg on the cleavage of proteins by pepsin. After an extended stay (1912–1921) as an assistant of Wolfgang Heubner [1877–1957] in Göttingen, Loewe became professor of pharmacology in Dorpat (Tartu). He moved to Mannheim in 1928, but was forced to leave Germany five years later; during the remainder of his life he worked in the United States at Cor-

---

[147] For Ellinger, see note 61, this chapter; for Lifschitz, see Laubender (1949).
[148] For Pick, see Molitor (1961).

nell Medical College and the University of Utah. Loewe's important contributions included studies on sex hormones and the active ingredients of hashish.[149]

Three other Hofmeister pupils also became noted professors of pharmacology. Julius Pohl had been a student and assistant of Hofmeister before succeeding him in Prague, where he did important work on the metabolism of alcohols, the action of bromoacetic acid on muscle and the toxicity of chloroform; from these studies Pohl drew conclusions that were accepted only many years later. From 1911 until 1928 he was professor of pharmacology in Breslau, where his research output slackened considerably. The second was Hermann Wieland, a brother of the chemist Heinrich Wieland, who worked with Hofmeister for his Dr.med. degree (1909) on the lipids of intestinal mucosa. After the First World War, Hermann Wieland became successively professor of pharmacology at Königsberg and at Heidelberg, but his career was cut short by his death in 1929. The third was the Belgian Edgard Zunz, who worked with Hofmeister in 1899 on the action of pepsin and (according to his biographer) "received from him the ideas that became the basis of his scientific orientation." Zunz was professor of pharmacology at Brussels from 1919 until 1939, and published many papers on blood coagulation, anaphylaxis and other biochemical and pharmacological topics.[150]

As was noted above, the biochemists Embden and Raper held professorships of physiology, and to these men should be added Albrecht Bethe. Although his association with the Hofmeister group was in one sense brief, through his work in 1905 on the staining of tissues, in another sense it was a prolonged and close one, because of his leading role in bringing together the younger members of various departments in the Strassburg medical faculty in a journal club that met frequently. Bethe received his Dr.med. degree (1898) in Strassburg for a study of the histology of nerve cells and, in succeeding years, won renown for his investigations on nerve conduction, the permeability of tissue membranes and the periodicity of physiological processes. He left Strassburg in 1911 to become professor of physiology at Kiel, and four years later he moved to Frankfurt, where he joined his former Strassburg colleagues Embden and Ellinger.[151]

[149] For Loewe, see Voss (1959, 1964) and Kattermann (1964).
[150] For Pohl, see Riesser (1931); for Hermann Wieland, see Oppenheimer (1929) and Diepgen (1960), pp.99–102. For Zunz, see La Barre (1939).
[151] For Bethe, see Thauer (1955).

Among the many Hofmeister pupils who became professors of internal medicine and reflected his influence in their clinical investigations were Friedrich Kraus and Gustav von Bergmann. Kraus became an assistant in Hofmeister's Prague institute after receiving his Dr.med. degree in 1882. After serving as professor of medicine in Graz, Kraus was appointed to the prestigious chair at the Charité hospital in Berlin, where he established a productive research school; his studies dealt largely with metabolic and cardiovascular disorders. In 1927, Kraus was succeeded by Bergmann, who had worked in Hofmeister's Strassburg laboratory for his Dr.med. degree (1903) on the metabolic conversion of cystine to taurine. According to his biographer Katsch, during his medical studies Bergmann had been "especially stimulated and influenced by Hofmeister." Bergmann then joined Kraus and, after holding posts in Hamburg, Marburg and Frankfurt, returned to Berlin, where he continued to make significant contributions to the study of metabolic dysfunctions. Another Hofmeister pupil who conducted research at the Charité hospital was Adolf Magnus-Levy, who had obtained Dr.med. and Dr.phil. degrees before coming to Strassburg, and in 1895 had discovered the relation of the thyroid to the respiratory exchange (basal metabolic rate). In Bernhard Naunyn's Strassburg clinic, Magnus-Levy also elucidated the cause of diabetic acidosis and found time to work in Hofmeister's laboratory on the crystallization of the Bence-Jones protein by the salting-out method. In 1901, Magnus-Levy moved to Berlin; after having been arrested by the Nazis, he succeeded in coming to the United States in 1940.[152]

As may be seen in Appendix 7, Hofmeister's progeny included other persons who became professors of internal medicine. Two of them, Wilhelm Stepp and Léon Blum, also merit special mention. Stepp came to Strassburg after his Dr.med. degree (1907) in Munich, at a time when Hofmeister was interested in the study of what he called "accessory nutrients," and Stepp's demonstration in 1909 of the nutritional requirement for a fat-soluble factor was one of the important events at the beginning of the vitamin era. Stepp later was professor of medicine at Jena, Breslau and Munich, where he conducted an active program of clinical research in the vitamin field. Blum was one

---

[152] For Kraus, see Bergmann (1936); for Bergmann, see Katsch (1955); for Magnus-Levy, see Zadek (1955) and Goldner (1955). In view of the role that Kraus and Bergmann played in the activities of the Charité hospital, it may be noted that, before 1933, this hospital (and the Moabit hospital) were important centers of biochemical research in Berlin; see Sauer et al. (1961).

of the several Frenchmen whom Hofmeister encouraged during the German occupation of Alsace-Lorraine. His Dr.med. dissertation (1901) dealt with the nutritional quality of albumoses and he also worked on the metabolism of cystine and of tyrosine. According to Spiro, "I do not remember any other medical student who performed quantitative analyses with such eagerness and such patience . . . [and] he later trained his students in the same exact methodology. . . . He was also a good synthetic chemist; for example, for his studies on alcaptonuria he synthesized for the first time homologues of tyrosine." Upon the reversion of Strasbourg to French rule, Blum became professor of medicine there and conducted a fruitful program of research on clinical aspects of mineral metabolism.[153] Other future professors of internal medicine who derived an impulse for their later clinical investigations from their association with Hofmeister were Georg Haas, Adolf Oswald, Paul Morawitz, Otto Porges and Friedrich Umber.[154]

Hofmeister's pupils in Prague and Strassburg also included several men who later became noted professors of pediatrics. The first of them was Adalbert Czerny, who received his Dr.med. at Prague (1888) and later held professorships in Breslau, Strassburg and (after 1913) Berlin. He made important contributions to the study of the clinical aspects of digestion and metabolism in children. Another student at Prague was Josef Langer, who worked in Hofmeister's laboratory on bee venom, and later became professor of pediatrics at Prague. Among the future pediatricians in Hofmeister's Strassburg laboratory was Leo Langstein, whose work there earned him a Dr.phil. degree (1902) in addition to his earlier Dr.med. degree from Vienna. Despite his promising start in physiological chemistry, Langstein decided to specialize in pediatrics, and in 1911 he became the head of the Kaiserin Auguste Victoria Haus, a state institute for the study and prevention of tuberculosis in children. Of Jewish parentage, Langstein chose to commit suicide in 1933. Another Hofmeister student at Strassburg was Karl Stolte, who received his Dr.med. degree there and remained as an assistant in physiological chemistry (1906–1909) before joining Czerny in the children's clinic. Stolte later held professorships in pediatrics at Breslau, Greifswald and Rostock. And to these men should be added Ernst Freudenberg, who spent a year (1912–1913) with Hofmeister after his Dr.med. degree in Munich, and worked on the metabo-

---

[153] For Stepp, see Kühnau (1964). For Blum, see Spiro (1930); the quotation is on p.379.
[154] For Haas, see Benedum (1982); for Oswald, see Grumbach (1956); for Morawitz, see Krehl (1936); for Porges, see Novak (1968); for Umber, see Störring (1941).

lism of fats. After military service during the First World War he
became professor of pediatrics at Marburg, but was dismissed in
1933 and moved to Basle, where he held the same post until
1954. Freudenberg's principal contributions were in the field of
metabolic disorders in children.[155] Apart from the clinicians
mentioned in the above paragraphs, Hofmeister's pupils also included the photobiologist Walter Hausmann (who had worked
in Strassburg on the analysis of proteins) and the pathologist
Hans Eppinger (see Appendix 7).

Finally, no list of Hofmeister's scientific progeny would be
complete without the inclusion of the botanist Friedrich Czapek,
who worked in Hofmeister's Prague laboratory. After receiving
his Dr.med, in 1892, Czapek began the study of plant physiology
with Wilhelm Pfeffer [1845–1920] in Leipzig, and in 1894 obtained his Dr.phil. in Vienna. Czapek later became professor of
botany in Prague and, shortly before his death, had been selected to be Pfeffer's successor. Czapek's many contributions to
plant biochemistry and physiology, as well as his book *Die Biochemie der Pflanzen* (dedicated to Hofmeister and Pfeffer), gave
evidence of his wide-ranging biochemical interests.[156]

Conclusion

During the period of the activity of their research groups, span-
ning the same four decades, Fischer had more than twice as
many junior associates than did Hofmeister. An examination of
the later scientific productivity of the members of the two
groups indicates that this numerical disparity is reduced if one
takes account of the fact that at least 55 per cent of the Fischer
people disappeared from the scientific, medical or technical literature, whereas the comparable figure for the Hofmeister
group is about 25 per cent. Such numbers, however, can only
serve as a background for the comparison of the influence that
Fischer and Hofmeister, each in his own way, exerted on the
later development of the chemical and biochemical sciences, not
as measures of their influence. Although Fischer's personal
achievements unquestionably outshone those of Hofmeister in
the field of their common scientific interest, the biographical

---

[155] For Czerny, see Schiff (1956) and Opitz (1963); for Langer, see Epstein (1937); for
Langstein, see Rietschel (1933); for Stolte, see Ocklitz (1951); for Freudenberg, see
Buchs (1977).
[156] For Czapek, see Boresch (1922).

sketches in the preceding two sections of this chapter suggest that the educational contributions of the two men to the advancement of the biochemical sciences were more nearly equal than is implied by the relative place accorded to them in the history of these sciences. There can be no doubt that, at the turn of the century, organic chemistry had a much more distinctive and highly specialized methodology than did physiological chemistry. This made for a more disciplined training of students, and tended to encourage great chemists like Fischer to direct the work of their junior associates rather autocratically. On the other hand, in an emergent field such as biochemistry, whose problems came largely from mammalian physiology, the methods and concepts of organic chemistry, while indispensable, were not sufficient, and required experimental approaches derived from biology and from other branches of chemistry. Thus, apart from the differences in the personal qualities of Fischer and Hofmeister, and in the institutional status of their respective disciplines, it would seem that, for the later development of the biochemical sciences, Hofmeister's liberal style of leadership was the more appropriate one.

Chapter Six

Modern Research Groups in the Biochemical Sciences

What relevance, if any, do the case histories recounted in the preceding chapters, and drawn from the period 1830–1914, have to the more recent development of the biochemical sciences? The accretion of knowledge in these sciences during the twentieth century has totally transformed the field championed by Hoppe-Seyler and Hofmeister, and the organic chemistry of today is rather different in conceptual structure and methodology from that of Baeyer and Fischer. Indeed, owing to the impact of biochemical discovery on medicine and agriculture, at mid-century the institutional status of biochemists began to match that of the chemists who provided manpower and new products to industrial firms. As a consequence, the large groups, led by senior scientists, in the organic-chemical institutes of pre-World War I Germany, became prominent features of biochemical research in other countries, especially the United States. Thus, although the body of knowledge that passed for either organic chemistry or biochemistry had undergone great change, the institutional forms of the education of future investigators in the biochemical sciences began to resemble those in Wilhelmine Germany, with the important difference that the infusion of sizable funds from governmental agencies, private foundations and industrial firms made it possible to accommodate several relatively large research groups in a single university department. The question posed at the beginning of this paragraph may therefore be rephrased: To what extent are the contrasts in scientific style, such as those of Hoppe-Seyler and Kühne, or of Fischer and Hofmeister, reflected in the leadership of more recent successful research groups in the biochemical sciences? In particular, have there been significant changes in such matters as the interaction of the group leader with his junior associates, his role in the education of the next generation of productive scientists, and the contributions of his students and assistants to the achievements for which he was given credit?

I hope that future historians will find these, and related, questions worthy of close examination. Regrettably, with several notable exceptions,[1] such studies do not appear to have attracted much interest among historians of the biochemical sciences. More attention has been given to the work and thought of individual scientists in an effort to discern the sources of their success. That this is an important historical enterprise cannot be doubted, for much mythology about "flashes of insight" has permeated the biographical, and especially the autobiographical, literature about famous scientists.[2] In writing about his successful career and of the people who participated in the work that brought him fame, too infrequently has a prominent leader of a large research group in the chemical or medical sciences given more than passing credit to the contributions of his junior associates, thus implying that the fruitful ideas came from his brain alone and that his students and assistants merely provided the hands needed to establish the validity of these ideas. Such an inference may be justified in the case of investigators who themselves worked in the laboratory, perhaps with the help of one or two assistants, but is open to question in the case of the leaders of relatively large research groups.

On the other hand, as I have noted in the first chapter, scientific research groups have been actively studied by sociologists, and a recent report[3] is of special interest in relation to the theme of this book. In particular, from an examination of Swedish academic groups active during the 1970s, it was concluded that the optimum group size is about five to seven scientists, and that when groups grow larger the quality of their performance suffers more than the quantity of their output of publications. Moreover, the data appear to confirm the conventional wisdom that the personal qualities of the leader, as well as his previous research experience, are decisive factors in the productivity of an academic group in the experimental sciences. Reports of

---

[1] See Morrell (1972), Geison (1978), Snelders (1984) and Klosterman (1985).
[2] Holmes (1986a).
[3] Andrews (1979); see especially pp.191–221.

such statistical studies on contemporary research groups, which of necessity cannot identify the group leaders and their associates by name, may be valuable as guides to governmental or institutional policy makers, but numbers alone do not illuminate sufficiently the special features of group research. Although it must be left to future historians of science to examine in detail important recent and contemporary research groups in the chemical and medical sciences, I believe that much more can be learned from the study of other groups, especially those active during the period 1860–1960, that were only mentioned briefly in the preceding chapters of this book.

Other Organic-chemical Research Groups

In focusing attention on the groups associated with Adolf Baeyer and Emil Fischer, it was not my intention to denigrate the role of their German contemporaries in the education of future leaders in chemical research. Despite less elegant facilities than those in Munich and Berlin, other university chemical institutes also attracted many promising students from German-speaking countries and from abroad during the period 1870–1914. Although much has been written about August Wilhelm Hofmann, his research group in Berlin (1865–1892) merits more detailed study, for his students there included many important figures in the later development of organic chemistry, such as Karl Friedrich von Auwers [1863–1939], Eugen Bamberger and Siegmund Gabriel. The same may be said of the research groups of Hermann Kolbe in Marburg (1851–1865) and Leipzig (1865–1884) and of Victor Meyer in Zurich (1872–1885) and Heidelberg (1888–1897). Still another research group of that time which deserves closer study is that of Kolbe's devoted pupil Adolf Claus [1838–1900] at Freiburg im Breisgau. Of particular interest is Rudolf Fittig [1835–1910], successively professor at Göttingen (1866–1870), Tübingen (1870–1876) and Strassburg (1876–1902). During his period of active research Fittig's students (approximately 200 persons) made many valuable contributions to the study of aromatic compounds, and included 17 Dr. phil. recipients from Great Britain (including William Ramsay) and 17 from the United States (including Ira Remsen). Indeed, Remsen occupies a special place in the development of American graduate education in chemistry, for he established the first large research group in the United States.

Between 1879 and 1913, of the 130 Ph.D.s in organic chemistry at Johns Hopkins, 105 were awarded to Remsen's students, who included William Albert Noyes and Elmer Peter Kohler, as well as the future biochemist Walter Jones.[4]

In all these laboratories, the emphasis was on the experimental elaboration of the conceptual structure of organic chemistry, on the development of new or improved methods for the synthesis of organic compounds (especially those of practical value as dyes or drugs) and on the application of the new concepts and methods to the elucidation of the structure of organic substances obtained from natural sources. The research of students, assistants and guests was largely directed along restricted lines determined by the head of the laboratory, although some group leaders (notably Hofmann) appear to have been somewhat more liberal in that regard. The ties of most of the individual leaders to the growing chemical industry were rather close, and it would seem that the majority of the students in all these groups later joined chemical firms or became commercial chemists. These general features of the organization of academic research in organic chemistry have persisted into the twentieth century. Of the many German organic-chemical groups whose closer study might further illustrate both continuity and individuality in styles of leadership, I offer as examples the ones associated with Theodor Curtius and with Ludwig Knorr.[5]

After his *Habilitation* (1886) in Erlangen, Curtius was *ordentlicher Professor* at Kiel (1890–1897), Bonn (1897–1898) and Heidelberg (1898–1924). During this period approximately 200 Dr.phil. candidates worked with him almost exclusively on the chemistry of organic hydrazides and azides. Although Curtius's pupils learned the importance of clean and careful laboratory technique (many of the azides are explosive), the limited scope of his research program may have deprived them of the opportunity to acquire a broader chemical education. It should be noted, however, that at Heidelberg, where there was a large enrollment in chemistry courses, Curtius had a sizable staff of teaching assistants, some of whom (notably Emil Knoevenagel [1865–1921]) directed the Dr.phil. work of a small number of students along other lines of research. Moreover, in 1901, Curtius set up in his institute a section of physical chemistry, with Georg Bredig [1868–1944] as its head; among Bredig's Dr.phil. pupils

---

[4] For Hofmann, see Chapter Two, note 91; for Meyer, see Chapter Four, note 12. For Claus, see Vis (1900); for Fittig, see Fichter (1911) and Remsen (1911); for Remsen, see Noyes and Norris (1932), Getman (1940) and Hannaway (1976).

[5] For Curtius, see Chapter Four, note 74; for Knorr, see Chapter Five, note 125.

were such later notables as Kasimir Fajans [1887–1975] and James William McBain [1882–1953].[6] Of the total of about 230 research students, 35 came from non-German-speaking countries (14 from Great Britain and nine from the United States). Curtius's own scientific progeny do not appear to have contributed significantly to chemical education for nearly all of his Dr.phil. students later joined industrial firms or became commercial or consulting chemists. Among them were the scions of prominent German industrial families Gustav von Brüning [1864–1913], Karl Otto Leverkus [1883–1957] and Heinrich Thyssen [1875–1947].[7]

Knorr's association with Emil Fischer, discussed in the previous chapter, was marked by Knorr's preparation of the pyrazolone *Antipyrine,* and the beginning of a thirty-year long collaboration with the Hoechst chemical firm. Although he remained in academic life, and in 1889 went from Fischer's institute to Jena as *ordentlicher Professor* of chemistry, Knorr's research largely continued along lines that were derived from or directed to problems of pharmaceutical interest. While still in Würzburg, Knorr had six Dr.phil. students, all of whom worked on pyrazoles and other heterocyclic compounds, and he also began studies on the structure of morphine. These two fields were the main parts of his research program until 1912, when he ceased to publish scientific papers. At Jena, Knorr's research group included about 150 persons, with a sizable contingent (22) from Great Britain and two from the United States; in addition, about ten students received their Dr.phil. degree for work with Paul Rabe [1869–1952], one of Knorr's chief teaching assistants. Several of Knorr's junior associates achieved distinction (Fritz Haber, Hermann O. L. Fischer and Friedrich Stolz have already been mentioned). As in the case of the Curtius group, however, except for a few other men who later enjoyed modest academic careers, the great majority joined chemical firms or became chemical consultants.[8]

---

[6] For Knoevenagel, see Jacobson (1921); for Bredig, see Kuhn (1952); for Fajans, see Schwab (1976) and Hurwic (1985); for McBain, see Rideal (1953). Both Bredig and Fajans were obliged to leave Germany during the 1930s, and came to the United States.

[7] Thyssen, who received his Dr.phil. (1899) at Heidelberg, was a famous art collector. In this connection, it may be noted that another of Curtius's pupils, Hermann Hille [1871–1962], went to the United States after his Dr.phil. (1900) at Heidelberg, and became a partner of Albert Coombs Barnes [1872–1951] in the manufacture of Argyrol; the profits made possible the acquisition of the remarkable Barnes art collection.

[8] During the course of my search for biographical data about Knorr's British student Frank Bernard Dehn [1886–1964], I learned from his son, Mr. Stanley G. Dehn (letter dated 23 March 1983) that his father's first employment after his doctorate was as a

After the First World War, the German university institutes and chemical firms specializing in organic chemistry were obliged to cede their position of world leadership to those in Britain, Switzerland and the United States. As was indicated in the previous two chapters, the scientific progeny of Baeyer and Fischer, in the persons of men such as Heinrich Wieland, were still prominent in organic-chemical research and education, but the eminence of their research groups was being matched in these other countries. Moreover, although the establishment of I. G. Farben served to maintain some of the prewar strength of the German chemical industry, its dominance was increasingly challenged by firms such as Imperial Chemical Industries, the group of chemical firms in Basle, and the DuPont company. The symbiosis between research in university laboratories and the growth of industrial chemistry, evident in Wilhelmine Germany, thus became, after 1920, a more international feature of the development of organic chemistry. Among the leading figures in that development were men such as Robert Robinson [1886–1975] in Manchester and Oxford, Leopold Ruzicka [1887–1976] in Zurich and Roger Adams [1889–1971] in Urbana.[9]

Robert Robinson began chemical research at Manchester during 1902–1912 with Baeyer's pupil William Henry Perkin, Jr. His first success was in the study of the structure of brazilin, and Robinson's lifelong interest in natural pigments may be traced to this early achievement. After holding professorships in organic chemistry at Sydney (1913–1915), Liverpool (1916–1919) and St. Andrews (1920–1922), as well employment by the British Dyestuffs Corporation (1919–1920), Robinson became professor at Manchester (1922–1928) and University College London (1928–1930) before succeeding Perkin at Oxford. During the 25 years of Robinson's professorship there, the Dyson Perrins Laboratory became a world center of organic chemistry. His research program reflected both adherence to the experimental tradition he had learned from Perkin and the application of the then-new electronic theory of valency to the theoretical approach he had learned from Lapworth. In the synthesis of natural products, Robinson's remarkable ability to discern from a structural for-

---

works chemist in a cement factory, and that he later established a successful firm of chartered patent agents. Mr. Dehn also noted that "I have no recollection of my father ever mentioning Knorr."

[9] For Robinson, see Robinson (1976) and Todd and Cornforth (1976). For Ruzicka, see his autobiographical article (1973) and Prelog and Jeger (1980, 1983). For Adams, see Tarbell and Tarbell (1981, 1982); the first-named source is especially valuable in providing an appendix listing the names and later occupations of Adams's Ph.D. and postdoctoral students, but regrettably their dates are not given.

mula a feasible route involving relatively simple reactants became evident in 1917 with his synthesis of tropinone, and his
later work on various alkaloids, anthocyanins and steroids gave
further proof of his keen chemical insight. His use of the electronic theory of valency to suggest a revision of Thiele's interpretation of the behavior of unsaturated organic compounds
and for the design of routes of synthesis was extremely fruitful,
but brought him into controversy with Christopher Kelk Ingold
[1893–1970].[10] Robinson also offered speculations about the
chemical pathways in the biosynthesis of natural products. Like
many organic chemists before him, he suggested the occurrence
of intermediate biochemical reactions analogous to those found
to be effective in the chemical laboratory. Although several of
his speculations, especially in the case of some alkaloids, were
later shown to be correct, others were disproved in biochemical
studies involving the use of isotopes or of mutant strains of
micro-organisms.[11] Robinson was the author or co-author of
over 600 research articles, most of them from his laboratories in
Manchester and Oxford. Many of his early papers were published jointly with Perkin, and he later collaborated with other
senior colleagues. From an examination of the list of his publications, it would appear that during the sixty-year period of his
independent research, Robinson had over 250 junior associates;
one of the most valued of them was his first wife Gertrude
Maud (Walsh) Robinson [1886–1954]. Among his scientific progeny are such notable chemists as John Warcup Cornforth [b.
1917], John Masson Gulland [1898–1947] and Alexander Rober-
tus Todd [now Lord Todd, b. 1907]. Robinson's research group
included many doctoral and postdoctoral students from over-
seas, especially from India; one of them was Tiruvenkata Rajen-
dra Seshadri [1900–1975]. According to his biographers,

Robinson did not pursue solid experimental studies in the way that his
predecessor Perkin did. He confined himself mainly to exploratory experiments usually conducted in test-tubes or boiling-tubes. . . . Once a
test-tube trial of a reaction indicated that something had occurred, the
follow-up was usually left to a junior colleague. Robinson's visits to in-

---

[10] For Ingold, and his controversy with Robinson, see Shoppee (1972) and Saltzman
(1980).

[11] Robinson's self-assurance about his theories was proverbial. When I first met him in
1948, I ventured the comment that he must have been pleased at the introduction of
the isotope technique into the study of metabolic pathways, so that the validity of his
hypotheses could be tested experimentally. As nearly as I can recall his reply, it was:
"Of course they are correct."

dividual research students, although inspiring, were erratic and depended largely on his somewhat mercurial interest in a particular topic at a given time. So it was that a research student might have spells when he had discussions once or twice a day and others when a week or two could pass without his speaking to Robinson at all.[12]

Robinson maintained his association with British chemical industry, especially Imperial Chemical Industries, to which he assigned several of his patents, and later (1955) he became a director of the Shell Chemical Company, which provided a laboratory after his retirement at Oxford.

Leopold Ruzicka's entry into organic chemistry was a consequence of his association with Hermann Staudinger [1881–1965] at the Karlsruhe Technische Hochschule where Staudinger was an *ausserordentlicher Professor*. A Dr.phil. student of Daniel Vorländer [1867–1941] at Halle, Staudinger had gone to the laboratory of Johannes Thiele in Strassburg, and there had discovered in 1907 the highly reactive ketenes. This achievement marked the beginning of a scientific career that is now chiefly known through the belated recognition of Staudinger's contributions during the 1920s to the establishment of the macromolecular concept of the structure of natural polymers such as rubber and cellulose.[13] In 1910, Ruzicka obtained his doctoral degree for work with Staudinger on the chemistry of ketenes, and when, two years later, Staudinger became professor of chemistry at the Zurich ETH, Ruzicka accompanied him there. It would seem that, in the fashion of the time, Staudinger chose to enter the field of natural product chemistry, and Ruzicka was assigned the task of studying the insecticidal pyrethrins. During the course of this work Ruzicka became interested in the chemistry of the terpenes but, according to his biographers,

Staudinger was an inspiring teacher, and yet at the same time his drive to further his own particular interests made him a hard master. After Ruzicka had informed him that he wanted to work in a field of research of his own he lost his assistantship [in 1916] and found his research facilities to be severely curtailed. Thus he came to regard his mentor with mixed feelings; admiration for his scientific achievements coupled with disappointment at his personal attitude. The outcome of

---

[12] Todd and Cornforth (1976), p.424
[13] For Staudinger, see his autobiography (1961); an English translation appeared in 1970 under the title *From Organic Chemistry to Macromolecules*. See also Yarsley (1967).

his decision to become his own master was that he was forced to turn to the chemical industry to find the financial wherewithal for his livelihood and laboratory expenses. It took some time before he and his partners in industry arrived at a mutually satisfactory framework for cooperation. Later this was to develop into a model for interaction between the Swiss chemical industry and the academic chemical community.[14]

Ruzicka was born in the Croatian region of pre-World War I Austria-Hungary, but in 1917 he obtained Swiss citizenship, and during the succeeding ten years he held unpaid appointments as *Privatdozent* and titular professor at the Zurich ETH. The most notable of his early experimental successes was the demonstration that the musk perfume civetone is a 17-membered ring compound, thus disproving the widely held view that such macrocyclic organic compounds could not exist. This work, and Ruzicka's initial efforts in the terpene field, were aided by financial support from several Swiss chemical firms, especially the Naef perfumery company in Geneva, whose research staff he joined in 1926. In the following year, however, he was appointed professor of organic chemistry at Utrecht, and in 1929 he returned to the Zurich ETH as professor of chemistry. There, with support from major Swiss pharmaceutical firms (especially CIBA and Sandoz), he assembled an outstanding group of colleagues, and made memorable contributions to the elucidation of the structure and to the synthesis of terpenes, steroids and sex hormones. It should be emphasized that these achievements involved mainly chemical operations, frequently with small amounts of material, and demanded exceptional skill; the Zurich laboratory became widely known for the superb training students received in chemical manipulations, especially through the instruction they were given by Ruzicka's colleague Tadeus Reichstein [b. 1897]. Also, in his research, Ruzicka used as a guiding principle the so-called "empirical isoprene rule," according to which the molecules of most of the terpenes could be considered to be derived from the condensation of a number of the 5-carbon isoprene units. Indeed, later biochemical studies, involving the use of the isotope technique, showed that the biosynthesis of steroids from acetate proceeds by the intermediate formation and assembly of isoprenoid compounds. Between 1920 and 1957, Ruzicka had about 250 junior associates, doc-

---

[14] Prelog and Jeger (1980), p.413.

toral students and postdoctoral guests. In addition to Reichstein, they included Vladimir Prelog [b. 1906], as well as the younger men Duilio Arigoni [b. 1928] and Albert Eschenmoser [b. 1925], all of whom achieved great distinction for their organic-chemical contributions.

After receiving his Ph.D. (1912) at Harvard, Roger Adams visited Emil Fischer's institute, where he was associated with Diels, but no publication emerged from his stay in Germany. In 1916, Adams joined William Albert Noyes in the chemistry department at the University of Illinois; ten years later he succeeded Noyes as head of the department, and he served in that capacity until 1954. At Urbana, Adams conducted an active research program for 40 years, and his group totalled about 250 persons. The major problems tackled in his laboratory included the structure of natural products (chaulmoogric acid, gossypol, alkaloids) and the stereochemistry of biphenyl derivatives. As head of the department, Adams gathered about him other professors of organic chemistry, among whom Carl Shipp Marvel [1894–1988] was especially outstanding. These associates conducted independent programs of research, but Adams's group was the largest in the department. Early in his career Adams became a consultant to chemical firms, including Abbott, DuPont and Cola-Cola, and well over half of his students became industrial chemists, some of whom rose to executive positions. The most famous of these men was Wallace Hume Carothers [1896–1937], at DuPont, where he invented nylon. More than any other organic chemist of his time, Adams provided scientific manpower for the growth of research in American chemical industry between the two World Wars. Approximately a quarter of his scientific progeny remained in academic life, and many of them rose to professorships of organic chemistry at American universities; among them were Samuel Marion McElvain [1897–1973] at Wisconsin, John Raven Johnson [1900–1983] at Cornell and Carl Robert Noller [1900–1980] at Stanford. Adams also played a significant, but insufficiently appreciated, role in the development of biochemistry in the United States. In 1922, Noyes had brought to Urbana William Cumming Rose [1887–1985] to head a section of physiological chemistry. Adams's encouragement of Rose's research program and the close collaboration of members of the organic and biochemical groups in the department made possible the education of a generation of biochemists soundly trained in organic chemistry. Moreover, Adams's own students included the later Nobel Prize winner Wendell Meredith Stanley [1904–1971], as well as other produc-

tive biochemists such as Armand James Quick [1894–1978] and
Hubert Orion Calvery [1897–1945]. Although the scientific
achievements of Adams's research group were less distinguished
than those of Robinson or of Ruzicka, between the two World
Wars Adams was unquestionably the leading American organic
chemist not only because of the eminent place his department
had come to occupy in the United States, but also because of his
energy and enthusiasm in promoting the advancement of his
discipline.

During the years between the two World Wars, much of the
organic-chemical research continued along traditional lines, but
increasingly followed the lead of Emil Fischer, Richard Willstätter and Heinrich Wieland in attacking problems that had come
from biochemical studies. These efforts overshadowed the emergence of research groups in what came to be called physical-
organic chemistry. In this development, British chemists such as
Ingold, Joseph Kenyon [1885–1961] and Nevil Sidgwick [1873–
1953] were especially prominent, and in the United States Morris Kharasch [1895–1957] and Louis Plack Hammett [1894–
1987] inspired a generation of chemists to bring the new ideas
derived from atomic physics and physical chemistry into the theoretical structure of organic chemistry.[15] The most brilliant organic chemist of that generation was Robert Burns Woodward
[1917–1979] at Harvard, where he led a research group that included about 400 junior associates, most of them postdoctoral
students. During the course of his scientific work, he contributed to the solution of the chemical structure of many alkaloids
and antibiotics, and his research group achieved the synthesis of
numerous natural products such as quinine and chlorophyll;
Woodward's paper in 1960 on the total synthesis of chlorophyll
listed 17 co-authors. In planning a route of synthesis, Woodward
brought to the task not only a phenomenal knowledge of the
chemical literature, but also new insights into the mechanisms of
organic-chemical reactions. His most notable contribution to
chemical theory was the one he made in collaboration with
Roald Hoffmann [b. 1937] on the conservation of orbital sym-
metry in organic reactions. Moreover, he was among the first organic chemists to use new physical methods, such as infrared
and nuclear magnetic resonance spectroscopy, for the determination of structure; the latter method, together with x-ray crystallography, are now the procedures of choice. Woodward
maintained close ties to industrial firms, at first the Polaroid

---

[15] Saltzman (1986).

Corporation and later CIBA in Basle. I hope that, in time, historians of chemistry will examine Woodward's research group more closely than has been done heretofore, because his style of leadership has left an indelible mark on the development of organic chemistry and of the biochemical sciences during the latter years of this century.[16]

The Impact of Physical Chemistry on the Biochemical Sciences

As was noted in the account of Franz Hofmeister's personal research, one of his major contributions was in the study of the "salting-out" of proteins, and in the application of his findings to their crystallization. The phenomena Hofmeister described were later studied fruitfully by scientists who identified themselves as physical chemists, but it would exaggerate the importance of Hofmeister's studies to include him among the pioneers in that branch of chemistry. During the period 1900–1914, however, many chemically minded physiologists, as well as some organic chemists who, like Hofmeister's associate Karl Spiro, had turned to biochemistry, were strongly influenced by the emergence in Germany of physical chemistry as an independent university discipline. Although the interplay of physics and chemistry had long been a prominent feature of the development of nineteenth-century chemistry, most notably in the achievements of Michael Faraday, it was not until the 1880s that the dominance of organic chemistry, with its firm connection to the burgeoning German chemical industry, was seriously challenged by those who called themselves physical chemists. The leader in their drive for recognition was Wilhelm Ostwald [1853–1932], who had studied at Dorpat (Tartu) with Liebig's pupil Carl Schmidt, and had served as professor of chemistry at Riga from 1881 until 1887, when Ostwald was appointed to the professorship of physical chemistry at Leipzig, then the only chair in this discipline at a German university.[17] While at Riga, Ostwald learned of the work in Stockholm of Svante Arrhenius [1859–1927] on the galvanic conductivity of electrolytes and in Amsterdam of Jacobus Henricus van't Hoff [1852–1911] on the

---

[16] Todd and Cornforth (1981).
[17] References to the extensive biographical literature about Ostwald are collected in Poggendorff (1971), pp.476–482. Ostwald's three-volume autobiography (1926–1927) is an essential source of information about his career and his prejudices. Among the scholarly accounts, that by Hiebert and Körber (1978) is especially valuable.

application of thermodynamics to the study of chemical equilibria. By 1887, Ostwald had established close connections with both men, and had founded a new journal (with van't Hoff as co-editor) entitled *Zeitschrift für physikalische Chemie, Stöchiometrie und Verwandtschaftslehre*.

Of this triumvirate, van't Hoff was clearly the foremost investigator. Despite the derision of Kolbe, the stereochemical concepts that van't Hoff and Le Bel had advanced during the 1870s had become indispensable components of organic-chemical theory. In his laboratory at Amsterdam, where he was a full professor from 1878 until 1896, van't Hoff led a sizable research group (about 50 persons), and during this period he made his principal contributions in physical chemistry, notably his thermodynamic equation for the relationship between the absolute temperature and the equilibrium constant of a chemical reaction, and his use of osmotic pressure data to formulate the theory that the behavior of dilute solutions accords with the general gas law ($PV = iRT$). After Arrhenius's enunciation of his theory of electrolytic dissociation, and Ostwald's application of the law of mass action to the equilibria between the dissociated and undissociated portions of an electrolyte (the so-called dilution law), van't Hoff became an adherent of Arrhenius's theory, and joined what came to known as the "ionist" club. In 1896, van't Hoff moved to Berlin where he held a research appointment in the Prussian Academy of Sciences. Although he attracted many junior colleagues (about 40) there, he is reported to have missed his homeland, and his research in Berlin was less spectacular than it had been in Amsterdam. His principal contribution after 1900 was in the application of Gibbs's phase rule to the study of the deposition of salt deposits; this work brought him into close contact with the German potash industry. In his final years, van't Hoff turned to the investigation of enzyme-catalyzed reactions, and two of his last papers dealt with the ability of hydrolytic enzymes to catalyze the reverse synthetic process.[18]

Most of van't Hoff's students at Amsterdam had been Netherlanders; among them were Ernst Julius Cohen [1869–1944], who became professor of chemistry at Utrecht, and Johannes van Laar [1860–1938]. The foreign guests included, apart from Arrhenius, the American Wilder Dwight Bancroft [1867–1953]. In Berlin, there were Frederick George Donnan [1870–1956]

---

[18] For van't Hoff, see Cohen (1912); this volume contains a list of van't Hoff's publications. See also Jorissen and Reicher (1912), Tenderloo et al. (1952) and Snelders (1984).

from Britain and Hans von Euler [1873–1964] from Sweden.[19] A distinctive feature of van't Hoff's style of research, evident from his inaugural lecture in 1878 on the importance of imagination in science, was that for him "Experiments were an illustration for a theory and tools for testing, not probes for investigating or discovery."[20] Although he could teach his students techniques, for he is reported to have been a meticulous experimenter, in his discussions with younger colleagues he did not tell them that he considered their opinions to be wrong, but rather asked "why do you think so?"[21] Some of his students, especially in Berlin, found the assignments that van't Hoff gave them to be uninspiring, and Euler later reported that he had asked van't Hoff whether he was not interested in more stimulating problems. According to Euler,

He [van't Hoff] maintained that a group of young scientists would possibly follow him in hypothetical paths on account of his authority and his former successes, and that he would have a heavy responsibility as an academic teacher concerning the scientific education of a new generation.[22]

Euler later turned from physical chemistry to the study of biochemical problems.

Svante Arrhenius's doctoral dissertation in 1884 was received less favorably in Stockholm than abroad, and the influence of Ostwald and van't Hoff made his theory of electrolytic dissociation more widely known.[23] Prominent physicists, especially Friedrich Kohlrausch [1840–1910] and Ludwig Boltzmann [1844–1906], gave their support, but the theory met strong resistance, especially in Britain and the United States, as well as from van't Hoff's former student van Laar.[24] After an extended absence from Sweden during 1886–1890, when Arrhenius visited the laboratories of Ostwald, van't Hoff, Kohlrausch and Boltzmann, and had refined and extended his theory, he suc-

---

[19] For Cohen, see Donnan (1948); Cohen died in the Nazi concentration camp at Auschwitz. For van Laar, see Scheffer et al. (1930), Van Klooster (1962) and Snelders (1984), pp.22–23. For Bancroft, see Findlay (1953) and Servos (1982); for Donnan, see Freeth (1957); for Euler, see Lynen (1965). Hoytink (1970) has discussed van't Hoff's influence on the later development of physical chemistry in Holland.
[20] Snelders (1984), p.23.
[21] D'Ans (1952), p.641.
[22] Euler (1952), p.644.
[23] For Arrhenius, see Walker (1928) and Riesenfeld (1931).
[24] Dolby (1976) has prepared an outstanding account of this debate.

ceeded in 1895 in obtaining a professorship in physics at the Technical High School in Stockholm. Arrhenius remained there until 1905, when he became head of a new Nobel Institute for Physical Chemistry in Stockholm, two years after he had been awarded a Nobel Prize in chemistry. He appears to have had relatively few co-workers in Stockholm; among them were Richard Abegg [1869–1910], Georg Bredig, Ernst Cohen and Hugh Stott Taylor [1890–1974]. Apart from his work on electrolytic dissociation, Arrhenius made another significant contribution in 1889 by applying van't Hoff's equation for the temperature dependence of chemical equilibria to the formulation of a theory that introduced the concept of "activation energy" in the consideration of the kinetics of chemical reactions.[25] By the end of the nineteenth century, however, Arrhenius had left research on these subjects and, during the remaining years of his life devoted his attention to cosmology and to immunology.

Arrhenius entered the field of immunology because he had come to believe that the new physical chemistry could illuminate biological problems, and he collaborated with the Danish bacteriologist Thorvald Madsen [1870–1957] in a study of the reactions between toxins and antitoxins. In a paper published in 1903, Arrhenius and Madsen described such reactions as chemical equilibria analogous to those attained in the interactions of weak acids and weak bases. This approach was quite different from that of Paul Ehrlich [1854–1915], who had proposed during the 1890s that living protoplasm elaborates antibody side chains which interact specifically and stoichiometrically with chemical groups of toxins in a kind of lock-and-key process analogous to that postulated by Emil Fischer for the specific binding of a substrate by an enzyme.[26] In 1906, shortly before he turned from the study of immunity to chemotherapy, Ehrlich wrote:

In view of the extraordinary success which physical chemistry has scored, it is readily understood how tempting it was for so eminent a representative of this science as Arrhenius to apply its principles to the new field of immunity. I have always emphasized the chemical nature of the reaction, and am glad therefore that the attempt to apply these principles has been made. It has been demonstrated anew that the

---

[25] See Logan (1982) and Laidler (1984).

[26] The Arrhenius-Ehrlich controversy has been discussed by Rubin (1980) and Fruton (1982d), pp.91–94. For Madsen, see Parnas (1981). The biographical literature about Ehrlich is immense, and frequently borders on hagiography; for recent articles, with references to older sources, see Bauer (1954), Witebsky (1954), Dolman (1971), Witkop (1981) and Parascandola (1981).

phenomena of animate nature represent merely the resultants of infinitely complex and variable actions, and that they differ herein from the exact sciences, whose problems can be treated mathematically. The formulas devised by Arrhenius and Madsen for the reaction of toxins and antitoxins explain absolutely nothing. . . . Neither do I believe that the phenomena observed in toxins and antitoxins bear any relation to the processes of colloid chemistry. . . .Structural chemistry, on the other hand, has not only served to explain all the phenomena in immunity studies, but has also proved a valuable guide in indicating the lines along which further progress might be made. The limitations of colloid chemistry have already manifested themselves, and enthusiastic advocates of this science have been compelled to assume the existence of specific atom groupings in accordance with my views. I therefore see no reason for abandoning the views expressed in my receptor theory, a theory in complete accord with the principles of synthetic chemistry.[27]

Some years afterward, Arrhenius took occasion to write:

I am convinced that biological chemistry cannot develop into a real science without the aid of the exact methods offered by physical chemistry. The aversion shown by biochemists, who have in most cases a medical education, to exact methods is very easily understood. . . . The physical chemists have found that the biochemical theories, which are still accepted in medical circles, are founded on an absolutely unreliable basis and must be replaced by other notions agreeing with the fundamental laws of general chemistry.[28]

As a member of the Nobel Prize committee on physics from 1900 until 1927, Arrhenius was also influential in the awards in chemistry.[29]

The member of the triumvirate of "ionists" who gained the greatest public attention was Wilhelm Ostwald. In many respects, Ostwald was a scientific entrepreneur in the style of Liebig.[30] He was an effective organizer and inspiring lecturer, he wrote over forty books (some of which were translated into many languages) and an immense number of articles and reviews, and he created an important scientific journal. Although

[27] Ehrlich (1910), p.578.
[28] Arrhenius (1915), p.vi.
[29] Crawford (1984b).
[30] See Chapter Two, note 6.

his personal research achievements did not bear the mark of originality that characterized the work of van't Hoff or Arrhenius, they were substantial contributions to the development of the physical chemistry of his time. Apart from Ostwald's "dilution law" of 1888, during the 1890s he formulated a theory of acid-base indicators that became an important part of analytical chemistry and his systematic investigation of catalytic reactions helped put the further study and industrial application of catalysis on a fruitful path. Also, Ostwald was influential in making known the importance of the free-energy functions of Gibbs and Helmholtz, rather than the heats of reaction, in the evaluation of equilibria in chemical reactions. However, his enthusiastic advocacy of thermodynamics, or "energetics" as he referred to it in his writings, led him to reject the atomistic and molecular concepts upon which the organic chemistry of his time was based.[31] It is difficult to escape the impression that this extreme view, which he abandoned in 1909, was part of Ostwald's campaign to counteract the indifference of leading organic chemists, notably Emil Fischer, as well as that of many prominent physicists, to his claim for the primacy of the concept of energy above that of substance. In 1906, Ostwald withdrew from his professorship at Leipzig to settle in his country home, where he devoted his last years to the study of colors and to writing extensively about internationalism, pacifism, monism, scientific genius and what he called "cultural energetics."

In his autobiography, Ostwald wrote that "in the present circumstances, I must consider the organizer as more important than the discoverer."[32] This statement suggests Ostwald's later recognition that his principal achievement was the creation, during the period 1887–1906, of the seedbed for the emergence of new talent in physical chemistry. Foremost among his German junior associates was Walther Nernst [1864–1941], the first in the succession of Ostwald's assistants, and of whom more will be said below. Others were (in alphabetical order) Max Bodenstein [1871–1942], Georg Bredig [1868–1944], Carl Drucker [1876–1959], Herbert Freundlich [1880–1941], Max Le Blanc [1865–1943], Paul Alwin Mittasch [1869–1953], Robert Luther [1868–1945] and Gustav Tammann [1861–1953], as well as many who did not attain the later distinction of these men. In addition to Arrhenius, there was a large foreign contingent, especially from the United States. Among the approximately thirty Americans who either received their Dr.phil. degree for work in Ostwald's

---

[31] Leegwater (1986).
[32] Ostwald (1926–1927), vol. 3, p.435.

Leipzig institute, or came as postdoctoral students, were Wilder Dwight Bancroft [1867–1953], Frederick Gardner Cottrell [1877–1948], Louis Kahlenberg [1870–1941], Arthur Becket Lamb [1880–1952], Samuel Colville Lind [1879–1965], Morris Loeb [1863–1912], Arthur Amos Noyes [1866–1936] and Theodore William Richards [1868–1928]. The British contingent was somewhat smaller, and included Frederick George Donnan [1870–1956], Alexander Findlay [1874–1966] and James Walker [1863–1935]. All these men played leading roles in the promotion of physical chemistry, whether in academic research and education, or in its industrial applications.

There can be little doubt that in the constellation of scientific talent in Ostwald's institute, that of Walther Nernst outshone the rest. Before coming to Leipzig in 1887, Nernst had acquired a Dr.phil. degree in physics, and soon demonstrated exceptional theoretical insight and technical skill.[33] During his four years with Ostwald, Nernst expanded the thermodynamic theory of solutions to include the relationship between electromotive force and ionic concentration (the so-called Nernst equation) and he also formulated an equation for the partition of a solute between two immiscible phases (the so-called distribution law). In 1891, Nernst moved to Göttingen, where he became professor of physics and directed the work of an effective research group on largely electrochemical problems of interest to him. He also displayed his technological skill and financial acumen by inventing an arc lamp in which Edison's carbon filament was replaced by a mixture containing cerium oxide, and sold his patent at a handsome price to the Allgemeine Elekrizitäts Gesellschaft.

As was noted in the previous chapter,[34] Nernst went to Berlin in 1905 to head what had been called the II. Chemisches Institut, to distinguish it from the one directed successively by Hofmann and Fischer. At Nernst's insistence and by ministerial decree, the name was changed to Physikalisch-chemisches Institut. Before he had arrived in Berlin, Nernst had formulated his most famous theoretical achievement which, with his characteristic lack of modesty, he termed "mein Wärmesatz." This contribution to thermodynamic theory offered a solution to the problem of determining the equilibrium constant of a chemical process from measurements of the changes in the heat content

---

[33] For Nernst, see the excellent biographical article by Hiebert (1978), which contains many references to other sources. See also Mendelssohn (1973). For a description of his laboratory in Göttingen, see Nernst (1896).
[34] See Chapter Five, note 102.

and the entropy accompanying the process.[35] Although the changes in heat content could be estimated readily by means of the calorimetric methods available at that time, the measurement of the entropy change presented greater problems. Nernst's heat theorem was based on the postulate that entropy differences between condensed phases of the reactants in a chemical process vanish at absolute zero; after its experimental verification, the theorem became known as the third law of thermodynamics. For several years after his enunciation of this postulate in 1906, one of the main tasks of Nernst's research group in Berlin was to study chemical reactions at very low temperatures, and for this purpose new apparatus for the measurement of thermal and electrical changes had to be constructed. These efforts involved the efforts of many of Nernst's junior associates, among whom Franz Eugen Simon [1893–1956] was especially gifted.[36] Among his other pupils were Max Bodenstein (who succeeded Nernst in Berlin), Niels Bjerrum [1879–1958], Arnold Eucken [1884–1950], Joel Henry Hildebrand [1881–1983] and Frederick Alexander Lindemann (later Lord Cherwell [1886–1957]). According to one of Nernst's students, " . . . in his lectures he related practically everything to his own work, or works of his pupils, so that a student who attended his lectures got the impression that most of physical chemistry and related fields were worked out by Nernst."[37] Moreover, in a report of the activities in Berlin, Nernst (and his assistant Sand) stated:

As the chief purpose of instruction we strive to encourage students to carry out independent physical-chemical research in accordance with the directions and aims given to them. It will therefore continue to be the custom in the institute to emphasize the personal responsibility of the students for their work and also in the designation of authorship of publications.[38]

It would seem that, along with his eccentricities and self-esteem, Nernst's style of leadership was somewhat more dictatorial than that of Ostwald but that he also frequently exhibited great generosity.

---

[35] See Hiebert (1978), pp.436–441, for a succinct discussion of the problem, and of Nernst's proposed solution.
[36] For Simon, see Kurti (1958).
[37] Harteck (1960), p.464.
[38] Lenz (1910), vol.3, p.310.

Before 1914, many biologists found in the new physical chemistry fruitful approaches to the study of physiological processes such as transport across biological membranes, muscular contraction, electrical conduction in nerve or the fertilization of egg by sperm. Particular emphasis was placed on the colloidal nature of protoplasm, and some investigators invoked the existence of giant protein molecules endowed with special physical properties.[39] One of the adherents of this view was the British physiologist William Maddock Bayliss [1860–1924]:

It is more than likely that chemical facts will sooner or later find their description in terms of molecular physics. The enormous molecules and aggregates of molecules which play so large a part in vital phenomena differ from small molecules in that they already begin to show the properties of matter in mass, especially those concerned with the development of surface.[40]

This physicalist attitude was expressed in the great interest among physiologists in the kind of adsorption phenomena studied intensively by Ostwald's pupil Herbert Freundlich.[41] Colloid chemistry, together with the "ionist" theory of solutions and the new ideas about the equilibria and kinetics of chemical processes, became prominent features of research in the biochemical sciences. To illustrate this impact of the emergence of physical chemistry as an independent discipline I have selected two twentieth-century research groups, those led by Jacques Loeb [1859–1924] and by Leonor Michaelis [1875–1949].

After receiving his Dr.med. degree (1884) at Strassburg for work on brain physiology, during the succeeding seven years Jacques Loeb continued to publish papers in this field, as well as on other physiological problems (muscle, heliotropism). His conviction that biological phenomena could be best understood through the application of physical and chemical theories and methods stemmed from these early years of his research, especially through his association with Adolf Fick [1829–1901] and Julius von Sachs [1832–1897] in Würzburg. Loeb went to the United States in 1891 and, after a year at Bryn Mawr College, was at the University of Chicago (1892–1902) and the University

---

[39] Fruton (1976), pp.329–330.
[40] Bayliss (1924), pp.19–20. For a biography of W. M. Bayliss, see L. E. Bayliss (1961). See also Hill (1969).
[41] For Freundlich, see Donnan (1942).

of California (1902–1911) before joining the staff of the Rockefeller Institute for Medical Research in New York.[42] During the
course of Loeb's physiological research in Chicago, his study of
Arrhenius's theory of electrolytic dissociation led him to investigate the effect of ions on the fertilization of sea urchin eggs and
to the discovery that an increase in the osmotic pressure could
induce artificial parthenogenesis. In Loeb's view this finding
had transferred the problem of fertilization from the realm of
morphology to that of physical chemistry, and he expanded on
this theme in his famous books *The Dynamics of Living Matter*
(1906), *The Mechanistic Conception of Life* (1912) and *The Organism
as a Whole, from a Physico-chemical Viewpoint* (1916), which attracted wide attention and stimulated lively controversy. In his
laboratory in New York Loeb continued to work on the physiological effects of ions, but in 1917 he turned his attention to the
study of the binding of ions by proteins. Loeb had touched on
this problem in 1900, when it was also being studied by Wolfgang Pauli, the pupil of Hofmeister, but since that time, largely
through the contributions of Søren Peter Lauritz Sørensen
[1868–1939], biologists had come to appreciate the importance
of the control of the hydrogen ion concentration in biochemical
experiments. Loeb's work on proteins from 1917 until his death
in 1924 led him to conclude that, as amphoteric electrolytes (a
term introduced by Bredig), they combine with ions stoichiometrically and not by mere adsorption on a surface, and that their
osmotic behavior accords with the theory advanced by Ostwald's
pupil Donnan.[43] Whatever may be said about the significance of
Loeb's contributions to biochemical knowledge, there can be no
doubt that the remarkable transformation of a student of brain
physiology into a protein chemist gives evidence not only of
Loeb's scientific imagination and philosophical outlook but also
of the attraction of the physical chemistry promoted by Ostwald
for some of the young biologists of his time. Nor can there be
question of the influence of Loeb's early writings on some of the
future leaders in the biochemical sciences. A notable example is
provided by the career of the botanist Winthrop John Vanleuven
Osterhout [1871–1964], whose physical-chemical studies on the
role of salts in plant physiology owed much to Loeb's earlier

---

[42] See Osterhout (1930) and Pauly (1987).
[43] Loeb (1922).
[44] For Osterhout, see Blinks (1974).

work.[44] Another example may be found in Otto Warburg's decision, after his Dr.phil. work with Emil Fischer, to turn from organic chemical research to the study of biological oxidations in the sea urchin egg.

In the United States, Loeb had approximately 70 co-workers. Among those in Chicago were Martin Henry Fischer [1879–1962] and Ralph Stayner Lillie [1875–1952]. During Loeb's stay in California, the group included Wolfgang Ostwald [1883–1943], Thorburn Brailsford Robertson [1884–1930] and Hardolph Wasteneys [1881–1965]. At the Rockefeller Institute, apart from his son Robert Frederick Loeb [1895–1973], later a distinguished physician, two of Loeb's co-workers, John Howard Northrop [1891–1987] and Moses Kunitz [1887–1978], merit special mention.[45]

Northrop joined Loeb in 1915, after receiving his Ph.D. at Columbia University for work with John Maurice Nelson on the enzyme invertase. After Loeb's death, Northrop moved to the Princeton branch of the Rockefeller Institute, and in 1949 he went to the University of California at Berkeley. Although a man of narrower scientific and philosophical outlook, Northrop was greatly influenced by Loeb's belief that the physical chemistry taught by Ostwald and Arrhenius provided the most reliable basis for the experimental study of biological problems; in particular, Northrop shared Arrhenius's view that one should employ as few hypotheses as possible.[46] Northrop's most notable scientific success came in 1930, when he prepared a crystalline form of pepsin. In providing strong evidence for the view that the catalytic activity of the enzyme is an intrinsic property of the protein that he had isolated, Northrop gave support to the claim of James Batcheller Sumner [1887–1955] that the enzymatic activity of the crystalline preparation of urease Sumner had isolated in 1926 is associated with its protein nature. Although, at the turn of the century, organic chemists considered enzymes to be proteins, by the 1920s preference was given to the view of the Nobel Prize winner Richard Willstätter that enzymes are reactive small molecules which are readily adsorbed on catalytically inac-

---

[45] For Northrop, see Herriott (1983); for Kunitz, see Herriott (1978, 1989). Paul de Kruif [1890–1971], a member of Simon Flexner's group during 1920–1922, also came under Loeb's influence. De Kruif's fame came largely from his best-seller *Microbe Hunters* (1926) and from his association with Sinclair Lewis [1885–1951] in the preparation of Lewis's novel *Arrowsmith* (1925); see de Kruif (1923, 1962).

[46] Northrop (1961).

tive colloidal carriers.[47] Northrop's work on pepsin was followed by a series of successes by his associate Moses Kunitz, who had come to Loeb's laboratory in 1913 as a technical assistant, and had obtained a Ph.D. at Columbia University in 1924 during his employment at the Rockefeller Institute. Between 1934 and 1952, Kunitz successively crystallized the inactive precursors of pancreatic trypsin and chymotrypsin, and prepared crystalline forms of the active enzymes, as well as pancreatic ribonuclease, deoxyribonuclease, yeast hexokinase and pyrophosphatase. Two other associates of Northrop also participated in this fruitful program: Mortimer Louis Anson [1901–1968] crystallized pancreatic carboxypeptidase and Roger Moss Herriott [b. 1908] crystallized pepsinogen.[48] The achievements of the Northrop group during the 1930s led not only to the confirmation of Sumner's claim, but also to the entry of Otto Warburg's group into the field of enzyme crystallization. By 1940, Willstätter's view had been largely rejected, and there was widespread agreement that enzymes are proteins. Moreover, in 1935 Wendell Meredith Stanley, then in the plant pathology section of the Princeton branch of the Rockefeller Institute, isolated tobacco mosaic virus in the form of crystals; his success owed much to his proximity to the Northrop group. Northrop's later research dealt largely with bacterial viruses. His group was relatively small; apart from the persons named above, there were, over the years, several foreign guests, among them John Alfred Valentine Butler [1899–1977].

The scientific career of Leonor Michaelis, like that of Jacques Loeb, offers striking evidence of the role of the new physical chemistry in shaping biochemical research during the first half of the twentieth century. Michaelis began scientific work as a medical student in Berlin, and received his Dr.med. degree (1897) for an embryological study in the laboratory of Oscar Hertwig [1849–1922]. His enthusiasm for communicating new knowledge to students, evident throughout his later life, led Michaelis to prepare a small textbook of embryology.[49] Then followed a one-year assistantship at Paul Ehrlich's laboratory in

---

[47] For Sumner, see Maynard (1958) and Dounce and Allen (1988); it should be noted that most of the historical accounts of the work of Sumner and of the Northrop group fail to mention the fact that the methods they employed were similar to the ones used by Osborne and by the Hofmeister group during the 1890s for the crystallization of proteins. For Willstätter's views on the nature of enzymes, see Fruton (1977).

[48] For a summary of the work on enzymes in Northrop's laboratory, see Northrop, Kunitz and Herriott (1948).

Frankfurt am Main; according to Michaelis's account, Ehrlich insisted that he enter clinical medicine because "only a man of sufficient wealth should stay permanently in fundamental scientific research."[50] During that year, Michaelis continued Ehrlich's histological work, and delved deeply into the organic chemistry of the dyes that had been developed for the textile industry. He also studied intensively the new physical chemistry, and the mathematics underlying its principles, in order to understand better the interaction of dyes with chemical constituents of biological tissues, and remained an omnivorous student of these subjects the rest of his life. While with Ehrlich, Michaelis discovered the specific staining by Janus Green of cellular granules later to be called mitochondria; this method of vital staining came to be widely employed.[51] In accordance with his agreement with Ehrlich, Michaelis then spent five years (1900–1905) in clinical work at Berlin hospitals, but managed to continue research on histological staining and also worked on immunological problems. At the end of that period, he received a professorial title at the University of Berlin, a post without salary, laboratory or funds for research. This dubious distinction convinced Michaelis that, as a Jew, his chances for an established academic post were negligible, and in 1905 he accepted a position as bacteriologist at the Berlin municipal hospital *am Urban*. Except for the years of the First World War, when he served as a physician in various hospitals, Michaelis remained at the *am Urban* until 1921, and there made many of his most distinguished contributions to biochemistry. Together with his friend Peter Rona [1871–1945], the chemist at the hospital, he established a small laboratory from which there emerged a series of studies on the role of the hydrogen ion concentration in determining the properties of proteins and enzymes.[52] There, Michaelis de-

---

[49] Michaelis (1898). This textbook was well received, and went through many editions; the ninth appeared in 1921. By that time, Michaelis's research interests were far removed from embryology. Three further editions appeared during 1927–1933, under the authorship of Richard Weissenberg [1882–1974].

[50] MacInnes and Granick (1958), p.284. This obituary notice contains a lengthy autobiographical sketch and a list of Michaelis's publications.

[51] For a history of histological staining, see Clark and Karsten (1983).

[52] Priority for some of Michaelis's work on hydrogen ion concentration belongs to Sørensen, whose memorable paper (Sørensen, 1909) presented a detailed description of the electrometric method for the determination of hydrogen ion concentration, the introduction of the term pH (negative logarithm of the hydrogen ion concentration), the preparation of buffer solutions of fixed pH, the use of dyes for the colorimetric determination of pH and the results of studies on the effect of pH on the catalytic activity of several enzymes. According to Michaelis, Sørensen's paper appeared when

veloped the theory of the dissociation of amphoteric electrolytes, such as amino acids and proteins, and formulated a
quantitative treatment of the concept of the isoelectric point, introduced in 1900 by William Bate Hardy [1864–1934]. In 1914,
he published an influential book on this subject.[53] Also, in keep-
ing with the scientific fashion of his time, Michaelis studied the
adsorption of small molecules on colloidal surfaces. The contribution for which he became best known was the outcome of his
work on the effect of substrate concentration on the initial rate
of reactions catalyzed by enzymes. Such measurements had been
made before by several investigators, notably Victor Henri
[1872–1940], who proposed that an intermediate enzyme-
substrate complex is formed in the catalytic process, but the validity of Henri's data and his theory were questioned, especially
by Bayliss.[54] Michaelis's work, initially conducted with the help
of a Canadian guest, Maud Leonora Menten [1879–1960], not
only removed many of the objections to Henri's data, but also
provided the definition of a constant for the affinity of an enzyme for its substrate.[55] Later authors termed the dissociation
constant of the assumed enzyme-substrate complex the "Michaelis constant" (or "Michaelis-Menten constant"), and the symbol
K_M to denote it has become a lasting part of the language of
modern biochemistry. It should be noted, however, that the recognition of the significance of Michaelis's mathematical treatment of enzyme kinetics did not come until the 1930s, and was
notably promoted by the appearance of the book on enzymes by
John Burdon Sanderson Haldane [1892–1964].[56] Michaelis also
extended Henri's treatment of the inhibition of enzyme action;
later, this aspect of enzymology was greatly refined because of
its importance in the study of enzyme mechanisms and in the
development of new chemotherapeutic agents.[57]

Between 1905 and 1921, Michaelis had attracted to his modest laboratory at the *am Urban* about 40 co-workers, most of
whom were German physicians who later entered medical prac-

---

his own work on the use of the hydrogen electrode to determine the effect of hydrogen ion concentration on enzymatic activity were nearing completion. For Sørensen,
see Linderstrøm-Lang (1939).

[53] Michaelis (1914). There was a second edition of this book in 1922, and an English
translation (by William A. Perlzweig) of this version appeared in 1926. Between 1905
and 1921, Michaelis also published *Dynamik der Oberflächen* (1909; English translation,
1915), *Einführung in die Mathematik für Biologen und Chemiker* (1912; later editions 1922,
1927), and *Praktikum der physikalischen Chemie insbesondere der Kolloidchemie für Mediziner
und Biologen* (1921; later editions in 1922, 1926, 1930).

[54] Bayliss (1908), pp.39,61.

[55] Michaelis and Menten (1913).

[56] Haldane (1930); see also Boyde (1980).

[57] See Segel (1975).

tice, but also included postdoctoral guests from abroad; among the latter was Albert Szent-Györgyi [1893–1986]. However, his achievements did not lead to an improvement in Michaelis's academic prospects. In 1921 he joined the staff of a Berlin firm that manufactured scientific apparatus, and in the following year he became a visiting professor at the Aichi Prefectural Medical School, which later became part of the University of Nagoya. Michaelis continued work on adsorption phenomena and electrochemistry, and also initiated a sustained effort on semipermeable membranes. He later wrote enthusiastically about his three-year stay in Japan, for he was given a good laboratory and had about 15 research students, among them Akiji Fujita [1894–1985].

In 1923, Jacques Loeb invited Michaelis to come to the United States for a lecture tour. Loeb arranged the tour, but died before Michaelis arrived in 1924. One of the stops was the Johns Hopkins School of Medicine, and on that occasion Michaelis accepted an invitation to come there as a resident lecturer. During his stay in Baltimore (1926–1929), Michaelis continued research on the permeability of artificial membranes, but also began electrometric studies of oxidation-reduction reactions; his co-workers in the latter effort included Eleazar Sebastian Guzman Barron [1898–1957] and Harry Eagle [b. 1905]. As was his wont, Michaelis wrote a book on the subject.[58]

At the age of 54, Michaelis finally attained his hope of securing a permanent position when, in 1929, he was appointed a Member of the Rockefeller Institute for Medical Research. He remained there until his death, and his research was not interrupted by his transfer to emeritus status in 1941. Shortly after moving to New York, Michaelis made his most important contribution to the study of the reversible oxidation-reduction of organic substances. With Ernst Friedheim [1899–1989] he found that, at acidic pH values, the oxidation-reduction of the natural pigment pyocyanine involves the stepwise transfer of one electron, with the intermediate formation of a free radical which Michaelis termed a "semiquinone."[59] At first, Michaelis's claim

---

[58] Michaelis (1929); an English translation (by L. B. Flexner) appeared a year later, and a second German edition was published in 1933.

[59] This discovery was made independently at about the same time by Bene Elema (b. 1901) when he was a graduate student at Delft. In a letter to me dated 9 May 1972, Dr. Elema wrote: "I remember that after my first titration experiments at lower pH, I proposed to Professor Kluyver publishing a short note in *Nature*. His answer was that he thought it practically impossible that someone should work on the subject elsewhere in the world simultaneously. So the paper by Friedheim and Michaelis came to us rather as a shock!" Albert Jan Kluyver [1888–1956] was a noted microbiologist who made many important biochemical contributions.

was greeted with skepticism, for it was widely believed that in such two-electron transfer processes no intermediates are formed, but he soon produced massive evidence in support of his conclusion. The pioneer in the electrometric measurement of the oxidation-reduction potentials of organic dyes, William Mansfield Clark [1884–1964], later wrote: "As one whose work occasionally overlapped that of Dr. Michaelis, I wish to say that he taught me much and that true to the instincts of a good teacher he generously ignored the fact that I had had an opportunity to find what he later discovered."[60] Apart from potentiometric studies showing that the concept of stepwise univalent oxidation-reduction applies to organic dyes other than pyocyanine, Michaelis also demonstrated that the semiquinones are paramagnetic, owing to the presence of an unpaired electron. During the course of this work, he acquainted himself thoroughly with the quantum theory underlying the principles of magnetochemistry and his enthusiasm as a teacher was again evident in the series of lectures he gave on quantum mechanics at the Rockefeller Institute during the 1930s.

During his stay at the Rockefeller Institute, the group associated with Michaelis made valuable contributions to other biochemical problems. Thus, studies on the reactions of the amino acid cysteine led to the finding by David Rockwell Goddard [1908–1985], later a prominent botanist, that the reduction of the cystine-rich hair keratin by means of thioglycollate gives a soluble form of the protein, digestible by proteolytic enzymes; this observation was later taken up by the cosmetic industry to produce a "cold permanent wave." Another area of investigation, developed by Sam Granick [1909–1977], was the study of the iron-containing protein ferritin. In addition to Granick, Michaelis's chief research associates were Maxwell Schubert [b.1902], Carl Vincent Smythe [b.1903] and Edgar Smith Hill [1907–1952]. Among the numerous guest workers, apart from Goddard, were such later notables as Jannik Bjerrum [b.1909], John Runnström [1888–1971] and Gerold Schwarzenbach [1904–1978]. In all, during his years in the United States, Michaelis had about 30 junior associates; when added to those in Berlin and Nagoya, the total comes to about 85 persons who participated in his various research programs. Michaelis did not seek honors, and received few, possibly because, though kindly toward younger scientists, he was not always tactful in his relationships with senior colleagues.

---

[60] Clark (1950); for Clark, see Vickery (1967). See also the book by Clark (1960).

There can be no doubt that Loeb and Michaelis, as well as others, contributed significantly to the introduction of some aspects of the new physical chemistry into the biochemical research of their time. It must be recalled, however, that the theory of solutions elaborated by the "ionists" soon came under sharp attack from other chemists (including some of the pupils of van't Hoff and Ostwald) on the ground that theory is valid only for solutions of weak electrolytes as such solutions approach infinite dilution.[61] Among the extensive efforts during 1900–1925 to formulate a more satisfactory theory, those of Gilbert Newton Lewis [1875–1946] were especially important in connection with the chemical thermodynamics developed by Josiah Willard Gibbs [1839–1903] during the 1870s. Although the elegance of Gibbs's mathematical treatment, and the significance of his concepts of free energy, entropy and the criteria for thermodynamic equilibrium, had been recognized by prominent theoretical physicists, at the turn of the century few chemists (and fewer biochemists) were able to master Gibbs's sophisticated formal language. Lewis's book (with his associate Merle Randall [1888–1946]) made Gibbs's ideas understandable even to students of elementary physical chemistry.[62] Apart from Lewis's notable contributions the theory of chemical valence, he revised Arrhenius's theory to include the ionic interaction, and introduced valuable concepts such as "activity" and "ionic strength." Another important figure in the formulation of a satisfactory theory of electrolytic dissociation was Niels Bjerrum [1879–1958], whose contributions included the demonstration that, at their isoelectric point, amino acids exist as dipolar ions.[63] A high point came in 1923, when Peter Debye [1884–1966] and his associate Erich Hückel [1896–1980] proposed an equation derived from the mathematical analysis of a model system in which an ion interacts with an ionic atmosphere of opposite charge, thus emphasizing the importance of the dielectric constant of the solution.[64] The Debye-Hückel theory accounted well for the

---

[61] See Wolfenden (1972) and Dolby (1976).

[62] Lewis and Randall (1923). For Lewis, see Hildebrand (1948) and Calvin (1977). The latter article contains a list of the Ph.D. recipients in chemistry during the period (1912–1945) of Lewis's activity at the University of California, Berkeley. In its time, Lewis's research group was one of principal seedbeds for the nurturing of future leaders in physical chemistry in the United States.

[63] For Bjerrum, see Guggenheim (1960), Bok (1974) and Kauffman (1980). It should be noted that in 1916 Lewis's student Elliot Quincy Adams [1888–1971] had advanced the idea that, in their electrically neutral state, amino acids bear positive and negative charges.

[64] For Debye, see Davies (1970) and Williams (1975); for Hückel, see Hückel (1975) and Hartmann and Longuet-Higgins (1982).

thermodynamic properties of electrolyte solutions, but not for their conductance, and was later subjected to critical study by many investigators, notably Charles August Kraus [1875–1967] and Lars Onsager [1903–1976].[65]

The biochemical field in which these new advances had their most immediate impact was the physical chemistry of proteins. Among the first applications of the Debye-Hückel theory were the studies of Kaj Ulrik Linderstrøm-Lang [1896–1959], in Sørensen's biochemical laboratory in Copenhagen, on the ionization of proteins. Linderstrøm-Lang succeeded Sørensen as head of the chemical department of the Carlsberg Laboratory and, after the Second World War, this laboratory attracted many foreign postdoctoral students, largely because of Linderstrøm-Lang's personal qualities and scientific acumen as the leader of his productive research group.[66] Before 1940, however, the most prominent biochemical laboratory in which the new theories of electrolyte solutions were applied to the study of proteins was that of Edwin Joseph Cohn [1892–1953] at Harvard.[67] As a student, Cohn had been influenced by Loeb and (especially) by Henderson, and had worked on proteins with Osborne and Sørensen. In 1922 Cohn joined the newly established Department of Physical Chemistry at the Harvard medical school, and shortly afterward began a lifelong association with the physical chemist George Scatchard [1892-1973] of the Massachusetts Institute of Technology.[68] Scatchard had just begun to apply the Debye-Hückel theory to salt solutions, and became interested in the behavior of proteins. As Scatchard later described it: "When I came to M.I.T., I was the only one who called himself a physical chemist in the area, perhaps in the country, who would listen to a physician or even a physiologist talking science. Cohn is probably responsible for my doing so."[69] Cohn's initial studies dealt largely with the solubility of the proteins, and especially the "salting-out" effect investigated by Hofmeister during the 1880s. Also, he attempted to correlate the binding of acids and bases by proteins with their amino acid composition, as it was known during the 1920s. In succeeding years, apart from a sustained effort to isolate the constituent in liver that Minot and Murphy had found to be effective in the treatment of pernicious anemia, the research in Cohn's laboratory dealt with many as-

---

[65] Kraus (1958). For Kraus, see Fuoss (1971); for Onsager, see Longuet-Higgins and Fischer (1978) and Lyons (1981).
[66] Holter (1976).
[67] Edsall (1955, 1961, 1981) and Scatchard (1969).
[68] Edsall and Stockmayer (1980).
[69] Scatchard (1969), p.39.

pects of the physical chemistry of proteins, peptides and amino acids, and the data were collected in an important book that he wrote with his principal associate John Tileston Edsall [b. 1902].[70] During the Second World War, and for several years afterward, Cohn led a massive effort on the fractionation of blood proteins, and many of the results of the 1930s proved to be invaluable in this program. From this program there also came Scatchard's important studies on the binding of ions by serum albumin, and the formulation of a widely applicable method for the determination of the number of binding sites and the magnitude of association constants. Moreover, during the 1930s, Scatchard's associate John Gamble Kirkwood [1907–1959] formulated a valuable theory to describe the interactions between ions and dipolar ions.[71]

An outstanding organizer but (like his mentor Henderson) not an enthusiastic experimenter, Cohn assembled a sizable research group. From the list of the publications it would seem that, between 1922 and 1953, approximately 130 persons had worked in his laboratory, many of them as members of subgroups directed by his principal associates, with whom Cohn conferred directly.[72] Apart from Edsall, who was with Cohn for twenty-six years, these associates included John Douglas Ferry [b. 1912], Jesse Philip Greenstein [1902–1959], Thomas Leroy McMeekin [1900–1979] and John Lawrence Oncley [b. 1910]. Although Cohn's style of leadership was that of a benevolent despot, the variety of talent he attracted to his laboratory and to its seminars made it an important center for the education of future leaders in the study of the physical chemistry of proteins. According to Edsall,

...yet with all his driving force and his tendency to dominate the situation, Edwin Cohn succeeded in aiding the development of gifted young investigators of very diverse talents. Sometimes by encouragement, sometimes by sharp criticism, he often evoked capacities and responses in his younger colleagues of which they had not known they were capable.[73]

---

[70] Cohn and Edsall (1943).
[71] For Kirkwood, see Scatchard (1960).
[72] The total given above does not include senior colleagues (such as Conant and Scatchard) nor the many persons in other laboratories whom Cohn had enlisted into the blood fractionation program. A list of investigators who worked in Cohn's laboratory between 1920 and 1950 is included in Edsall (1950).
[73] Edsall (1955), pp.27–28.

In his final years, Cohn offered speculations about the detailed structure of proteins; although received respectfully, his ideas were not accepted, and are now forgotten.

The introduction into biochemical thought of the concepts of the new physical chemistry was accompanied by the appearance in biochemical laboratories of new physical instruments for the study of proteins. A stimulus came from William Bate Hardy's report in 1900 on the migration in an electric field of heat-coagulated egg albumin (which he termed an "irreversible hydrosol"). This pioneering contribution to colloid chemistry was soon followed by the elaboration of Hardy's apparatus by Wolfgang Pauli and by Leonor Michaelis, and a high point in this development was reached in 1936, when Arne Tiselius [1902–1971] described his electrophoresis apparatus.[74] Despite the cost of its construction and maintenance, this instrument quickly became a much-desired addition to the equipment of pre-World War II biochemical laboratories. Tiselius had been a student in Uppsala of The Svedberg [1884–1971], a colloid chemist who invented an optical ultracentrifuge for the measurement of the rates of sedimentation of colloidal particles. Svedberg's demonstration in 1925 that proteins sediment as homogeneous substances, rather than as polydisperse particles, brought chemists back to the view, espoused by Hofmeister and rejected by Emil Fischer, that proteins are molecules of high molecular weight.[75] During the 1930s, the Svedberg centrifuge was improved by Jesse Wakefield Beams [1898–1977], and his student Edward Greydon Pickels [b. 1911] later brought out a commercial model. Similarly, the Tiselius apparatus was extensively modified by Lewis Gibson Longsworth [1904–1981], an associate of Duncan Arthur MacInnes [1885–1965] at the Rockefeller Institute, and less expensive versions were widely used by biochemists.

At that time, other major physical instruments began to appear in biochemical laboratories. For example, the use of stable isotopes (deuterium, ^{15}N) as labels in metabolic studies and of mass spectrometers for their determination brought into biochemistry physical chemists who had worked with Harold Clayton Urey [1893–1981]. After 1945, when radioactive isotopes became available for civilian research, Geiger counters were re-

---

[74] For Tiselius, see Kekwick and Pedersen (1974) and Kay (1988).

[75] For Svedberg, see Claesson and Pedersen (1972) and Kerker (1976, 1986). As in the case of other distinguished chemists (see Chapter Five, note 129), Svedberg succumbed to the hypnotic power of numerology, and concluded from his molecular weight data that proteins are composed of multiples (2, 4, 8, 24, 48, 192, 384) of units of 17,600 molecular weight.

placed by scintillation counters. Moreover, commercial spectro-
photometers, notably the Beckman DU model, and reliable
instruments for the determination of infrared and fluorescence
spectra also became normal tools of biochemical research. The
polarimeters used by Louis Pasteur and Emil Fischer for the
measurement of optical activity were replaced by recording instruments, and apparatus became available for studies on optical
rotary dispersion and on circular dichroism. By the 1960s, two
other powerful physical techniques had been added to biochemical methodology. After the achievements of John Cowdery Ken-
drew [b. 1917] and Max Ferdinand Perutz [b. 1914] in their
studies on the structure of myoglobin and hemoglobin, many
major biochemical departments sought to establish research
units for the X-ray crystallographic determination of the structure of biological materials. Also, the emergence of high-
resolution nuclear magnetic resonance (NMR) spectroscopy as
the method of choice for the determination of organic-chemical
structure brought into biochemistry specialists in this field. The
importance of the biochemical knowledge gained by the use of
physical instruments cannot be exaggerated but, as Edwin Cohn
suggested,

. . . the very specialized nature of the tools might well result in the
future in the training of men who knew the tools more intimately than
the substances they were investigating. Indeed, each tool begins to
have so interesting a history and to involve so intricate a theory that
few not trained in its use can interpret the important investigations
that are being reported and critically appraise the evidence that may
be deduced from them.[76]

Among the tools of modern biochemistry few are more important that those used to perform the separation and analysis of
chemical substances by chromatographic procedures. The roots
of these methods may be found in the early efforts of Friedlieb
Ferdinand Runge [1794–1867], Christoph Friedrich Goppel-
sroeder [1837–1919], Carl Engler [1842–1925] and David Talbot
Day [1859–1925].[77] During the first decade of the twentieth century, Mikhail Semenovich Tswett (Tsvet) [1872–1919] described

---

[76] Cohn (1941), pp.79–80. For a critical view of the interpretation of X-ray crystallographic data, see Jones (1984). See also Hill (1956) on the interaction of biochemistry
and biophysics.
[77] Williams and Weil (1952); Weil and Williams (1953).

the use of columns of calcium carbonate to separate plant pigments in a process he termed "chromatographic adsorption analysis," but his procedure was not widely adopted.[78] Although employed sporadically by organic chemists before 1941, it was not until the work of Archer John Porter Martin [b. 1910] and Richard Laurence Millington Synge [b. 1914] on the separation of amino acids by partition chromatography that such methods began to be explored intensively. In addition to the studies of Arne Tiselius on adsorption chromatography, the systematic investigation of the separation of amino acids with solid supports having charged groups (ion-exchange chromatography) led to the development of the first reliable methods for the quantitative analysis of the amino acid composition of protein hydrolysates. These advances were followed by the introduction of gas-liquid chromatography and, more recently, of high pressure liquid chromatography (HPLC) and, at each stage, instruments of higher selectivity and precision became available for biochemical investigation. Moreover, procedures were developed for the electrophoretic separation of such substances as proteins and nucleic acids on polyacrylamide gels. Then, the introduction of minicomputers made possible the control of the instruments and the automatic collection of data. In short, during the period 1945–1985, the availability of new physical instruments had a profound impact on biochemical research and education.

Other Biochemical Research Groups

Among the sizable research groups not discussed in the previous chapters the one led by the Polish biochemist Marceli Nencki [1847–1901] merits further study.[79] After his association with the Baeyer group in Berlin in 1870–1872, Nencki was at Berne, where he became *ordentlicher Professor* of medical chemistry, and in 1891 he moved to St. Petersburg to head the chemical department of a newly established institute of experimental medicine. In his two laboratories, Nencki had 147 students, assistants and postdoctoral guests (84 in Berne, 63 in St. Petersburg). Among his foreign associates in Berne were John Jacob Abel [1857–1938] from the United States, Piero Giacosa [1853–1928] from

---

[78] For Tswett, see Robinson (1960).
[79] For Nencki, see Chapter Four, note 21 and Appendix 5. For the collected papers from his laboratories, see Nencki (1904).

Italy, Martin Hahn [1865–1934] from Germany and Allan Macfadyen [1860–1907] from Great Britain. An indication of his attitude toward his junior associates is suggested by the fact that,
of the approximately 450 papers which appeared from his laboratories, only about one-half bear Nencki's name. His manifold
research interests included topics in organic chemistry, largely
derived from his stay in Baeyer's laboratory, and in physiological
chemistry and bacteriology. In the latter two areas, Nencki and
his associates published papers on a variety of problems, notably
the products of bacterial putrefaction, the metabolic oxidation
of aromatic compounds, and on the chemistry of the porphyrins. Nencki also wrote in support of the speculations offered by
Eduard Pflüger and Oscar Loew about "living proteins." He was
highly regarded by some of his scientific contemporaries and
Polish compatriots, as is indicated by the later establishment in
Warsaw of the Nencki Institute of Experimental Biology. Although the influence which Nencki exerted, through his students, on the development of the biochemical sciences in Europe
is not comparable to that of Hoppe-Seyler or Hofmeister,
Nencki's pupil Abel later made important contributions to the
advancement of biochemistry in the United States.

   Abel worked in Nencki's laboratory during 1889–1890, after
two years of medical study and research in Ludwig''s institute in
Leipzig, and a further two years in Strassburg, where Abel received his Dr.med. degree in 1888, and in Würzburg. Shortly before his return to the United States, Abel worked with the noted
biochemist Edmund Drechsel [1843–1897] at Ludwig's institute.
From Abel's stay in Berne there emerged three papers, dealing
with melanins, derivatives of biuret, and molecular weight determinations of several biochemical substances. This relatively
lengthy exposure in Europe to physiology, pharmacology and
biochemistry led Abel to the conviction that the future of medical research lay in the application of chemistry to the study of
biological problems. In 1893, he became professor of pharmacology at the Johns Hopkins School of Medicine, and continued
research there after his retirement in 1932.[80]

   One of Abel's first independent scientific efforts was in the
purification of the adrenal pressor substance, which he named
epinephrin.[81] Later (1917–1924), he attempted to isolate what
appeared to be the single hormone of the posterior pituitary.
Abel's greatest success came in 1925, when he succeeded in ob-

---

[80] For Abel, see Voegtlin (1939), Lamson (1941), MacNider (1946) and Parascandola
(1982).
[81] See Chapter Five, note 56.

taining crystals of insulin, but at that time his evidence for the protein nature of this hormone was not widely accepted. As in the case of Sumner's claim for his crystalline urease, the prevalent view among chemists was that "bioactive" substances such as hormones, enzymes and vitamins are small organic molecules which can be adsorbed on inert colloidal carriers.[82] During the course of these researches, Abel had a series of notable junior associates. Thomas Bell Aldrich [1861–1938], a doctoral student of Ludwig Knorr, played an important role in the work on epinephrine, as did Eugene Maximilian Karl Geiling [1891–1971] in the studies on the pituitary hormone, and among those who worked with Abel on the chemistry of insulin were Vincent du Vigneaud [1901–1978], Hans Friedrich Jensen [1896–1959] and Oskar Wintersteiner [1898–1971]. Although the research in Abel's relatively small department was largely channeled along the lines of his own scientific interests, he encouraged independent thought and shared his ideas freely with his junior associates. The list of Abel's personal publications is a modest one; most of the experimental papers on insulin from his laboratory do not bear his name.

Apart from Abel's considerable influence on the institutional development of pharmacology in American medical schools, he also played a leading role in the promotion of biochemistry in the United States. Abel enlisted the help of Christian Archibald Herter [1865–1910], professor of pharmacology at Columbia University (and a man of some wealth), in founding the *Journal of Biological Chemistry;* the two men were the first editors (with Alfred Newton Richards as associate editor) of the new journal, which began to appear in 1905. Moreover, in the following year Abel initiated the establishment of the American Society of Biological Chemists.[83]

If, by the mid-1920s, there was more professors of physiological chemistry at medical schools in the United States than in Germany, the number of American research groups which contributed significantly to the study of central problems in the biochemical sciences was relatively small. In many of the departments led by these professors, apart from their responsibility for the formal instruction of first-year students, whatever research was done dealt largely with the improvement of analytical methods in clinical chemistry. It has been suggested that if American biochemists had "tried instead to build a discipline on the basis of more highbrow programs of general biochemis-

---

[82] Murnaghan and Talalay (1967).
[83] Chittenden (1945).

try . . . it is doubtful that they would have done so well."[84] This suggestion may have merit, if one only considers the number of medical school professors who were acceptable to the heads of the dominant clinical departments. If account is also taken of the quality of the contributions these professors made to the advancement of clinical chemistry, they had not "done so well" as did Donald Dexter Van Slyke, the pupil of Gomberg and Emil Fischer, in his work at the hospital of the Rockefeller Institute. It must also be noted that the emergence of "highbrow" biochemistry in American universities, with great impact on medical practice in the United States, was markedly furthered by the influx of biochemists who came there from Germany and Austria during the 1930s. Moreover, despite their relatively small number, productive research groups in American universities, such as those of Abel at Johns Hopkins and of Mendel at Yale, carried on the tradition to which their leaders had been exposed in Germany. Also, by the 1920s, the Rockefeller Institute, directed by the bacteriologist Simon Flexner [1866–1959], had become a leading center of such "highbrow" biochemical research, with separate laboratories, each

ruled over by the distinguished scientist for whom it was organized. How egocentric the command, how strict the direction, varied with the character and temperament of the Member in charge and with the nature and range of his interests. Flexner, with a group of wide scope including several mature people, left his senior men free to pursue programs of their own. The others, accustomed to European methods, kept their staffs at work on problems in which the laboratory directors were personally interested . . . Levene, too, used his young chemists largely for his own experiments. Flexner did not conceal from Levene his fear that they were not being trained for independent work. Against this charge Levene defended himself, with some justification, by pointing out that Van Slyke and Jacobs were leading independent research in the Institute, and three other former assistants were filling responsible positions elsewhere.[85]

It should be added, however, that in contrast to Flexner's own research group, the number of staff positions in the separate biochemical laboratories of the Institute was relatively small, and

---

[84] Kohler (1982), p.252.
[85] Corner (1964), pp.152–153; see Fruton (1966). For Levene, Van Slyke and Jacobs, see Chapter Five, notes 131 and 132.

their size was only increased by the temporary stay of guest investigators.[86] For example, Levene's group also included such later notables as Fritz Lipmann [1899–1986]. Also, despite the small size of the individual groups, the increasingly chemical orientation of the Institute as a whole was reflected in the achievements of several of the biological laboratories. The most famous of these laboratories was that of Oswald Theodore Avery [1877–1955], whose studies on the pneumococcus led to the discovery of the immunological specificity of polysaccharides and of the role of its DNA in the transformation of bacterial types.[87] Before 1950, however, the Rockefeller Institute was not strongly represented in the important area of biochemistry dealing with the chemical pathways of cellular metabolism.

Indeed, with the prominent exception of the medical school department at Columbia University, headed by Hans Thacher Clarke,[88] who during the 1930s welcomed many of the outstanding young biochemists from Germany and Austria, at that time the research programs in such biochemistry departments of American medical schools were largely determined by the scientific interests of their chief professors, and independent efforts by their junior associates were generally not encouraged. Among these departments, the ones headed by Vincent du Vigneaud at the George Washington School of Medicine (1932–1938) and at the Cornell University Medical College (1938–1967) provide a notable example of a closely directed biochemical research group. After his association with Abel in Baltimore and with Max Bergmann in Dresden, du Vigneaud spent three years (1929–1932) in Adams's department in Urbana, where he began independent research.[89] Between 1930 and 1975 (after his retirement he transferred his group to the Cornell Ithaca campus), du Vigneaud had about 170 junior associates. During the course of these years, his research group made distinguished contributions to the study of the chemistry and metabolism of sulfur-containing amino acids (1930–1946), as well to the elucidation of the chemical structure of the vitamin biotin (1940–1945) and of the peptide hormones oxytocin and vasopressin of the posterior pituitary (1949–1975). An outstanding organizer, alert to new developments in instrumentation and

---

[86] Before the Second World War, the Rockefeller Institute did not accept direct research grants for the support of the research of individual investigators. After the war, this policy was changed, with large increases in the size of its biochemical research groups.

[87] See Dubos (1976) and McCarty (1985).

[88] See Chapter Three, note 119.

[89] See du Vigneaud (1952) and Hofmann (1987).

in the methods of peptide synthesis, du Vigneaud was always in command.

In Great Britain, the pattern of the institutional development of the biochemical sciences before the First World War was rather different from that in the United States. Although there had been a long tradition of the association of notable chemists with the London hospital medical schools, between 1870 and 1914 the dominant feature of the emergence of biochemistry as a university discipline had become its identification as a part of animal physiology. This attitude, which reflected the opposition to Hoppe-Seyler's aspirations by German physiologists such as Pflüger and Kühne, led to the inclusion of what was termed "chemical physiology" as subdivisions of departments of physiology or, as in the case of Halliburton in London and (later) of Raper in Manchester to the formation of closely directed biochemical research groups in British departments of physiology. The first appointment to a professorship of biochemistry in Britain, at Liverpool in 1902, was given to the physiologist Benjamin Moore [1867–1922],. who proved to be a rather undistinguished scientist, despite his later designation (1920) as Whitley Professor of Biochemistry at Oxford.[90] It was rather from the physiology department in Cambridge that Frederick Gowland Hopkins emerged as the most important figure in the development of biochemistry in Britain before the Second World War.[91]

Before he became lecturer in physiology at Cambridge in 1898, Hopkins had worked for many years at Guy's Hospital, had obtained a medical degree in 1894, and had done chemical research on butterfly pigments, protein chemistry and porphyrins (the last in collaboration with Archibald Edward Garrod [1857–1936]). As Hopkins later wrote in his autobiography, "I went to Cambridge without any training as a specialized biochemist. I had never paid the then orthodox visit to a German laboratory and, indeed, had had no contact with any master of the subject."[92] In 1902, Hopkins was promoted to a readership and, in succeeding years, his income improved markedly through appointments at Emmanuel College and Trinity College. By 1914, when he became professor of biochemistry (with only a modest increase in salary), Hopkins had made several

---

[90] For Moore, see Hopkins (1926) and Morton (1972). Edward Whitley [1879–1945], Moore's associate in Liverpool, was a wealthy man. He helped to endow the professorship of biochemistry at Oxford and to launch the publication of the *Biochemical Journal.*

[91] See Chapter Three, note 103.

[92] Needham and Baldwin (1949), p.21.

contributions that attracted wide notice. With his assistant
Sydney William Cole [1877–1951], he used the Denigès reagent
(mercuric sulfate in sulfuric acid) to isolate tryptophan, and thus
explained the long-known appearance of indole upon the digestion of proteins.[93] Then, he and Edith Gertrude Willcock [1879–
1953] showed that young mice fed zein (which lacks tryptophan)
as the sole source of protein did more poorly than the comparable animals to whose diet tryptophan had been added; these
experiments marked Hopkins's entry into the study of what he
called "accessory food factors" in nutrition. Also, Hopkins collaborated with Walter Morley Fletcher [1873–1933] in studies on
the formation of lactic acid during anaerobic muscular contraction and its disappearance during recovery in air.[94]

There can be little doubt that Hopkins's 16 years as a member of the Cambridge department of physiology were decisive in
molding his style of scientific leadership. Archibald Vivian Hill
[1886–1977] later wrote:

The Cambridge laboratory, in those years before the war in 1914, was
a wonderful place—because of the people in it; the laboratory itself
was pretty awful by present standards. There were probably more great
physiologists there to the square yard than in any other place, before
or since; and not only because there were so few square yards. . . . How
it all happened nobody knows, though I suppose it was an after-effect
of Michael Foster; he died in 1907, but Langley had succeeded him in
1903. It certainly was not a "school" like a school of herring, in which
everyone works on the same subject as the head of the department.
We worked on all sorts of things; Langley never kept that kind of
school. Anyhow it did happen. So, when I am asked why I became a
physiologist, the complete answer really is the extraordinary interest
and variety of work that was going on in that laboratory and the sort
of people who were doing it; also the kindness and encouragement we
youngsters received from them all.[95]

Among the many people Hill listed as workers in the laboratory
before 1914 were, apart from Hopkins, Fletcher and himself,
Edgar Douglas Adrian [1889–1972], Joseph Barcroft [1872–
1947], William Bate Hardy [1864–1934], Keith Lucas [1879–
1916] and Rudolph Albert Peters [1889–1982].

---

[93] See Chapter Three, note 35.
[94] Fletcher and Hopkins (1907).
[95] Hill (1970), pp.35–36. For Hill, see Hill (1965) and Katz (1978); for Langley, see
Chapter Three, note 97.

In 1913, Fletcher became secretary of the newly formed
Medical Research Committee (it was renamed a Council in
1919), and in that capacity played a leading role in securing
funds from the Dunn Trust to endow the professorship of
biochemistry and to construct a building for Hopkins's
department.[96] The new Dunn Institute was opened in 1924, and
during the ensuing 15 years it became a world center of biochemical research and instruction. Hopkins's objectives were
clear:

The greatest need of biochemistry at the moment in my opinion is
equipment which shall make possible the study under one roof (of
course, from its own special standpoint alone) of all living material. No
full understanding of the dynamics of life as a whole, no broad and
adequate views of metabolism can be obtained save by studying with
equal concentration the green plant and micro-organisms as well as the
animal. . . . Such institutes of General Biochemistry would have to be
equipped for teaching as well as for research, for we have to prepare a
future generation to bear burdens greater than our own. Modern
physical chemistry and modern organic chemistry . . . are to provide
entirely new concepts for biochemistry. There must be experts to understand and apply them; a task which will be impossible for those
who must continue to concern themselves with the rapid growth on
other sides of physiology and biology, as it will be for the equally pre-
occupied pure chemist. I do not undervalue the difficulties of equip-
ping and staffing an institute of general biochemistry such as I am
picturing, but I must not stop to discuss them. I will only state that we
have, in not too ambitious fashion, attempted the task in Cambridge.[97]

Between 1924 and 1943, when Hopkins retired from his professorship, several hundred people had worked in the Cambridge
biochemical laboratory.[98] Those of his junior associates who
were there for many years included (in alphabetical order)
such notable biochemists as Malcolm Dixon [1899–1985], John
Burdon Sanderson Haldane [1892–1964], Dorothy Moyle Needham [1896–1987], Joseph Needham [b. 1896], Norman Wingate
Pirie [b. 1907], Marjory Stephenson [1885–1948] and Muriel

---

[96] Kohler (1978). For Fletcher, see Elliott (1934).
[97] Hopkins (1926), pp.35–36. See Fruton (1951) for an echo of this statement.
[98] A lengthy roster of Hopkins's collaborators and colleagues was included in Baldwin
and Needham (1949), pp.333–353. The list provides an essential starting point for
any future study of Hopkins's research school, but omits most of his associates before
1910, and many of the persons who were included do not appear to have published
papers from the Cambridge laboratory.

Wheldale [1880–1932]. Their research interests were in separate fields: enzymes, biochemical genetics, chemical embryology, bacterial metabolism, immunology, the biochemistry of muscle, and much else. Among the many guest investigators from abroad were David Ezra Green [1910–1983], Luis Federico Leloir [1906–1987] and Albert Szent-Györgyi [1893–1986]. In particular, Hopkins extended a welcome to Hans Adolf Krebs [1900–1981] and to other émigrés from Nazi Germany, such as Ernst Friedmann [1877–1946].

Clearly, between 1924 and 1943, the Cambridge laboratory was a beehive of intense activity, reminiscent of the scientific atmosphere of the Hofmeister institute in Strassburg between 1896 and 1914. Although of different background and temperament, both men guided the research of their junior associates in similar fashion. According to Pirie, Hopkins "did not direct research but influenced it by his obvious interest in some aspects rather than in others."[99] After 1924, Hopkins's personal research dealt largely with the study of the substance identified in 1888 by Joseph de Rey-Pailhade [1850–1935], which Hopkins called glutathione and which for a time was considered to be an important participant in biological oxidation processes.

To this brief account of Hopkins's achievements in Cambridge must be added some words about his colleague David Keilin [1887–1963], who headed the Molteno Institute of Parasitology.[100] Through his work on the cytochromes and their role in cellular oxidation processes, Keilin had gained, despite the opposition of Otto Warburg, wide recognition as one of the leading biochemists of his time. A man of broad biological perspective, scientific acumen and exceptional personal qualities, Keilin drew to him many of the younger investigators in the biochemical sciences at Cambridge, not only in Hopkins's department but also in the Cavendish physics laboratory.

Among the others who made important contributions to the development of the biochemical sciences in Britain during the first decades of the twentieth century was William Maddock Bayliss [1860–1924] at University College London. Through the successive editions of his book *Principles of General Physiology* he influenced a generation of biologists and biochemists. Bayliss's most famous experimental achievement came during the course of his collaboration with Ernest Henry Starling [1866–1927] in studies on pancreatic secretion. In 1902, they discovered secretin, a chemical agent elaborated in the intestinal mucosa, which

---

[99] Pirie (1983), p.124.
[100] For Keilin, see Mann (1964), Dixon and Tate (1966) and Keilin (1966).

stimulates the flow of pancreatic juice. Starling introduced the name "hormone" to denote such chemical messengers. Later, Bayliss worked on the physical chemistry of enzymes.

It should also be noted that, during the decade before the First World War, the Lister Institute of Preventive Medicine, established in 1898, had become an important center of biochemical research.[101] Among its members were Allan Macfadyen [1860–1907], John Beresford Leathes [1864–1956], Arthur Harden [1865–1940], Charles James Martin [1866–1955], Henry Drysdale Dakin [1880–1952] and Robert Robison [1883–1941]. Moreover, in 1914, Fletcher's committee established the National Institute for Medical Research; after the First World War, it was headed by Henry Hallett Dale [1875–1968] and by Charles Robert Harington [1897–1972], who encouraged programs of biochemical research.[102] Fletcher was succeeded by Edward Mellanby [1884–1955], who had worked with Hopkins in Cambridge.[103] After the Second World War, few actions of the Medical Research Council were as decisive for the development of the biochemical sciences in Britain as the establishment, in 1947, of the Unit for Molecular Biology at the Cavendish Laboratory in Cambridge, with Max Ferdinand Perutz [b. 1914] as its head. After its transfer to a new building in 1961, the unit became a leading center of research from which there came not only great discoveries but also many of the present-day leaders in the biochemical sciences. I hope that, in time, future historians of science will examine more closely than has been done heretofore the role this research group has played in the development of the biochemical sciences during the latter part of the twentieth century. In the context of the theme of this book, however, I venture to suggest that no small part of the success of the group came from Perutz's style of scientific leadership, in the tradition carried forward in Cambridge by John Newport Langley and Frederick Gowland Hopkins. Indeed, that spirit was evident at other major British universities, most notably at Oxford, where Rudolph Albert Peters [1889–1982] and Hans Adolf Krebs [1900–1981], both of whom had been associated with Hopkins, successively held the professorship in biochemistry.

Since most of this book has dealt with chemical and biochemical groups in pre-World War I Germany, it should be recalled that, between 1919 and 1933, despite military defeat,

---

[101] Chick, Hume and Macfarlane (1971).
[102] Thompson (1973). For Dale, see Feldberg (1970); for Harington, see Himsworth and Pitt-Rivers (1972).
[103] For Mellanby, see Dale (1955) and Platt (1956).

economic collapse and political crisis, Germany retained much
of its past eminence in biochemical investigation. Most of the
leading research groups have been mentioned before, but two
more that should be added are those headed by Carl Neuberg
[1877–1956] and Otto Meyerhof [1884–1951].[104] Neuberg received his Dr.phil. (1900) in Berlin for work with Alfred Wohl,
who introduced him to carbohydrate chemistry, and in succeeding years Neuberg made many valuable contributions in this
field. In 1913 he became head of the biochemical division of the
newly established Kaiser-Wilhelm Institute for Experimental
Therapeutics, directed by the immunologist August von Wassermann. After Wassermann's death in 1925, the institute was
renamed the Kaiser-Wilhelm Institute of Biochemistry, with
Neuberg as its chief but, as a Jew, nine years later he was
obliged by the Nazis to resign this post. Neuberg did not leave
Germany until 1939, and eventually came to New York, where
he continued his research. His sizable group in Berlin published
many papers, the most important of which dealt with the pathways of the fermentation of sugars. Among his more notable pupils was Friedrich Franz Nord [1889–1973].

Otto Meyerhof's scientific career was more illustrious than
that of Neuberg. After his Dr.med. degree (1909) at Heidelberg,
and a brief collaboration with Otto Warburg, Meyerhof joined
the institute of physiology headed by Rudolf Höber [1873–1953]
at Kiel, where he remained until 1924. There, Meyerhof began
his famous work on the chemistry of muscular contraction, and
demonstrated that the disappearance of lactic acid during the
aerobic recovery phase is accompanied by the formation of carbohydrate. This discovery won him a Nobel Prize in 1923, which
he shared with Archibald Vivian Hill. Although Warburg later
claimed Meyerhof (along with Krebs) as his pupil, it would seem
that Hill's paper in 1912 on heat production in muscle exerted a
greater influence.[105] After 1924, Meyerhof worked successively
in Kaiser-Wilhelm Institutes in Berlin and Heidelberg. Although
his position, as a Jew, became increasingly precarious after 1933,
he did not leave Germany until five years later, when he went to
Paris. The German invasion of France forced him to flee to the

---

[104] For Neuberg, see Nachmansohn (1979), pp. 311–327 and Nordwig (1984); to my
knowledge, a complete list of the publications from Neuberg's laboratory (stated to
have been about 900) has not been published. For Meyerhof, see Muralt (1952),
Nachmansohn (1979), pp.268–311 and Schweger (1986). A brief article by Gemmill
(1966) is valuable in giving a glimpse of Meyerhof's style of leadership.

[105] Hill (1912). Apart from his experimental skill, Hill's mathematical knowledge was
extensive, and enabled him to appreciate the significance of Nernst's thermodynamics for biological studies; see pp.507–513 of this paper.

United States, and he continued research at the University of Pennsylvania.[106] Meyerhof's program of research during the 1920s, based on the view that the energy for muscular contraction is derived from the formation and removal of lactic acid, was rudely shaken by what Hill termed the "revolution in muscle physiology."[107] The decisive events in that revolution came from work in several laboratories, notably that of Hofmeister's pupils Embden and Parnas, of Einar Lundsgaard [1899–1968], of Cyrus Hartwell Fiske [1890–1978] and his student Yellapragada SubbaRow [1895–1948], and of Meyerhof's chemical associate Karl Lohmann [1898–1978].[108] During the course of his research in Berlin and Heidelberg, Meyerhof had about 80 junior colleagues among whom, apart from Lohmann, were such later notables as Fritz Lipmann [1899–1986], David Nachmansohn [1899–1983], Severo Ochoa [b. 1905] and George Wald [b. 1906].

Conclusion

At the beginning of this chapter, several questions were posed concerning the relevance of the case histories treated in the preceding chapters to the development of the chemical and biochemical sciences in the recent past. Although definitive answers require more detailed examination of twentieth-century research groups such as those sketched above, there can be little doubt regarding the transmission of the contrasts in the styles of scientific leadership. As during the 1870–1930, but in much larger numbers, after the Second World War there have been benevolent despots who achieved notable success in the study of chemical and biochemical problems through the efforts of their closely directed research groups. The eminent biochemist Hans Adolf Krebs, who had worked in the laboratory of the autocratic Otto Warburg, later wrote:

. . . although the scene has changed some fundamental principles governing successful research are still the same, and will always remain the same—the recognition of leadership in research, the value of long

---

[106] It should be noted that Rudolf Höber also had to leave Germany during the 1930s, and also ended his days in Philadelphia.
[107] Hill (1932).
[108] For an account of these developments, see Fruton (1972), pp.364–372.

training, the need for hard work and dedication, an attitude of humility. . . . Today a different basis of the relations between seniors and juniors has evolved. The master may still rule, and rule firmly, but the basis of his authority is now a natural respect, a natural mixture of admiration and affection which he has earned by his work and conduct; in a good laboratory authority is no longer based on the power invested in a head of a laboratory.[109]

A longer historical perspective will, I think, be required to discern what effect, if any, these changes in the personal relationships between seniors and juniors will have had in molding the next generation of leaders in the biochemical sciences. It is, however, difficult to escape the impression that, in many successful present-day research groups, the admiration (and perhaps affection) of its members for their leaders may have more to do with entrepreneurial skill than with scientific genius. Entrepreneurship is not a new feature of the scientific life, at least since the time of Liebig, and is not inconsistent with outstanding achievement in research, but it may at times, as in the case of Emil Abderhalden,[110] give only the public appearance of scientific distinction. To a greater extent than ever before, the essential ingredients for recognition now appear to be co-authorship of as many papers as possible and large financial support from government agencies, private foundations and chemical firms, as well as prominence in scientific societies and visibility at national and international meetings.

Among the other features of present-day biochemical research by closely directed academic groups is competition in fields deemed to have possible commercial importance for medical and agricultural practice. The elucidation of the structure of nucleic acids and proteins, and the identification of enzymes that act upon them, have led to a biochemical technology that has attracted the interest of leading chemical firms in the United States, West Germany and Japan, as well as to the establishment of smaller enterprises in this field. This development appears to be strikingly similar to the impact of the new organic chemistry of the 1860s on the activities of chemical firms in Wilhelmine Germany, in their search for profitable dyes and drugs. The data presented in the preceding chapters suggest that the principal social function of leading organic-chemical groups, such as that of Emil Fischer, was to provide manpower

---

[109] Krebs and Lipmann (1974), pp.15–16.
[110] See Chapter Five, note 92.

for a burgeoning German chemical industry. I hope that future historians will find it of interest to determine the extent to which this analogy applies to the state of the biochemical sciences in the 1980s.

As will have been evident from the emphasis in this book on the influence of leaders of research groups in the chemical and biochemical sciences on the later contributions of their scientific progeny, I consider the role of the group leader in the education of his junior associates no less important than the research achievements for which he gained fame. In this respect, the styles of scientific leadership exhibited by Hoppe-Seyler, Hofmeister or Hopkins were, in my opinion, more effective than those of their equally distinguished contemporaries Kühne and Kossel. Apart from the differences in their personal qualities, the first three men had a much broader view of the scope of their discipline, and a greater interest in encouraging their junior associates to explore lines of research other than their own. Although it is clear that the closely directed group, perhaps ruled less autocratically than before, has remained a dominant form of the organization of chemical and biochemical research, there are still research laboratories in which broad scientific education and independent initiative are regarded as highly as the faithful performance of assigned tasks. It would be inappropriate to mention such laboratories by name, but I am confident that future historians will readily identify them through the achievements of the scientific progeny of the leaders of these groups, and recognize styles of leadership akin to those of men such as Adolf Baeyer and Frederick Gowland Hopkins.

Appendix 1*

The Liebig Research Group

Allan, James [1825–1866] *(UK)*. Giessen mat. 1844 chem.; Dr.phil. 1846 (worked on zinc salts and uric acid). Later conducted analytical practice in Manchester (1849–54), then in Sheffield, where he taught chemistry at several schools.

ANDERSON, THOMAS [1819–1874] *(UK)*. Edinburgh M.D. 1841; Giessen mat. 1843 chem. (also studied with Berzelius in Stockholm). Regius Prof. chem. Glasgow 1852–74. Published many important chemical articles, especially on the destructive distillation of bone oil (discovered picoline, etc.) and on alkaloids; also wrote extensively on agricultural chemistry.

BABO, LAMBERT von [1818–1899]. Heidelberg Dr.med. 1842; Giessen mat. 1843 chem. (worked on arsenic analysis with Fresenius). Freiburg i.B. Pv.Dz.; ao.Prof. 1854–59; o.Prof. chem. 1859–83. Published extensively on many topics: vapor pressure of salt solutions, plant chemistry, photography.

---

*In this and the subsequent appendices, the names in capital letters denote individuals who achieved a measure of scientific distinction in their later careers. The nationality is only indicated for persons from countries other than the German and Austrian states. Universities are usually denoted in terms of the cities of their location, and Technische Hochschulen as TH. The dates and period of association with the respective research group are approximate, especially when only one date is given. The nature of the work done as a member of the research group is mentioned briefly, only to indicate the subjects under investigation. The statement that "no further publications" appeared only means that no references were found in the standard bibliographical sources cited in Chapter Five, note 115.

BAUMERT, FRIEDRICH MORITZ [1818–1865]. Berlin Dr.med. 1842; medical practice 1842–47; Giessen mat. 1847 chem. (worked on gentianine and brucine). Also studied with Bunsen and Redtenbacher. Breslau Pv.Dz. 1853; Bonn ao.Prof. chem. 1855–. Wrote on various chemical topics (ozone, etc).

Bensch, Friedrich August [1817–?]. Giessen mat. 1844 pharm.; Dr.phil. 1845 chem. (worked on uric acid, sulfur content of bile). Later joined chemical factory in Ringkühl (near Kassel).

Bernays, Albert James [1823–1892] (UK). Giessen mat. 1841 chem.; Dr.phil. 1853. London private laboratory 1845–55; Lecturer chem. St. Mary's Hosp. 1855–60, St. Thomas's Hosp. 1860–92. Published papers on plant chemistry, water and food analysis, hygiene.

Blanchet, Rodolphe [1807–1864] (SWITZ). Giessen mat. 1832 pharm. (worked on analysis of oils, camphor, solanin). Later publications dealt largely with natural history of Vevey region.

Bleibtreu, Hermann [1821–1881]. Giessen mat. 1845 chem.; Dr.phil. 1846 (worked on coumarin). In 1853 founded near Stettin the first German factory for the manufacture of Portland cement.

Blyth, John [1814–1871] (UK). Edinburgh M.D. 1839; Giessen mat. 1843 chem. (worked on narcotine). Also studied in London (Graham), Paris and Berlin (Rose, Magnus). At Royal Coll. Chem. London 1845–47; Royal Agric. Coll. Cirencester 1847–49; Prof. chem. Queen's Coll. Cork 1849–72. After 1845 paper on styrol, devoted his efforts largely to teaching.

Böckmann, Emil [1811–?]. Giessen mat. 1836 chem. (worked on double salts, eugenol). Later chemist in Fries factory in Heidelberg.

Bopp, Friedrich [1824–1849]. Giessen mat. 1844 chem. Asst. of Liebig (worked on hydrolysis of proteins; isolated tyrosine). Died after participation in the uprising in Baden.

BRODIE, BENJAMIN COLLINS, Jr. [1817–1880]. (UK). Giessen mat. 1844 chem.; Dr.phil. 1850 (worked on analysis of beeswax). Research in private laboratory London 1847–55; Waynfleete Prof. chem. Oxford 1855–73. Apart from his experimental work on waxes and other chemical subjects, he is best known for his unsuccessful attempt to introduce a new chemical "calculus of operations."

Bromeis, Johann Conrad [1820–1862]. Giessen mat. 1839 chem. (worked on action of nitric acid on fatty acids); Marburg Dr.phil. 1841. Teacher chem. and physics Realschule Hanau 1842–51. Marburg Pv.Dz. chem. and technol. 1851–57; ao.Prof. 1857–62. His later writings dealt with mineralogy and thermal springs.

BUCHNER, LUDWIG ANDREAS [1813–1897]. Paris studied pharm. and chem. 1834–?; Munich Dr.phil. 1839; Dr.med. 1842. Giessen ca. 1843 (not in mat. list; not evident what experimental work he did). Munich ao.Prof. physiol. and pathol. chem. 1847–52; o.Prof. pharmacy and toxicology 1852–. Made many contributions in plant chemistry and in physiological chemistry.

BÜCHNER, PHILIPP THEODOR [1821–1890]. Giessen mat. 1841 pharm.; Dr.phil. 1842 chem. (worked on maleic acid, tannins, etc.). Chem. teacher Realschule Mainz 1845–55; Darmstadt 1855–63. Darmstadt Polytechnicum ao.Prof. chem. 1863–69; o.Prof. 1869–80. Published many papers on the chemistry of natural products.

Buff, Heinrich [1805–1878]. Geissen mat. 1826 math.; Dr.phil. 1827 chem. (worked on indigo). Chemist in Kestner's factory in Alsace; studied in Paris (Gay-Lussac). Teacher of physics and technol. Kassel 1834 (worked with Bunsen). Giessen o.Prof. physics 1838–. After his initial chemical publications, he only reported on researches in physics.

Buff, Heinrich Ludwig [1828–1872]. Giessen mat. 1851 chem. (worked on analysis of iron compounds). London asst. of Stenhouse (1853) and Hofmann (1854). Started chem. factory Osnabrück 1859–61; Göttingen Dr.phil. 1863. Prof. chem. German Polytechnicum Prague 1869–72.

Campbell, Robert Corbett [1817–1840] *(UK)*. Giessen mat. 1838 chem. (worked on ferrocyanides).

Clemm-Lennig, Karl [1817–1887]. Giessen mat. 1839 pharm.; Dr.phil. 1845 chem. (worked on fatty acids). In 1854 established in Mannheim the first sizable German factory for the manufacture of artificial fertilizers. From 1853 he was a U.S. citizen.

Crasso, Gustav Ludwig [1810–?]. Giessen mat. 1840 chem. (worked on citric acid). Later chem. inspector Royal Porcelain Works in Meissen.

Demarçay, Horace Marc [1813–1866]. *(FR)*. Giessen mat. 1832 philosophy; worked in Liebig's laboratory on fumaric acid, bile. Later became politician.

Döpping, Otto [1814–1863]. Giessen mat. 1842 pharm.; Dr.phil. 1844 (worked on analysis of plant materials). Later chemist Royal Porcelain Factory in St. Petersburg and then (1857) chemist in Nevsky stearin factory.

Dollfus, Charles [1828–1907]. *(FR)*. Giessen mat. 1846 chem.; Dr.phil. 1846 (worked on alkaloids, hippuric acid in blood). Entered family chemical firm. After 1870 went to France, opened hotels in Cannes and Switzerland. In 1881 founded Dollfusville near Oran, Algeria.

Enderlin, Karl Friedrich [1819–1893]. Giessen mat. 1842 chem. (worked on gastric acids, bile). Later established silk dyeing factory in Basle.

Engelhardt, Heinrich Hermann. Giessen mat. 1846 philosophy (worked on muscle lactic acid). Later entered chemical industry.

ERLENMEYER, EMIL [1825–1909]. Giessen mat. 1845 med.; 1847 chem.; Dr.phil. 1851 (worked on lead cyanide with Will). Heidelberg

Pv.Dz. 1857; ao.Prof. 1863–68. Munich Polytechnicum o.Prof chem. 1868–83. One of the leading experimental and theoretical organic chemists of his time.

Ettling, Karl Jakob [1806–1856]. Giessen mat. 1831 pharm.; Dr.phil. 1846; asst. of Liebig, principally in the program of laboratory instruction (worked on creosote, beeswax). In 1846 became Pv.Dz. mineralogy Giessen; ao.Prof. 1849–56.

FEHLING, HERMANN [1811–1885]. Heidelberg 1835–37 pharm.; Dr.phil. 1837; Giessen mat. 1837 chem. (worked on fulminic acid, aledhydes and hippuric acid). In Paris with Dumas 1838. Prof. chem. and technol. Stuttgart Polytechnicum 1839–85. Apart from his many valuable contributions to organic chemistry he invented the "Fehling solution" for sugar analysis.

Fellenberg-Rivier, Ludwig Rudolf von [1809–1878]. (SWITZ). Giessen Dr.phil. 1841 (not in mat. list; thesis on mineral analysis). Director of family paper factory 1835–36; Prof. chem. and mineralogy Lausanne 1841–46; private laboratory in Berne 1846–. Wrote many papers on mineralogy, archeology and chemistry.

Fleitmann, Thoedor [1828–1904]. Giessen mat. 1845 chem.; Dr.phil. 1850; asst. of Liebig 1849–51 (worked on ash analysis of plants, sulfur in proteins, pyrophosphate). In 1851 entered industry and later developed an important method of nickel manufacture.

FOWNES, GEORGE [1815–1849]. (UK). Giessen mat. 1838 chem.; Dr.phil. 1841 (on the equivalence of carbon). Became Prof. chem. Birkbeck Laboratory, University Coll. London. A very promising chemist who made valuable contributions before his untimely death.

Francis, William [1817–1904]. (UK). Giessen mat. 1841 chem.; Dr.phil. 1842 (worked on cocculus indicus). Later principally an editor of chemical journals and book publisher.

FRANKLAND, EDWARD [1825–1899]. (UK). London laboratory of Playfair 1845–47; collaborated with Kolbe. Teacher at Queenswood Coll. 1847 (began work on alcohol radicals). Marburg Dr.phil. 1849 (with Bunsen); Giessen autumn 1849 (not in mat. list). Prof. chem. Owens Coll. Manchester 1851–57; Lecturer chem. St. Bartholomew's Hosp. London 1856–64; Prof. Royal Coll. of Mines 1865–. Made many important contributions, notably on metal-alkyl compounds.

FRESENIUS, KARL REMIGIUS [1818–1897]. After training in pharmacy, in 1840 was in private laboratory of L.C. Marquart in Bonn, and wrote book on qualitative chemical analysis. Giessen mat. 1841 chem.; Dr.phil. 1842; Pv.Dz. 1843; asst. of Liebig. Prof. analytical chem. Agricultural Institute Wiesbaden 1845–48. In 1848 he established his own laboratory. The leading German analytical chemist of his time; his books went through many editions and were translated into many languages.

Gay-Lussac, Jules [1810–?]. *(FR)*. Giessen mat. 1831 chem. (worked on paraffin). Later worked on salicin and lactic acid with Pelouze. Subsequent activity not determined; in 1886 he appears to have resided in Cairo.

GENTH, FRIEDRICH AUGUST (FREDERICK AUGUSTUS) [1820–1893]. Heidelberg 1839–41; Giessen mat. 1841 philosophy (worked in Liebig laboratory on masopine); Marburg mat. 1844 chem.; Dr.phil. 1845; asst. of Bunsen 1845–48. Went to U.S. in 1848; established an analytical laboratory in Philadelphia; for a time (1872–88) Prof. chem. and mineralogy U. of Penna. Published (with O. W. Gibbs) valuable papers on cobalt-ammonia compounds, but most of his numerous publications dealt with mineralogy and analytical chemistry.

GERHARDT, CHARLES [1816–1856] *(FR)*. Schoolmate of Wurtz; Karlsruhe Polytechnicum 1831; after brief service in French cavalry went to Leipzig to study chemistry; first paper (1835) on silicates. Giessen mat. 1836 chem. (stayed about 6 months; did analysis of picric acid; no publication; Liebig asked him to translate Berzelius and Liebig writings into French). Went to Paris 1838 (collaborated with Laurent; Ph.D. 1841). Montpellier Chargé des Cours chem. 1841–44; Prof. 1844–48. Private laboratory in Paris 1848–55. Prof. chem. Strasbourg 1855–56. His experimental and theoretical contributions had a profound influence on the mid-nineteenth-century development of organic chemistry (see TEXT).

GLADSTONE, JOHN HALL [1827–1902]. *(UK)*. Studied in London with Graham; Giessen mat. 1847 chem.; Dr.phil. 1848 (worked on formation of urea from fulminic acid). London Lecturer St. Thomas's Hosp. 1848–50; Fullerian Prof. chem. Royal Institution 1874–77; had private laboratory. Best known for experimental work on chemical equilibria and on refractivity of solutions, and for his efforts to improve British technical education.

GREGORY, WILLIAM [1803–1858]. *(UK)*. Edinburgh M.D. 1828; Giessen 1835, 1841 (not in mat. list; worked on manganates and uric acid). Prof. chem. Anderson Institution Glasgow 1837–38; Aberdeen 1839–44; Edinburgh 1844–58. Published many chemical papers of secondary importance; best known for his English translations of Liebig's most popular books.

GRIEPENKERL, FRIEDRICH [1826–1900]. Giessen mat. 1847 philosophy; Dr. phil. chem. 1848 (worked on role of minerals in potato disease). Also studied with Wöhler. Göttingen ao.Prof. agriculture 1850–57; o.Prof. 1857–. Continued to publish papers in agricultural chemistry.

Guckelberger, Carl Gustav [1820–1902]. Giessen mat. 1845 chem.; Dr.phil. 1848; asst. of Liebig 1847–49 (worked on volatile decomposition products of albumin). Became technical director first of a paper factory, then of a soda factory (retired 1867). Made important improvements in soda manufacture.

Heldt, Wilhelm [1823–1865]. Giessen mat. 1842 chem. (worked on citric acid, santonin); Berlin Dr.phil. 1846. Later papers dealt with bleaching, cement, metallurgy, etc.

Hempel, Karl Wilhelm [1820–1898]. Giessen mat. 1844 chem.; Dr.phil. 1848 (worked on oxidation of fennel oil). Became pharmacist in Giessen; introduced new analytical methods.

HENNEBERG, WILHELM [1825–1890]. Giessen mat. 1846 chem. (worked on ash analysis of blood and on phosphates); Jena Dr.phil. 1849. Director of agricultural station near Göttingen; published mainly on animal nutrition.

Hodges, John Frederick [1815–1899]. *(UK)*. Dublin M.D.; medical practice in Newcastle and Downpatrick; Giessen Dr.phil. 1843 (thesis on Peruvian matico). Prof. chem. Belfast 1845–99; wrote on agricultural subjects.

Hoffmann, (Gustav) Reinhold [1831–1919]. Giessen mat. 1849 philosophy, then chem. (worked on leucine and tyrosine). London 1854 (asst. of Williamson); Heidelberg Dr.phil. 1856 (studied with Kekulé). Became director of ultramarine factory Marienberg, then of Kalle chem. factory Wiesbaden. Wrote papers on chemical technology.

HOFMANN, AUGUST WILHELM [1818–1892]. Giessen mat. 1836 law; Dr.phil. 1841 chem.; asst. of Liebig 1843–45 (worked on aniline, indole). Bonn Pv.Dz., ao.Prof. chem. 1845. Director Royal College of Chemistry London 1845–65. o.Prof. chem. Berlin 1865–92. Made outstanding contributions in organic chemistry, notably on aniline, amines and dyes (see TEXT).

HORSFORD, EBEN NORTON [1818–1893] *(US)*. Giessen mat. 1844 chem. (worked on glycine, nitrogen content of foods). Rumford Prof. Harvard 1847–63; founded Rumford Chem. Co. (baking powder). Many diverse publications.

Hruschauer, Franz [1807–1858]. Vienna M.D. 1831; Giessen 1843 (not in mat. list; worked on albumin). Graz Prof. chem. 1851–58; published papers on agricultural chem.

Ilienkov, Pavel Antonovich [1821–1877] *(RUSS)*. Giessen mat. 1844 chem. (worked on casein, volatile acids in cheese). St. Petersburg ao.Prof. technology 1850–60; director sugar factory 1860–65. Moscow Prof. organic and agronomic chem. Agricultural Acad. 1865–75. Minor publications on various chemical and technical subjects.

Ilisch, Friedrich [1822–1867] *(RUSS)*. Giessen mat. 1841 chem.; Dr.phil. 1844 (worked on acid in potatoes). Also studied in Kharkov and Moscow. Chemist in govt. service 1849–63; wrote on manufacture of vinegar and on food preservation.

JONES, HENRY BENCE [1813–1873] *(UK)*. Studied with Graham and Fownes 1839–40; London M.D. 1841. Giessen mat. 1841 chem.; Dr.phil. 1843 (worked on analysis of plant proteins). London physician St. George's Hosp. 1842–. Many publications on urine analysis (e.g., Bence-Jones protein) and other aspects of medical chemistry.

KANE, ROBERT JOHN [1809–1890] *(UK)*. Lecturer (then Prof.) natural philosophy Royal Dublin Soc. 1834–47. Giessen 1836 (not in mat. list; worked on sulfomethylic acid). Cork President Queen's Coll. 1847–. In 1833 suggested that alcohol, ether and some esters contain the same radical (Liebig named it "ethyl"); in 1837 discovered mesitylene. Also published on other chemical topics and was an active educator.

KEKULÉ, AUGUST [1829–1896]. Giessen mat. 1847 architecture; Dr.phil. chem. 1852 (worked on amyloxysulfates with Will). Studied in Paris 1851; asst. of Planta (1852–53) and of Stenhouse (1853–54). Heidelberg Pv.Dz. chem. 1856–58. o.Prof. Ghent 1858–67; Bonn 1867–92. The most important nineteenth-century German organic chemist (see TEXT).

Keller, Wilhelm [1818–?]. Giessen mat. 1840 chem. (worked on hippuric acid); also studied with Wöhler. Became physician, moved to Philadelphia ca. 1848.

Kerndt, (Carl Huldreich) Theodor [1821–?]. Giessen Dr.phil. 1846 (thesis on analysis of geochronite). Chemist Kuhnheim factory Berlin 1846–47. Leipzig Pv.Dz. 1849–52; teacher at agricultural institute. Published several papers on mineralogy.

Kersting, Richard Georg [1821–1875]. Giessen mat. 1848 chem.; Dr.phil. 1850 (worked on analysis of wines). Also studied in Leipzig and Munich. Became head of a mineral water plant in Riga.

KHODNEV, ALEKSEI IVANOVICH [1818–1883] *(RUSS)*. Giessen mat. 1843 chem. (worked on pectin); also studied in Leipzig. Kharkov Prof. chem. 1848–; wrote on thermochemistry and physiological chemistry; one of the first Russian supporters of Laurent and Gerhardt.

Kleinschmidt, Johann Ludwig. Giessen mat. 1843 pharm. (worked on ash analysis). Later wrote geological papers.

KNAPP, FRIEDRICH LUDWIG [1814–1904]. Giessen mat. 1835 chem.; Dr.phil. 1837 (worked on formation of cyanuric acid from melam). Studied in Paris 1837–38; married Liebig's sister 1841. Giessen ao.Prof. technology 1841–47; o.Prof. 1847–53. o.Prof. technical chem. Munich (also Royal Porcelain Works Nymphenburg) 1854–63; Braunschweig 1863–89. Published many papers, mostly in technical chemistry.

Kodweiss, Friedrich [1803–1866]. Giessen Dr.phil. 1830 (not in mat. list; worked on uric acid). Entered sugar industry; published on manufacture of beet sugar.

KOPP, HERMANN [1817–1892]. Heidelberg mat. 1836 philology; Marburg mat. 1838 chem.; Dr.phil. 1838. Giessen mat. 1839 (worked on decomposition of mercaptans by nitric acid); Pv.Dz. 1841; ao.Prof. 1843–52; o.Prof. 1852–63. Heidelberg o.Prof. chem. 1863–90. Except for one paper (1844) in organic chemistry, his many publications dealt with the physical properties (specific gravity, specific heat, etc.) of chemical substances. He also wrote an important history of chemistry (1843–47).

Kosmann, Constant Philippe [1810–1881] *(FR)*. Giessen mat. 1835 chem.; Dr.phil. 1854 (thesis on Bonleu resin). Published many papers on pharmaceutical chemistry.

Kremers, Peter [1827–?]. Giessen mat. 1848 chem. (worked on sulfurous chloride); Dr.phil. Berlin 1851. Had private laboratory first in Bonn, then in Cologne; published many papers on various aspects of organic chemistry.

KROCKER, EUGEN OTTO FRANZ [1818–1891]. Giessen mat. 1845 chem.; Dr.phil. 1845 (worked on ammonia content of soils, starch, bile). Prof. chem., physics and technol. at Agricultural Academy in Proskau (Silesia) 1847–. Wrote extensively on agriculture.

Laskowski (Lyaskovsky), Nikolai Erastovich [1816–1871] *(RUSS)*. Giessen mat. 1844 chem. (worked on sulfur in proteins); also studied in Berlin and Paris 1843–46. Moscow M.D. 1849; adjunct prof. chem. 1855–. Published several papers on chemical elements.

Lehmann, Julius Alexander [1825–1894]. Giessen mat. 1849 chem.; Dr.phil. 1851 (worked on coffee). Became chief chemist at Agricultural Experiment Station Weidlitz. Wrote extensively on agricultural chemistry and on nutrition.

Lenoir, Georg [1824–1909]. Giessen mat. 1846 chem. (worked on pentathionic acid). Became owner of a chemical firm in Vienna.

Löwe, Julius [1823–1909]. Giessen mat. 1847 chem.; Dr.phil. 1852 (worked on hippuric acid). Established commercial analytical laboratory in Frankfurt a.M.; published many papers in analytical chemistry.

Luck, Eduard [1819–1889]. Giessen mat. 1842 pharm.; Dr.phil. 1845 (worked on acids of Artemesia). Became analytical chemist in Hoechst; published many papers in analytical chemistry.

Macadam, Stevenson [1829–1901] *(UK)*. Giessen Dr.phil. 1853 (not in mat. list; worked on iodine in plants). Edinburgh chemical consultant, lecturer at medical school, active in chemical societies. Published papers on water supply and geology.

MARIGNAC, JEAN CHARLES GALLISARD de [1817–1894] *(SWITZ)*. Giessen mat. 1840 chem. (worked on nitric acid oxidation of naphthalene). Geneva Prof. chem. 1841–78; private laboratory 1878–84. Except for the one organic chemical paper from Giessen, all of his publications dealt with important problems in inorganic chemistry, especially the rare earths, and in physical chemistry.

Marsson, Theodor Friedrich [1816–1892]. Giessen mat. 1841 chem. (worked on laurel fat; discovered lauric acid). Took over father's pharmacy, which he gave up in 1870 to pursue full-time work in botany (he had obtained a Dr.phil. in botany at Greifswald in 1856).

MATTHIESSEN, AUGUSTUS [1831–1870] (UK). Giessen mat. 1852 chem.; Dr.phil. 1853 (no publications?). Heidelberg 1853–57 (Bunsen). London private laboratory 1857–61; Lecturer chem. St. Mary's Hosp. 1862–68; St. Bartholomew's Hosp. 1869–70. Published important papers on the organic chemistry of alkaloids and on the electrical conductivity of metals.

Mayer, Wilhelm [1827–1891]. Giessen mat. 1851 chem.; Dr.phil. 1852; asst. of Liebig (worked on Jalappa resin). Munich Pv.Dz. 1856–57; published many organic-chemical papers before 1858. Then became director of chemical factory in Heufeld.

MELSENS, LOUIS HENRI [1814–1886] (BELG). Giessen Dr.phil. 1841 (not in mat. list; thesis on action of chlorine on stagnant water). Asst. of Dumas (showed that trichloroacetic acid is converted to acetic acid by nascent hydrogen). Brussels Prof. physics and chem. school of veterinary medicine. Wrote on various topics in organic and physiological chemistry and in technology.

Merck, Georg Franz [1825–1873]. Studied chem. in London (Hofmann) 1845. Giessen mat. 1847 chem.; Dr.phil. 1848 (worked on opium; discovered papaverine). After his father's death (1855) he and his brothers assumed management of the family chemical firm.

Meyer, (Hermann Christian) Wilhelm. Giessen mat. 1845 chem. (worked on volatile acids in Angelica). Later activity not determined.

MUSPRATT, JAMES SHERIDAN [1821–1871] (UK). Giessen mat. 1843 chem.; Dr.phil. 1844 (worked on indigo and sulfites). With A. W. Hofmann in London 1845–48. In 1848 he founded Liverpool College of Chemistry. In addition to his chemical papers, he wrote books (see TEXT).

Namur, Joseph François Pierre [1823–1892] (LUXEMBOURG). Giessen mat. 1845 chem. (worked on ash analysis of leaves). Became pharmacist and teacher at Echternach Progymnasium; wrote on mineralogical and agricultural subjects.

NICKLÈS, (FRANÇOIS JOSEPH) JÉRÔME [1820–1869] (FR). Giessen mat. 1845 chem. (worked on tartaric acid fermentation). Prof. chem. Nancy 1854–69; wrote many papers on various chemical topics (fluorine, crystallography, etc.).

Noad, Henry Minchin [1815–1877] (UK). After publishing articles on electricity, studied chemistry with Hofmann in London 1845–47. Prof. chem. St. George's Hosp. London 1847–77. Giessen Dr.phil.

1851 (no evidence of what, if anything, he did in Liebig laboratory). Later wrote extensively on chemistry and electricity.

Nöllner, Carl [1808–1877]. Giessen mat. 1836 pharm. (worked on tartaric acid fermentation). Partner in chemical factory Zoeppritz & Co. Freudenstadt 1840–48; director nitrate factory in Harburg 1854–. Wrote extensively on technical chemistry.

OPPERMANN, CHARLES FRÉDÉRIC [1805–1872] (FR). Giessen mat. 1829 chem.; Dr.phil. 1830 (worked on analysis of waxes, turpentine and naphthalene). Prof. (1835) and Director (1848) of École Supérieure de Pharmacie Strasbourg. Published many papers on organic and pharmaceutical chemistry.

Ortigosa, Vicente [1817–1877] (MEXICO). Giessen mat. 1842 chem.; Dr.phil. 1842 (worked on nicotine and coniine). Later wrote on various subjects, especially nutrition.

OTTO, FRIEDRICH JULIUS [1809–1870]. Jena Dr.Phil. 1832; Giessen 1838 (not in mat. list; worked on solanin). Prof. chem. and pharm. (1842) and Director (1866) Braunschweig Polytechnicum. Many publications in toxicology and technical chem. (described guncotton in 1846). Especially well known for the Graham-Otto textbook of chemistry.

Paul, Benjamin Horatio [1827–1917] (UK). Giessen mat. chem. 1847; Dr.phil. 1848 (worked on alkaloids). Editor of *Pharmaceutical Journal* 1870–1912.

PENNY, FREDERICK [1816–1869] (UK). Giessen Dr.phil. 1842 (not in mat. list; thesis on action of nitric acid on salts). Prof. chem. Anderson's Coll. Glasgow 1839–69. His most important chemical contribution was his 1839 paper on the combining weights of several elements; later papers dealt with analytical topics.

PETTENKOFER, MAX JOSEF von [1818–1901]. Munich Dr.med. & pharm. 1843; Würzburg mat. 1843 chem. (with Scherer; isolated creatinine from urine). Giessen mat. 1844 chem. (worked on meat extract). Munich ao.Prof. med. chem. 1847–52; o.Prof. 1852–65; o.Prof. hygiene 1865–94. A versatile physiologist and chemist; best known for his work with Voit on respiration and for his writings on public health, but also published valuable papers in pure and applied chemistry. Played a significant role in bringing Liebig to Munich in 1852 (see TEXT).

Peyrone, Michele [1814–1885] (ITAL). Turin Dr.med. 1835; hospital physician. Giessen mat. 1842 chem. (worked on action of ammonia on platinum chloride). Turin ao.Prof. chem.; published a few chemical and agricultural papers.

PLANTA, ADOLF von [1820–1895] (SWITZ). Heidelberg mat. 1843 chem.; Dr.phil. 1845. Giessen mat. 1846 chem. (worked on alkaloids). In 1851 set up private laboratory in Reichenau (Kekulé was his assistant 1852–53); published a few papers on alkaloids, but largely wrote about agricultural chem., mineral springs and especially apiculture.

PLANTAMOUR, PHILIPPE [1816–1898] *(SWITZ)*. Giessen mat. 1838 chem.; Dr.phil. 1839 (worked on Peru balsam, acetone, nitration of benzene); also studied with Berzelius in Stockholm. Did research in private laboratory; wrote papers on chemical topics and on limnology, mineralogy and meteorology.

PLAYFAIR, LYON [1818–1898] *(UK)*. Giessen mat. 1839 chem.; Dr.phil. 1840 (worked on myristic acid and caryophyllene). During 1842–58, he held various academic and government posts, including that of adviser to Prince Albert on the 1851 Exhibition. Edinburgh Prof. chem. 1858–68; elected M.P. and spent the rest of his life in politics. Published many chemical papers; the most important later ones dealt with the characterization of nitroprussides.

Poleck, Theodor [1821–1906]. Giessen mat. 1843 chem. (worked on analysis of seeds); Halle Dr.phil. 1849. Managed family pharmacy in Neisse and taught chem. in Realschule 1853–67. Breslau o.Prof. pharm. chem. 1867–1902. Wrote many papers on toxicology, water analysis and public health.

Posselt, Louis [1817–1880]. Heidelberg Dr.phil. 1840; Giessen mat. 1841 chem. (worked on ferrocyanide compounds, analysis of seeds). Heidelberg Pv.Dz. pharm. 1842; ao.Prof. 1847–49; then went to Mexico and California. His later publications (until 1849) dealt with analytical chemistry.

Ragsky, Franz. Vienna Dr.med.; Giessen mat. 1844 chem. (worked on urea analysis). Later published several analytical papers.

REDTENBACHER, JOSEPH [1810–1870]. Vienna Dr. med. 1834; Giessen 1840 (not in mat. list; worked on fatty acids). o.Prof. chem. Prague 1840–49; Vienna 1849–70. Continued to do significant work on fatty acids; also wrote on analysis of mineral waters and on mineralogy.

REGNAULT, HENRI VICTOR [1810–1878] *(FR)*. Giessen 1835 (not in mat. list; worked on action of alkali on ethylene chloride). Paris Prof. chem. École Polytechnique 1840–41; Prof. physics Collège de France 1841–54; Director Sèvres Porcelain Factory 1854–70. After his meticulous work in organic chemistry (1835–39) he conducted equally outstanding studies on specific heats, the physical properties of gases and, with Reiset, on animal respiration.

Richardson, Thomas [1816–1867] *(UK)*. Giessen mat. 1836 chem. (worked on composition of coal, use of lead chromate in organic analysis); also studied in Paris (Pelouze). Became important industrial chemist in Newcastle; among his enterprises was the manufacture of superphosphates. Also published some papers on other chemical subjects, such as the one with R. D. Thomson on emulsin.

Rieckher, Theodor [1818–1888]. Giessen mat. 1842 chem.; Dr.phil. 1844 (worked on fumaric acid). Owner of pharmacy Marbach am Neckar 1845–86.

Riegel, Emil [1817–1873]. Giessen 1840 (not in mat. list; worked on oil of Madia sativa); Karlsruhe Dr.phil. 1845. Opened Pharmazeutisches Institut Carlsruhe at which he taught pharmacists (lectures and laboratory). Published many papers on pharmaceutical chemistry.

ROCHLEDER, FRIEDRICH [1819–1874]. Vienna Dr.med. 1842; Giessen mat. 1842 chem. (worked on camphor, casein, legumin). o.Prof. chem. Lemberg 1845–49; Prague 1849–70; Vienna 1870–74. Made important contributions in theoretical organic chemistry and in the study of the constitution of plant substances.

Ronalds, Edmund [1819–1889] (UK). Giessen mat. 1842 chem.; Dr.phil. 1842 (worked on nitric acid oxidation of wax); also studied in Jena, Berlin, Heidelberg, Zurich and Paris. Prof. chem. Queen's College Galway 1849–56; then moved to Edinburgh, where he had a private laboratory and was associated with the Bonnington Chemical Works.

ROWNEY, THOMAS HENRY [1817–1894] (UK). Giessen Dr.phil. 1852 (not in mat. list; worked on sebacic acid). Prof. chem. Queen's College Galway 1856–. Wrote papers on fats and oils, and on topics in analytical chemistry.

SACC, FRÉDÉRIC [1819–1890] (SWITZ). Giessen mat. 1843 chem.; Dr.phil. 1844 (worked on linseed oil). Prof. chem. Neuchâtel 1845–48, 1866–75; during 1848–66 chemist in factory of Gros et al. in Wesserlingen. Prof. chem. Santiago, Chile 1875–90. Wrote many papers on agricultural chemistry and nutrition.

SANDBERGER, FRIDOLIN [1826–1898]. Giessen mat. 1845 philosophy; Dr.phil. 1846 chem. (thesis on lake minerals). Director Natural History Museum Wiesbaden 1849–55; o.Prof. geology Karlsruhe Polytechnicum 1855–63; o.Prof. mineralogy Würzburg 1863–96. Published many important papers in geology and mineralogy.

SCHERER, JOHANN JOSEPH von [1814–1869]. Würzburg Dr.med. 1836; medical practice 1836–38. Munich mat. 1838 chem.; Giessen 1840–41 (not in mat. list; worked on analysis of proteins). Würzburg ao.Prof. chem. medical faculty 1842–47; o.Prof. 1847–69. Published many important papers in clinical chemistry; discovered inositol and hypoxanthine; led a productive research group at Würzburg.

Schiel, Jakob Heinrich Wilhelm [1813–1889]. Heidelberg Dr.phil. 1842; Giessen mat. 1842 chem. (worked on sanguinain). Heidelberg Pv.Dz. 1845–49, 1859–? (in U.S. 1849–58); moved to Baden-Baden. Wrote on organic chemistry, electricity, geology and philosophy.

Schlieper, Adolf [1825–1887]. Giessen mat. 1844 chem. (worked on decomposition products of uric acid, nitric acid oxidation of cholic acid). After stay in U.S. (1848–51), joined textile firm Schlieper & Baum, founded by his father, and later developed valuable new dyeing methods.

SCHLOSSBERGER, JULIUS EUGEN [1819–1860]. Tübingen Dr.med. 1840; Giessen mat. 1843 chem. (worked on analysis of muscle). Edinburgh 1845–46 (asst. of Gregory). Tübingen Prof. chem. 1847–60. In addition to a valuable textbook of organic chemistry, he published many papers in physiological and analytical chemistry.

Schnedermann, Georg Heinrich Eberhard [1818–1881]. Giessen Dr.phil. 1845 (not in mat. list; worked on cetrarin, ash analysis of oats). Science teacher commerce school Leipzig 1845–47; Prof. chem. (1847–50) and Director (1850–66) technical school Chemnitz.

Schoedler, Friedrich [1813–1884]. Giessen mat. 1834 pharm.; Dr.phil. 1835; asst. of Liebig (worked on fumaric acid, combustion of wood). Science teacher Realschule Worms (1842–54); Rector Realschule Mainz (1854–83). Wrote books and general articles on science.

SCHMIDT, CARL [1822–1894]. Giessen mat. 1843 philosophy; Dr.phil. chem. 1844 (worked on plant mucins, introduced term *Kohlenhydrat*); also studied in Berlin (Rose) and Göttingen (Wöhler). Göttingen Dr.med. 1845. Dorpat Pv.Dz. 1847–50; o.Prof. pharm. 1850–52; o.Prof. chem. 1852–85. Made many valuable contributions in physiol. chem.; also published papers on geochemistry. Created an important research school at Dorpat (one of his students was Wilhelm Ostwald).

SCHUNCK, (HENRY) EDWARD [1820–1903] *(UK)*. Giessen mat. 1840 chem.; Dr.phil. 1841 (worked on nitric acid oxidation of aloe and on lichens); also studied in Berlin (Rose, Magnus). In 1842 became chemical manager of the family textile firm in Belfield, but continued research in his private laboratory. Did important research on plant dyes, especially alizarin, indigo and chlorophyll (see TEXT).

Sell, Ernst [1808–1854]. Pharmacist; Giessen mat. 1832 chem.; Dr.phil. 1834 (worked on analysis of oils). In 1837 partner in chemical firm; in 1842 set up in Offenbach important factory for distillation of tar.

SMITH, JOHN LAWRENCE [1818–1883] *(US)*. South Carolina M.D. 1840; Giessen mat. 1841 chem. (worked on products of distillation of spermaceti); also studied in Paris (Dumas, Orfila). After return to U.S. in 1843, briefly practiced medicine, then became gold assayer, and worked on agricultural chemistry in U.S. and in Turkey. Prof. chem. Louisiana Univ. 1850–52; Univ. Virginia 1852–54. Prof. med. chem. and toxicol. Univ. Louisville 1854–66. Later papers dealt largely with meteorites and mineralogy.

SOBRERO, ASCANIO [1812–1888] *(ITAL)*. Turin Dr.med. 1840; studied in Paris (Pelouze) 1840–43; Giessen mat. 1843 chem. (worked on dry distillation of guaiac resin). Turin Technical Institute Lecturer chem. (1845–49); Prof. (1849–). In 1846 descovered nitroglycerine, which he did not patent; in 1863 Nobel began its manufacture, and in 1867 invented dynamite. Sobrero also published extensively on the chemistry of plant products.

SOKOLOV, NIKOLAI NIKOLAEVICH [1826–1877] *(RUSS)*. Giessen mat. 1850 chem. (worked on occurrence of creatinine in urine); also studied in Paris (Gerhardt, Regnault). Successively Prof. chem. Novosibirisk, St. Petersburg, Odessa. Wrote on various organic-chemical topics; discovered glyceric acid.

Souchay, August. Giessen mat. 1842 chem. (worked on ash analysis of seeds). Later published papers on analytical chemistry.

Spirgatis, (Johann Julius) Hermann [1822–1899]. Jena Dr.phil. 1849; Giessen mat. 1850 chem. (no publication?). Königsberg Pv.Dz. pharm. chem. 1855–61; ao.Prof. 1861–68; o.Prof. 1868–96. Published papers on various topics in pharmaceutical chemistry.

Stähelin, Christoph [1804–1870] *(SWITZ)*. Giessen mat. 1842 chem. (worked on analysis of plant rinds). Basle Dr.phil. physics 1848; Prof. physics 1853–. Later papers only in physics.

Stammer, Karl [1828–1893] *(LUXEMBOURG)*. Giessen mat. 1848 chem. (worked on ash analysis of cabbage); Berlin Dr.phil. 1850. Teacher industrial school Münster 1857; then head of a sugar factory near Breslau. Wrote mostly on technical subjects, especially sugar manufacture.

Stein, (Heinrich) Wilhelm [1811–1889]. Giessen mat. 1839 pharm. (worked on action of acids on sugars; was Liebig's secretary). Prof. techn. and pract. chem. Dresden Polytechnicum 1850–79. Published many papers on assorted chemical topics.

STENHOUSE, JOHN [1809–1880] *(UK)*. Giessen mat. 1839 chem.; Dr.phil. 1840 (worked on hippuric acid, plant oils). Lecturer chem. St. Bartholomew's Hosp. Med. School 1851–57; worked in private laboratory 1860–. Wrote extensively on chemistry of plant products, as well as on technical subjects, notably the use of charcoal filters for disinfecting and deodorizing purposes (see TEXT).

Stoelzel, Carl [1826–1896]. Heidelberg Dr.phil. 1849; Giessen mat. 1850 philosophy (worked in Liebig's laboratory on analysis of inorganic components of ox blood and meat). Teacher at industrial schools in Kaiserslautern and Nürnberg; after 1868 successively ao.Prof. and o.Prof. techn. chem. in Munich. Wrote many papers, mostly on technical subjects.

STRECKER, ADOLPH [1822–1871]. Giessen mat. 1840 chem.; Dr.phil. 1842. After teaching at Realschule Darmstadt 1842–45, returned to Giessen as asst. to Liebig 1846–48 (worked on atomic weight of silver and carbon, and on fibrin and glycoholic acid). o.Prof. chem. Christiania 1851–60; Tübingen 1860–70; Würzburg 1870–71. A gifted experimenter; made important contributions to the study of amino acids and uric acid derivatives, as well as to the development of the periodic table (see TEXT).

Sullivan, William Kirby [1821–1893] *(UK)*. Giessen mat. 1842 chem. (no publication?). At Museum for Irish Industry 1847–73 (Prof. 1854–73); President Queen's College Cork 1873–93. Published on various topics in chemistry, mineralogy and agriculture (beet sugar manufacture).

Thaulow, (Moritz Christian) Julius [1812–1850] *(NOR)*. Giessen mat. 1837 chem. (worked on analysis of rhodizonic acid, cystine, citric acid). Became Prof. chem. Christiania.

Thiel, Karl Eugen [1830–1915]. Giessen mat. 1849 chem.; Dr.phil. 1852 (worked on ash analysis of meat). Darmstadt Lecturer techn. chem. and mineral. 1864–71; ao.Prof. 1871–75; o.Prof. 1875–. Wrote extensively on food production.

THOMSON, ROBERT DUNDAS [1810–1864] *(UK)*. Glasgow M.D. 1831; studied chem. with Thomas Thomson (his uncle). After voyage to India, had medical practice in London. Giessen mat. 1842 chem. (worked on lichens, pine resin, digestion). Glasgow Deputy Prof. and asst. to Thomson 1842–52. London Lecturer chem. St. Thomas's Hosp. 1852–56; later active in public health affairs. Wrote many papers on nutrition and other aspects of physiological chemistry.

Thomson, Thomas, Jr. [1817–1878] *(UK)*. Glasgow M.D. 1839; Giessen 1839 (not in mat. list; worked on pectic acid in carrots). Went to India as a physician and became Prof. botany Calcutta Med. School. 1854–61. Wrote many botanical papers.

Tilley, Thomas George [?–1849] *(UK)*. Giessen mat. 1840 chem.; Dr.phil. 1841 (worked on berberine, nitric acid oxidation of castor oil); also studied in Edinburgh, Paris, Berlin. Prof. chem. Queen's College Birmingham 1845–49. Published papers on chemistry of plant materials (in 1848 with Redtenbacher in Prague).

Unger, Julius Bodo [1819–1885]. Giessen mat. 1845 chem. (worked on uric acid, xanthine, guanine, fibrin). Became owner of soap factory in Hannover.

Varrentrapp, Franz [1818–1877]. Studied pharmacy Lausanne 1832–35; chemistry Berlin 1837–39. Giessen mat. 1839 chem.; Dr. phil. 1840 (worked on fatty acids and, with Will, on new method of nitrogen analysis). Prof. chem. Braunschweig med. school 1844–68; Director Aachen Polytechnicum 1868–77. He was also partner in the Vieweg publishing firm in Braunschweig.

Verdeil, François [1826–?] *(SWITZ)*. Giessen mat. 1845 med.; Dr.phil. chem. 1848 (worked on sulfur determination of organic compounds, ash analysis of blood, bile). In Paris after 1850; with Robin wrote treatise on anatomical and physiological chemistry (1852–53).

Vogel, Julius [1814–1880]. Munich Dr.med. 1838; Giessen mat. 1838 chem. (wrote on theoretical chemistry). Göttingen Pv.Dz. pathology 1839–42; ao.Prof. 1842–46; o.Prof. 1846–. Published extensively on general pathology.

Vohl, Hermann [1823–1878]. Giessen 1845 chem. (worked on chromium analysis). Established private technical chem. laboratory in Bonn (then in Cologne) and gave practical instruction to students. Published many papers on technical subjects, especially fossil fuels and gas lighting.

Voskressensky, Aleksandr Abramovich [1809–1880] (RUSS). Giessen mat. 1837 chem. (worked on analysis of naphthalene, quinic acid). St. Petersburg ao.Prof. chem. 1843–48; o.Prof. 1848–73. Published valuable paper on theobromine (1841).

Wallace, William [1832–1888] (UK). Giessen mat. 1849 chem.; Dr.phil. 1857 (thesis on chloroarsenious acid). Became analytical and consulting chemist in Glasgow, where he served as City Analyst. Published largely on technical and public health problems such as sugar refining, gas manufacture and sewage disposal.

Weidenbusch, Valentin. Giessen mat. 1845 chem.; Dr.phil. 1847 (worked on analysis of albumin, action of acids and alkalis on aldehydes, ash analysis of bile). Established a chemical factory in Odenwald.

WETHERILL, CHARLES MAYER [1825–1871] (US). Giessen mat. 1847 chem.; Dr.phil. 1848 (worked on organic sulfur compounds); also studied in Paris (Pelouze). Set up in Philadelphia a chemical laboratory for commercial analysis and private instruction 1851–53. Chemist Dept. of Agriculture and Smithsonian Institution 1862–65. Prof. chem. Lehigh Univ. 1866–71. Published extensively on technical topics, mineralogy and nutrition.

Whitney, Josiah Dwight [1819–1896] (US). Giessen mat. 1846 chem. (no publication?). Geologist in Michigan, Iowa, California 1847–74. Prof. geol. and mineral. Harvard 1875–. Published many important geological papers and reports.

WILL, HEINRICH [1812–1890]. Giessen 1837 chem.; Dr.phil. 1839 (worked on composition of chelidonin and jervine); asst. of Liebig in research, teaching and editorial work; Pv.Dz. 1844; ao.Prof. 1845–52; o.Prof. 1852–82. Published many papers on analytical chemistry (ash analysis, nitrogen analysis) and on the chemistry of plant products (see TEXT).

WILLIAMSON, ALEXANDER WILLIAM [1824–1904] (UK). Giessen mat. 1844 chem.; Dr.phil. 1845 (worked on bleaching salts, ozone, oenanthol, Prussian blue). Private laboratory in Paris 1846–49 (studied mathematics with Comte, associated with Dumas, Gerhardt, Laurent). Prof. chem. Univ. College London 1849–87. His theory of etherification (1850) was an important component in the nineteenth-century development of organic chemistry.

Winkelblech, Karl Georg [1810–1865]. Giessen mat. 1832 chem. (worked on cobalt oxides); Marburg Dr.phil. 1835; ao.Prof. chem. 1836–38. Prof. chem. Kassel 1839–53, 1861–65 (tried for treason after 1848 revolution).

Wolff, Julius August [1830–1898]. Giessen mat. 1848 chem.; Dr.phil. 1850 (worked on aspartic acid, styracine, madder). Became partner in family dye works in Barmen.

WURTZ, CHARLES ADOLPHE [1817–1884] *(FR)*. Strasbourg M.D. 1843 (thesis on fibrin and albumin); Giessen mat. 1842 chem. (worked on hypophosphorous acid). Paris asst. of Dumas 1844–48; School of Med. Lecturer organic chem. 1849–53; Prof. 1853–66; Sorbonne Prof. chem. 1874–84. Made many important contributions to organic chemistry, and developed valuable synthetic methods (see TEXT).

ZININ, NIKOLAI NIKOLAEVICH [1812–1880] *(RUSS)*. After completing studies in Kazan (1836), worked with Mitscherlich and Rose in Berlin (1837–38). Giessen 1838–39 (not in mat. list; worked on benzoyl compounds, decomposition products of oil of almonds). St. Petersburg Med. Acad. Prof. chem. 1847–74. Published many important papers in organic chemistry, some of them continuations of his work at Giessen.

Appendix 2

Selected Other Liebig Pupils

Amend, Bernard Gottwald [1820–1911]. Stated to have been student and asst. of Liebig ca. 1845 (not in mat. list; no publication). Went to U.S. 1847; set up chemical firm in New York (it later became Eimer & Amend).

Archinard, Jean Jacques François [1819–1890] *(SWITZ)*. Giessen mat. 1844 chem. (no publication). Later activity not determined.

Baist, Ludwig [1825–1899]. Giessen mat. 1848 chem. (no publication); was asst. of Will. Worked in several factories; in 1856 set up large plant to manufacture sulfuric acid, soda, artificial fertilizer, etc. (it later became Chemische Fabrik Griesheim-Elektron in Frankfurt a.M.).

Baldamus, Alfred Ferdinand [1820–1886]. Giessen mat. 1841 chem. (no publication). Became Kommerzienrat, landowner in Gerlebogk and member of the Reichstag.

Bastick, William [1818–1903] *(UK)*. Giessen mat. 1842 philosophy (no publications before 1848). Later pharmaceutical chemist in Buckingham; published on pharmaceutical and chemical subjects.

Beauclair, Louis Theodor de [1813–1846]. Giessen mat. 1832 pharm. (no publication). Became pharmacist in Usingen.

Benckiser, Edmund [1818–1836]. Giessen mat. 1835 chem. (no publication). From chemical manufacturing family.

BERLIN, WILLEM [1825–1902] *(NETH)*. Giessen mat. 1844 chem.; Heidelberg mat. 1846 med.; Leiden Dr.med. (no publications before

1853). Became Prof. anatomy Amsterdam; published valuable papers on hemoglobin crystals.

Bernouilli, Friedrich [1824–1913] *(SWITZ)*. Giessen mat. 1844 chem. (no publication). Became pharmacist.

Bichon, Gerhard Wilhelm *(NETH)*. Giessen mat. 1843 chem.; Dr.phil. 1844 (worked on ash analysis of cereals; translated *Chemische Briefe* into French). Later activity not determined.

Binder, August [1818–?]. Giessen mat. 1837 chem. (no publication). Became pharmacist in Worms.

Bindewald, Hugo [1820–?]. Giessen mat. 1841 pharm. (no publication). Died while pharmacist's apprentice in Worms.

Blank, Hugo [1824–1898]. Giessen mat. 1843 chem. (no publication). Later activity not determined.

Böttinger, (Wilhelm) Heinrich [1820–1874]. Giessen mat. 1843 chem.; Dr.phil. 1844 (worked on ash analysis of wood). Went to England 1847; became factory director in Boston.

Breed, Daniel *(US)*. Dr.; Giessen ca. 1850 (not in mat. list; worked on ash analysis of human brain, phosphate in urine). Translated books by Löwig and Will.

Breidenbach zu Breidenstein, Eberhard [1803–1872]. Giessen mat. 1845 chem. (no publication). Became landowner.

Brill, Louis [1814–1876]. Giessen mat. 1838 pharm. (no publication). Became pharmacist in König.

Brommer, Paul. Giessen mat. 1850 chem.; Dr.phil. 1851 (no publication?). In 1857 published two papers on chemistry of wine.

Buch, Friedrich. Giessen mat. 1843 pharm. (worked on ash analysis of herbs). No further publications found; became pharmacist in König 1850–.

Buchka, Franz Anton [1828–1896]. Giessen mat. 1850 pharm. (no publication). Became owner of Kopfapotheke in Frankfurt a.M..

BUCKLAND, FRANCIS TREVELYAN [1826–1880] *(UK)*. Giessen mat. 1845 chem. (does not appear to have done advanced laboratory work). Later many publications in natural history.

Büchner, Louis Wilhelm [1816–1892]. Giessen mat. 1837 pharm. (no publication). Became pharmacist.

Bujard, Benjamin Louis [1824–1862] *(SWITZ)*. Giessen mat. 1848 chem. (no publication). Became pharmacist in Yverdon.

Bull, Buckland W. *(US)*. Giessen mat. 1848 chem. (worked on emulsin). No further publications found; later activity not determined.

Bullock, John Lloyd [1812–1905] *(UK)*. Giessen mat. 1837 chem. (no publication). Became pharmacist and chemical manufacturer in London; made business arrangement with Liebig and Hofmann to produce quinine.

Caesar, Karl [?–1891]. Giessen mat. 1841 pharm. (no publication). Became pharmacist in Katzelnbogen.

Cameron, William [1822–1855] *(UK)*. Giessen mat. 1838 nat. sci. (no publication). Became medical officer in British army; died on service in India.

Clemm, Gustav [1814–1866]. Giessen ca. 1840 (not in mat. list; worked on analysis of sea water). Joined his brother Karl (see Appendix 1) in their chemical firm.

Cohen, Jacob [1822–?]. Giessen mat. 1850 chem. (worked on ash analysis of seeds); Heidelberg mat. 1852 chem. Later activity not determined.

Conn, Franz Karl Friedrich. Giessen mat. 1847 chem. (no publication). In 1859 bought Elephantenapothek in Hamburg.

Conrad, Friedrich Ferdinand [1826–1857]. Giessen mat. 1850 pharm. (no publication). Became pharmacist in Gernsheim.

Crichton, James [?–1868] *(UK)*. Edinburgh M.D. 1835; Giessen mat. 1846 chem. (no publication). Became physician in Glasgow.

Crum, Alexander [1828–1893] *(UK)*. Giessen mat. 1844 chem. (worked on solubility of calcium phosphate in acid). No further scientific publications; became merchant in Glasgow and was M.P. 1880–85. Son of Walter Crum [1796–1867].

Curtze, Philipp Heinrich [1809–?]. Giessen mat. 1832 pharm. (no publication). Became pharmacist in Worms.

Darby, Stephen [1825–1911] *(UK)*. Giessen mat. 1847 chem. (worked on analysis of ammonium chromate and mustard oil). Later wrote on diastases and on the history of Cookham.

Denecke, Ferdinand. Giessen mat. 1846 chem.; Dr.phil. 1851 (worked on analysis of mineral waters). No further scientific publications; later activity not determined.

Dieffenbach, Ernst Johann [1811–1855]. Giessen mat. 1828 med. (no publication from Liebig laboratory); Zürich Dr.med. 1835. Explored New Zealand 1839–41; in England 1845–46 as Liebig's agent to promote his artificial fertilizer. Giessen Pv.Dz. mineralogy 1849–50; ao.Prof. 1850–55. Published papers on geological subjects.

Drevermann, August. Giessen mat. 1846 chem.; Dr.phil. 1857 (worked on crystallization of minerals). Later activity not determined.

Dunlop, Charles J. *(UK)*. Giessen mat. 1842 chem. (no publication). Later held patents for a chlorine process and for the recovery of manganese by bleachers.

Eatwell, William [1819–1899] *(UK)*. Giessen mat. 1837 chem. (no publication). Glasgow M.D. 1840. Surgeon in India 1841–57; Principal Calcutta Medical College 1857–61.

Ehrhardt, Wilhelm [1825–?]. Giessen mat. 1847 pharm. (no publication). Became pharmacist in Darmstadt.

Elbers, (Johann) Christian [1824–1911]. Giessen mat. 1850 chem.; Dr.phil. 1852 (worked on molybdic acid). No further scientific publications; joined family textile factory.

Engelmann, Christian Gotthold [1819–1884]. Giessen mat. 1844 chem. (worked on ash analysis of plant materials). Became pharmacist in Basle.

Ettling, Friedrich Karl [?–1889]. Giessen mat. 1841 chem. (no publication). Became pharmacist in Kirchheimbolanden.

Faber, Karl [1822–?]. Giessen mat. 1843 pharm. (no publication). Became pharmacist in Crumstadt.

Faber, William Leonard (US). Giessen mat. 1850 chem. (no publication). In 1852 was metallurgist and mining engineer.

FEILITZSCH, FABIAN KARL OTTOKAR von [1817–1885]. Bonn Dr.phil. 1841; Giessen mat. 1842 chem. (no publication). Greifswald ao.Prof. physics 1848–54; o.Prof. 1854–. Later papers dealt solely with magnetism.

Feyen, Franz. Giessen mat. 1848 pharm. (no publication). Became pharmacist in Lorsch.

Fink, Alexander. Giessen mat. 1848 pharm. (no publication). In 1855 wrote a botanical paper; in 1862 became pharmacist in Lorsch.

Fleck, Wilhelm Hugo [1828–1896]. Giessen mat. 1850 pharm. (no publication); Dresden Dr.phil. 1857. o.Prof. chem. and physics Dresden TH 1862–71; Director Center of Public Health Dresden 1871–94. Wrote on technical subjects and on sanitation.

Gaedechens, Julius Heinrich [1820–1862]. Giessen mat. 1844 chem. (no publication). Later activity not determined.

Gail, Georg [1819–1882]. Giessen mat. 1843 pharm. (no publication). Later joined family textile firm.

Gardner, John [1804–1880] (UK). Giessen Dr.med. 1843 (not in mat. list; does not appear to have worked in Liebig's laboratory). London Prof. chem. and materia medica Apothecaries Co. Translated Liebig writings; played role in foundation of Royal College of Chemistry.

Geiger, Gustav [1819–1900]. Giessen mat. 1843 pharm. (worked on analysis of lymph). Became manufacturer of dyes, then malt extract, in Stuttgart.

Geromont, Karl. Giessen mat. 1830 pharm; 1835 med. (worked on alcohol content of wine from Bingen). No further scientific publications; became physician.

GIBBS, OLIVER WOLCOTT [1822–1908] (US). M.D. 1845 Coll. Phys. and Surg. N.Y.; studied in Berlin (Rose, Rammelsberg); Giessen mat. 1846 chem. (does not appear to have done research in Liebig's laboratory). New York Free Academy Prof. physics and chem. 1849–63; then Rumford Prof. Harvard 1871–87. Published extensively on topics in inorganic and analytical chemistry, notably on cobalt-ammonia compounds (with Genth).

GILBERT, JOSEPH HENRY [1817–1901] *(UK)*. Giessen mat. 1840 chem. (no publication). In 1843 joined Lawes at Rothamsted, where they conducted important agricultural research; by 1851 they had shown that some of Liebig's views on soil nutrition were incorrect (see TEXT).

Gindroz, Théophile [1813–1872] *(SWITZ)*. Giessen mat. 1838 chem. (no publication). Became pharmacist in Morges.

Giulini, Lorenz [1824–1898]. Giessen mat. 1842 chem. (no publication); Heidelberg Dr.phil. 1845. Became head of chemical factory in Ludwigshafen.

Glasson, Karl Eduard *(RUSS?)*. Giessen mat. 1846 chem.; Dr.phil. 1847 (worked on theobromine, ash analysis). No further publications found; later activity not determined.

Glogau, Henrik Moritz [1821–1877] *(NOR)*. Schooling in Germany; Giessen mat. 1844 chem. (no publication). After medical studies at Jena turned to geography and economics. Secretary Chamber of Commerce Frankfurt a.M. 1863–77.

Gravelius, Georg [1808–?]. Giessen mat. 1837 pharm. (no publication). Became pharmacist in Babenhausen.

Groll, Karl. Giessen mat. 1848 pharm. (no publication). Later activity not determined; in 1863 published paper in analytical chemistry.

Gros, James [1817–?]. Giessen mat. 1837 chem. (worked on platinum salts). No further scientific publications; joined family textile factory in Wesserlingen.

Gundelach, Karl [1821–1878]. Giessen mat. 1838 chem; Dr. phil. 1846 (worked on bile with Strecker). No further chemical publications; became owner of chemical factory in Luisenthal.

Haeffely, Edouard *(FR)*. Giessen mat. 1848 chem. (worked on pigment from sandalwood). Became dye chemist in Mulhouse.

Haidlen, (Paul) Julius [1818–1883]. Giessen mat. 1842 pharm.; Dr.phil. 1843 (worked on analysis of milk). No further chemical publications; took over family pharmacy in Stuttgart.

Hardy, Edmund [1816–1878]. Giessen mat. 1835 pharm. (no publication). Became pharmacist.

Hartmann, Jules Albert [1823–1905] *(FR)*. Giessen mat. 1841 chem. (no publication). Worked in several textile dyeing factories in Mulhouse and elsewhere.

Hautz, (Friedrich) Oswald. Giessen mat. 1845 chem. (worked on double salts). No further chemical publications; later activity not determined.

Hegmann, Friedrich [1813–1860]. Giessen mat. 1837 pharm. (no publication). Went to U.S.; became pharmacist in New York City.

Helmolt, August von [1829–?]. Giessen mat. 1849 chem. (no publication). Became physician.

Helmolt, Otto von [1829–1901]. Giessen mat. 1847 chem. (no publication). Became physician.

Henry, William Charles [1804–1892] (UK). Edinburgh M.D. 1827; studied chem. in Berlin (Rose, Mitscherlich); Giessen 1836 (not in mat. list; does not appear to have done research). Became landowner in Surrey.

Hering, Édouard [1814–1893] (FR). Giessen mat. 1838 chem. (worked on reactions of sufurous acid). Became pharmacist in Barr.

Hertwig, Carl [1820–1896]. Giessen mat. 1842 chem. (worked on ash analysis of plants). Became cigar manufacturer in Mühlhausen.

Hess, Isidor. Giessen mat. 1850 chem.; Dr.phil. 1853 (no publication). Later activity not determined.

Heydenreich, Eduard [?–1885]. Giessen mat. 1845 chem. (no publication). Became pharmacist in Strassburg.

Heyl, Adolph. Giessen mat. 1846 pharm.; Dr.phil. 1847 (worked on analysis of bell metal and coal). No further publications; later activity not determined.

Hofstetter, Johann Josef (SWITZ). Giessen mat. 1842 chem. (worked on analysis of fruit rinds and nitrate from Peru). No further publications; later activity not determined.

Jacobi, Bernhard. Giessen Dr.phil. 1843 (not in mat. list; thesis on effect of soil nutrition on plant growth). Later wrote botanical papers.

Jamieson, Alexander John (UK). Giessen mat. 1845 chem. (worked on sulfur cyanide compounds). No further chemical publications; later activity not determined.

Janosi, Ferenc [1819–1879]. Giessen mat. 1846 chem. (no publication). Became teacher, journalist and writer on popular science in Hungary.

Jobst, Karl [1816–1896]. Giessen mat. 1836 chem. (worked on analysis of sarsparilla). Entered family chemical firm in Stuttgart.

Johnson, Carl (US). Giessen mat. 1848 chem.; Dr.phil. 1851 (worked on ash analysis of cheese). No further scientific publications; later activity not determined.

Kayser, Gustav Adolf [1817–1878]. Giessen mat. 1843 chem.; Dr.phil. 1844 (worked on double salts, jalap resin). A further paper (1864) on meteorology; became pharmacist in Hermannstadt.

Keller, Franz. Dr.phil.; Giessen mat. 1848 chem. (worked on analysis of plant materials, ash analysis of meat). Later activity not determined.

Kessler, Georg [1828–1873]. Giessen mat. 1848 chem.; Dr.phil. 1849 (worked on tartrates). No further chemical publications; later activity not determined.

Koch, Karl Jakob Wilhelm [1827–1882]. Giessen mat. 1849 chem. (no publication.) In 1853 became head of iron works in Dillenberg; in 1873 member of Geological Institute Wiesbaden. Wrote papers on mineralogy.

Kocher, Rudolph Friedrich [1811–1875] *(SWITZ)*. Giessen mat. 1838 chem. (no publication). Became pharmicist.

Koechlin, Jean Albert [1818–1889] *(FR)*. Giessen mat. 1844 chem. (worked on ash analysis of wood). No further scientific publications; joined the family textile firm in Mulhouse.

Krüger, Rudolf [1815–1846]. Giessen mat. 1838 pharm. (no publication). Became pharmicist in Korbach.

Kühnert, Ernst [1818–?]. Giessen mat. 1838 chem. (worked on analysis of coal). No further scientific publications; later activity not determined.

Kugler, Ludwig [1827–1894]. Giessen mat. 1845 chem. (worked on analysis of lead cyanide). Became pharmacist in Gnesen.

Kyd, John *(UK)*. Giessen mat. 1848 chem. (worked on nitroprusside compounds). No further scientific publications; later activity not determined.

Lade, Friedrich Gustav [1821–1856]. Giessen mat. 1844 chem. (worked on water analysis, glycyrrhizin). No further scientific publications; later activity not determined.

Langsdorff, Wilhelm [1827–1898]. Giessen mat. 1846 public affairs; Dr.phil. chem. 1852 (thesis on silver as unit in measurement of electrical resistance). No further chemical publications; became geologist and botanist in Clausthal.

Leers, Heinrich Gustav. Giessen mat. 1850 chem.; Dr.phil. 1852 (worked on composition of quinidine). Became pharmacist.

Leers, Ludwig [1812–1860]. Giessen mat. 1832 pharm. (no publication). Became pharmacist.

Lehmann, Johann [1823–1899]. Giessen mat. 1847 chem. (no publication). Became pharmacist in Rensburg.

Lehr, Gustav [?–1892]. Giessen mat. 1843 pharm. (no publication). Became pharmacist in Bensheim; then Dr.med. and director of sanatorium in Wiesbaden.

Lengerke, Ernst August Karl von [1823–1870]. Giessen mat. 1842 chem. (no publication). Became pharmacist.

Lenz, August. Giessen mat. 1840 chem. (worked on double salts). No further scientific publications; later activity not determined.

Leuchtweiss, Alexander. Giessen mat. 1843 pharm. (worked on ash analysis of seeds, manna). No further scientific publications; later activity not determined.

Leverkus, Carl Friedrich Wilhelm [1804–1889]. Studied in Paris 1827–28; Berlin 1829; Giessen Dr.phil. 1830 (thesis on silver; not in mat. list and probably did not work in Liebig's laboratory). Partner in soda factory in Barmen 1830–33. Obtained patent (1838) for production of ultramarine; in 1860 established first large dye factory in Prussia. It was bought by the Bayer Co. in 1890, and later (1925) incorporated into the I.G. Farbenindustrie.

Liebig, (Georg) Karl [1818–1870]. Brother of Justus Liebig. Giessen mat. 1842 pharm. (no publication). Became pharmacist in Darmstadt.

Liesching, Franz [1818–1903]. Giessen mat. 1843 pharm. (no publication). Later with Chamber of Commerce and Industry Stuttgart.

Linck, Christian. Giessen mat. 1842 chem.; Dr.phil. 1845 (no publication). Went to U.S. and published on various chemical topics there.

Lindsay, Thomas (UK). Giessen mat. 1847 chem. (no publication). Later wrote popular science articles.

Loew, Wilhelm Christian [1818–1908]. Giessen 1841 pharm. (no publication). Became pharmacist in Markt-Redwitz. Father of Oscar Loew [1844–1941].

Mackenzie, Kenneth Smith [1832–1900] (UK). Giessen mat. 1850 chem. (no publication). Landed gentry in Rossshire.

Maddrell, Robert (UK). Giessen mat. 1845 chem. (worked on phosphates, lactic acid). No further chemical publications; later activity not determined.

Mallinckrodt, Gustav [1829–1904]. Giessen mat. 1847 chem. (no publication). Joined family chemical firm.

Mangold, Friedrich Wilhelm [1827–1898]. Giessen mat. 1849 pharm. (no publication). Became pharmacist in Darmstadt.

Marty, Rudolph [1829–1909] (SWITZ). Giessen mat. 1849 chem. (no publication). Became pharmacist.

Mayer, Ferdinand [?–1869]. Giessen mat. 1850 chem. (no publication). Went to U.S.; became pharmacist in New York City.

Meidinger, Johann Hermann [1831–1905]. Giessen mat. 1849 chem. and physics; Dr.phil. 1853; also studied Heidelberg 1853–55, Paris and London 1855–56. Heidelberg Pv.Dz. technology 1857–69; Karlsruhe Polytechnicum o.Prof. technical physics 1869–. All his publications were on physics and on technical subjects.

Mertzdorff, Charles [1818–1883] (FR). Giessen mat. 1836 chem. (no publication). Became important Alsatian industrialist.

Meyer, Hermann. Giessen mat. 1839 chem.; Dr.phil. 1840 (worked on elaidic acid). Became pharmacist.

Meyer, Johann Ludwig [1819–1894] *(SWITZ)*. Giessen mat. 1843 pharm. (no publication). Became pharmacist.

MILLER, WILLIAM ALLEN [1817–1870] *(UK)*. Giessen mat. 1840 chem. (no publication). London King's College M.D. 1842; Prof. chem. 1845–. Made important contributions to spectroscopy.

Mitchell, Alexander [1822–1874] *(UK)*. Giessen mat. 1839 chem. (no publication). Became partner in coal firm in Glasgow.

Möricke, Emil [1822–1897]. Giessen mat. 1845 pharm.; Dr.phil. 1846 (no publication). Became pharmacist in Wimpfen.

Möricke, Martin [1824–1881]. Giessen mat. 1849 pharm. (no publication). Became pharmacist in Winnenden.

Mohr, Philipp [?–1885]. Giessen mat. 1850 chem.; Dr.phil. 1853 (no publication). Became pharmacist in Würzburg.

Müller, Christian [1816–?]. Giessen mat. 1836 pharm. (no publication). Became pharmacist in Berne.

Muspratt, Edmund Knowles [1833–1923] *(UK)*. Giessen mat. 1850 chem. (no publication). Joined family chemical firm.

Muspratt, Frederick [1825–1872] *(UK)*. Giessen mat. 1843 chem. (no publication). Joined family chemical firm.

Muspratt, Richard [1822–1885] *(UK)*. Giessen 1840 (not in mat. list; no publication). After initial partnership in alkali company (1841–52), set up a new such company with two of his brothers; entered politics.

Papon, Jakob [1827–1860] *(SWITZ)*. Giessen mat. 1846 (no publication). Later wrote articles on geological subjects.

Parkinson, Robert [1831–1913] *(UK)*. Giessen Dr.phil. 1853 (not in mat. list; worked on valeraldehyde). Became partner in chemical firm in Bradford.

Petersen, (Daniel) Christian. Giessen mat. 1834 pharm. (worked on analysis of several drugs). No further chemical publications; became pharmacist in Apenrode.

Petry, Philipp [?–1896]. Giessen mat. 1850 pharm. (no publication). Became pharmacist in Frankfurt a.M.

Pistor, Hermann [1822–1883]. Giessen mat. 1846 chem. (no publication). Became pharmacist in Mainz.

Polunin, Aleksei Ivanovich [1820–1888] *(RUSS)*. Giessen mat. 1845 chem. (no publication). Moscow Dr.med. 1848; Prof. pathol. anat. and physiol. 1849–.

Porter, John Addison [1822–1866] *(US)*. Giessen mat. 1847 chem. (worked on ash analysis of feces, action of nitric acid on wood fibers). Yale Univ. Prof. analytical and agricultural chem. 1852–56; Prof. organic chem. 1856–64; Dean Sheffield Scientific School

1861–64. In 1855 married Josephine Sheffield, daughter of donor after whom the school was named in 1861. Porter published several textbooks but few experimental papers.

Price, David Simpson [1823–1888] *(UK)*. Giessen mat. 1846 (no publication). Later became superintendent of the Technological Museum at the Crystal Palace; wrote papers on topics in technical chemistry.

Quentel, Eduard [1823–1865]. Giessen mat. 1839 chem. (no publication). Became agriculturist; emigrated to Parana, Brazil.

Radcliff, William *(UK)*. Giessen mat. 1841 chem. (worked on nitric acid oxidation of spermaceti). No further chemical publications; became physician.

Rehe, Johann August [?–1892]. Giessen mat. 1843 chem. (no publication). Became pharmacist in Cologne.

Reissig, Wilhelm [1829–1901]. Giessen mat. 1850 pharm. (no publication). Later wrote papers on topics in analytical chemistry.

Reuling, Ludwig [1811–1879]. Giessen mat. 1836 pharm. (worked on chelidonic acid). Became physician in Wöllstein.

Reuling, Robert [1808–1852]. Giessen mat. 1830 pharm. (no publication). Became pharmicist in Darmstadt.

Reynier, Henri Frédéric [1824–1902] *(SWITZ)*. Giessen mat. 1842 chem. (no publication). Later activity not determined.

Ricker, Albin Heinrich [1811–1852]. Giessen mat. 1835 pharm. (no publication). Became pharmacist in Kaiserslautern.

Rittershausen, Friedrich [?–1875]. Giessen mat. 1839 pharm. (no publication). Became pharmacist in Herborn.

Römheld, Julius [1827–1901]. Giessen mat. 1849 chem. (no publication). Went to U.S.; became pharmacist in Chicago.

Rogers, John Robinson *(UK)*. Giessen mat. 1846 chem.; Dr.phil. 1848 (worked on ash analysis of feces). No further scientific publications; later activity not determined.

Rosengarten, Samuel George [1827–1908] *(US)*. Giessen mat. 1847 chem. (worked on nitric acid oxidation of benzene). No further scientific publications; joined family chemical firm in Philadelphia.

Roser, Gustav [1823–1860]. Giessen mat. 1848 chem. (worked on ash analysis of blood, phloridzin). No further scientific publications; later pharmacist in Schwäbisch Hall.

Rubach, Wilhelm. Giessen mat. 1849 chem.; Dr.phil. 1850 (worked on analysis of bar iron). No further scientific publications; later activity not determined.

Rübsamen, Karl [1826–1902]. Giessen mat. 1848 pharm. (no publication). Became pharmacist in Frankfurt a.M.

Rüling, Eduard [1811–1875]. Giessen mat. 1845 chem.; Dr.phil. 1846 (worked on sulfur analysis of plant and animal materials; ash analysis of plants). No further scientific publications; later activity not determined.

Saalmüller, Louis Eduard. Giessen mat. 1845 chem. (worked on fatty acids of castor oil). No further scientific publications; later activity not determined.

Sander, Wilhelm [1812–1881]. Giessen mat. 1832 pharm. (no publication). Became pharmacist and botanist; wrote on flora of Hamburg region.

Sandmann, Friedrich [1818–1876]. Giessen mat. 1850 pharm.; Dr.phil. 1853 (worked on analysis of galena). No further scientific publications; later activity not determined.

Schaffner, Ludwig Friedrich Carl. Giessen mat. 1843 pharm. (worked on magnesium phosphate, composition of hydrates). No further scientific publications; later activity not determined.

Schedel, Henry Edward [1804–1856] (UK). Giessen mat. 1847 chem. (no publication). Published books on various non-chemical topics.

Schild (Schilt), Josef [1824–1866] (SWITZ). Giessen mat. 1846 chem. (no publication). Chemistry teacher in several cantonal schools 1854–66. Founder and first President Swiss Alpine Society; published on agricultural chemistry, as well as on Alpine geology and meteorology.

Schlenther, Emil [?–1892]. Giessen mat. 1842 chem. (no publication). Became pharmacist in Insterburg.

Schlienkamp, Christian. Giessen mat. 1849 chem.; Dr.phil. 1849 (worked on ash analysis of cabbage and asparagus). No further scientific publications; later activity not determined.

Schlosser, Theodor [1822–1907]. Giessen mat. 1843 chem. (worked on bile). Vienna Dr.phil. 1848 (thesis on history of chemistry). Later activity not determined.

Schmid, Wilhelm. Giessen mat. 1847 pharm.; Dr.phil. 1854 (worked on mangosteen). No further chemical publications; later activity not determined.

Schulthess, Edmund [1826–1906] (SWITZ). Giessen mat. 1846 chem. (no publication). Became landowner in Aargau.

Schwarzenberg, Adolf Emil. Giessen mat. 1846 chem. (worked on pyrophosphates, bismuth salts). No further scientific publications; later activity not determined.

Scriba, Emil [1814–1886]. Giessen mat. 1836 pharm. (no publication). Became pharmacist; later published on identification of blood spots.

Scriba, Theodor [?–1886]. Giessen mat. 1844 pharm. (no publication). Became pharmacist in Schotten.

Siebold, Georg von [1812–1873]. Giessen mat. 1831 pharm. (no publication). Became pharmacist in Mainz.

Silber, Gustav [1826–1904]. Giessen mat. 1850 (no publication). Later chemist in Stuttgart.

SMITH, ROBERT ANGUS [1817–1884] *(UK)*. Giessen mat. 1840 chem.; Dr.phil. 1841 (studied German language and philosophy; apparently no chemical publications). Manchester Royal Institution (asst. of Playfair) 1842–45; consulting chemist 1845–. Became a leading authority on sanitary chem.; wrote important papers and reports on urban air and water.

Soldan, Friedrich [1817–1881]. Giessen mat. 1846 chem. (no publication). Became pharmacist.

Stein, James *(UK)*. Giessen mat. 1848 chem. (worked on arsenious acid salts). No further scientific publications; later activity not determined.

Steiner, Cäsar Heinrich [1813–?] *(SWITZ)*. Giessen mat. 1837 chem.; Dr.phil. 1838 (no publication). Became pharmacist.

Sthamer, (Johann Georg) Bernhard [1817–1903]. Giessen mat. 1842 chem. (worked on analysis of wax); Rostock Dr.phil. 1845. Head of chemical pathology laboratory in Rostock hospital; then set up chemical factory in Hamburg.

Strecker, Hermann. Giessen mat. 1850 chem. (no publication). Asst. of his brother Adolph in Christiania; later head of analytical laboratory of Meister, Lucius & Brüning in Hoechst; published several chemical papers.

Summer, Thomas Jefferson *(US)*. Giessen mat. 1846 chem. (worked on analysis of cotton plant and seed; results published posthumously in 1852 by Wetherill).

Tenner, Alfons [1829–1898]. Giessen mat. 1850 pharm. (no publication). Became pharmacist in Darmstadt.

Theyer, Joseph. Giessen mat. 1843 chem. (worked on bile). No further scientific papers; later activity not determined.

THUDICHUM, LUDWIG [1829–1901]. Giessen mat. 1847 med.; Dr.med. 1851 (no publications from Liebig's laboratory). London physician 1853–; Prof. chem. St. George's Hosp. medical school 1858–65; Director pathol.-chem. laboratory St. Thomas's Hosp. 1865–71. Made important discoveries on brain lipids and on animal pigments (see TEXT).

Thurn, Georg Wilhelm [1813–?]. Giessen mat. 1830 pharm.; Dr.med. 1835 (no publication). Became physician in Babenheim.

Tillmanns, Heinrich [1831–1907]. Giessen mat. 1850 chem. (worked on mineral water analysis). No further chemical publications; became a chemical manufacturer in Krefeld.

TRAUBE, MORITZ [1826–1894]. Giessen mat. 1844 philosophy; Berlin Dr.phil. 1847 (thesis on chromium compounds). Managed family wine concern in Ratibor; set up private laboratory. Made outstanding contributions, notably in the study of biological oxidation and semipermeable membranes (see TEXT).

TRAUTSCHOLD, HERMANN von [1817–1902]. Giessen mat. 1844 chem.; asst. of Liebig 1845–46; Dr.phil. 1847 (no publication). Went to Russia, was teacher in Moscow, and became a mineralogist (Dorpat Dr.phil. 1871). Moscow Prof. mineralogy and geology at Agricultural School 1868–88. Wrote important articles on the geology and paleontology of Russia.

Treupel, Ernst Wilhelm [1828–1871]. Giessen mat. 1847 chem. (no publication). Later activity not determined.

Tribolet, Georges de [1830–1873] (SWITZ). Giessen mat. 1850 chem. (no publication); Heidelberg Dr.phil. 1853. A wealthy man; traveled extensively; later wrote mostly on geological and paleontological topics.

Tschudi, Joachim [1822–1893] (SWITZ). Giessen mat. 1842 chem. (no publication). Entered family dye firm.

Vogel, August [1817–1889]. Munich Dr.med. 1839; Giessen 1840 (not in mat. list; no publication); Erlangen Dr.phil. 1848. Munich ao.Prof. agric. chem. 1848–69; wrote articles on many subjects, especially plant chem.

VOGT, KARL [1817–1895]. Giessen mat. 1833 med. (worked in Liebig's laboratory; no publication); Berne Dr.med 1839. Giessen Prof. zoology 1872–95. Became well-known zoologist and also wrote on anthropology and philosophy (see TEXT).

Wernher, Karl Christian [1830–1889]. Giessen mat. 1849 pharm. (no publication). Became pharmacist.

WILHELMY, LUDWIG FERDINAND [1812–1864]. Heidelberg Dr.phil. 1846; Giessen mat. 1846 philosophy (studied with Liebig; no publication). Heidelberg Pv.Dz. 1849–54; then private life in Berlin. Best known for his 1850 paper on chemical kinetics.

Wilkens, Hermann [1816–1886]. Giessen mat. 1838 pharm.; Dr.phil. 1842 (no publication). Became director of ultramarine factory in Kaiserslautern.

Wimmer, Johannes [?–1890]. Giessen mat. 1838 pharm. (no publication). Became pharmacist in Kraiburg.

Winkler von Mohrenfels, Wolf Karl Rudolf [1820–1888]. Giessen mat. 1843 chem. (no publication). Became forester.

Wrightson, Francis Trippe (UK). Giessen mat. 1843 (worked on ash analysis of wood); Marburg Dr.phil. 1853 (Kolbe). Later activity not determined.

Wydler, Ferdinand [1821–1873] *(SWITZ)*. Giessen mat. 1840 chem. (no publication); Heidelberg mat. 1846 med.; also studied in Berlin and Würzburg. Became physician in Aarau.

Wydler, Franz Wilhelm [1818–1877] *(SWITZ)*. Giessen mat. 1840 chem. (no publication). Later activity not determined.

Wydler, Rudolf *(SWITZ)*. Giessen mat. 1840 chem. (no publication). Later activity not determined; in 1847 published paper (with Bolley) on pigment of *Anchusa tinctoria*.

Zedeler, Adolf Johann *(DEN)*. Giessen Dr.phil. 1851 (not in mat. list; worked on ash analysis of cocoa beans). Later activity not determined.

Zoeppritz, Johann Friedrich [1814–1861]. Giessen mat. 1835 pharm. (no publication). Acquired chemical factory in Freudenstadt; in 1850 went to U.S. and established pharmacy in New York City.

Zwenger, Constantin [1814–1884]. Giessen mat. 1835 med.; Marburg Dr.phil. chem. (worked on catechin); Pv.Dz. pharm. chem. 1841–44; ao.Prof. 1844–52; o.Prof. 1852–84. Published papers on the chemistry of natural products.

Appendix 3

The Hoppe-Seyler
Research Group

Adrian, Carl [?–1937] *(FR)*. Strassburg Dr.med. 1893 (worked on nutrition). Became physician in Strasbourg.

Amthor, Carl [1853–1939]. Strassburg 1878 (worked on wine chemistry); Freiburg Dr.phil. 1880. Head of Alsace-Lorraine chemical research station Strassburg 1881–1918. Published many papers on chemistry of wine, honey, etc.

ARAKI, TORASABURO [1853–1939] *(JAP)*. Strassburg Dr.med. 1891 (worked on hemoglobin, lactic acid formation, tissue oxidation). Kyoto Univ. Prof. med. chem. 1899–1915; President 1915–29. Published on many topics in physiological chemistry.

Aronheim, Felix [1843–1913]. Göttingen Dr.med 1868; Tübingen 1868 (worked on effect of salts on blood flow). Became physician in Braunschweig.

Astashevski, Pavel Petrovich [1845–?]. *(RUSS)*. Kazan Dr.med. 1873; Strassburg 1880 (worked on lactic acid in muscle). Became physician in Tomsk.

Bary, Johann Jakob de [1840–1915]. Tübingen Dr.med. 1864 (worked on chondrin, digestion of proteins). Became prominent physician in Frankfurt a.M.; published extensively on clinical subjects.

BAUMANN, EUGEN [1846–1896]. Tübingen Dr.rer.nat. 1872 (asst. of Hoppe-Seyler). Strassburg 1872–77 (Pv.Dz. 1876). Berlin Head of chem. divn. Institute of Physiology 1877–83 (ao. Prof. 1882).

Freiburg i.B. o.Prof. chem. 1883–96. Published many important chemical and biochemical papers on sulfur compounds, homogentisic acid, thyroid, etc. (see TEXT).

Bayer, Heinrich [1853–1926]. Strassburg Dr.med. 1879 (worked on acids in bile); Pv.Dz. obs. and gyn. 1885; ao.Prof. 1893–. Published many papers on biochemical aspects of obs. and gyn.

Bókay, Arpad [1856–1919]. Strassburg 1877 (worked on lecithin and nuclein). Budapest Dr.med. 1879. Later o.Prof. pharmacology Klausenburg 1883–89; Budapest 1889–19. Published extensively on toxicology, mineral waters, etc.

BOTKIN, SERGEI PETROVICH [1832–1889] (RUSS). St. Petersburg Dr.med. 1855; Berlin 1857 (worked on action of salts on erythrocytes). St. Petersburg Prof. physiol. 1861–89. Published extensively on physiological and clinical subjects; among his pupils was Ivan Pavlov [1849–1936].

Bruns, Paul von [1846–1916]. Tübingen Dr.med. 1870 (worked on chemistry of the cornea); Pv.Dz. surgery 1875–77; ao.Prof. 1877–82; o.Prof. 1882–1916. Published many papers on surgical topics, especially military surgery.

Bubnov, Nikolai Aleksandrovich [1851–1884] (RUSS). St. Petersburg Dr.med 1880; Strassburg 1880 (worked on chemistry of thyroid, action of iron compounds on digestion); also worked with Rudolf Heidenhain in Breslau.

Buliginsky, Aleksandr Dmitrievich [1838–1907] (RUSS). Moscow Dr.med. 1860; Tübingen 1867 (worked on urinary phenol compounds); also studied with Bunsen and Erlenmeyer. Moscow Prof. med. chem. 1878–1907. Published on various biochemical topics, especially bile acids and urine analysis.

Cahn, Arnold [1859–1927]. Strassburg Dr.med. 1881 (worked on chemistry of connective tissue and the eye). Became head of municipal hospital in Strassburg 1906–19; published on many medical topics.

Chandelon, Théodore [1851–1921] (BELG). Strassburg 1876 (worked on muscle glycogen); Liège Dr.med. 1878. Became physician in Liège; published on topics in physiological chemistry and toxicology.

Chevalier, Josephine (US). Dr.?; Strassburg 1886 (worked on chemistry of nerves). Later activity not determined.

Cohn, Felix [1869–1942?]. Strassburg Dr.med. 1889 (worked on effect of gastric juice on fermentation). Became physician in Munich; died in Nazi concentration camp near Minsk.

Dähnhardt, Christian Johann [1844–1892]. Tübingen 1867 (worked on oxidation in blood); Dr.med. Kiel 1868. Became physician in Kiel.

Demant, Bernhard *(RUSS)*. Dr.med.; Strassburg 1879 (worked on chemistry of muscle, secretion of intestinal juice). Later in Kharkov Institute of Physiology (Anrep).

Diakonov, Konstantin Sergeevich [1839–1868] *(RUSS)*. Dr.med. Kazan 1866; Tübingen 1867 (worked on lecithin). Died in Tübingen.

Disqué, Ludwig [1854–1928]. Strassburg Dr.med. 1878 (worked on urobilin). Became physician in Alt Thann, Chemnitz and Potsdam; wrote books on nutrition.

Dreyfuss, Isidor. Strassburg Dr.med. 1893 (worked on cellulose in micro-organisms). Became physician in Ottweiler.

Drozdov, Viktor Ivanovich [1846–1899] *(RUSS)*. St. Petersburg Dr.med. 1872; Strassburg 1877 (worked on blood analysis). Became Prof. physiol. milit. med. sch. St. Petersburg.

Duncan, C. *(UK?)*. Strassburg 1893 (worked on respiration of fish). No biographical data found.

Dybkovski, Vladimir Ivanovich [1830–1870] *(RUSS)*. St. Petersburg Dr.med. 1861; Tübingen 1866 (worked on hemoglobin and on phosphorous poisoning). Kiev Prof. pharmacology 1868–70.

Ernst, Carl. Strassburg Dr.med. 1893 (worked on chemistry of bile). Became physician in Wiesbaden.

ERRERA, LÉO ABRAM [1858–1905] *(BELG)*. Brussels Dr.sci. 1879; Strassburg 1880 (worked on glycogen in invertebrates). Brussels Prof. botany 1894–1905. Made outstanding contributions to plant physiology.

Esov, Ivan *(RUSS)*. Dr.med.; Strassburg 1876 (worked on urobilin). Later activity not determined.

Fischer, Charles Sumner [1866–1926] *(US)*. Long Isl. Coll. Med. M.D. 1887; Strassburg Dr.phil. 1894 (worked on glycine content of gelatin). Became physician in New York City; published papers on clinical chemistry.

Flückiger, Max [1863–1887] *(SWITZ)*. Strassburg Dr.med. 1885 (worked on urine analysis). Son of Friedrich August Flückiger [1828–1894], Prof. pharmacy at Strassburg. Died soon after receiving degree (suicide).

Fox, Wilson [1831–1887] *(UK)*. London M.D. 1855; Berlin 1857 (worked on gastric secretion). University College London Prof. pathol. anat. 1861–67; Prof. medicine 1867–87; published extensively on medical subjects.

FREDERICQ, LÉON [1851–1935] *(BELG)*. Ghent Dr.sci. 1871; Strassburg 1877 (worked on blood and muscle of invertebrates). Liège Prof. physiology 1879–. Founded important school of comparative biochemistry and physiology.

Froriep, August von [1849–1917]. Tübingen 1870 (worked on connective tissue in invertebrates); Leipzig Dr.med. 1874. Tübingen Prof. anatomy 1884–1917. Published many papers on anatomical topics.

Fudakowski, Herman Boleslaw [1834–1878] *(RUSS/POL)*. Dorpat Dr.med. 1859; Tübingen 1867 (worked on lactose); also studied with Bernard and Kühne. Warsaw Pv.Dz. medical chem. 1869–78. Wrote many papers on physiological chemistry.

Gähtgens, Karl [1839–1915]. Dorpat Dr.med. 1866; Tübingen 1867 (worked on urinary creatinine and uric acid). o.Prof. pharmacology Rostock 1875–80; Giessen 1880–98. Published extensively on topics in physiological chem. and toxicology.

Gaertner, Frederick [1861–1929] *(US)*. St. Louis Med. Coll. M.D. 1882; Strassburg Dr.med. 1885 (worked on pigments in animal tissues). Became physician in Pittsburgh, Pa.

Garcia-Valenzuela, Adeodato [1864–1936] *(CHILE)*. Dr.med.; Strassburg 1892 (worked on ptomaines). Became Prof. physiol. chem. Santiago, Chile.

Geoghegan, Edward George *(UK)*. Strassburg Dr.med. 1879 (worked on cerebrin). Became physician in Dublin.

Giacosa, Piero [1853–1928] *(ITAL)*. Turin Dr.med. 1876; Strassburg 1878 (worked on chemistry of frog's egg, urinary phenol); also studied with Nencki in Berne. Turin o.Prof. pharmacology 1895–. Apart from many contributions in physiological chemistry, he wrote extensively on the history of medicine.

Gilkinet, Alfred [1845–1926] *(BELG)*. Strassburg Dr.phil. 1872 (worked on plant chemistry). Liège Prof. pharmacy 1882–1919; published principally on fossil plants.

Gilson, Eugène [1862–1908] *(BELG)*. Strassburg Dr.phil. 1890 (worked on lecithin). Prof. pharmacy Ghent 1901–08; published on plant chemistry.

Gottwalt, Eduard *(RUSS)*. Strassburg 1880 (worked on filtration of protein solutions through animal membranes). Later activity not determined.

Gusserow, Adolf Ludwig Sigismund [1836–1906]. Würzburg Dr.med. 1859; Berlin 1860 (worked on urine analysis); o.Prof. obs. and gyn. Utrecht 1867; Zurich 1867–72, Strassburg 1872–78; Berlin 1878–1906.

Härlein, Julius [1835–?]. Tübingen 1862 (worked on paralbumin). Became pharmacist.

Härter, E. Tübingen 1871 (worked on guanine). Later activity not determined.

Harkavy, Alexander *(RUSS)*. Strassburg Dr.med. 1877 (worked on putrefaction of yeast). Later activity not determined.

Hasebroek, Karl [1860–1941]. Kiel Dr.med. 1886; Strassburg 1887 (worked on lecithin, digestion, etc.). Became head of Zander medical institute in Hamburg; in addition to medical papers, wrote extensively on entomology.

Hermann, Maximilian [1834–?]. Berlin Dr.med. 1861 (worked on urinary secretion, hippuric acid). Later Pv.Dz. physiology Vienna.

Herter, Erwin [1849–1908]. Tübingen Dr.phil. 1870; Strassburg 1872–81 (asst. of Hoppe-Seyler; collaborated with Baumann in work on ethereal sulfates). Berlin Pv.Dz. 1881–; set up private laboratory (in Naples zool. exp. sta. 1890–92). Published few experimental papers; prepared many literature reports for abstract journals.

Hirschfeld, Eugen. Strassburg Dr.med 1889 (worked on acetic and lactic fermentation). Became physician in Australia.

Hirschler, Agoston [1861–1911]. Budapest Dr.med. 1886; Strassburg 1887 (worked on effect of carbohydrates on protein putrefaction). Became chief physician St. Istvan Hosp. Budapest; published papers on clinical topics.

Hoffmann, Arthur. Strassburg Dr.med. 1877 (worked on hippuric acid formation). Became physician in Darmstadt.

Horvath (Khorvat), Aleksei Nikolaevich [1836–?] *(RUSS)*. Dr.med. Kiev 1876; Strassburg 1877 (worked on respiration). Kazan ao.Prof. pathology 1882–.

Irisawa, Tatsukichi [1867?–1935] *(JAP)*. Tokyo Dr.med. 1889; Strassburg 1890 (worked on lactic acid in blood and urine). Tokyo Prof. medicine Univ. 1901–.

Jacobj, Carl [1857–1944]. Strassburg Dr. med. 1887 (worked on urea determination, excretion of iron). Tübingen o.Prof. pharmacology 1897–1927; published extensively on physiological and pharmacological topics.

Jacobson, John (Fritz Emil) [1859–?]. Berlin Dr.med. 1891; Strassburg 1892 (worked on soluble ferments). Became physician in Berlin.

Jaksch, Rudolf von [1855–1947]. Prague Dr.med. 1878; Strassburg 1880 (worked on urease). Graz ao.Prof. pediatrics 1887–89; Prague o.Prof. medicine 1889–25. Made important contributions to the study of intermediary metabolism, especially in regard to formation of urea and of ketone bodies.

Jüdell, Gustav [1847–1876]. Tübingen Dr.med. 1869 (worked on blood analysis, disinfection, etc.). Became Pv.Dz. physiology Erlangen shortly before his death.

Kistiakovski, Vasili Fedorovich [1841–1901] *(RUSS)*. Kiev Dr.med. 1873; Strassburg 1874 (worked on pancreas peptones). Kiev o.Prof. med. chem.

Klüpfel, Richard [1848–1917]. Dr.med.; Tübingen 1868 (worked on effect of muscular activity on acidity of urine). Became physician in Urach.

Koch, Paul [1844–1911]. Tübingen Dr.rer.nat. 1869 (worked on optical rotation of organic compounds). Became pharmacist in Neuffen.

Komanos, Anton (GREECE). Dr.med. Strassburg 1875 (worked on digestion of inulin). Became physician in Athens.

KOSSEL, ALBRECHT [1857–1927]. Rostock Dr.med. 1878; Strassburg 1876–83 (worked on peptones, nuclein). Berlin head chem. divn. Institute of Physiology 1883–93; o.Prof. physiology Marburg 1893–1901; Heidelberg 1901–24. His outstanding work on nucleic acids, protamines and histones won him a Nobel Prize (see TEXT).

Krauss, Ernst. Dr.med.; Strassburg 1893 (worked on nutrition). Later activity not determined.

Kriege, Hermann. Strassburg Dr.med. 1885 (worked on treatment of diphtheria with papain). Became physician in Barmen.

Kukol-Yasnopolski, Vladislav Aleksandrovich (RUSS). St. Petersburg Dr.med. 1873; Strassburg 1876 (worked on formation of indole in liver). Later activity not determined.

Laas, Rudolph. Strassburg Dr.med. 1894 (worked on effect of fats on the utilization of proteins). Became physician in Strassburg.

Lahousse, Émile [1850–?] (BELG). Ghent Dr.med. 1876; Strassburg 1883 (worked on blood gases). Ghent Prof. physiology 1890–; published on various topics in physiology.

Landwehr, Hermann Adolf. Strassburg Dr.med. 1881 (worked on mucin of bile and of submaxillary gland). Würzburg Pv.Dz. physiology 1885–87.

Lange, Gerhard. Tübingen Dr.phil. 1887; Strassburg 1889 (worked on lignin and cellulose). Became commercial chemist in Hannover.

Lapchinski, Mikhail Demyanovich [1841–1889] (RUSS). St. Petersburg Dr.med. 1875; Strassburg 1876 (worked on chemistry of lens tissue). Warsaw ao.Prof. pathology 1885–89.

Laves, Ernst [1863–1927]. Studied with Baumann in Freiburg 1886–88; Würzburg Dr.phil. 1891; Strassburg 1892 (asst. of Hoppe-Seyler; worked on chemistry of milk and on respiration physiology). Head of chemical laboratory of the municipal hospital in Hannover 1895–1913. In 1908 he set up a firm to exploit his patents on several pharmaceutical preparations.

Ledderhose, Georg [1855–1925]. Strassburg Dr.med. 1880 (worked on glucosamine). Later ao.Prof. surgery Strassburg 1891–; then in Munich.

Levy, Ludwig [1864–?]. Strassburg Dr.med. 1888 (worked on muscle pigments). Became physician in Stephansfeld.

Levy, Maurice [?–1934]. Strassburg Dr.med. 1895 (worked on chemistry of bone). Became physician in Illkirch-Graffenstaden (Alsace).

LEYDEN, ERNST von [1832–1910]. Berlin Dr.med. 1853; Berlin 1863 (worked on chemistry of sputum). o.Prof. medicine Königsberg 1865–72; Strassburg 1872–76; Berlin 1872–1907. Made many important contributions to clinical medicine.

Liebreich, Oskar [1839–1908]. Berlin Dr.med. 1865; Tübingen 1865 (worked on chemistry of brain). Berlin o.Prof. pharmacology 1871–1907. Continued research on lipids; demonstrated efficacy of chloral as a hypnotic.

Limbourg, Philipp Maria [1860–?]. Marburg Dr.med. 1887; Strassburg 1888 (worked on blood proteins). Became physician in Strassburg.

Lindenmeyer, Oskar [1839–1889]. Dr.rer.nat. Tübingen 1863 (worked on cholesterol). Became pharmacist in Heilbron (1866–74), then chemical manufacturer and owner of pharmacy in Stuttgart (1875–89).

Loebisch, Wilhelm Franz [1839–1912]. Vienna Dr.med. 1863; Tübingen 1870 (worked on urine analysis). Innsbruck ao.Prof. applied med. chem. 1878–82; o.Prof. 1882–1910. Published on various topics in clinical chem.

Lubavin, Nikolai Nikolaevich [1845–1918] (RUSS). Tübingen 1869 (worked on products of protein digestion); also studied in Leipzig (Kolbe), Berlin (Baeyer) and Heidelberg (Bunsen). St. Petersburg Dr.phil. 1887. Moscow Prof. applied chem. 1890–. Published on a variety of chemical and biochemical subjects, notably proteins and colloids.

Lücke, Albert [1829–1894]. Göttingen Dr.med. 1854; Berlin 1862 (worked on mucins). o.Prof. surgery Berne 1865–72; Strassburg 1872–94.

Luedeking, Robert [1853–1908] (US). Dr.med. Strassburg 1876 (worked on regeneration of muscle fibers). Prof. pathol. anat., then pediatrics, and dean, St. Louis School of Medicine.

Lukyanov, Sergei Mikhailovich [1855–?] (RUSS). Strassburg 1880 (worked on bile). Warsaw Prof. pathology 1886–94; St. Petersburg Dr.med. 1883; Prof. experimental medicine 1894–.

Makris, Constantinus (GREECE). Strassburg Dr.med. 1876 (worked on milk proteins). Became physician in Athens.

Manasse, Paul [1866–1927]. Strassburg Dr.med. 1890 (worked on lecithin and cholesterol in eryrthrocytes). Prof. otolaryngology Strassburg 1902–18; Würzburg 1919–27.

Manassein, Vyacheslav Avksentievich [1841–1901] (RUSS). Tübingen 1865 (worked on gastric juice, antipyretic agents). Moscow Dr.med. 1866; St. Petersburg Prof. medicine 1877–1901.

Mauthner, Julius [1852–1917]. Vienna Dr.med. 1879; Strassburg 1881 (worked on chemistry of amino acids). Vienna Pv.Dz. med. chem. 1881; ao.Prof. 1885–13 (set up private laboratory); o.Prof. 1913–17. Published many papers on organic-chemical and biochemical subjects.

Mayer, Siegmund [1842–1910]. Tübingen Dr.med. 1865 (worked on blood coagulation). Prague ao.Prof. physiology and histology 1872–87; o.Prof. 1887–1910.

MERING, JOSEPH von [1849–1908]. Strassburg Dr.med. 1873 (worked on chemistry of cartilage); Pv.Dz. pharmacology 1879–86; ao.Prof. 1886–94; Halle o.Prof. medicine 1894–1908. Among his many contributions, the best known are the introduction of veronal (with Emil Fischer) and his induction of experimental diabetes by means of phloridzin or (with Minkowski) by surgical removal of the pancreas.

MIESCHER, FRIEDRICH [1844–1895] *(SWITZ)*. Basle Dr.med. 1868; Tübingen 1868 (worked on chemical composition of pus cells; isolated nuclein); also studied with Ludwig in Leipzig. Basle Prof. physiology 1871–95. His later work on the nuclein of the sperm of the Rhine salmon was especially outstanding (see TEXT).

Möhlenfeld, J. *(RUSS)*. Dr.med.; Tübingen 1871 (worked on formation of peptones from fibrin). Became physician in St. Petersburg.

Munk, Immanuel [1852–1903]. Berlin Dr.med. 1873; Strassburg 1873 (worked on urea formation in liver). Berlin Pv.Dz. physiology and physiol. chem. 1883–99; ao.Prof. 1899–1903. Published extensively on topics in physiological chemistry.

Musculus, Frédéric Alphonse [1829–1888] *(FR)*. A pharmacist, and a student of Boussingault, after the Franco-Prussian war he became chief pharmacist at the Strassburg municipal hospital, and collaborated in research with junior members of Hoppe-Seyler's laboratory. Among his notable contributions were studies on urease and on amylase.

Obolensky, Ivan Nikolaevich [1840–?] *(RUSS)*. St. Petersburg Dr.med. 1868; Tübingen 1870 (worked on mucin of submaxillary gland). Later ao.Prof. pathology Kharkov.

Oeffinger, Heinrich Carl [1840–?]. Tübingen 1865 (asst. of Hoppe-Seyler; no publication). Became pharmacist in Stuttgart.

Onsum, Ivar [1834–1881] *(NOR)*. Christiania Dr.med. 1860; Tübingen 1865 (no publication?). Became physician at Christiania (Oslo).

Otto, Jacob Gottfried [1859–1888] *(NOR)*. Strassburg 1883 (worked on digestion of proteins by pancreatin); Dr.phil. Christiania 1887.

Parke, John Latimer [1825–1907] *(UK)*. M.D.; Tübingen 1866 (worked on taurocholic acid). Became physician in Tideswell.

Pashutin, Viktor Vasilievich [1845–1901] *(RUSS)*. St. Petersburg
Dr.med. 1868; Strassburg 1874 (worked on butyric fermentation;
gastrointesinal digestion). Prof. pathology Kazan 1874–75; St. Petersburg 1875–1901.

Petrovski, D. Y. *(RUSS)*. St. Petersburg Dr.med. 1872; Strassburg 1873
(worked on the chemistry of the brain). Kharkov Pv.Dz. physiology.

Pickardt, Max [1868–?]. Berlin Dr.med. 1891; Strassburg 1892 (worked
on chemistry of cartilage, blood sugar). Became physician in Berlin.

Plósz, Pal [1844–1902]. Budapest Dr.med. 1867; Tübingen 1869
(worked on nuclein, paralbumin, pigments in liver and spleen).
Klausenberg ao.Prof. physiol. chem. 1872–74 (worked in Kühne's
laboratory); Budapest ao.Prof. 1874–82; o.Prof. 1882–1902. Published on many topics in physiological chemistry, especially serum
proteins, peptones, etc.

Popov, Lev Vasilievich [1845–1906] *(RUSS)*. St. Petersburg Dr.med.
1868; Strassburg 1872 (worked on pathology of muscle fibers). St.
Petersburg Prof. pathology 1890–1906.

Putzeys, Félix [1847–1932] *(BELG)*. Liège Dr.med. 1871; Strassburg
1874 (worked on repetition of Huizinga's expts. on abiogenesis).
Liège ao.Prof. hygiene 1879–82; o.Prof. 1882–1919.

Raevski, Arkadi Aleksandrovich [1848–1916] *(RUSS)*. Dr.med. St. Petersburg 1871; Strassburg 1875 (worked on alcohol metabolism, determination of hemoglobin content of blood). Became Prof. and
Director of Veterinary Institute Kharkov.

RECKLINGHAUSEN, FRIEDRICH von [1833–1910]. Berlin Dr.med.
1855 (worked on mineral constituents of bone); asst. in Virchow's institute of pathology 1858–64. o.Prof. pathological anatomy Königsberg 1865–67; Würzburg 1867–72; Strassburg 1872–. Made many
important contributions to the anatomical and pathological aspects
of clinical medicine.

Riesell, Albert [1844–1889]. Tübingen 1868 (worked on urinary excretion of phosphates); Leipzig Dr.med. 1869. Became physician in
Echte.

Sakharin, Grigori Antonovich [1829–1897] *(RUSS)*. Moscow Dr.med.
1852; Berlin 1856 (worked on blood analysis). Moscow Prof. medicine 1862–96. Published extensively on many clinical subjects.

SALKOWSKI, ERNST LEOPOLD [1844–1923]. Königsberg Dr.med.
1867; Tübingen 1868 (worked on bilirubin); also briefly with Kühne
1872. Berlin head chem. divn. Institute of Pathology 1880–1923.
Published very many papers on a large variety of topics in physiological chemistry, such as pentosuria, tissue autolysis, the formation of
urea and hippuric acid, etc.

Schadow, Gottfried [?–1885]. Strassburg Dr.med. 1876 (worked on physiological action of nitropentane). Became ophthalmologist in Strassburg.

SCHMIDT, ALEXANDER [1831–1894]. Dorpat Dr.med. 1858; Berlin 1861 (worked on fibrin ferment). Dorpat Pv.Dz. physiology 1862–69; o.Prof. 1869–94. Made many important biochemical contributions, especially in the study of blood coagulation; also led a productive research group.

Schmidt, August [1844–1907]. Dr.rer.nat. Tübingen 1871 (worked on emulsin and legumin). Became pharmacist in Sulzbach.

Schrader, Max [1860–1892]. Strassburg Dr.med. 1886 (worked on cardiac physiology); later Asst. Physiol. Inst. Strassburg (Goltz).

Schulze, Ernst. Strassburg Dr.med. 1889 (worked on nitrogenous components of urine). Later activity not determined.

Schwarz, Hugo. Budapest Dr.med. 1891; Strassburg 1893 (worked on hydrolysis and oxidation of proteins). Became physician in Budapest, then St. Louis.

SECHENOV, IVAN MIKHAILOVICH [1829–1905] (RUSS). St. Petersburg Dr.med. 1860; Berlin 1861 (worked on metabolism of alcohol). Prof. physiology St. Petersburg 1861–70, 1876–88 (Odessa 1870–76); Moscow 1891–1901. One the leading nineteenth-century Russian physiologists (see TEXT).

Sertoli, Enrico [1842–1910] (ITAL). Pavia Dr.med. 1865; Tübingen 1867 (worked on CO_2 exchange in blood and lungs). Milan Prof. physiology 1870–1910.

Severi, Domenico (ITAL). Dr.med.; Tübingen 1867 (worked on effect of gastric juice on fermentation). Later activity not determined.

Shchelkov, Ivan Petrovich [1833–1909] (RUSS). Kharkov Dr.med. 1857; Berlin 1860 (worked on changes in volatile fatty acids in muscle during tetanus). Kharkov Prof. physiology 1863–90.

Sokolov, Nil Ivanovich [1846–1894] (RUSS). St. Petersburg Dr.med. 1874; Strassburg 1874 (worked on liver secretion, composition of bile). St. Petersburg ao.Prof. pathology 1878–94.

Sotnichevski, Roman Amfilokhievich (RUSS). Kiev Dr.med. 1876; Strassburg 1879 (worked on glycerophosphoric acid in urine, chemistry of lung tissue). Later activity not determined.

Spiro, Petr Antonovich [1844–1893] (RUSS). St. Petersburg Dr.nat.sci. 1867 (with Mechnikov); Odessa Pv.Dz. biol. 1874–78; Strassburg 1878 (worked on bile, metabolism of lactic acid); Kharkov Dr.med. 1881; Odessa Prof. physiology 1884–93.

Stolnikov, Yakov Yakovlevich [1850–?] (RUSS). Strassburg 1877 (worked on effect of bile on digestion of protein and fat); St. Petersburg Dr.med. 1880. Warsaw Prof. medicine 1888–.

Strauss, Leopold [1873–1944]. Strassburg Dr.med. 1893 (worked on fat absorption in diabetes). Became physician in Berlin; died in Theresienstadt concentration camp.

Strogonov, Nikolai Alekseevich [1842–?] *(RUSS)*. St. Petersburg Dr.med. 1873; Strassburg 1876 (worked on O_2 uptake by hemoglobin). Became physician at Odessa municipal hospital.

Sundvik, Ernst [1850–1918] *(FIN)*. Helsingfors Dr.med. chem. 1875; Strassburg 1880 (worked on chitin). Helsingfors ao.Prof. physiol. chem. and pharmacy 1879–86; o.Prof. 1886–1915.

Szabó, Denes [1856–1918]. Budapest Dr.med. 1879; Strassburg 1878 (worked on gastric juice). Klausenburg Prof. gynecology 1892–1918.

Takacs, Endre (Andreas) [1848–1895]. Dr.med.; Strassburg 1878 (worked on biological oxidation). Became psychiatrist in Budapest.

Tarkhanov, Ivan Romanovich [1846–1908] *(RUSS)*. St. Petersburg Dr.med. 1869; Strassburg 1873 (worked on formation of bile pigments). St. Petersburg Prof. physiology 1875–95. Published extensively on various physiological topics.

THIERFELDER, HANS [1858–1930]. Rostock Dr.med. 1883; Strassburg 1884–87 (asst. of Hoppe-Seyler; worked on glucuronic acid). Berlin Pv.Dz. 1887; ao.Prof. 1896 (head of chem. divn. Institute of Physiology 1895–1908); Tübingen Prof. physiol. chem. 1908–30. Made important contributions to the study of brain lipids, especially cererosides (see TEXT).

Tolmachev, Nikolai Aleksandrovich [1823–1901] *(RUSS)*. Tübingen 1868 (worked on chemistry of milk, effect of blood loss); Kazan Dr.med. 1875; ao.Prof. pediatrics 1881–1901.

Treskin, Fedor Vasilievich [1836–?] *(RUSS)*. St. Petersburg Dr.med. 1869; Strassburg 1872 (worked on urea determination, chemistry of testicular tissue). Became physician in Revel (Tallinn).

Trifanovski, Dmitri Semenovich [1845–?] *(RUSS)*. Strassburg 1874 (worked on composition of human bile); Moscow Dr.med. 1875. Became physician in Moscow.

Udránszky, László von [1862–1914]. Budapest Dr.med. 1883; Strassburg 1885 (worked on pigments in urine). o.Prof. med. chem. Klausenburg 1892–1909; Budapest 1909–14. Published extensively on topics in physiological chem. and in sensory physiology.

Vahlen, Ernst [1865–1941]. Berlin Dr.med. 1890; Strassburg 1890–95 (asst. of Hoppe-Seyler; worked on bile acids and on urea formation). Halle ao.Prof. pathological chem. 1903–37. Wrote many papers on a variety of biochemical topics.

Vandevelde, Guillaume *(BELG)*. Dr.med., Dr.phil.; Strassburg 1884 (worked on chemistry of B. subtilis). Later in Institute of Physiology at Ghent.

Vasiliev, Nikolai Petrovich [1852–1891] *(RUSS)*. St. Petersburg Dr.med. 1880; Strassburg 1880 (worked on action of calomel on intestinal bacteria). St. Petersburg Pv.Dz. medicine 1885–91.

Vincenzi, Livio *(ITAL)*. Dr.med.; Strassburg 1887 (worked on chemical composition of molds). Later became Prof. hygiene in Sassari; wrote many papers on topics in bacteriology.

Vogt, Paul [1844–1885]. Tübingen 1864 (worked on salamander poison); Greifswald Dr.med. 1865; Pv.Dz. surgery 1869–73; ao.Prof. 1873–82; o.Prof. 1882–85.

Voroshilov, Konstantin Vasilievich [1842–1899] *(RUSS)*. St. Petersburg Dr.med. 1871; Strassburg 1875 (worked on protein digestion). Kazan ao.Prof. physiology 1876; o.Prof. 1885–99.

Vvedenski, Nikolai Evgenievich [1852–1922] *(RUSS)*. St. Petersburg Dr.med. 1879; Strassburg 1881 (worked on carbohydrates in urine). St. Petersburg Pv.Dz. physiology 1884–89; ao.Prof. 1889–95; o.Prof. 1895–. Made important contributions to neurophysiology.

Weintraud, Wilhelm [1866–1920]. Strassburg Dr.med. 1889 (worked on carbohydrate metabolism); Pv.Dz. medicine 1893. Wiesbaden head of divn. in municipal hospital 1898–1920 (suicide). Published many papers on carbohydrate metabolism.

WELCH, WILLIAM HENRY [1850–1934] *(US)*. Coll. Phys. and Surg. N.Y. M.D. 1875; Strassburg 1876 (worked on analytical methods); also studied in Leipzig (Ludwig) and Breslau (Cohnheim). Johns Hopkins Prof. pathology 1885–1916 (Dean 1893–98). Apart from his scientific work (e.g., gas gangrene bacillus), he played a leading role in the reform of medical education in the United States.

Weyl, Theodor [1851–1913]. Strassburg Dr.med. 1877 (worked on crystalline seed proteins). Berlin (with Baumann) chemistry of amino acids and proteins). Erlangen Pv.Dz. physiology 1879–80; Naples zool. exp. station (electrical organ of the Torpedo); Berlin 1884 set up private biochemical laboratory from which came many publications, especially on amino acid chemistry. Then turned to public health problems (Pv.Dz. hygiene Charlottenburg TH 1895), but continued biochemical research.

White, Thomas Philip [1855–1901] *(US)*. Strassburg Dr.med. 1880 (worked on physiological effect of zinc compounds). Became physician in Cincinnati.

Winternitz, Hugo [1868–1934]. Strassburg 1890 (worked on chemistry of milk, protein in urine, tryptophan); Vienna Dr.med. 1894. Halle Pv.Dz. medicine 1908; chief physician Elisabeth Hospital. Published many papers in physiological chemistry, especially on lipid metabolism.

Zaleski, Nikolai Lavrentievich [1835–?] *(RUSS)*. St. Petersburg Dr.med. 1865; Tübingen 1866 (worked on chemistry of bone, uremia, toad poisons). Kharkov Prof. pharmacology 1883–.

Zapolski, Nikolai Vasilievich [1835–1883] *(RUSS)*. Moscow Dr.med. 1860; Tübingen 1869 (worked on action of phenol on proteins and ferments). Moscow Pv.Dz. obstetrics and gynecology 1873–83.

Zillessen, Hermann [1871–?]. Strassburg 1891 (effect of cyanide on carbohydrate metabolism); Dr.med. Jena 1899. Became physician in Berlin.

Zweifel, Paul [1848–1927]. Zurich Dr.med. 1872; Strassburg 1874 (worked on hemoglobin, meconium). o.Prof. obstetrics and gynecology Erlangen 1876–87; Leipzig 1887–1921.

Appendix 4

The Kühne Research Group

Anderson, Richard John [1848–1914] *(UK)*. Belfast M.D. 1872; Heidelberg 1875 (worked on histology of myoneural junction). Prof. biology, mineralogy and geology Queens College, Galway 1883–1914.

Arthus, Maurice [1862–1945] *(FR)*. Paris Dr.med. 1886; Heidelberg 1888 (worked on coagulation of milk). Prof. physiology Marseille 1903–07; Lausanne 1907–. Extensive studies on anaphylaxis, blood coagulation, enzymes.

Asher, Leon [1865–1943]. Leipzig Dr.med. 1890; Heidelberg 1892 (worked on muscular contraction, resorption through blood vessels). Berne Pv.Dz. physiology 1895–06; ao.Prof. 1906–14; o.Prof. 1914–36. Published extensively on endocrinology.

Ayres, William C. [1853–1896] *(US)*. Tulane M.D. 1876; Heidelberg 1878 (worked on regeneration of rhodopsin). After 1880 practiced ophthalmology in New York City and in New Orleans.

Biernacki, Edmund [1866–1911] *(RUSS/POL)*. Warsaw Dr.med. 1889; Heidelberg 1890 (worked on effect of temperature on digestive enzymes). Lemberg ao.Prof. pathology 1907–11. Contributed to hematology, nutrition and neurophysiology.

Brunton, Thomas Lauder [1844–1916] *(UK)*. Edinburgh M.D. 1868; Amsterdam 1869 (worked on blood chemistry, digestion); also studied in Vienna (Brücke) and Leipzig (Ludwig). Physician at St. Bartholomew's Hosp. London 1871–16. Best known for his introduction of amyl nitrite for the treatment of angina pectoris; also worked on digestive physiology.

Calberla, Ernst [1848–1878]. Freiburg i.B. Dr.med. 1874; Heidelberg 1874 (worked on resorption in the aqueous humor); asst. in anatomy.

CHITTENDEN, RUSSELL HENRY [1856–1943] *(US)*. Heidelberg 1878 (worked on xanthine in albumin, histochemical studies on retinal epithelium and sarcolemma). Yale Ph.D. 1880; Prof. physiol. chem. 1882–1922 (Director Sheffield School 1898–1922); collaborated with Kühne by correspondence (1883–90) in work on albumoses and peptones. Also wrote about toxicology and human nutrition (see TEXT).

COHNHEIM, JULIUS [1839–1884]. Berlin Dr.med. 1861 (worked on cleavage of starch by saliva). Prof. pathology Kiel 1868–72; Breslau 1872–'78; Leipzig 1878–84. One of the leading pathologists of his time.

COHNHEIM (after 1917, KESTNER), OTTO [1873–1953]. Heidelberg Dr.med. 1896 (worked on peptones); Pv.Dz. physiology 1898; ao.Prof. 1903–13. After military service (1914–18) became Prof. physiology Hamburg 1919–34 (dismissed by Nazis). In 1901 discovered erepsin; later published on nutrition and protein chemistry (see TEXT).

DANILEVSKI, ALEKSANDR YAKOVLEVICH [1838–1923] *(RUSS)*. Kharkov Dr.med. 1860; Berlin 1871 (worked on myosin). Prof.med. (or physiol.) chem. Kazan 1863–85; Kharkov 1885–92; St. Petersburg 1892–. Published extensively on protein and enzyme chemistry.

Dreser, Heinrich [1860–1924]. Heidelberg Dr.med. 1884 (worked on chemistry of retinal rods, histochemical studies on kidney tissues). Pv.Dz. pharmacology Tübingen 1890–93; Bonn 1893–96; Göttingen ao.Prof. 1897–1904; Bayer & Co. Elberfeld head of pharmacological divn. 1904–14; o.Prof. pharmacology Düsseldorf medical academy 1914–24.

Ewald, August [1849–1924]. Bonn Dr.med. 1873; Heidelberg 1874–78 (asst. of Kühne; worked on rhodopsin, muscle-nerve physiology); Pv.Dz. physiol. 1880; ao.Prof. 1883–1914; o.Prof. (hon.) 1914–25. Continued histological studies.

FLEMMING, WALTHER [1843–1905]. Rostock Dr.med. 1868; Amsterdam 1869–71 (asst. of Kühne; worked on histology of fat cells and connective tissue). Rostock Pv.Dz. anatomy 1872; Prague ao.Prof. 1873–75; Kiel o.Prof. 1876–1902. One of the leading cytologists of his time; made important contributions to the study of the cell nucleus and of nuclear division.

Fudakowski, Herman Boleslaw [1834–1878] *(RUSS/POL)*. Berlin 1867 (worked on action of pancreatic juice). Also studied with Hoppe-Seyler (see Appendix 3).

Gamgee, Arthur [1841–1909] *(UK)*. Edinburgh M.D. 1862; Heidelberg 1871 (worked on hemoglobin?). Manchester Prof. physiology Owens

College 1873–82; Royal Institution 1882–85. After medical practice in Switzerland until 1904, continued research in England. Wrote on various biochemical topics, especially hemoglobin (see TEXT).

Hanau, Arthur [1858–1900]. Bonn Dr.med. 1881; Heidelberg 1882 (worked on intestinal secretion). Zurich Pv.Dz. pathology 1887, then (1892–1900) prosector at hospital at St. Gallen.

Jani, Curt. Heidelberg 1882 (worked on histology of nerve tissue). A medical student who appears to have died before receiving his Dr.med. degree.

Kaiser, Karl [1861–1933]. Heidelberg Dr.med. 1888 (worked on muscle physiology); Pv.Dz. physiology 1893–96; ao.Prof. 1896–1902. Berlin ao.Prof. 1902–. Continued work in muscle physiology.

Kijlstra, Heinrich Johann [1862–1929?] (NETH). Heidelberg Dr.med. 1888 (worked on albuminuria). Became physician in Amsterdam.

Kistiakovski, Vasili Fedorovich [1841–1901] (RUSS). Heidelberg 1873 (worked on blood coagulation). Also studied with Hoppe-Seyler (see Appendix 3).

Klug, Nándor (Ferdinand) [1845–1909]. Budapest Dr.med. 1870; Pv.Dz. physiol. 1874. Heidelberg 1880 (worked on rhodopsin). o.Prof. physiol. Klausenburg 1878–90; Budapest 1890–1909. Published extensively on many topics in physiology.

Knies, Max [1851–1917]. Heidelberg Dr.med. 1874 (asst. of Kühne; worked on ageing of lens). Freiburg i.B. ao.Prof. ophthalmology 1888–1917.

KRONECKER, HUGO [1839–1914]. Berlin Dr.med. 1863 (worked with Kühne on muscle). Leipzig Pv.Dz. physiology 1871–75; ao.Prof. 1875–77 (Berlin 1877–84). Berne o.Prof. 1884–. Made many contributions to physiology; noted for his generous help to younger colleagues, especially foreign guests.

Krukenberg, Carl Friedrich Wilhelm [1852–1889]. Heidelberg Dr.phil. 1883 (worked on comparative physiology of digestive and contractile tissues). Jena ao.Prof. physiol. chem. 1884–89 (see TEXT).

Krukenberg, Richard [1863–1924]. Heidelberg Dr.med. 1888 (worked on gastric hydrochloric acid). Became physician in Braunschweig.

Lahousse, Émile [1850–] (BELG). Heidelberg 1883 (worked on histology of nerve fibers). Also studied with Hoppe-Seyler (see Appendix 3).

LANGLEY, JOHN NEWPORT [1852–1925] (UK). Heidelberg 1877 (worked on salivary secretion in the cat); Cambridge M.A. 1878; Lecturer in physiology 1883–1903; Prof. 1903–25. Together with his teacher, Michael Foster, one of the founders of the great twentieth-century British school of physiology (see TEXT).

Lea, Arthur Sheridan [1853–1915] (UK). Heidelberg 1880 (worked on histological study of pancreatic secretion). Cambridge Sc.D. 1886;

Lecturer in physiology 1881–1915. Published valuable papers on digestive enzymes (see TEXT).

Leber, Theodor [1840–1917]. Heidelberg Dr.med. 1862; Berlin 1867 (worked on muscle). o.Prof. ophthalmology Göttingen 1871–90; Heidelberg 1890–1910.

MAGNUS, RUDOLF [1873–1927]. Heidelberg Dr.med. 1898 (worked on measurement of blood pressure); Pv.Dz. pharmacology 1900–04; ao.Prof. 1904–08. Utrecht o.Prof. pharmacology 1908–27. Made outstanding contributions to pharmacology and neurophysiology.

Maslov, A. A. (RUSS). Dr.med.; Heidelberg 1878 (worked on intestinal juice). Later in chemical-microscopic laboratory Kharkov.

Mays, Karl. Heidelberg Dr.med. 1877 (asst. of Kühne; worked on eye pigments, motor nerve endings, action of trypsin and pepsin on proteins); continued to publish from Heidelberg institute of physiology until 1903; later activity not determined.

Meyer, Arthur [1874–1942?]. Heidelberg Dr.med. 1897 (worked on products of protein digestion). Became physician in Berlin; died in Auschwitz concentration camp.

Morokhovets, Lev Zakharovich [1848–1918] (RUSS). Heidelberg Dr.med. 1876 (worked on action of silver nitrate on nerve fibers). Moscow ao.Prof. physiology 1884–.

Neumeister, Richard [1854–1905]. Rostock Dr.phil. 1882; Heidelberg Dr.med. 1887 (worked on albumoses). Würzburg Pv.Dz. physiol. chem.; Jena head of chem. section of institute of physiology 1892–97; then entered medical practice. Continued work on protein digestion and wrote valuable textbook of physiological chemistry.

PALLADIN, VLADIMIR IVANOVICH [1859–1922] (RUSS). Warsaw Dr.phil. 1889; Heidelberg 1895 (worked on plant proteins). Prof. plant physiology Kharkov 1889–1901; St. Petersburg 1901–14; laboratory at Academy of Sciences 1914–21. Many important contributions to the study of plant respiration.

Plósz, Pal [1844–1902] Heidelberg 1872 (worked on sugar-forming ferment of blood, protein-like substances in liver). Also studied with Hoppe-Seyler (see Appendix 3).

Pollitzer, Sigmund [1859–1937] (US). Coll. Phys. and Surg. M.D. 1884; Heidelberg 1885 (worked on peptones). New York Postgraduate Med. School Prof. dermatology 1906–15.

Rudnev, Mikhail Matveevich [1837–1878] (RUSS). St. Petersburg Dr.med. 1863; Berlin 1865 (worked on amyloid). St. Petersburg ao.Prof. pathology 1867–70; o.Prof. 1870–76.

Rumpf, Theodor [1851–1934]. Heidelberg Dr.med. 1877 (worked on histology of nerve fibers). Bonn Pv.Dz. medicine 1882; ao.Prof. 1887–88. Head of hospital Marburg 1888–92; Hamburg 1892–1901; Bonn 1901–22. Published on many clinical subjects.

Saake, Wilhelm [1865–?]. Heidelberg Dr.med. 1893 (worked on glycogen). Became physician in Thedinghausen.

SALKOWSKI, ERNST LEOPOLD [1844–1923]. Heidelberg 1872 (worked on effect of phenol on animal organism, determination of uric acid). Also studied with Hoppe-Seyler (see Appendix 3).

Sachs, Carl [1853–1878]. Berlin cand.med. (with du Bois-Reymond); Heidelberg 1875 (worked on innervation of tendons).

Sasse, Hendrik Frederick August (NETH). Heidelberg 1878 (worked on "Descemet" membrane of the eye); Utrecht Dr.med. 1886. Became physician in Amsterdam.

Scholz, G. Dr.? Berlin 1865 (worked on determination of ozone in blood). Later activity not determined.

SCHWALBE, GUSTAV ALBERT [1844–1916]. Berlin Dr.med. 1866; Amsterdam 1868 (worked on histology of taste organs). Halle Pv.Dz. anatomy 1870; o.Prof. Jena 1873–81; Königsberg 1881–83; Strassburg 1883–1914. Made many contributions to neuroanatomy and neurophysiology, then turned to anthropology.

SEWALL, HENRY [1855–1936] (US). Johns Hopkins Ph.D. 1879; Heidelberg 1880 (worked on histology and chemistry of retinal epithelium). Univ. of Michigan Prof. physiology 1882–89; Denver M.D. 1891; practiced medicine in Denver 1891–. During his relatively brief academic career, he made important research contributions to physiology.

Smith, Herbert Eugene [1857–1933] (US). Univ. of Penna. M.D. 1882; Heidelberg 1883 (worked on chemistry of bone). Yale Prof. chem. and dean medical school 1885–1910. His later publications dealt largely with public health.

Stadelmann, Ernst [1853–1941]. Königsberg Dr.med. 1878; Heidelberg 1888 (asst. of Kühne; worked on protein digestion). Berlin chief physician Urban hospital 1895–1921. Published extensively on many biochemical topics.

Steiner, Isidor [1849–1914]. Berlin Dr.med. 1873; Heidelberg Pv.Dz. physiology 1878–86; ao.Prof. 1886–88 (worked on nerve histology and physiology). Set up medical practice (neurology, electrotherapy) in Köln.

Strauch, H. Dr.? Berlin 1864 (worked on the occurrence of ammonia in blood). Later activity not determined.

Tiegel, Ernst [1849–1889]. Heidelberg Dr.med. (worked on ferments in blood). Tokyo Lecturer physiology 1876– (helped found the first dept. of physiology in Japan); later went to the United States?

Uexküll, Jakob [1864–1944]. Heidelberg 1898 (wrote on neurophysiology). A private scholar in Heidelberg who was associated with Kühne's group; later wrote extensively on philosophical aspects of biology.

Umbach, Carl [1856–1926]. Heidelberg Dr.med. 1887 (worked on the effect of antipyrine on nitrogen excretion). Became a physician in Tamm.

Van Syckel, Benjamin Miller [1857–1903] *(US)*. Bellevue Med. School. M.D. 1883; Heidelberg 1884 (worked on histology of neuromuscular junctions). Became physician in New York City.

Wenz, Joseph [1862–?]. Dr.med.?; Heidelberg 1885 (worked on albumoses). Later activity not determined.

Wolffhügel, Gustav [1845–1899]. Würzburg Dr.med. 1869; Heidelberg 1872 (worked on products of protein digestion by pepsin). Munich Pv.Dz. hygiene 1875; Berlin head of chem. lab. Dept. of Health 1879–87; Göttingen ao.Prof. hygiene 1887–99.

Zangemeister, Wilhelm [1871–1930]. Heidelberg Dr.med. 1895 (worked on colorimetric determination of hemoglobin). Königsberg Pv.Dz. obs. and gyn. 1904–08; ao.Prof. 1908–10; Marburg o.Prof. 1910–25; Königsberg 1925–30. Published extensively on topics in obstetrics and gynecology.

Appendix 5

The Baeyer Research School

ADOR, ÉMILE [1845–1920] (*SWITZ*). Berlin 1868–71 (Baeyer, V. Meyer; 6 papers on benzene derivatives); also studied in Edinburgh, London 1865–66. Zurich Dr.phil. 1872. Geneva suppl. Prof. org. chem. 1874–78; private research laboratory 1878–. Later published on a variety of topics in organic chemistry.

Aickelin, Hans [1885–1944]. Munich Dr.phil. 1907 (Dimroth; worked on oxytriazole, triphenylmethane and xanthone compounds); res. asst. of Baeyer. Then in U.S. with General Aniline & Film Co.

Alibegov, Georg (*RUSS*). Dr.phil.? Munich 1886 (Zimmermann; 2 papers on uranium compounds). Later at Kazan; no chemical publications 1897–1926.

Althausse, Max [1861–?]. Munich Dr.phil. 1888 (Bamberger, Krüss; 2 papers on tetrahydronaphthylamine, absorption spectra of organic compounds). No chemical publications 1897–1926; later activity not determined.

ANGELI, ANGELO [1864–1931] (*ITAL*). Bologna Dr.chem. 1891 (Ciamician); Pv.Dz. 1893–97; Munich 1894 brief stay (no publication). Palermo Prof. pharm. chem. 1897–1915; Florence Prof. chem. and director School of Pharmacy 1915–31. Published extensively on many aspects of organic chem.

Arnoldi, Heinrich. Munich 1906 (Hofmann; worked on diazonium perchlorates); Munich TH Dr.phil. 1907; later consulting chemist to leather industry.

Aronheim, Berthold [1850–1881]. Göttingen Dr.phil. 1873 (Wöhler); Munich 1879 (4 papers in inorganic chemistry). Later with Gebr. Dittler Hoechst.

ASCHAN, OSSIAN [1860–1939] (*FIN*). Helsingfors Dr.phil. 1884; Munich 1891 (Baeyer; worked on naphthenic acids). Helsingfors ao.Prof. chem. 1904–08; o.Prof. 1908–27. Published extensively on many topics in organic chemistry.

Ascher, Max [?–1908]. Berlin 1869–72 (Baeyer; 3 papers on benzene derivatives); Dr.phil.? Later set up Max-Ascher Co. Berlin-Tempelhof.

Astié, Hermann [1860–1903] (*SWITZ*). Munich Dr.phil. 1889 (Baeyer; worked on reduction products of phthalic acid). Became head of the state bacteriological service (Vaud).

Bagh, Aleksandr von [1882–?] (*RUSS*). Munich Dr.phil. 1909 (Einhorn; worked on derivatives of salicylic acid). Later activity not determined.

Bailey, James Robinson [1868–1941] (*US*). Munich Dr.phil. 1897 (J. Thiele; worked on derivatives of propionic acid). Univ. of Texas (Austin) Prof. chem. 1901–; published on various topics in organic chemistry, also wrote book (1933) on petroleum.

Baltzer, Otto [1861–?]. Erlangen Dr.phil. 1890; Munich 1891 (Pechmann; worked on pyridine and triazole derivatives). No further chemical publications 1891–1926; later activity not determined.

BAMBERGER, EUGEN [1857–1932]. Berlin Dr.phil. 1881 (Liebermann); Munich 1882–93 (Pv.Dz 1885, ao.Prof. 1891; published 115 papers, 84 of them with 37 students); Zurich ETH o.Prof. chem. 1893–1905 (resigned because of neurological problem that left his right arm paralyzed, continued research with aid of an assistant). Made outstanding contributions to many branches of organic chemistry (see TEXT).

Bammann, Johannes [1866–1910]. Munich Dr.phil. 1889 (Bamberger; worked on naphthylamines). Later became deputy director Bayer & Co. Elberfeld.

Banzhaf, Eugen. Munich 1894 (Besthorn; worked on quinoline derivatives). No further chemical publications 1897–1926; later activity not determined.

Barlow, William (1869–?] (*UK*). Munich Dr.phil. 1896 (J. Thiele; worked on condensation of aminoguanidine with quinones). Later activity not determined.

Basler, Adolph. Munich 1883 (Baeyer; published 3 organic-chemical papers). Dr.phil.?; later became commercial chemist, specialized in photography.

Bassett, Henry, Jr. [1881–1965] (*UK*). Munich Dr.phil. 1904 (Baeyer; worked on aminotriphenylcarbinol). Univ. of Reading Prof. chem.

1912–46; Dir. research Peter Spence Co. Widnes 1947–50; with geological survey Tanganyika 1950–54; Reading Natl. Institute of Dairying 1954.

Bauer, Édouard [1879–1915] (*FR*). Munich 1901 (Einhorn; worked on reduction of aminobenzoic acid). Nancy Dr.phil. 1904; research assistant of Albin Haller 1904–15.

Bauer, Hugo [1883–1968]. Munich Dr.phil. 1907 (Wieland; worked on aliphatic azo and nitroso compounds). Frankfurt a. M. Georg-Speyer-Haus 1909–; head of chem. section 1922–35. Went to U.S.; at Natl. Inst. Hth. 1936–53. Published many organic-chemical papers, especially in relation to chemotherapy.

Baumann, Artur [1885–?]. Munich Dr.phil. 1912 (Wieland; worked on polymerization of fulminic acid). Later res. asst. of Wieland; then with BASF Ludwigshafen.

Baumeister, Eduard [1865–?]. Munich Dr.phil. 1898 (Einhorn; worked on caffeine). No further chemical publications 1907–26; later activity not determined.

Baumgärtel, Konrad [1868–?]. Munich Dr.phil. 1897 (Baeyer; worked on oxycarone). Became consulting chemist in Langenfeld im Vogtland.

BAUR, EMIL [1873–1944]. Munich Dr.phil. 1897 (Muthmann; worked on conductivity of organic compounds). o.Prof. physical chem. Braunschweig 1907–11; Zurich ETH 1911–42. Made many contributions to the study of the spectroscopic properties of chemical substances.

Bayer, Joseph [1889–?]. Munich Dr.phil. 1914 (Kalb; worked on indolone). Later with BASF Ludwigshafen.

Beck, Ludwig [1880–?]. Munich Dr.phil. 1907 (Vanino; worked on synthesis of dyes from furfurol and dicarbonyl compounds). No further chemical publications 1908–56; later activity not determined.

Becker, Paul [?–1925]. Freiburg i. B. Dr.phil. 1881; Munich 1883 (Baeyer; worked on parabenzaldehyde, nitrodiphenylmethane). Later at Karlsruhe TH, then commercial dye chemist in Moscow.

Bedall, Karl [1858–1930]. Munich Dr.phil. 1882 (O. Fischer; worked on quinoline derivatives). Later took over family pharmacy in Munich.

Behr, Arno [1846–1920]. Heidelberg Dr.phil. 1869; Berlin 1870–72 (Baeyer; worked on tetraphenylethylene). Went to U.S.; with sugar refining company Jersey City 1874–81; supt. Chicago sugar refining company 1881–92; then consultant to sugar industry.

Bell, Charles John [1854–1903] (*US*). Munich 1879 (Koenigs; worked on sugar chemistry). Dr.phil.?; Univ. of Minnesota Prof. chem. 1890–1903.

Bellenot, Gustave [1858–1935] (*SWITZ*). Munich Dr.phil. 1885 (Perkin; worked on nitrobenzoylacetic acid). Prof. pharmacy school and school of commerce Neuchâtel 1894–1931.

Benack, Julius. Munich Dr.phil. 1896 (Thiele; worked on aminophenyl-triazole). Later activity not determined.

Bender, Georg [1838–1918]. Munich 1886 (published 5 papers on inorganic chem. topics). Dr.phil.; later established chemical firm in Munich.

Bentheim, Alfons von [1862–?]. Munich Dr.phil. 1907 (Koenigs; worked on condensation of trimethylpyridine with benzaldehyde); was army officer; no further chemical publications 1908–26; later activity not determined.

Beran, Alfred. Munich 1879 (Wurster; worked on bromomethyl-aniline). Zurich Dr.phil. 1884; no further chemical publications 1897–1926; later activity not determined.

Berend, Max [?–1888]. Berlin 1863–65 (Baeyer; worked on acetylene, formamide); Dr.phil. Set up chemical factory on Neu-Schönfeld (Leipzig).

Berlé, Bernhard [1866–?]. Munich Dr.phil. 1891 (Bamberger; worked on imidazole derivatives). No further chemical publications 1897–1926; later activity not determined.

Bernhart, Carl [1851–1920]. Munich 1884 (Perkin; worked on dehydroacetic acid); Dr.phil.?; later research asst. of Koenigs.

Besemfelder, Eduard [1863–1929]. Munich Dr.phil. 1890 (Baeyer; worked on reduction products of naphthoic acid). Later active in sugar industry, on editorial staff *Chemiker-Zeitung*.

BESTHORN, EMIL [1858–1921]. Freiburg i. B. Dr.phil. 1882; at Meister, Lucius & Brüning 1882–90 (accident in lab.); Munich private lab. 1892–1921 (worked on quinoline derivatives; had 5 research students).

Beyer, Carl [1859–1891]. Rostock Dr.phil. 1886; Munich 1886–89 (published 4 papers; 2 with Claisen); died soon after joining Meister, Lucius & Brüning Hoechst.

Bihan, Richard [1867–?]. Munich Dr.phil. 1894 (J. Thiele; worked on condensation of aminoguanidine with aldehydes). No further chemical publications 1897–1936; later activity not determined.

Bijvanck, Hendrik [1875–?] (*NETH*). Munich Dr.phil. 1898 (Besthorn; worked on lepidine derivatives). No further chemical publications 1899–1936; later activity not determined.

Bischkopff, Eduard [1875–1932]. Halle Dr.phil. 1898; Munich 1901 (Koenigs; worked on quinoline compounds). Became chemist at I. G. Farben.

Bishop, Arthur Wright [1867–1920] (*UK*). Munich Dr.phil. 1890 (Claisen; worked on formylcamphor). Later chemical consultant; died in India.

Blau, Fritz [1865–1929]. Vienna Dr.phil. 1886; Munich 1895 (Baeyer; worked on terpinol). Vienna Pv.Dz. chem. 1890–1902; entered electrical industry (became director at Osram) and held ca. 185 patents, mostly in lamp technology.

Bleyer, Benno [1885–1945]. Munich Dr.phil. 1910 (Prandtl; worked on atomic weight of molybdenum); Pv.Dz. pharm. and food chem. 1913. ao.Prof. Weihenstephan 1923; o.Prof. 1925; Munich TH 1929–45; published on various aspects of food chemistry.

Bloch, Ignaz [1878–1942]. Munich TH Dr.ing. 1901; Munich 1903 (Wieland; worked on diazo compounds). Scientific editor *Chemisches Zentralblatt* 1904–28; Berlin Pv.Dz. org. chem. (worked on polysulfides); then conult. chemist in Dessau. Died in Nazi concentration camp near Lodz.

Bloch, Siegfried [1882–?]. Munich Dr.phil. 1905 (Wieland; worked on addition reactions of unsaturated ketones). Later activity not determined.

Bloem, Friedrich [1860–1915]. Munich Dr.phil. 1882 (Baeyer; worked on synthesis of indigo from aminoacetophenone). Became owner of chemical explosives factory in Düsseldorf.

Blömer, Alfred [1888–1967]. Munich Dr.phil. 1913 (Piloty; worked on synthesis of pyrroles). Later with Bayer & Co. Leverkusen.

Bockmühl, Max [1882–1949]. Munich Dr.phil. 1909 (Einhorn; worked on eugenol and isoeugenol series). Became head pharmaceut. divn. I. G. Farben Hoechst.

Bode, Adolf. Munich Dr.phil. 1902 (Willstätter; worked on synthesis of cocaine). No further chemical publications 1907–36; later activity not determined.

Boeckmann, Otto [1860–?]. Erlangen Dr.phil. 1884; Munich 1887 (Bamberger; worked on naphthalene aldehyde). Became consulting chemist in Darmstadt.

Boesler, Magnus [1852–1924]. Freiburg i. B. Dr.phil. 1881; Munich 1882 (E. Fischer; worked on cuminoin, tolylhydrazine). Became consulting chemist in Munich.

Böttinger, Carl Conrad [1851–1901]. Tübingen Dr.phil. 1873; Munich 1880–81 (independent investigator; published 15 papers on various topics in organic chemistry). Later had private research lab. in Darmstadt.

Bordt, Friedrich [1851–?]. Munich Dr.phil. 1888 (Bamberger; worked on tetrahydronaphthalene derivatives). No further chemical publications 1892–1926; later activity not determined.

Borgmann, Eugen [1843–1895]. Berlin 1867–70 (Graebe; worked on toluquinone); Leipzig Dr.phil. 1868. After 1877 with Fresenius lab., Wiesbaden, where he specialized in food and wine chemistry, and published extensively on these subjects.

Born, Oscar. Berlin 1866 (Graebe; worked on phthalic acid). Dr.phil.; became commercial chemist in Darmstadt.

Bornhardt, Carl [1886–?]. Munich Dr.phil. 1913 (Schlenk; worked on triarylmethyl). No further chemical publications 1917–56; later activity not determined.

Brandis, Ernst [1861–1921]. Munich Dr.phil. 1889 (Baeyer; worked on condesations with naphthalene aldehyde). Later with aniline factory Griesheim a.M.

Brandl, Josef [1856–1925]. Munich Dr.phil. 1882 (E. Fischer; worked on minerals of the cryolite series); Dr.med. 1890; ao.Prof. pharmacol. veterinary school 1899–1905; o.Prof.1905–25.

Brantl, Josef. Munich Dr.phil. 1898 (Einhorn; worked on diethylbenzyl-amino compounds). No further chemical publications 1907–56; later activity not determined.

Breest, Fritz [1879–1923]. Munich Dr.phil. 1906 (Dieckmann; worked on acetoacetic ester). Became chemist at fisheries exper. station Munich.

Brettauer, Erwin [1884–?]. Munich Dr.phil. 1909 (Sand; worked on radioactive substances). Became head of a firm for utilization of brewery wastes.

Brömme, Eduard [1860–1893]. Munich Dr.phil. 1889 (Claisen; worked on reaction of oxalic ester on acetophenone). Died shortly after doctorate.

Brüning, Gustav von [1864–1913]. Munich Dr.phil. 1888 (Baeyer; worked on phenylhydrazine derivatives of succinyl compounds). Later became general director of Meister, Lucius & Brüning chemical firm.

Bruns, Wilhelm [1864–1945]. Munich 1890 (v.d. Pfordten; worked on mercuric oxide). Later with Bayer & Co. Elberfeld.

Buchka, Karl von [1856–1917]. Göttingen Dr.phil. 1881; Munich 1881 (Baeyer; worked on gallein and coerulein). Göttingen Pv.Dz. chem. 1881–91; ao.Prof. 91–96. Berlin successively in patent office, health ministry, govt. bureau of testing 1896–1916. Wrote on nutrition, also on historical topics.

BUCHNER, EDUARD [1860–1917]. Munich Dr.phil. 1888 (Curtius; worked on diazoacetic ester); Pv.Dz. 1891–93 (had 4 research students). Kiel 1893–96 (ao.Prof. 1895); Tübingen Prof. pharm. chem. 1896–98; Berlin Prof. chem. Agricult. Sch. 1898–1909; Breslau Prof. physiol. chem. 1909–11; Würzburg 1911–. He received the 1907 Nobel Prize in chemistry for his discovery of cell-free alcoholic fermentation (see TEXT).

Buchner, Karl [1868–?]. Munich Dr.phil. 1909 (Hofmann; worked on hexanitrocobaltic acid). Became consulting chemist, specialized in water purification.

Bührig, Heinrich. Munich 1876 (Baeyer; worked on aminophthalic acid). Later activity not determined.

Buff, Heinrich Ludwig [1828–1872]. Berlin 1868 (guest in Baeyer's lab; published on hexylene and amylene). See Appendix 1.

Bugge, Günther [1885–1944]. Munich Dr.phil. 1908 (Hofmann; worked on compounds of metal salts with nitriles). After various chemical assistantships, was head of library and patent divn., Holzverkohlung-Industrie Frankfurt a.M. 1918–43. Wrote on history of chemistry.

Buhlmann, Otto Ludwig [1877–?]. Munich Dr.phil. 1902 (Einhorn; worked on anthranil). Became commercial chemist in Leipzig.

Bull, Benjamin Samuel [1866–1910] (UK). Munich Dr.phil. 1897 (Einhorn; worked on hexahydroanthranilic acid, etc.). With Wilkinson, Heywood & Clark (paint manufacturers) 1897–1910.

Burg, Otto [?–1884]. Berlin 1864 (Baeyer; worked on oleic acid). Dr.phil.; became head of Russo-German Naphtha Import Co.

Burgdorf, Christian [1868–?]. Munich Dr.phil. 1891 (Bamberger; worked on chrysene). Later with Bayer & Co. Elberfeld.

Burger, Oskar [1883–?]. Munich Dr.phil. 1907 (Hofmann; worked on molybdenum and chromium complexes). No further chemical publications 1917–56; in 1907 moved to London; later activity not determined.

Burkhardt, Johann Baptist [?–1920]. Munich 1878 (Baeyer; worked on phthaleins). Dr.phil.?; later chemist with Weiler & Co., BASF.

Burton, Beverly Scott [1846–1904] (US). Prof. agricult. chem. Tennessee 1873–77; Würzburg Dr.phil. 1881 (Wislicenus); Munich 1883 (Pechmann; worked on derivatives of benzil and acetone dicarboxylic ester). Prof. chem. Univ. of Penna. 1888–1904.

Busch, Albert [1867–1902] (SWITZ). Munich Dr.phil. 1891 (Koenigs; worked on lepidine and quinaldine). Went to U.S.; died in Buffalo, N.Y.

Calman, Albert [1862–?] (US). Munich 1885 (Perkin, Bamberger; worked on benzoyl acetic ester and azo compounds); Erlangen Dr.phil. 1885. Became consulting chemist in New York City.

Carl, Richard Waldemar [1868–1933]. Munich Dr.phil. 1891 (Koenigs; worked on condensation reactions of phenols). Later with G. Jaeger company (printing inks; subsidiary of I. G. Farben).

CARO, HEINRICH [1834–1910]. After technical training in Germany and England in calico printing, ca. 1860 began research on artificial dyes. In 1866 was in Heidelberg (Bunsen). Director BASF Ludwigshafen 1868–1889. Collaborated with Baeyer and Graebe; Berlin 1870 (acridine); Strassburg 1874 (aniline dyes). Made many important contributions to organic chemistry (see TEXT).

Chattaway, Frederick Daniel [1860–1944] (*UK*). Munich Dr.phil. 1893 (Bamberger; worked on chrysene and picene). London Lecturer chem. St. Bartholomew's Hosp. 1893–1906; Oxford Tutor (then Praelector) Queen's College 1906–44. Published many valuable papers in organic chemistry.

Cherbuliez, Émile [1891–1985] (*SWITZ*). Munich Dr.phil. 1915 (Pummerer; worked on dehydromethylnaphthol). Geneva Pv.Dz. pharm. chem. 1920–22; ao.Prof. 1922–28; o.Prof. 1928–52; o.Prof. chem. 1952–66. Made many contributions to chemistry of proteins and organic phosphates.

Chojnacki, Kasimir [?–1904] (*POL*). Berlin 1871 (Liebermann; worked on opianic acid and rufiopine). Later director dye works in Poland.

CLAISEN, LUDWIG [1851–1930]. Bonn Dr.phil. 1875 (Kekulé); also studied in Göttingen (Wöhler); Pv.Dz. 1879. Manchester Owens College (Roscoe, Schorlemmer) 1882–85. Munich 1886–90 (Pv.Dz. 1887; published 40 papers, had 13 research students). o.Prof. chem. Aachen TH 1890–97; Kiel 1897–1904. After a stay (1904–07) in Fischer's institute in Berlin, continued research in his private laboratory in Godesberg 1907–26. A gifted chemist who made many important experimental contributions to the development of organic chemistry (see TEXT).

Clever, August [1869–?]. Munich Dr.phil. 1896 (Muthmann; worked on selenium and sulfur compounds). Later joined family chemical factory near Essen.

Cobliner, Jesaiah [1878–?] (*POL*). Munich Dr.phil. 1903 (Einhorn; worked on pyrogallol). No further chemical publications 1907–36; later activity not determined.

Coblitz, Franz Peter [1868–?]. Munich Dr.phil. 1896 (Einhorn; worked on reduction of hydroxybenzoic acid). No further chemical publications 1907–36; later activity not determined.

COHEN, JULIUS BEREND [1859–1935] (*UK*). Munich Dr.phil. 1884 (Pechmann; worked on reaction of phenols with ethyl acetoacetate). Leeds Lecturer chem. 1892–1904; Prof. 1904–24. Although productive in chemical research, his chief contribution was as a teacher (see TEXT).

Cohn, Lassar [1858–1922]. Königsberg Dr.phil. 1880; Pv.Dz. chem. 1888–94; Prof. 1894–97; Munich Pv.Dz. 1897–98 (worked on bile acids); Königsberg Prof. 1902–09. At various times, associated with chemical firms. Made contributions to the study of bile acids, the electrolysis of organic compounds, analytical chemistry and wrote several popular books on chemistry.

Comstock, William James [1860–1922] (*US*). Munich 1883 (Baeyer; worked on oxindole); 1886 (Koenigs; worked on quinine alkaloids). No doctoral degree; Yale instructor chem. 1896–1917.

Conrad, Ludwig [1878–?]. Munich Dr.phil. 1907 (Baeyer; worked on derivatives of eugenol, thymol, guiacol). No further chemical publications 1908–36; later activity not determined.

Corleis, Ehrenfried [1855–1919]. Dr.phil. Erlangen 1885; Munich 1886 (v.d. Pfordten; worked on sulfur compounds of tungsten). Became chief chemist Krupp Works.

Cornelius, Hans [1863–1947]. Munich Dr.phil. 1886 (Pechmann; worked on synthesis of orcine); Pv.Dz. chem. 1887–88. Left chemistry, Dr.phil. philosophy; Munich ao.Prof. 1903–10; Frankfurt a. M. Acad. Soc. Sci. 1910–.Wrote on many philosophical subjects, especially the philosophy of art.

Corti, Arnold [1873–1932] (*SWITZ*). Munich Dr.phil. 1899 (J. Thiele; worked on condensation products of cyanoacetamide and cyanoacetic acid). Later head chemical factory Flora in Dübendorf; also active entomologist.

CURTIUS, THEODOR [1857–1928]. Leipzig Dr.phil. 1882 (Kolbe); Munich 1882–86 (worked on diazo compounds, synthesis of hippuric acid derivatives). Erlangen Pv.Dz. 1886–90; o.Prof. Kiel 1890–97; Bonn 1897–98; Heidelberg 1898–1924. The leader in the development of the chemistry of azides and their use in organic synthesis; had many research students (see TEXT).

Dahl, Friedrich [1856–1929]. Kiel Dr.phil. 1884 (biology); Munich 1890 (Pechmann; worked on reduction of diketones). Later o.Prof. zoology Berlin.

Damm, Gustav. Berlin 1864 (Baeyer; worked on uric acid). Dr.phil.; later at chemical technology lab. Stuttgart TH. No chemical publications 1892–1926.

Deichsel, Theodor. Berlin 1864 (Baeyer; worked on uric acid). Dr.phil.?; later activity not determined.

Dent, Frankland [1869–1929] (*UK*). Munich Dr.phil. 1898 (J. Thiele; worked on urethane derivatives). Chemist Sierra Co. Burgos (Spain) 1900–05; Govt. analyst Singapore 1905–24.

Dessauer, Hans [1869–1926]. Basle Dr.phil. 1892; Munich 1892–94 (Buchner; worked on derivatives of triphenylmethylene, pyrazole). Became chemist in paper factory.

Dieckmann, Walter [1869–1925]. Munich Dr.phil. 1892 (Bamberger; worked on tetrahydroisoquinoline); Pv.Dz. 1898–1905; ao.Prof. 1905–26. Published many valuable papers on organic-chemical topics; before 1915 he had 7 research students.

Diehl, Claus [1879–?]. Munich Dr.phil. 1906 (Baeyer; worked on alkylated aminobenzophenone, tetraphenylmethane derivatives). After 1910 with Merck Darmstadt.

Dienstbach, Oskar [1883–1914]. Munich Dr.phil. 1908 (Dimroth; worked on reactions of phenyloximidotriazolone). Later with BASF Ludwigshafen; died in action World War I.

Diesbach, Henri de [1880–1970] (*SWITZ*). Munich Dr.phil. 1907 (Einhorn; worked on derivatives of substituted malonic acids). With BASF Ludwigshafen 1907–19; Fribourg (Switz.) o.Prof. chem. 1920–55. Published on synthesis of dyes.

DIMROTH, OTTO [1872–1940]. Munich Dr.phil. 1895 (J. Thiele; worked on nitrobenzyl chloride). Tübingen Pv.Dz. 1900–05; Munich ao.Prof. 1905–13 (did research on heterocycles, diazo compounds, physical-organic chem.; had 14 research students). o.Prof. Greifswald 1913–18; Würzburg 1918–37. An outstanding chemical experimenter, and one of the pioneers in the development of physical-organic chemistry in Germany (see TEXT).

Dörr, Gustav [1874–?]. Munich Dr.phil. 1901 (J. Thiele; worked on cinnamylidene malonic acid). Became consulting food chemist in Hamburg.

Dorfmüller, Gustav. Munich TH Dr.phil. 1909; Munich 1912 (Pummerer; worked on isophthalanil). Later associated with sugar industry.

Dormann, Edmund [1884–1959]. Munich Dr.phil. 1916 (Piloty; worked on synthesis of pyrroles). Became commercial chemist in Munich.

Dorp, Willem Anne van [1847–1914] (*NETH*). Heidelberg Dr.phil. 1871; Berlin 1871 (Liebermann; worked on cochineal dye). Continued research in private laboratory; published extensively on organic chemistry of natural products.

Dralle, Eduard [1870–1940]. Munich Dr.phil. 1898 (J. Thiele; worked on condensation of aminoguanidine with aldehydes and ketones). Later joined family chemical firm in Hamburg-Altona.

Drewsen, Viggo Beutner [1858–1930] (*DAN*). Munich Dr.phil. 1881 (Baeyer; worked on synthesis of indigo). Trondheim technical school Pv.Dz. 1882–87; in cellulose factory 1887–93. Went to U.S. and was active in cellulose industry.

Ducca, Wilhelm [1880–?]. Munich Dr.phil. 1904 (Hofmann; worked on phosphorescent materials). Went to U.S.; became consulting chemist in Babbitt, N.J.

Dünschmann, Max [1858–1923]. Erlangen Dr.phil. 1886; Munich 1891 (Pechmann; worked on acetone dicarboxylic acid). Later chemist at I. G. Farben.

DUISBERG, KARL [1861–1935]. Jena Dr.phil. 1882; Munich 1883 (Pechmann; worked on reaction of phenols with ethyl acetoacetate).

With Bayer & Co. Elberfeld 1883–; general director 1912–25; then head of I. G. Farben (see TEXT).

Ebert, Georg von [1885–1956]. Munich Dr.phil. 1908 (Hofmann; worked on reaction of phenyl magnesium bromide with inorganic chlorides). Left chemistry; Würzburg Dr.jur. 1913; became economist and headed institute for Bavarian economy.

Eckert, Fritz [1888–?]. Berlin Dr.phil. (physics) 1913; Munich 1914 (Pummerer, worked on quinoid salts of thiazines). Berlin Pv.Dz. 1923, and associated with the glass industry.

Ehret, Hermann [1869–1913]. Munich Dr.phil. 1897 (Einhorn; worked on cocaine). Later active in textile industry.

Ehrhardt, Ernest Francis [1866–1929] (UK). Munich Dr.phil. 1889 (Claisen; worked on acetyl acetone). Became consulting chemist in Manchester.

Ehrhardt, Oskar [1888–?]. Munich Dr.phil. 1912 (Hofmann; action of hydrazine on dicyanamide). Later with Deutsche Gold- und Silber-Anstalt Frankfurt a. M.

Ehrhardt, Wilhelm. Dr.phil.; Munich 1878–79 (E. Fischer; worked on derivatives of phenylhydrazine). No further chemical publications; later activity not determined.

Eichwald, Egon [1883–1943]. Munich 1900 (Hofmann; worked on mercarbide); Halle Dr.phil. 1906 (Vorländer). After World War I became head of scientific lab. Ossag Hamburg; in 1927 moved to Amsterdam, central lab. of Royal Dutch Shell. Died in Nazi concentration camp in Holland.

Eichwede, Heinrich [1875–1956]. Munich Dr.phil. 1900 (J. Thiele; worked on action of ethyl nitrite on phenols). Later chemist with I. G. Farben.

EINHORN, ALFRED [1857–1917]. Tübingen Dr.phil. 1880; Munich 1882–85 (Pv.Dz.); Aachen TH 1886–91 (Prof. 1890); Munich 1891–1917 (Pv.Dz.; published ca. 100 papers and had 48 research students, including Willstätter). A remarkably gifted and versatile chemist and teacher (see TEXT).

Ekstrand, Åke Gerhard [1846–1933] (SWED). Uppsala Dr.phil. 1875; Munich 1878 (Baeyer; worked on hydroquinone phthalein). Later with Swedish finance ministry and editor of Svensk Kemisk Tidskrift.

Emmerling, Adolf [1842–1906]. Freiburg i.B. Dr.phil. 1865; Berlin 1869 (Baeyer; worked on synthesis of indole and indigo). Kiel head of chem. lab. Agric. Exp. Station 1871–; Pv.Dz. agric. and physiol. chem. 1874–82; Prof. 1882–1906; published extensively on agricultural chemistry.

Eppens, August [1864–?]. Munich Dr.phil. 1892 (Koenigs; worked on camphor). Later consulting chemist sugar industry.

Eppinger, Paul [1879–?]. Munich Dr.phil. 1907 (Piloty; worked on chemistry of hemoglobin). No chemical publications 1917–56; later activity not determined.

Erwig, Emil [?–1900]. Munich 1889 (Koenigs; worked on sugar chemistry). Dr.phil.?; later pharmacist in Kalk.

Escales, Richard [1863–1924]. Würzburg Dr.phil. 1886; Munich 1901 (Einhorn, J. Thiele; worked on hydrazides, dinitrotoluene). In 1898 patented ammonal; set up lab. in Munich for research on explosives 1902–07.

Ettlinger, Friedrich [1877–?]. Munich Dr.phil. 1902 (Willstätter; worked on pyrrolidine compounds). No further chemical publications 1907–56; later activity not determined.

Ewan, Thomas [1868–1955] (*UK*). Munich Dr.phil. 1890 (Claisen; action of oxalic ester on dibenzylketone). Research chemist Aluminium Co. 1896–1900; chief chemist Cyanide Chem. Co. Glasgow 1900–27 (director 1924–27); after working at I.C.I. (1927–33). took over family oil refinery firm; later chief chemist Cyanide Chem. Co. Glasgow.

Fabinyi, Rudolf [1849–1920]. Budapest Dr.phil. 1868; Munich 1878 (Baeyer; worked on diphenol ethane). Prof. chem. Klausenberg 1878–1919.

Faust, Edwin Stanton [1870–1928]. Munich Dr.phil. 1893 (Einhorn; worked on cocaine); Strassburg Dr.med. 1898; Pv.Dz., ao.Prof. pharmacology 1898–1907; Würzburg o.Prof. pharmacology 1908–20; Basle CIBA 1920–23.

Feer, Adolf [1862–1913] (*SWITZ*). Munich Dr.phil. 1886 (Koenigs; worked on derivatives of pyridine and quinoline). With Geigy Basle 1887–91; Scheurer-Kestner Mulhouse 1891–99; then set up textile-printing firm in Brombach.

Feibelmann, Richard [1883–1948]. Munich Dr.phil. 1907 (Einhorn; worked on oxyacid amides. quinoline carboxylic acid). Later with Heyden A. G. Dresden; after 1935 in U.S. (Activin divn. N.Y.).

Feigel, Heinrich [1877–?]. Munich Dr.phil. 1905 (Hofmann; worked on reaction of heavy metal compounds with polysulfides). No further chemical publications 1907–56; later activity not determined.

Fester, Gustav [1886–?]. Munich Dr.phil. 1910 (Dimroth; worked on trizoles and tetrazoles). Frankfurt a.M. Pv.Dz. chem. 1919–21; ao-.Prof. 1921–24. Went to Argentina, o.Prof. Univ. Nat. del Litoral Santa Fe 1924–.

Finckh, Carl [1878–1941]. Munich Dr.phil. 1903 (Piloty; worked on uric acid derivatives). Later chief chemist with Osram (lamp technology).

Fink, Heinrich. Munich 1912 (Piloty; worked on molecular size of hemin and hemoglobin); Freiburg i.B. Dr.phil. 1920. Later with I. G. Farben.

FISCHER, EMIL [1852–1919]. Strassburg Dr.phil. 1874 (Baeyer; worked on fluorescein and phthalein-orcin); Munich Pv.Dz. 1878; ao.Prof. 1879–82 (published 41 papers; had 5 research students). o.Prof. chem. Erlangen 1882–85, Würzburg 1885–92, Berlin 1892–1919. Baeyer's most distinguished pupil and the leading German organic chemist at the turn of the century (see TEXT).

Fischer, Louis [1862–1921] (US). Munich Dr.phil. 1891 (Claisen; worked on benzoyl acetaldehyde. Also worked with Einhorn on tropine derivatives). Became chemist at the U.S. Bureau of Standards.

FISCHER, OTTO [1852–1932]. Strassburg Dr.phil. 1874 (Baeyer; worked on action of chloral on toluene); Munich Pv.Dz. 1878–84 (published 50 papers; had 14 research students; collaborated with his cousin Emil in research on triphenylmethane dyes); o.Prof. Erlangen 1885–1925. Made contributions to several areas of organic chemistry.

Fitz, Albert [1842–1885]. Strassburg 1875 (Baeyer; worked on resorcinol). Dr.phil.; conducted research on mold fermentation in private lab. in Strassburg.

Forrer, Carl [1856–1921] (SWITZ). Zurich Dr.phil. 1882; Munich 1883–84 (Baeyer?; published 3 papers on organic chemical topics). Became a patent attorney in Basle.

Foucar, Georg [1870–?]. Munich Dr.phil. 1898 (J. Thiele; worked on semi-carbazones). Later in metal smelting industry.

FOURNEAU, ERNEST [1872–1949] (FR). Paris Dr.pharm. 1898; Munich 1902 (Willstätter; worked on lupinine); also studied with Emil Fischer and with Curtius. Director pharm. chem. lab. Poulenc Ivry-sur-Seine 1902–10; Paris chef de service Pasteur Institute 1911–1944. A talented chemist; developed an anaesthetic (stovaine) and contributed to the study of other chemotherapeutic agents and of alkaloids (see TEXT).

Frank, Christian [1872–?]. Munich Dr.phil. 1898 (J. Thiele; worked on hydrazinoisobutyric acid). No further chemical publications 1900–36; later activity not determined.

Frankfurter, Fritz [1890–?]. Munich Dr.phil. 1914 (Pummerer; worked on organic radicals). Continued as research asst. of Pummerer at Munich, Greifswald and Erlangen.

Fraude, Georg [1848–1899]. Munich 1879–81 (Baeyer; worked on phthaleins). Dr.phil.?; later became owner of pharmacy in Bühl (Alsace).

Freer, Paul Caspar [1862–1912] (US). Rush Med. College M.D. 1883; Munich Dr.phil. 1887 (Perkin; worked on synthesis of cyclopropane derivatives). Prof. chem. Univ. of Michigan 1890–1901; Supt. Govt. lab. Philippines 1901–12.

Fressel, Hans [1887–?]. Munich Dr.phil. 1912 (Wieland; worked on aliphatic hydrazines and tetrazenes). No further chemical publications 1917–56; later activity not determined.

Frew, William [1870–1910] (*UK*). Munich Dr.phil. 1893 (Bamberger; worked on isocoumarin and isoquinoline). Later associated with UK breweries or cider makers.

Frey, Karl [1868–1947] (*SWITZ*). Munich Dr.phil. 1893 (Einhorn; worked on eugenol and isoeugenol). Became bacteriologist in Munich.

Friedländer, Albert [1869–1942?]. Munich Dr.phil. 1893 (Einhorn; worked on ecgonine). Later commercial chemist in Berlin; died in Nazi concentration camp near Lodz.

FRIEDLÄNDER, PAUL [1857–1923]. Munich Dr.phil. 1878 (Baeyer; worked on phenanthrene quinone); asst. of Baeyer 1879–81 (worked on isatin); Pv.Dz. 1883–84 (published 20 papers; had 8 research students). Head of research lab. Oehler Co. Offenbach 1884–87. Karlsruhe TH ao.Prof. 1888–95; Vienna technical-industrial museum 1895–1911; Darmstadt TH o.Prof. 1911–23. The leading dye chemist of his time; more than 200 patents and many papers on dyes of the indigo series.

Frisoni, Erich [?–1905]. Munich 1905 (Dimroth; worked on condensation of diazophenylimid with ketones); died during doctoral studies.

Fritsch, Martin [1868–?]. Munich Dr.phil. 1893 (Buchner; worked on pyrazole derivatives). Later with I. G. Farben.

Fritsch, Paul [1859–1913]. Munich Dr.phil. 1888 (Baeyer; worked on hydroxyphenylacetic acid). Assistantships at Breslau and Rostock, and work in chemical industry 1888–93. Marburg Pv.Dz. pharm. chem. 1893; Prof. 1899–1913; published valuable papers in pharmaceutical chemistry.

Frobenius, August Ludwig [1866–1926]. Munich Dr.phil. 1894 (Pechmann; worked on aromatic diazo compounds). Became teacher Realgymnasium Augsburg.

Fuchs, Friedrich [?–1926]. Strassburg Dr.phil. 1876 (Baeyer; worked on nitrosonaphthol). Later with Central Exp. Station, Vienna, then commercial chemist in Jena.

Gambaryan, Stepan Pavlovich [1879–1948] (*RUSS*). Munich Dr.phil. 1907 (Wieland; worked on tetraphenylhydrazine and diarylhydroxylamine). Later Prof. chem. Erivan (Armenia).

Garben, Eduard [1873–1940]. Munich Dr.phil. 1902 (Besthorn; worked on reaction of acetone dicarboxylic ester with aromatic amines). Became commercial chemist in Hannover.

Gassner, Sebastian. Munich 1914 (Pummerer; worked on quinoid salts of thiazines). Dr.phil.; later with I. G. Farben.

Gehrenbeck, Clemens [1859–1909]. Erlangen Dr.phil. 1888; Munich 1888 (Einhorn; worked on nitrocinnamic aldehyde). Became owner of commercial chemistry lab. in Halle.

Geigy, Rudolf [1862–1943] (*SWITZ*). Munich Dr.phil. 1885 (Koenigs; worked on derivatives of pyridine and benzophenone). Joined family chemical firm, and later headed it.

Genssler, Otto [1881–?]. Munich Dr.phil. 1904 (Sand; worked on cobaltamine complexes). No further chemical publications 1907–56; later activity not determined.

Gerber, Niklaus [1850–1914] (*SWITZ*). Strassburg 1873 (no publication?); Zurich Dr.phil. 1874. Established famous milk company in Zurich.

German, Ludwig [1855–?]. Erlangen Dr.phil. 1883; Munich 1883 (O. Fischer; worked on triphenylmethane derivatives, skatole). No further chemical publications 1885–1936; later activity not determined.

Gernsheim, Alfred [1870–1931]. Munich Dr.phil. 1895 (Einhorn; worked on nitrophenylglycidic acids). Became commercial chemist in Worms.

Giese, Oskar [1873–1917]. Munich 1902 (J. Thiele; worked on condensation products of dihydroterephthalic acid). Strassburg Dr.phil. 1903; later chemist at Hoechst.

Gilbody, Alexander William [1870–?] (UK). Munich Dr.phil. 1893 (Einhorn; worked on action of chloral on pyridine bases). Later research asst. of W. H. Perkin in Manchester, then at Bradford Technical College Antigua.

Gmelin, Erwin [1883–1961]. Munich Dr.phil. 1909 (Wieland; worked on furoxanes). Later with Boehringer Ingelheim.

Göhring, Carl Friedrich [1857–?]. Munich Dr.phil. 1884 (Baeyer; worked on aromatic aldehydes). With Spindler textile dyeworks 1888–1922.

Goes, Bruno [?–1900]. Munich 1880 (Baeyer; worked on diphenyl diaminonaphthol). Dr.phil.?; became pharmacist in Frankfurt a.O.

Göttler, Maximilian [1882–?]. Munich Dr.phil. 1908 (Einhorn; worked on halogenacetylamides, isatin). Later with Boehringer Ingelheim.

Goldschmidt, Carl [1867–?]. Munich Dr.phil. 1891 (Bamberger; worked on ethyl naphthylamine). Later commercial chemist in Geneva.

Goldschmidt, Martin [1834–1915]. Berlin 1871 (Baeyer; worked on pyrrole). Dr.phil. Heidelberg; later established chemical factory in Köpenick (near Berlin).

GOLDSCHMIDT, STEFAN [1889–1971]. Munich Dr.phil. 1912 (Dimroth; worked on pigment of stick-lac). Würzburg Pv.Dz. 1919–23;

Karlsruhe TH ao.Prof. 1923–27; o.Prof. 1927–35. With Organon
Co. in Oss, Holland, 1938–47; Munich TH o.Prof. 1947–. Published
many valuable papers on a great variety of organic-chemical topics.

GOLDSCHMIEDT, GUIDO [1850–1915]. Heidelberg Dr.phil. 1872;
Strassburg 1872–74 (Baeyer; worked on diphenylethane, compounds
of bromal and chloral with benzene). Vienna Pv.Dz. 1875, ao.Prof.
1890. o.Prof. Prague 1891–1911; Vienna 1911–15. Many important
contributions to the study of the structure of natural products.

GOMBERG, MOSES [1866–1947]. (*US*). Michigan Ph.D. 1894; Munich
1896–97 (J. Thiele; worked on isonitramine- and nitroso-isobutyric
acid). Also studied with V. Meyer in Heidelberg, where Gomberg began the famous work that led to his discovery of triphenylmethyl.
Michigan Instr. chem. 1893–99; Asst. Prof. 1899–1904; Prof. 1904–
36. Created at Michigan one of the leading American research
groups in organic chemistry (see TEXT).

Gonder, Karl Ludwig [1880–?]. Erlangen Dr.phil. 1905; Munich 1905
(Hofmann; worked on induced radioactivity); later at Chem. Inst.
Univ. Kiel.

Grabfield, Gustave Philip [1861–1935] (*US*). Munich Dr.phil. 1887
(Einhorn; worked on methoxyphenylacrylic acid). Chief chemist
Morris meat packing company 1888–1929.

Grabowski, Julian [1848–1882] (*POL*). Warsaw Dr.phil. 1870; Berlin,
Strassburg 1871–73 (Baeyer; worked on naphthol derivatives);
Dr.phil. 1874. Lwow Pv.Dz. 1875–77; Cracow Prof. chem. technol.
1877–82.

GRAEBE, KARL [1841–1927]. Heidelberg Dr.phil. 1862; Berlin 1865–
69 (asst. of Baeyer; published 26 papers, including the reports, with
Liebermann, on the conversion of alizarin to anthracene and on the
synthesis of alizarin). Leipzig Pv.Dz. 1869; Königsberg o.Prof. 1870–
78; Geneva o.Prof. 1878–1906. Wrote history of organic chemistry
(see TEXT).

Gräter, Adolf [1874–1946]. Munich Dr.phil. 1898 (J. Thiele; worked on
nitramines of carbonic acid). No further chemical publications 1899–
1936; later activity not determined.

Grammling, Franz [1881–?]. Munich Dr.phil. 1908 (Sand; complex
chromium compounds). No further chemical publications 1917–56;
later activity not determined.

Greiff, Philipp [?–1885]. Munich 1879 (O. Fischer; worked on aniline
dyes). Dr.phil.?; later commercial dye chemist in Frankfurt a.M.

Grimm, Ferdinand [1845–1919]. Strassburg 1873 (Baeyer; worked on
phthalein). Dr.phil. Erlangen 1873; later Dr.med., specialized in
tropical medicine.

Groeneveld, Anton [1871–?]. Munich Dr.phil. 1900 (Dieckmann; worked on reactions of methyladipic acid ester). No further chemical publications 1907–36; later activity not determined.

Gruber, Wolfgang [1886–?]. Munich Dr.phil. 1914 (Baeyer; worked on reduction of dimethylpyrone). Baeyer's last research asst.; later chief chemist Wacker electrochemical firm Burghausen.

Gruhl, Woldemar [1880–?]. Tübingen Dr.phil. 1906; Munich 1907 (Dimroth; worked on diazoamino compounds). Later head of mining company in Essen.

Gruyter, Paul de [1866–?]. Munich Dr.phil. 1893 (Bamberger; worked on cyanophenylhydrazines and formazylmethylketone). Later chemical consultant Frankfurt a. M.

Günther, Oscar [1876–1917]. Munich Dr.phil. 1902 (J. Thiele; worked on derivatives of phenylnaphthalene). Later with Bayer & Co. Elberfeld.

Haeckel, Siegfried [1877–1931]. Munich Dr.phil. 1901 (J. Thiele; worked on derivatives of nitrophenylethylene). Later with I. G. Farben.

Hahnenkamm, Wilhelm [1875–?]. Munich Dr.phil. 1902 (J. Thiele; worked on condensation products of dinitrotoluene and dinitrobenzaldehyde). Later with I. G. Farben.

Haiss, August [1856–1905]. Freiburg i. B. Dr.phil. 1882; Munich 1882 (Pechmann; worked on ditolylpropionic acid). Later commercial chemist in Munich.

Hallensleben, Richard [1877–?]. Munich Dr.phil. 1904 (Baeyer; worked on dibenzalacetone and triphenylmethane). No further chemical publications 1907–36; later activity not determined.

Hamburger, Alexander [1880–1914]. Munich Dr.phil. 1907 (Einhorn; worked on N-methylol derivatives of aliphatic amides). Later with Bayer & Co. Elberfeld; died in action World War I.

Happe, Gustav Heinrich [1876–1967]. Munich Dr.phil. 1903 (Koenigs; worked on reactions of pyridine derivatives). Later commercial chemist in Alsfeld.

Harrison, Leonard Hubert [1885–?] (*UK*). Munich Dr.phil. 1910 (Hofmann, worked on diffusion in molten salts). Later consulting chemist in Manchester.

Hartl, Ferdinand [1880–1907]. Munich Dr.phil. 1906 (Vanino; worked on colloidal solutions); died shortly after doctorate.

Hartmann, Max [1884–1952] (*SWITZ*). Munich Dr.phil. 1908 (Dimroth; worked on diazo compounds). Became head of the pharmaceutical division of CIBA.

Hauser, Otto [1877–1915]. Munich Dr.phil. 1902 (Vanino; worked on bismuth compounds). Berlin Pv.Dz. chem. 1905–15.

Haussknecht, Otto [1844–1914]. Berlin 1866 (Baeyer; worked on derivatives of erucic acid); Breslau Dr.phil. 1869. Later teacher at technical school Gleiwitz.

Hecht, Hans [1890–?]. Munich Dr.phil. 1916 (Prandtl; worked on vanadium and tungsten compounds). Later owner of lab. for cement chem. in Berlin.

Heide, Carl von der [1872–1935]. Munich Dr.phil. 1896 (Hofmann; worked on reactions of molybdenum compounds). Later head of wine research station in Giesenheim.

Heidenreich, Karl Friedrich [1871–1924]. Munich 1893 (J. Thiele; worked on triazole derivatives); Kiel Dr.phil. 1894. Later with Bayer & Co.

Heidepriem, Wilhelm [1876–1945]. Munich Dr.phil. 1901 (Hofmann; worked on analysis of bröggerite). No further chemical publications 1907–36; later activity not determined.

Heine, Otto [1876–?]. Munich Dr.phil. 1903 (Hofmann; worked on ferrocyanides). No further chemical publications 1907–36; later activity not determined.

Heintzel, Carl. Berlin 1866–67 (Baeyer; worked on uric acid); Leipzig Dr.phil. 1867. Later commercial chemist in cement industry.

Heinz, Robert [1865–1924]. Breslau Dr.med. 1891; Munich 1897 (Einhorn; worked on orthoform). Erlangen Pv.Dz. pharmacology 1898–1904; ao.Prof. 1904–10; o.Prof. 1910–24. Published extensively on hematology, dermatology, metallic colloids.

Helwig, Hermann [1864–1935]. Munich Dr.phil. 1888 (Bamberger; worked on derivatives of naphthylamines). Government industrial inspector 1894–.

Hemilian (Gemilian), Valerian Aleksandrovich [1851–1914] (RUSS). Göttingen Dr.phil. 1873; Strassburg 1873 (Baeyer; worked on synthesis of triphenylmethane). Warsaw Pv.Dz. techn. chem. 1877–86; Kharkov technical institute ao.Prof. 1887–89; o.Prof. 1889–1903.

HENRICH, FERDINAND [1871–1945]. Heidelberg Dr.phil. 1894 (Jacobson); Munich 1895 (Baeyer; worked on pulegone and carvone). Graz Pv.Dz. 1897; Erlangen ao.Prof. 1912–26; o.Prof. 1926–41. Made many valuable contributions to organic and analytical chemistry, and also to the history of chemistry.

Henriques, Robert [1857–1902]. Strassburg Dr.phil. 1881; Munich 1881–82 (Friedländer; worked on nitrobenzaldehyde). Later in Oehler dye factory Offenbach and in Kunheim & Co. Berlin; then set up own commercial lab.

Henry, Paul [1866–1917] (BELG). Louvain D.Sc. 1890; Munich 1893 (Pechmann; worked on action of nitrous acid on acetone dicarboxylic ester). Louvain Prof. physical chem. 1896–1917.

Hepp, Eduard [1851–1917]. Strassburg Dr.phil. 1875 (Baeyer; worked on compounds of monochloraldehyde with aromatic hydrocarbons). Later with Meister, Lucius & Brüning Hoechst.

Herb, Josef [1861–?]. Erlangen Dr.phil. 1890; Munich 1890 (Baeyer; worked on reduction products of terephthalic acid). Later went to U.S.; became pharmaceutical chemist in Wisconsin.

Herzenstein, Anna. Munich 1910 (Schlenk, worked on triarylmethyls); Zurich Dr.phil. 1911. Later research assoc. of Cherbuliez in Geneva.

Herzog, Georg. Berlin 1865 (Baeyer; worked on hydantoic acid). Dr.phil.; later commercial chemist in Oppeln.

Hess, Hermann [1882–1915]. Munich Dr.phil. 1909 (Wieland; worked on nitroso and azo compounds). Later with refrigeration company Strassburg; died in action World War I.

Hess, Ludwig Conrad [1882–1956]. Munich Dr.phil. 1913 (Prandtl; worked on vanadium compounds). Later head of factory Riedel A.G. Berlin-Friedenau.

Hess, Wilhelm [1862–1934]. Munich 1884 (Einhorn; worked on isopropylnitrophenyl lactic acid); Göttingen Dr.phil. 1887. Later with Edeleanu Chem. Co. (oils) in Berlin-Lichterfelde.

Hesse, Julius [1864–?]. Berlin Dr.med. 1887; Munich Dr.phil. 1897 (Baeyer; worked on acetals, derivatives of pyrocatechol). Later in medical practice.

Hessert, Julius [1852–1898]. Munich 1878 (Baeyer; worked on phthalaldehyde). Later teacher Realschule in Speyer.

Heuser, Karl [1869–?]. Munich Dr.phil. 1896 (J. Thiele; worked on semicarbazide and hydrazine derivatives of isobutyric acid). No further chemical publications 1907–36; later activity not determined.

Heymann, Bernhard [1861–1933]. Munich Dr.phil. 1887 (Koenigs; worked on oxidation of phenol homologues, lepidine compounds). Later with Bayer & Co. Elberfeld.

Hiendlmaier, Heinrich [1878–1967]. Munich Dr.phil. 1907 (Hofmann; worked on perchromates). Later owner of pharmacy in Munich.

Hirsch, Paul [1887–1942?]. Munich Dr.phil. 1912 (Piloty; worked on synthesis of pyrroles). Later consulting chemist in Berlin-Charlottenburg; died in Auschwitz concentration camp.

His, Hans [1866–1915] (SWITZ). Munich Dr.phil. 1894 (Einhorn; worked on cocaine). Later cantonal food inspector in Chur.

HJELT, EDVARD [1855–1921]. (FIN). Munich 1879 (Baeyer; worked on camphoronic acid); Helsingfors Dr.phil. 1879; Pv.Dz. chem. 1880; Prof. 1882–1907. Published on stereochemistry, natural products and history of chemistry; later was active in Finnish politics.

Hocheder, Ferdinand [1881–1945]. Munich Dr.phil. 1907 (Willstätter; worked on chlorophyll). Later consulting chemist in Leipzig.

Hock, Heinrich [1887–1971]. Munich Dr.phil. 1912 (Hofmann; worked on diazotetrazole). Clausthal Bergakademie ao.Prof. organic chem. and fuel chem. 1932–47; o.Prof. 1947–55.

Höbold, Kurt [1886–?]. Munich Dr.phil. 1911 (Hofmann; worked on salts of perchloric acid). Later with Fabrik F. Raschig Ludwigshafen, then director factory Lüneberg.

Höchtlen, Friedrich [1878–1951]. Munich Dr.phil. 1904 (Hofmann; worked on sulfur compounds of heavy metals). Later head of Stockhausen chemical factory in Krefeld.

Hoeppner, Max [1872–1922]. Munich Dr.phil. 1898 (Koenigs; worked on quinine alkaloids). Later head of chem. factory Goldenburg & Geromont Winkel.

Hoerlin, Julius [1869–1955]. Munich Dr.phil. 1893 (Koenigs; worked on sulfocamphylic acid). Later with dye factory Graf & Co. Neubabelsberg.

Hofe, Christian von [1871–1954]. Munich Dr.phil. 1894 (Einhorn; worked on eugenol derivatives); Berlin Dr.phil. 1898 (physics; E. Warburg). With Zeiss (1903–07) and Goertz (1907–35) optical firms; Charlottenburg TH o.Prof. physics 1935–41.

Hoffmann, Felix [1868–1946]. Munich Dr.phil. 1893 (Bamberger; worked on derivatives of anthrol). Later at Bayer & Co. Elberfeld, then consulting chemist in Breslau.

Hoffmann, Ludwig. Munich 1883 (Koenigs; worked on tetrahydroquinoline); Erlangen Dr.phil. 1887 (Krüss; worked on atomic weight of gold). Later commercial chemist in Leipzig (fats and oils).

HOFMANN, KARL ANDREAS [1870–1940]. Munich Dr.phil. 1892 (Krüss; worked on rare earths); Pv.Dz. 1895–98; ao.Prof. 1898–1910 (published about 100 papers; had 32 research students). Charlottenburg TH o.Prof. inorganic chem. 1910–35. An extremely productive chemist, whose principal contributions were in inorganic chemistry.

Holch, Ludwig [1887–1965]. Munich Dr.phil. 1913 (Dimroth; worked on carminic acid). Later with Feldmühle, Papier & Zellstoff A. G. Troisdorf.

Hollander, Charles Samuel [1877–1962] (US). Munich Dr.phil. 1902 (Willstätter; worked on ecgonine and hygrine). With Mallinckrodt St. Louis 1904–09 and eastern chem. works 1910–14; then director research Röhm and Haas Co. Philadelphia.

Hollandt, Friedrich [1868–1949]. Munich Dr.phil. 1898 (Einhorn; worked on acylation of alcohols and phenols). Became pharmacist in Güstrow.

HOLLEMAN, ARNOLD FREDERIK [1859–1953] (NETH). Leiden Dr.phil. 1887; Munich 1887 (Baeyer; worked on phenylacetylene and diphenylacetylene). Groningen head agricult. chem. clab. 1889–93;

Prof. chem. 1893–1905; Amsterdam Prof. chem. 1905–24. Published extensively on many aspects of organic chemistry.

Holzinger, Otto [1869–?]. Munich Dr.phil. 1897 (J. Thiele; worked on diaminodibenzyl). Later became pharmacist in Bamberg.

Homolka, Benno [1860–1925]. Munich 1884–86 (Baeyer; worked on quinisatin, cantharidin, aromatic aldehydes); Erlangen Dr.phil. 1884. Later with Meister, Lucius & Brüning Hoechst.

Hooker, Samuel Cox [1864–1935] (*UK*). Munich Dr.phil. 1885 (Bamberger; worked on retene). Went to U.S.; associated with sugar refining industry 1885–1915; then set up private research lab. in Brooklyn. Did important research on lapachol, which was published posthumously.

Hoppe, Johannes [1872–1949]. Munich Dr.phil. 1902 (Dieckmann; worked on reactions of dicarbonyl compounds). Later owner of private commercial chem. lab. in Munich.

Hori, Etznoyu (*JAP*). Munich 1891 (Claisen; worked on aconitic acid). Dr.phil.?; later activity not determined.

Hoskyns-Abrahall, John Leigh [1865–1891] (*UK*). Munich 1888 (Bamberger; worked on tetrahydronaphthylene diamine); Dr.phil. 1889 (Groth; crystallography); died soon after doctorate.

Hottenroth, Valentin [1878–1954]. Munich Dr.phil. 1905 (Willstätter; worked on bromonitromalonic acid ester). Later with Zellstoff Fabrik (artificial silk) in Waldhof.

Hütz, Hugo [1871–?]. Berlin Dr.phil. 1895 (E. Fischer); Munich 1900 (Einhorn; worked on aminooxybenzoic acid ester). Later with Hoechst (1899–1904), then with other chemical firms.

Hütz, Rudolf [1877–?]. Munich Dr.phil. 1902 (J. Thiele; worked on dimethylene quinone). No further chemical publications 1907–36; later activity not determined.

Husmann, August [1872–?]. Munich 1896 (Koenigs; worked on cinchonine); Bonn Dr.med. 1898. Later in medical practice.

Ibele, Josef [1877–1956]. Munich Dr.phil. 1905 (Besthorn; worked on synthesis of dyes from quinoline carboxylic acids). Later became agricultural chemist in Munich.

Iglauer, Fritz [1877–?]. Munich Dr.phil. 1901 (Willstätter; worked on tropinone). Later head of dye factory in Nürnberg.

Ilosvay de Nagyilosva, Lajos [1851–1936]. Budapest Dr.phil. 1875; Munich 1881 (Baeyer; no publication?); also studied with Bunsen and Berthelot. Budapest TH Prof. chem. 1883–1914, 1917–34.

Ingle, Harry [1869–1921] (*UK*). Munich Dr.phil 1893 (J. Thiele; worked on derivatives of tetrazole). Chief chemist linoleum manufacturer in Kirkcaldy 1896–1909, then consulting chemist in Leeds.

IPATIEFF, VLADIMIR NIKOLAEVICH [1867–1952] (*RUSS*). Munich 1896 (Baeyer; worked on caronic acid and isoprene). St. Petersburg Prof. chem. Artillery Academy 1898–1930 (Dr.phil. 1908). In 1930 went to U.S.; with Universal Oil Products Co. 1930–52 (Prof. chem. Northwestern 1930–35). Made important contributions to chemical technology through the use of controlled catalytic reactions.

Irschick, Alfred [1885–?] (*RUSS*). Jena Dr.phil. 1913; Munich 1914 (K. H. Meyer; worked on coupling of phenols with diazo compounds). Later research asst. of Paul Rabe in Hamburg.

Jackson, Louis Lincoln [1861–1935] (*US*). Munich 1891 (Baeyer; worked on phenylhydrazine derivatives of succinyl compounds). Dr.phil.?; later with Squibb, then Dissoway Chem. Corp.

Jackson, Oscar Roland [1855–1916] (*US*). Munich 1880 (Baeyer; worked on synthesis of methylketole and homologues of hydrocarbostyril and quinoline). Dr.phil.?; later with DuPont.

Jaeger, Carl [1850–1928] (*SWITZ*). Strassburg 1874 (Baeyer; worked on nitrosophenol); Zurich Dr.phil. 1876. Head of chem. factory in Offenbach a.M. 1876–1896; accident caused bad health, later lived in retirement.

Jaeger, Carl [1872–?]. Munich Dr.phil. 1902 (J. Thiele; worked on dioxyfluorescein, derivatives of oxyhydroquinone). Later commercial chemist in Düsseldorf.

Jaeger, Emil [1842–1922]. Strassburg Dr.phil. 1875 (Baeyer; worked on compound of chloral with thymol). Later co-owner chemical firm Carl Jaeger in Barmen.

Jaeger, Erich. Munich 1888 (Krüss; worked on chromium, determination of CO_2). Dr.phil.?; later activity not determined.

Jaeglé, Georges [1872–1931] (*FR*). Munich Dr.phil. 1894 (Besthorn; worked on hydroxyphenylquinoline). Later manager of family bleaching firm in Alsace.

Jaffé, Benno [1840–1923]. Berlin Dr.phil. 1865 (Baeyer; worked on bromoangelic acid, rufigallic acid). In 1867 established in Berlin a chemical factory that later became Vereinigte Chemische Werke A. G.

Jahn, Stephan [1876–1911]. Munich Dr.phil. 1902 (Einhorn; worked on aminocamphor, glycine esters of menthol and borneol). Charlottenburg TH Pv.Dz. physical chem. 1905–11.

Jay, Rudolf [1865–1928]. Erlangen Dr.phil. 1888; Munich 1891 (Baeyer; worked on phenylhydrazine derivatives of succinyl compounds). Later with Langbein-Pfanhauser Werke Leipzig.

Jehl, Paul. (*FR*). Strassburg Dr.phil. 1901; Munich 1901 (J. Thiele; worked on reduction of vinylacrylic acid). Later commercial chemist in Mulhouse.

Jenisch, Karl Albert [1868–1936]. Munich Dr.phil. 1891 (Pechmann; worked on reactions of acetone dicarboxylic acid). Later with family publishing firm in Stuttgart.

Jenny, Alexander [1871–1942]. Munich Dr.phil. 1902 (Hofmann; worked on cobalt tetramines). Later with Siemens-Halske A. G.

Jonas, August [1866–1927]. Erlangen Dr.phil. 1890; Munich 1891 (Pechmann; worked on triazole derivatives). Later with I. G. Farben Leverkusen.

Kämmerer, Heinrich [1881–?]. Munich Dr.phil. 1906 (Dieckmann; reaction of HCN with phenylisocyanate). Later commercial chemist in Mannheim.

Käswurm, August [1859–?]. Erlangen Dr.phil. 1885; Munich 1886 (Pechmann; worked on condensation of aromatic bases with aldehydes). Later landowner in Darmstadt.

Kahn, Anselm [1878–?]. Munich Dr.phil. 1903 (J. Thiele; worked on derivatives of dinitrophenylethylene). No further chemical publications 1907–36; later activity not determined.

Kahn, Walter Ernst [1878–?]. Munich Dr.phil. 1904 (Willstätter; worked on betaines). Later commercial chemist in Berlin.

Kalb, Ludwig [1879–1958]. Munich Dr.phil. 1905 (Willstätter; worked on quinoid derivatives of benzidine); Pv.Dz. 1912, ao.Prof. chem. technol. 1918–50. Later published extensively on plant chemistry.

Kappelmeier, Paul [1887–?] (NETH). Munich Dr.phil. 1911 (Wieland; worked on morphine). Later had private chem. lab. in Amsterdam.

Keller, Hugo [1882–?]. Munich Dr.phil. 1911 (Schlenk; worked on derivatives of biphenyl). Later with Nutrition Institute Frankfurt a. M.

Kerkovius, Berthold Woldemar [1882–?] (RUSS). Munich Dr.phil. 1911 (Dimroth; worked on dioxynaphthoquinones, coccinin). Later with family factory in Riga.

Kiesewetter, Paul [1862–?]. Munich 1889 (G. Krüss; worked on absorption spectra of rare earths). Later owner of Münchener Verbandstoff Fabrik.

Kimich, Carl [1852–1937]. Strassburg 1874 (Baeyer; worked on action of aromatic amines on nitrosophenol and nitrosodimethylaniline); Zurich Dr.phil. 1876. Later commercial chemist in Deidesheim.

KIPPING, FREDERICK STANLEY [1863–1949] (UK). Munich Dr.phil. 1887 (Perkin; worked on derivatives of xylene and of nephthalene). Nottingham Prof. chem. 1897–1936; did important research on organosilicon compounds (see TEXT).

Kirmreuther, Heinrich [1884–1961]. Munich Dr.phil. 1909 (Hofmann; worked on halogen derivatives of ethylene and acetylene). Later with Feldmühle Papier- und Zellstoff A. G. Düsseldorf.

Kitschelt, Max [1866–1939]. Munich Dr.phil. 1890 (Bamberger; worked on naphthalene derivatives). Later with Bayer & Co., I. G. Farben.

Klages, Ludwig [1872–1956]. Munich Dr.phil. 1901 (Einhorn; worked on derivatives of methylpimelic acid). Later became graphologist and philosopher.

Kliegl, Alfred [1877–1953]. Munich Dr.phil. 1903 (Baeyer; worked on condensation of benzaldehyde with toluene, phenylfluorene). Tübingen Pv.Dz. pharm. chem. 1909–14; ao.Prof. 1914–36; o.Prof. 1936–47.

Knell, Wilhelm [1876–1916]. Munich Dr.phil. 1902 (J. Thiele; worked on diphenylhexatriene caboxylic acid). Later activity not determined.

Knop, Conrad Alexander [1828–1873]. Berlin 1864–66 (Baeyer; worked on isatin); Leipzig Dr.phil. 1866; later became pharmacist in Berlin.

Knorr, Angelo [1882–1932]. Munich Dr.phil. 1909 (Schlenk; worked on quininoid compounds). Later with I. G. Farben.

Knorr, Eduard [1867–1926]. Jena Dr.phil. 1896; Munich 1900–02 (Koenigs; worked on sugar chemistry). Later commercial chemist in Munich.

KNORR, LUDWIG [1859–1921]. Munich 1880 (E. Fischer; worked on piperylhydrazine); Erlangen Dr.phil. 1882. Würzburg ao.Prof. chem. 1888–94; Jena o.Prof. 1894–1921. A productive organic chemist (see TEXT).

Koch, Franz (Ferenc) [1853–?]. Munich Dr.phil. 1885 (Curtius; worked on derivatives of diazosuccinic acid). Klausenburg Pv.Dz. chem. 1889; ao.Prof. 1891–.

Kochendoerfer, Ernst [1863–1918]. Munich 1889 (Baeyer; worked on action of phenylhydrazine on phloroglucinol and resorcinol, phthalein of pyrocatechol); Erlangen Dr.phil. 1889. Later with Deutsche Gold- & Silber Anstalt.

Koenig, Theodor [1862–?]. Munich Dr.phil. 1888 (v.d. Pfordten; worked on titanium). Later became pharmacist in Munich.

KOENIGS, WILHELM [1851–1906]. Bonn Dr.phil. 1875 (Kekulé); Munich 1876–1906 (Pv.Dz. 1881, ao.Prof. 1892; published 79 papers, had 28 research students). An outstanding experimenter, who made important contributions to organic chemistry; independent wealth, declined o.Prof. Aachen (see TEXT).

Körner, Georg. Munich 1883 (O. Fischer; worked on acridine, chrysaniline, triphenylmethane dyes). Dr.phil.?; no further chemical publications 1886–1936; later activity not determined.

Kohlschütter, Volkmar [1874–1938]. Munich Dr.phil. 1898 (Hofmann; worked on hydroxylamine compounds); Pv.Dz. 1900–02 (published 6 papers, had 1 research student). Strassburg Pv.Dz. 1902–09; Berne o.Prof. chem. 1909–38; published extensively on topics in inorganic chemistry.

Konek-Norwall, Frigyes [1867–1945]. Munich Dr.phil. 1892 (Einhorn; worked on action of bromine on dihydrobenzaldoxime). Director Central Control Experiment Station Budapest 1896–44 (ao.Prof. organic chem. 1908–44).

Kopp, Adolphe (*FR*). Strassburg 1874 (Baeyer; worked on nitrosodiethylaniline). Zurich Dr.phil. 1875; later became pharmaceutical chemist in Strassburg.

Korff, Joseph von [1843–1876] (*RUSS*). Berlin Dr.phil. 1865 (Baeyer; worked on meconic acid). Member of titled family in St. Petersburg.

Korn, Adolf [1877–?]. Munich 1901 (Hofmann; action of cathode rays on radioactive substances); Tübingen Dr.med. 1902. Later physician in Weiblingen.

Kornick, Erich [1886–?]. Munich Dr.phil. 1911 (Dimroth; synthesis of optically active allene compounds, tetrazolium bases). Later with Merck Darmstadt.

Krannich, Walter [1890–?]. Munich 1914 (Piloty; worked on diphenylmethane derivatives); Dr.phil. 1919. Later with I. G. Fraben Ludwigshafen.

Kranzfeld, Jacob (*RUSS*). Munich 1885 (Bamberger; worked on chrysene). Dr.phil.?; later commercial chemist in Odessa.

Kretschmer, O. Berlin 1871 (Liebermann; worked on propargyl ester); Dr.phil.?; no further chemical publications 1882–26; later activity not determined.

Krüger, Gerhard [1878–1965]. Munich Dr.phil. 1901 (J. Thiele; worked on action of alkali on esters of unsaturated acids). Later with I. G. Farben.

Krüss, Gerhard [1859–1895]. Munich Dr.phil. 1883 (Zimmerman; worked on sulfur compounds of molybdenum); Pv.Dz. 1886–90; ao.Prof. 1890–95 (published 48 papers; had 15 research students). An outstanding inorganic chemist whose career was cut short.

Küspert, Franz [1875–1929]. Munich Dr.phil. 1898 (Hofmann; compounds of hydrocarbons with metal salts). Later teacher and rector Realgymnasium in Nürnberg.

Kuhlemann, Friedrich [1864–1926]. Munich Dr.phil. 1894 (Bamberger; worked on diformazyl). Later director rubber factory Phönix A. G. Harburg.

Kuhn, Otto [1847–1919]. Berlin 1868 (Baeyer; worked on indigo); no Dr.phil. Later with Fratelli metal factory, Turin.

Lachman, Arthur [1873–1957] (*US*). Munich Dr.phil. 1895 (J. Thiele; worked on nitrourea derivatives). Univ. Oregon Prof. chem. 1897–

1902; commercial chemist San Francisco 1902–19; consulting chemist (oil industry) 1919–55.

Ladisch, Carl [1860–1925]. Munich Dr.phil. 1899 (Einhorn; worked on reduction of benzylaminocarboxylic acid); later research asst. of Einhorn, then teaching asst. of Baeyer.

Landsberg, Ludwig [1858–1923]. Munich Dr.phil. 1882 (Baeyer; worked on syntheses with phenylacetylene). In 1889 established oil refinery that became the A. G. Petroleumindustrie which prospered with the growth of the automobile industry.

LANDSTEINER, KARL [1868–1943]. Vienna Dr.med. 1891; Munich 1892–93 (Bamberger; worked on action of permanganate on diazobenzene); also did organic chemical work with E. Fischer (Würzburg) and Hantzsch (Zurich). One of the great medical scientists of the twentieth century (see TEXT).

Lauch, Richard [1860–1925]. Munich Dr.phil. 1887 (Einhorn; worked on action of hypochlorite on quinoline derivatives). Later dye chemist with Bayer & Co. Elberfeld 1888–95, Weiler & ter Meer 1899–1903, Bauer & Co. Berlin 1911–25, with intermittent work in university labs.

Lazarus, Maurice Julius [1863–?] (UK). Munich Dr.phil. 1884 (Friedländer; worked on nitration of cinnamic acid derivatives). Later commercial chemist in London.

Lecher, Hans [1887–1970]. Munich Dr.phil. 1913 (Wieland; worked on divalent nitrogen); Pv.Dz. 1920–22; Freiburg i. B. ao.Prof. 1922–27. Head of rubber divn. I. G. Farben 1927–34, then in U.S. with Calco Chem. Co. and Am. Cyanamid Co. 1934–52.

Lehmann, Ludwig [1858–1939]. Munich Dr.phil. 1880 (Baeyer; worked on indigo synthesis). Later with BASF Ludwigshafen, then I. G. Farben.

Lehne, Adolf [1856–1930]. Munich Dr.phil. 1880 (Baeyer; worked on condensation of benzhydrol and naphthalene). With BASF Stuttgart 1880–88; in Govt. Patent Office 1891–1917; o.Prof. Karlsruhe TH 1920–25. Specialized in textile manufacture and dyeing.

Lengfeld, Felix [1863–1938] (US). Johns Hopkins Ph.D. 1890; Munich 1890 (Bamberger; worked on reduction of quinoline). Held various teaching posts in U.S. 1891–1901; terminated because of failing eyesight. Set up pharmacy in San Francisco 1901–30; also chemical consultant to refrigeration industry.

Lenhardt, Sigismund [1889–?]. Munich 1913 (K. H. Meyer; worked on reactions of enols); Marburg Dr.phil. 1913. No further chemical publications 1917–46; later activity not determined.

Leser, Georges (FR). Munich 1884 (Baeyer; worked on derivatives of xylol). Dr.phil.?; later published organic-chemical papers from the chemistry institute at the University of Lyon.

Lessing, Rudolf [1878–1964]. Munich Dr.phil. 1902 (Willstätter; worked on benzene sulfonamides, methyl pyrrolidine dicarboxylic acid). Later consulting chemist in London for coal industry.

Leuchs, Robert Friedrich [1881–?]. Munich Dr.phil. 1908 (Piloty; worked on reduction of nitroaniline; synthesis of porphyrexin). Later with I. G. Farben.

Leuckart, Rudolf [1854–1889]. Leipzig Dr.phil. 1879; Munich 1882 (Baeyer; worked on action of sulfuric acid on bromocinnamic acid). Göttingen Pv.Dz. chem. 1883–89.

LIEBERMANN, CARL [1842–1914]. Berlin Dr.phil. 1865 (Baeyer; worked on propargyl compounds); 1867–69 work with Graebe on alizarin; Pv.Dz. 1869–72; ao.Prof. 1872–73; o.Prof. 1873–1914 (in 1879 the Gewerbeakademie became part of newly established Charlottenburg TH). Made many valuable contributions in organic chemistry (see TEXT).

Liebig, Hans von [1874–1931]. Munich Dr.phil. 1898 (Baeyer; worked on adipic acid). Giessen Pv.Dz. chem. 1908–14; ao.Prof. 1914–21. Grandson of Justus von Liebig.

Liebrecht, Arthur [1862–?]. Kiel Dr.phil. 1886; Munich 1886 (Einhorn; worked on action of chloral on picoline). Later with Chemisch-Pharmazeutische A. G. Bad Homburg.

Liebreich, Oskar [1839–1908]. Berlin 1867 (Baeyer; worked on protagon); see Appendix 3.

Lindenberg, Eugen [1872–?]. Munich Dr.phil. 1898 (Einhorn; worked on carbonates of dioxybenzenes). Later went to Brazil; Prof. chem. Univ. São Paolo until 1937.

Link, Gustav. Munich 1880 (Baeyer; worked on phthaleins). Dr.phil.?; later commercial chemist in Wiesbaden.

Lintner, Carl Joseph [1855–1926]. Munich Dr.phil. 1882 (Baeyer; worked on indigo). Munich TH Pv.Dz. applied chem. 1884–98; o.Prof. 1898–1926; published extensively on fermentation chemistry.

Locquin, René [1876–1965] (FR). Paris Dr.sci. 1904; Munich 1908 (Baeyer; no publication?). At Lyon 1909–47 (Prof. chem. 1929–47).

Lodter, Wilhelm [1864–1895]. Munich Dr.phil. 1887 (Bamberger; action of sodium alcoholate on aromatic nitriles); continued as research asst. of Bamberger.

Löhr, Richard [1863–?]. Munich Dr.phil. 1890 (Baeyer; worked on aminotriphenylcarbinol). No further chemical publications 1900–36; later activity not determined.

Löw, Wilhelm [1862–1940]. Munich Dr.phil. 1885 (Baeyer; worked on indigo, terephthalaldehyde). Later with I. G. Farben.

Loo, Henri van [1859–?] (BELG). Munich Dr.phil. 1885 (O. Fischer; worked on diquinolyline). No further chemical publications 1890–1936; later activity not determined.

Loose, Anton [1867–1933]. Munich Dr.phil. 1892 (Krüss; worked on rare earths). Later head of guncotton factory (Düren), then wool bleaching plant (Köln).

Lorenzen, Julius [1864–1945]. Munich Dr.phil. 1891 (Bamberger; worked on benzimidazole). Later with Bayer & Co. Elberfeld, then set up chemical factory in Berlin.

Lossow, Emil [1875–?]. Munich Dr.phil. 1900 (Koenigs; worked on derivatives of cinchoninic acid). Later co-owner wholesale yarn company in Glauchau.

Lovén, Johann Martin [1856–1920] (*SWED*). Lund Dr.phil. 1882; Munich 1885 (Baeyer; worked on action of nitrous acid on sulfoacetic acid, synthesis of sulfur-substituted cinnamic acid). Lund Prof. chem. 1900–20.

Lowman, Oscar [1861–1939] (*US*). Munich Dr.phil. 1887 (Claisen; worked on benzoylacetic acid ester, benzoylacetone). Later set up wholesale drug company in Detroit.

Lubavin, Nikolai Nikolaevich [1845–1918] (*RUSS*). Berlin 1869 (Baeyer; worked on pyrrole, cinchonine); see Appendix 3.

LUDWIG, ERNST [1842–1915]. Vienna Dr.phil. 1864; Berlin 1869 (Graebe; worked on naphthalene derivatives). Vienna o.Prof. medical chem. 1874–1915; made many valuable biochemical contributions.

Lüdecke, Karl [1880–1955]. Munich Dr.phil. 1905 (Willstätter; worked on lecithin). Later with Vereinigte Chem. Werke A. G., then Glanzfilm A. G.

Lumsden, John Scott [1867–1950] (*UK*). Munich Dr.phil. 1895 (Einhorn; worked on reduction of phenol carboxylic acids). Later head of Dundee Technical College 1912–29.

Lustig, Fritz [1879–?]. Munich Dr.phil. 1906 (Prandtl; compounds of vanadic and selenic acids). No further chemical publications 1917–56; later activity not determined.

Maas, Johanna [1885–?]. Munich Dr.phil. 1908 (Sand; worked on molybdenum hexarhodanates). Later chemist at Municipal Hospital, Charlottenburg.

Mähly-Eglinger, Jacob [1850–1920] (*SWITZ*). Munich Dr.phil. 1883 (Friedländer; worked on dintrocinnamic acid, isoindole). Later with Geigy A. G. Basle.

Mai, Karl [1865–?]. Munich Dr.phil. 1893 (Koenigs; worked on condensation of unsaturated hydrocarbons with phenols). Later (until 1921) Director food institute Munich.

Mair, Leopold [1885–?]. Munich Dr.phil. 1911 (Schlenk; worked on triphenylmethyl compounds). Later at Physical-chem. Institute, Berlin.

Manasse, Otto [1861–1942]. Berlin Dr.phil. 1886; Munich 1894–1902 (Claisen; worked on nitrosation of camphor and menthone). Later became musical composer (pseud. Thomas Aston).

Manchot, Wilhelm [1869–1945]. Munich Dr.phil. 1895 (J. Thiele; worked on derivatives of triazole). Würzburg ao.Prof. chem. 1903–14; Munich TH o.Prof. 1914–35. Published valuable papers on metal-catalyzed oxidation and synthesis of chemotherapeutic agents.

Manck, Philipp [1872–1929]. Munich Dr.phil. 1895 (Pechmann; worked on diazomethane disulfonic acid). Later became landowner in Edenkoben.

Manz, Hermann [1888–?]. Munich Dr.phil. 1912 (Prandtl; worked on vanadium methyl). Later commercial chemist (metal industry, water purification).

Marasse, Siegfried [1844–1896]. Berlin Dr.phil. 1869 (Baeyer; worked on creosote, fatty acids). In 1873 set up chemical factory in Berlin.

Marburg, Eduard Carl [1874–1925]. Munich Dr.phil. 1899 (Hofmann; worked on mercuric nitrogen compounds). Later with Griesheim-Elektron, then I. G. Farben.

Marshall, Joseph (UK). Munich 1906 (Dimroth; worked on condensation of diazophenylimide with ketones). Dr.phil.?; later in Chem. Dept. Leeds, then Boots Drug Co. Nottingham.

Marx, Wilhelm. Munich 1904 (Willstätter; worked on lupinidine and sparteine); Zurich Dr.phil. 1905. Later with Losenhausen A. G. (metal industry).

Mason, Frederick Alfred [1888–1947] (UK). Munich Dr.phil. 1912 (Dimroth; worked on influence of solvent on reaction rate and equilibrium). With British Dyestuffs Corp. 1916–26; Lecturer Manchester College of Technology 1926–31; H. M. Inspector of Technical Schools 1931–47.

Maull, Carl [1870–?]. Munich Dr.phil. 1894 (Rupe; worked on derivatives of camphor). Later in Chem. Dept. Charlottenburg TH.

Mawrow, Franz (BULG). Munich 1896 (Muthmann; worked on determination of copper and bismuth). Dr.phil.?; later at state chemical institute Sofia.

Mayer, Erwin Wilhelm [1885–?]. Munich 1904 (Willstätter; worked on quinone diimide); Dr.phil. Zurich ETH 1908. Later active in metal industry (flotation).

Mayr, Ernst [1872–1930]. Munich Dr.phil. 1897 (J. Thiele; worked on phenacylbromocinnamic acid). Later at Agricultural Experiment Station Hohenheim (specialized in wine chemistry).

Meer, Edmund ter [1852–1931]. Strassburg Dr.phil. 1875 (Baeyer; worked on nitrosophenol). In 1877 established factory in Uerdingen for manufacture of synthetic dyes; in 1896 it became the Weiler-ter Meer A. G., and in 1904 he became the sole head of the company.

Meimberg, Franz [1866–?]. Munich Dr.phil. 1894 (Bamberger; worked on conversion of aniline into nitrobenzene). No further chemical publications 1904–36; later activity not determined.

MEISENHEIMER, JAKOB [1876–1934]. Munich Dr.phil. 1898 (J. Thiele; worked on reactions of cinnamic acid derivatives). o.Prof. chem. Agricult. School Berlin 1909–18; Greifswald 1918–22; Tübingen 1922–34. Made many important contributions, especially to the study of the stereochemistry of organic nitrogen compounds.

Mengel, Alfred [1873–?]. Munich Dr.phil. 1903 (Koenigs; worked on derivatives of dimethylquinoline and trimethylpyridine). Later became pharmacist.

Merckle, Elsa [1874–?]. Munich Dr.phil. 1909 (Dieckmann; worked on desmotropic triazole compounds). No further chemical publications 1917–56; later activity not determined.

Merling, Georg [1865–1939]. Marburg Dr.phil. 1881; Munich 1894–95 (J. Thiele; no publication). With Schering Berlin 1895–98; Meister, Lucius & Brüning Hoechst 1898–1906; Bayer & Co. Elberfeld 1909–.

Merzbacher, Siegfried [1883–?]. Munich Dr.phil. 1908 (Dimroth; worked on condensation products of azides). Later commercial chemist in Berlin; in 1935 went to Ankara.

Mettler, Karl [1877–?] (SWITZ). Munich Dr.phil. 1902 (Einhorn; worked on action of phosgene and pyridine on oxyacids and amides). Later with Geigy Basle.

Metzeler, Karl [1865–1919]. Munich Dr.phil. 1889 (Baeyer; worked on iodo derivative of quinone). Later joined family chemical firm.

Metzener, Walther [1880–?]. Munich Dr.phil. 1906 (Hofmann; worked on amphoteric metallic oxides). Later with Knöfler & Co. Berlin; in 1934 in Shanghai.

Metzler, August [1883–1916]. Munich Dr.phil. 1910 (Hofmann; worked on secondary valences in organic compounds). Later with mining industry; killed in action World War I.

Meyenberg, Alexander [1871–?]. Munich Dr.phil. 1895 (Einhorn; worked on reduction of aminobenzoic acid derivatives). No further chemical publications 1907–36; later activity not determined.

Meyer, Carl [1872–1922]. Munich Dr.phil. 1895 (J. Thiele; worked on reduction of methyl- and ethylnitramine). In 1899 established chemical firm in Hamburg.

MEYER, KURT HEINRICH [1883–1952]. Leipzig Dr.phil. 1907 (Hantzsch); Munich 1909–14 (Pv.Dz. 1911; published 22 papers, had 8 research students), ao.Prof. 1918–21. Research director BASF Ludwigshafen 1921–32 (after 1926 I. G. Farben); o.Prof. chem. Geneva 1932–52. A remarkably versatile and productive chemist, who made important contributions to basic organic chemistry, and to the study of polymers, polysaccharides and enzymes.

Meyer, Richard Emil [1846–1927]. Göttingen Dr.phil. 1868 (Wöhler); Munich Pv.Dz. 1886 (no publications). o.Prof. chem. Braunschweig TH 1889–1918.

MEYER, VICTOR [1848–1897]. Heidelberg Dr.phil. 1867 (Bunsen); Berlin 1868–71 (published 11 papers; had 2 research students); Stuttgart Polytechnicum ao.Prof. 1871–72. o.Prof. Zurich ETH 1872–85; Göttingen 1885–89; Heidelberg 1889–97. An exceptionally gifted chemist, whose contributions include many important organic chemical discoveries, the development of the vapor-density method, and the preparation (with Paul Jacobson) of the first textbook of organic chemistry based on the theoretical advances made during the latter half of the nineteenth century (see TEXT).

Meyerowitz, Louis [1861–?] (RUSS). Munich Dr.phil. 1889 (Claisen; worked on ketoaldehydes). No further chemical publications 1892–1936; later activity not determined.

MILLER, WILLIAM LASH [1866–1940] (CAN). Munich Dr.phil. 1890 (Baeyer; worked on acetone oxalic ester); Leipzig Dr.phil. 1892 (Ostwald). Toronto assoc.Prof. chem. 1900–08; Prof. 1908–37. Made important contributions in physical chemistry, especially in the application of Gibbs's thermodynamic theories to experimental systems.

Mizerski, K. Berlin 1871 (Baeyer; worked on action of HI on hydrophthalic acid). Dr.phil.?; no further chemical publications 1882–1926; later activity not determined.

Mohs, Rudolf [1839–?]. Halle Dr.med. 1864; Berlin 1867 (Baeyer; worked on hydroterephthalic acid). No further chemical publications; later activity not determined.

Montmollin, Guillaume de [1884–?] (SWITZ). Munich Dr.phil. 1910 (Dimroth; worked on synthesis of pentazole). Later with CIBA Basle.

Moore, Charles Watson [1879–1956] (UK). Munich Dr.phil. 1907 (Willstätter; worked on emeraldin and aniline black). With Welcome research lab. 1907–11 and with Crossfield & Sons Ltd. 1911–44.

Moraht, Hermann [1864–1901]. Munich Dr.phil. 1890 (Krüss; worked on iron rhodanides). Continued research in private lab. in Munich.

Morley, Henry Forster [1855–1943] (UK). Munich 1880 (Wurster; worked on tetramethylphenyleneamine). London D.Sc. 1883; asst.Prof. chem. 1883–87; Prof. Queen's College 1888–1901. Later devoted energy to writing chemical books and managing the publication of the International Catalogue of Scientific Literature.

Moscheles, Robert [1863–?]. Munich 1888 (Cornelius; worked on tetrinic and pentinic acids); Erlangen Dr.phil. 1888. Later at Chem.-techn. Versuchsanstalt Berlin.

Mothwurf, Arthur [1878–?]. Munich Dr.phil. 1904 (Baeyer; worked on triarylcarbinols). Later with Bayer & Co. Elberfeld; after 1915 in U.S.

Müller, Alfred [1876–?]. Munich 1904 (Koenigs; worked on quinolyl-acrylic and quinolylpropionic acids); Giessen Dr.phil. 1905. Later in textile dyeing industry.

Müller, Carl [1889–?]. Munich Dr.phil. 1914 (Wieland; worked on organic radicals, diphenylhydroxylamine). Later with I. G. Farben.

Müller, Ferdinand [1864–?]. Munich Dr.phil. 1887 (Friedländer; worked on derivatives of pseudocarbostyril). Later set up chemical firm in Trier.

Müller, Hermann [1862–1938]. Munich Dr.phil. 1889 (Pechmann; worked on aromatic diketones). Later became director of chemical factory in Bayreuth.

Müller, Jens [1867–?]. Munich Dr.phil. 1893 (Bamberger; worked on reaction of diazobenzene with acetaldehyde and pyruvic acid). Later with Cassella & Co.

Müller, Rudolf [1860–1915]. Munich 1887 (Bamberger; worked on naphthylamine, carbazole blue); Erlangen Dr.phil. 1888. Later with Heine & Co. Leipzig.

Müller, Sebald [1887–?]. Munich Dr.phil. 1912 (Wieland; worked on sugar chemistry, hydrazones). Later with Gesellschaft für Teerver-wertung.

Müller, Wilhelm [1866–1915]. Munich Dr.phil. 1898 (Willstätter; worked on tropylamine, ecgonine). Later with Schimmel & Co. Leipzig; died in action World War I.

Murschhauser, Hans [1878–?]. Munich Dr.phil. 1907 (Prandtl; worked on vanadate salts).Later chemist at pediatric clinic Univ. of Freiburg.

Muthmann, Friedrich Wilhelm [1861–1913]. Munich Dr.phil. 1886 (Zimmermann; worked on oxides of molybdenum); Pv.Dz. 1894–99 (published 19 papers, had 8 research students), o.Prof. chem. Munich TH 1899–1913; made valuable contributions in inorganic chemistry.

Nagel, Wilhelm [1872–1916]. Munich Dr.phil. 1899 (Muthmann; worked on molybdates). Later owner of a sugar factory.

Narbutt, Johannes von [1879–1937] (RUSS). Dorpat Dr.phil. 1905; Munich 1907 (Hofmann; worked on compounds of platinum chloride with dicyclopentadiene). Dorpat 1908–24 (Pv.Dz. chem. 1916; ao.Prof. 1917; o.Prof. 1919–24).

Nauen, Otto [1857–1942] Dr.phil. Würzburg 1880; Munich 1884 (O. Fischer; worked on triphenylmethylamine). No further chemical publications 1886–1936; later activity not determined; died in Theresienstadt concentration camp.

NEF, JOHN ULRIC [1862–1915] (US). Munich Dr.phil. 1886 (Baeyer; worked on durene, benzoquinone carboxylic acids). Purdue asst.Prof.

chem. 1887–89; Clark assoc.Prof. 1889–92; Univ. of Chicago Prof. 1892–1915. Made significant contributions to theoretical and experimental organic chemistry (see TEXT).

Neger, Franz Wilhelm [1868–1923]. Munich Dr.phil. 1892 (Pechmann; worked on action of acetic anhydride on acetone dicarboxylic acid). Became noted botanist; successively Prof. at the forestry academies at Eisenach (1902–05) and Tharandt (1905–20) and at Munich (1920–23).

NENCKI, MARCELI [1847–1901] (POL). Berlin Dr.med. 1870 (Schultzen); 1870–72 (Baeyer; worked on uric acid). Berne Pv.Dz. 1872–77; ao.Prof. 1877–78; o.Prof. physiol. chem. 1878–90; St. Petersburg o.Prof. 1890–1901. One of the leading nineteenth-century biochemists (see TEXT).

Neresheimer, Julius [1880–1918]. Munich Dr.phil. 1908 (Piloty; worked on derivatives of malonic acid). Later with BASF Ludwigshafen; died in action World War I.

Neufville, Rudolf de [1866–?]. Munich Dr.phil. 1890 (Pechmann; worked on diphenyltriketone). With Physikalisches Verein Frankfurt a. M. 1891–95; later in Frankfurt city politics.

Nieme, Alexander [1863–1930]. Erlangen Dr.phil. 1888; Munich 1888 (Pechmann; worked on citracumalic acid). Later with I. G. Farben Leverkusen.

NOYES, WILLIAM ALBERT [1857–1941] (US). Johns Hopkins Ph.D. 1883; Munich 1889 (Baeyer; worked on succinyl compounds). After brief appointments at Tennessee, the Rose Polytechnic Institute and the Bureau of Standards, became Prof. chem. at Univ. of Illinois (Urbana) 1907–26, where he built one of the great centers of chemical research and instruction in the United States. Although primarly an organic chemist, he also made important contributions to other significant chemical problems.

Oeconomides, Spiridon [?–1894] (GREECE). Munich 1882 (Baeyer; worked on isatin). Dr.phil.?; no further chemical publications 1884–94; later activity not determined.

Oehler, Eugen [1871–1950]. Munich Dr.phil. 1896 (Baeyer; worked on menthone and tetrahydrocarvone). Later commercial chemist, book dealer, etc. in Offenbach.

Offenbächer, Mortiz. Munich 1914 (Wieland; worked on diphenyl nitrogen oxide). Dr.phil.?; no further chemical publications 1917–46; later activity not determined.

Ohnmais, Karl [1865–1916]. Erlangen Dr.phil. 1890; Munich 1890 (Krüss; worked on vanadium compounds). Later food chemist in Stuttgart.

Oppenheimer, Hugo [1866–1887]. Munich Dr.phil. 1886 (Baeyer; worked on terephthalaldehyde). Died soon after his doctorate.

Oppenheimer, Max [1875–1918]. Munich Dr.phil. 1900 (Einhorn; worked on glycine derivatives of aromatic amino acids). Later commercial chemist in Frankfurt a.M.

Osborne, Wilhelm [1871–?]. Munich Dr.phil. 1898 (J. Thiele; worked on prozane, triazane, aliphatic diazo compounds). Later with Chem.-Pharm. Lab. Sahir, then Holzverkohlung-Industrie A. G.

Ostermaier, Hermann. Munich 1881 (Friedländer; worked on carbostyril). Dr.phil.?; later commercial chemist in Munich.

Ott, Karl [1885–?]. Munich Dr.phil. 1908 (Hofmann; worked on sulfides and solfones). Later with I. G. Farben.

Otte, Rudolf [1858–1944]. Munich Dr.phil. 1890 (Pechmann; worked on homologues of diacetyl). Later a director at I. G. Farben.

Overbeck, Otto [1842–1919]. Berlin 1865 (Baeyer; worked on derivatives of oleic acid). Dr.phil.; later became director of brewery.

Papastavros, Stavros (GREECE). Munich 1900 (Einhorn; worked on reduction of diethylbenzylamino carboxylic acid). Dr.phil.?; later activity not determined.

Pape, Carl [1857–1926]. Kiel Dr.phil. 1882; Munich 1884 (Baeyer; worked on derivatives of xylene). Later commercial chemist in Nürnberg.

Papendieck, August [1867–?]. Munich Dr.phil. 1892 (Buchner; worked on synthesis of pyrazole derivatives). No further chemical publications 1897–1926; later activity not determined.

PECHMANN, HANS VON [1850–1902]. Greifswald Dr.phil. 1874; Munich 1877–95 (Pv.Dz. 1883–86, ao.Prof. 1886–95; published 86 papers; had 24 research students). Tübingen o.Prof. 1895–1902 (suicide). A brilliant organic chemist who made many original contributions (see TEXT).

PERKIN, WILLIAM HENRY, Jr. [1860–1929] (UK). Würzburg Dr.phil. 1882 (Wislicenus); Munich 1883–86 (Pv.Dz. 1884; published 20 papers; had 2 research students). Prof. chem. Heriot-Watt College Edinburgh 1887–92; Owens College Manchester 1892–1912; Oxford 1912–24. Made many important contributions, notably in his work on organic ring compounds and on natural products (see TEXT).

Perkowski, Zygmunt [1884–?] (POL). Munich Dr.phil. 1912 (Prandtl; worked on vanadium and molybdenum compounds). No further chemical publications 1917–36; later activity not determined.

Pfannenstiel, Adolf [1880–1957]. Munich Dr.phil. 1905 (Willstätter; worked on quinones). Later took over family factory for preserves.

Pfeiffer, Hermann [1874–1915]. Munich Dr.phil. 1898 (Einhorn; worked on disalicylide, butylbenzylamines). Later at chemical factory in Prague.

Pfister, Karl [1881–?]. Munich Dr.phil. 1909 (Dimroth; worked on triazenes). Later in U.S.; BASF 1911–19, Seigle Corp. 1919–24, Röhm & Haas 1924–27, General Dyestuffs Corp. 1927–50.

Pfordten, Otto von der [1861–1918]. Munich Dr.phil. 1883 (Zimmermann; worked on reduction of molybdenum and tungsten compounds); Pv.Dz. 1886–89 (published 14 papers; had 2 research students). Left chemistry for philosophy.

Pfyl, Balthasar [1873–1933] (*SWITZ*). Munich Dr.phil. 1898 (Einhorn; worked on aromatic oxyamino and oxynitro esters). Later with German Ministry of Health.

Philip, Max [1861–1918]. Zurich Dr.phil. 1885; Munich 1886–87 (Bamberger; worked on pyrene). Later in private commercial chem. lab. (Hundeshagen & Philip) in Stuttgart.

Piccard, Jean Félix [1884–1963] (*SWITZ*). Zurich ETH Dr.phil. 1909; Munich Pv.Dz. chem. 1914. Went to U.S.; after working in industrial firms, became Prof. aeronautical engineering Univ. Minnesota 1937–52.

Pickard, Robert Howson [1874–1949] (*UK*). Munich Dr.phil. 1898 (J. Thiele; worked on hydroxamic acids, indigo oxime, benzalphenylhydrazone). Principal Municipal Technical School Blackburn 1899–1919; Battersea Polytechnic 1920–27. Director Shirley Institute (textiles) 1927–43.

Piloty, Oskar [1866–1915]. Würzburg Dr.phil. 1889 (E. Fischer); Munich 1900–1914 (ao.Prof. 1900–14; published 34 papers; had 17 research students). Made valuable contributions to pyrrole chemistry; died in action World War I (see TEXT).

Platz, Ludwig Wilhelm [1878–1914]. Munich Dr.phil. 1905 (Dieckmann; worked on chloromalonic aldehyde). Later with Riebeck Montan A. G.; died in action World War I.

Prandtl, Wilhelm [1878–1956]. Munich Dr.phil. 1901 (Hofmann; worked on euxenite). Pv.Dz. chem. 1906–10; ao.Prof. 1910–37 (dismissed by Nazis; reinstated 1946). After research in inorganic chem., became historian of chemistry.

Prausnitz, Gotthold. Munich Dr.phil. 1884 (Einhorn; nitrophenyllactic acid). Later at Chem. Inst. Univ. Breslau.

Prentice, Bertram [1867–1938] (*UK*). Munich Dr.phil. 1895 (Baeyer; worked on pulegone). Head Chem. Dept. Royal Technical College Salford 1896–1932.

Prettner, August [1875–1931]. Munich Dr.phil. 1903 (Einhorn; worked on triethyltrimethylene triamine). Later with Vereinigte Aluminumwerke, Innwerk Töging.

PUMMERER, RUDOLF [1882–1973]. Munich Dr.phil. 1905 (Willstätter; worked on pyrone, acetone dioxalic ester); 1908–23 (Pv.Dz. 1911–21; ao.Prof. 1921–23; up to 1915 published 10 papers, had 7 research students); o.Prof. Greifswald 1923–25; Erlangen 1925–52. A gifted organic chemist who made many important contributions.

Quitmann, Eugen [1884–?]. Munich Dr.phil. 1910 (Piloty; worked on hematoporphyrin). Later in hygienic divn. of Berlin Health Dept.

Rabe, Wilhelm Otto [1873–?]. Munich Dr.phil. 1898 (Hofmann; worked on action of halogen alkyls on mercaptides). Later in chem. lab. of Mineralogical Institute Munich.

Racky, Georg [1889–?]. Munich Dr.phil. 1912 (Schlenk; worked on valence of carbon, bivalent arsenic). Later with Weiler & ter Meer Uerdingen.

Rassow, Berthold [1866–1954]. Leipzig Dr.phil. 1890 (Wislicenus); Munich 1892 (Einhorn; worked on dihydroxyanhydroecoginine). Leipzig ao.Prof. chem. technol. 1903–39 (Gen. Sec. Verein Deutscher Chemiker 1906–21). A prolific writer on chemical technology.

Rée, Alfred [1863–1933] (UK). Munich 1883 (Baeyer; no publication?); Berne Dr.phil. 1886. In 1890 set up dye factory in Manchester; retired 1907.

Reinsch, Sigmund [1874–?]. Munich Dr.phil. 1898 (Hofmann; worked on cobalt tetramines). No further chemical publications 1907–36; later activity not determined.

Reisenegger, Curt [1888–?]. Munich Dr.phil. 1913 (Wieland; worked on nitrogen dioxide, isocyanylic acid). After 1921 Prof. techn. chem. Santiago, Chile (during 1930s, member NSDAP organization in Chile).

Renning, Julius [1886–1964]. Munich Dr.phil. 1912 (Schlenk; worked on valence of carbon and silicon). Later commercial chemist in Munich (fermentation industry).

Renouf, Edward [1848–1934] (US). Freiburg i. B. Dr.phil. 1880; Munich 1885 (O. Fischer; worked on derivatives of quinoline and pyridine). Johns Hopkins assoc.Prof. 1886–92; Prof. 1892–1911.

Resenscheck, Friedrich [1879–?]. Erlangen Dr.phil. 1904; Munich 1905 (Hofmann; worked on blue ferrocyanide compounds). Later in chem. lab. Agricult. School Berlin.

Reverdy, Arno. Munich 1915 (Wieland; worked on triphenylmethylhydrazine). Dr.phil.?; no further chemical publications 1917–46; later activity not determined.

Richter, Victor von [1841–1891] (RUSS). St. Petersburg Dr.chem. 1872; Munich 1883 (Baeyer; worked on cinnoline derivatives). Breslau Pv.Dz. chem. 1875–79; ao.Prof. 1879–91. Published extensively on aromatic compounds, and wrote chemical textbooks.

Riedel, Carl [1856–1943]. Munich 1879–83 (O. Fischer; worked on derivatives of quinoline and pyridine). Dr.phil.?; later with azo dye divn. I. G. Farben.

Riemerschmid, Carl [1860–1915]. Munich Dr.phil. 1883 (O. Fischer; worked on pyridinesulfonic acid). Later commercial chemist in Munich.

Roelig, Hermann [1871–?]. Munich Dr.phil. 1898 (Muthmann; worked on cerium compounds). Later with I. G. Farben Leverkusen.

Römer, Adolf [1859–1924]. Tübingen Dr.phil. 1885; Munich 1886 (Baeyer; no publication?). Later with BASF, Deutsch-Kolonial Gerb- & Farbstoff Ges.; active in the development of the tanning industry.

Rössner, Heinrich [1876–1962]. Munich Dr.phil. 1899 (J. Thiele; worked on dibromide of phenylcinnamenylacrylic acid). Later with I. G. Farben.

Roseeu, Alexander [1886–?]. Munich Dr.phil. 1912 (Wieland; worked on diarylhydroxylamines). Later with Weiler & ter Meer Uerdingen.

Rosenthal, Otto [1881–?]. Munich Dr.phil. 1906 (Prandtl; compounds of zinc with vanadium, phosphorous and arsenic). Later activity not determined.

Roser, Ludwig, Munich 1881 (O. Fischer; worked on aminotrophenylmethane). Dr.phil.?; later commercial chemist in Wiesbaden.

Rossi, Heinrich [1872–1938]. Munich Dr.phil. 1903 (Kohlschütter; worked on uranium oxalic acid). Later commercial chemist in Hannover.

Roth, Rudolf [1887–]. Munich Dr.phil. 1910 (Hofmann; worked on perchlorates). Later Pv.Dz. Techn. Inst. Berlin; then with Boehringer Mannheim.

Rothlauf, Leo [1877–1942]. Munich Dr.phil. 1910 (Einhorn; worked on mixed carboxylic esters, alkylates of phenols). Later commercial chemist in Munich.

Rothmund, Viktor [1870–1927]. Munich Dr.phil. 1894 (for work done in Leipzig with Ostwald); Munich Pv.Dz. physical chem. 1898 (worked on miscibility of solutions). Prague ao.Prof. 1902–11; o.Prof. 1911–27.

Rudolph, Christian [?–1922]. Munich 1882 (O. Fischer; worked on flavaniline). Dr.phil.?; later dye chemist in Hoechst.

Rule, Harold Gordon [1887–1943] (UK). Munich Dr.phil. 1913 (Dimroth; worked on intramolecular rearrangements). Edinburgh Lecturer organic chem. 1915–43; wrote valuable chemical papers, especially on optical activity.

Runge, Paul [1869–1953]. Munich Dr.phil. 1895 (Pechmann; worked on formazyl compounds). Later owner of pharmacy in Hamburg.

RUPE, HANS [1866–1951] (SWITZ). Munich Dr.phil. 1889 (Baeyer; worked on reduction of dichloromuconic acid). Basle Pv.Dz. chem. 1899–1903; ao.Prof. 1903–17; o.Prof. 1917–37. A productive organic chemist who made valuable stereochemical contributions.

Ruppert, Eduard [1877–?]. Munich Dr.phil. 1902 (Einhorn; worked on aminooxybenzoic acid methyl ester). In 1904 went to Buffalo, N.Y.; later activity not determined.

Sachs, Paula [1886–1970]. Munich Dr.phil. 1913 (Vanino; worked on chemistry of silver); Dr.med. 1923. Later at Hospital Munich-Schwabing, then private medical practice.

Samuel, Ernst [1871–1934]. Munich Dr.phil. 1899 (Manasse; worked on camphor quinone). Later commercial chemist in Berlin.

Sand, Julius [1878–1917]. Munich Dr.phil. 1900 (Hofmann; worked on reaction of mercuric salts with olefins); Pv.Dz. 1903–08 (published 16 papers, had 5 research students). Berlin Pv.Dz. physical chem. 1908–17.

Sander, Albert [1887–1963]. Munich Dr.phil. 1913 (K. H. Meyer; worked on anthracene compounds). Later with Th. Goldschmidt A. G. Essen.

Schäfer, Josef [1867–?]. Munich Dr.phil. 1893 (Muthmann; worked on selenium). Later owner of pharmacy in Essen.

Schäffer, Alfred [1879–?]. Munich Dr.phil. 1906 (Hofmann; worked on hydrazine and mercapto derivatives in organic analysis). Later director nutrition lab. Brazil.

Scheibe, Anton [1856–1948]. Munich Dr.phil. 1879 (Wurster; worked on bromodimethylaniline). Later agricultural chemist at Munich TH.

Scherks, Emil. Munich 1885 (Baeyer; worked on hydrindonaphthene caboxylic acid). Strassburg Dr.phil. 1885. No further chemical publications 1892–1936; later activity not determined.

Scheurer, Wilhelm [1887–?]. Munich Dr.phil. 1912 (Dimroth; worked on kermes dyes). Later with I. G. Farben.

Schieffelin, William Jay [1856–1955] (*US*). Munich Dr.phil. 1889 (Bamberger; worked on hydrogenation of aromatic amines). Later head of family chemical firm.

Schiff, Robert [1854–1940]. Strassburg 1874 (Baeyer; worked on nitrosothymol); Zurich Dr.phil. 1876. o.Prof. chem. Modena 1879–92; Pisa 1892–1924; published on organic chem. (e.g., camphor) and on physical chem.

Schillinger, Albin [?–1919]. Munich 1880 (Baeyer; worked on derivatives of diphenylphthalide); Freiburg i. B. Dr.phil. 1881. Later commercial chemist in Heidelberg.

Schinner, Andreas [1885–]. Munich Dr.phil. 1914 (Vanino; worked on reactivity of formaldehyde, benzoyl peroxide and hexamethylene tetramine). No further chemical publications 1917–56; later activity not determined.

SCHLENK, WILHELM [1879–1943]. Munich Dr.phil. 1905 (Piloty; worked on metallo-isobutyryl compounds); Pv.Dz. 1910–13 (published 16 papers during 1905–13, had 9 research students). ao.Prof. Jena 1913–16; o.Prof. Vienna 1916–21; Berlin 1921–35 (successor of E. Fischer); Tübingen 1935–43. Made many important contributions in organic chemistry, especially on organic free radicals.

Schleussner, Karl [1862–1928]. Munich Dr.phil. 1897 (J. Thiele; worked on diaminophenylisotriazole). Later with I. G. Farben.

Schlösser, Hans Rudolf [1888–?]. Munich Dr.phil. 1914 (K. H. Meyer; worked on anthracene compounds). Continued as research asst. of K. H. Meyer.

Schmaedel, Wolfgang von [1877–1950]. Munich Dr.phil. 1905 (Dimroth; worked on sulfonation in presence of mercury). Later factory owner in Munich.

Schmidlin, Julius [1880–1962]. Dr.phil.; Munich 1902 (Einhorn; worked on carbonylsalicylamide). Zurich ETH Pv.Dz. 1906–12; tit. Prof. 1912–; known as a co-discoverer of triphenylmethyl.

Schmidt, Carl. Munich 1884 (O. Fischer; worked on condensation of aromatic bases with aldehydes); Dr.phil.? Later activity not determined.

Schmidt, Friedrich Wilhelm [1866–?]. Munich Dr.phil. 1887 (Krüss; worked on compounds of gold). Berne Pv.Dz. chem. 1893–97; Heidelberg Dr.med. 1909. later physician in Frankfurt a. M.

Schneider, Heinrich [1886–1971]. Munich Dr.phil. 1910 (Prandtl; worked on influence of solvent on reaction rate). Later director Boehringer factory at Ingelheim.

Schobig, Eugen [1856–1941]. Heidelberg Dr.phil. 1878; Munich 1879 (Wurster; worked on oxidation of tetramethylphenylene diamine). Later head of analytical and control lab. Schering Berlin 1883–1931.

Schoder, Robert [1866–?]. Munich Dr.phil. 1891 (Baeyer; worked on reduction of naphthoic acid). Later at Agricultural Experiment Station Möckern.

Schönewald, Hans [1874–?]. Munich Dr.phil. 1902 (Koenigs; worked on reaction of sulfurous acid with quinine). No further chemical publications 1907–36. Later activity not determined.

Schraube, Conrad [1849–1923]. Strassburg Dr.phil. 1875 (Baeyer; worked on nitrosodimethylaniline). With BASF Ludwigshafen 1877–1911, then private life in Munich.

Schröder, Ernst [1872–1914]. Munich Dr.phil. 1898 (Muthmann; worked on tellurium). Later with mining company Wittelsbach; died in action World War I.

Schröder, Hermann. Berlin 1866 (Baeyer; worked on palmitoleic acid). Dr.phil.; later director chemical firm in Karlsruhe.

SCHULTZEN, OTTO [1837–1875]. Berlin Dr.med. 1862; 1867 (Graebe; worked on methoxybenzoic acid; metabolism of aromatic acids). Dorpat Prof. med. 1871–75. A pioneer in the application of organic chemistry to the study of metabolic processes (see TEXT).

Schwarz, Karl Rudolf [1887–?]. Munich Dr.phil. 1913 (Wieland; worked on oxidation of cyclopentadiene). Later with chemical factory Dölau; then with government food and drug lab. Berlin.

Schwerin, Botho (Graf von) [1865–1917]. Berlin Dr.jur.; Munich 1901 (Piloty; worked on nitroso isobutyric acid nitrile, compounds of tetravalent nitrogen). Later established Elektro-osmose A. G. Berlin for separation and purification of colloidal materials.

Seeberger, Ludwig [1867–?]. Munich Dr.phil. 1893 (Bamberger; worked on dicyanodiamides). Later commercial chemist in Landau/Pfalz.

Seemann, Lorenz [1877–1957]. Munich 1900 (Vanino; worked on purification and analysis of gold); Erlangen Dr.phil. 1909; later teacher in Realgymnasium in Weilheim.

Seiler, Karl [1882–?]. Munich Dr.phil. 1907 (Hofmann; worked on compounds of mercuric chloride and alcohols with cyclopentadiene). Later commercial chemist in Basle.

Seitter, Eduard [1870–1953]. Munich Dr.phil. 1897 (Muthmann; worked on nitrogen sulfide). Later head of govt. chemical research bureau in Ulm.

Semper, Leopold [1882–1915]. Munich Dr.phil. 1907 (Wieland; worked on glyoxime peroxides). Later Pv.Dz. chem. Agricult. School Berlin; died in action World War I.

Sendtner, Rudolf [1853–1933]. Erlangen Dr.phil. 1877; Munich 1879 (Wurster; worked on dimethylphenylenediamine). Later director state food research institute Munich.

Seuffert, Otto [1875–1952]. Munich Dr.phil. 1900 (Baeyer; worked on bromination of menthone). With Merck Darmstadt 1901–30.

Seuffert, Rudolf Wilhelm [1884–?]. Munich Dr.phil. 1910 (Einhorn; worked on derivatives of aminobenzoic acid). Ao.Prof. physiol. chem. Veterinary School Berlin 1928–35, then went to Ankara.

Sherman, Penoyer Levi, Jr. [1867–?] (*US*). Munich Dr.phil. 1895 (Einhorn; worked on quinoline derivatives). Later activity not determined.

Sherndal, Alfred Einar [1884–1958] (*SWED*). Munich Dr.phil. 1912 (Dimroth; worked on kermes wax). Went to U.S.; after association with various firms, with Winthrop-Stearns Chem. Co., Rensselaer, N.Y. 1934–.

Sicherer, Walther von [1876–?]. Munich Dr.phil. 1901 (Willstätter; worked on pyrrolidine carboxylic acids). No further chemical publications 1907–36; later activity not determined.

Sinclair, William [1865–?] (*UK*). Munich Dr.phil. 1893 (Claisen; worked on oxymethylene camphor). No further chemical publications 1894–1936; later activity not determined.

Singer, Fritz [1879–1974]. Munich Dr.phil. 1903 (Sand; worked on reaction of mercuric salts with organic compounds). Later with Griesheim-Elektron Frankfurt a. M., then consultant metal industry.

Smith, Alexander [1865–1922] (*UK*). Munich Dr.phil. 1889 (Claisen; worked on diketones). Went to U.S.; Prof. chem. Univ. of Chicago 1894–1911, Columbia 1911–19. Worked on forms of sulfur and vapor pressure methods; best known for contributions to chemical education in the United States.

Solereder, Hans [1860–1920]. Munich Dr.phil. botany 1885; 1886 (Krüss; worked on reduction of inorganic sulfur compounds). Later Prof. botany Erlangen 1901–20.

Spiegel, Adolf [1856–1938]. Munich Dr.phil. 1881 (Baeyer; worked on vulpinic acid). With Meister, Lucius & Brüning Hoechst 1882–84; then with Messel Co. Düsseldorf 1884–1921.

Stange, Otto [1870–1941]. Munich Dr.phil. 1894 (J. Thiele; worked on semicarbazide). Later with I. G. Farben.

Stein, Richard [1877–1922]. Munich Dr.phil. 1904 (Dieckmann; worked on dicarbonyl compounds, resorcinol derivatives). Later in photochemical factory in Munich.

Steinbock, Hermann [1873–?]. Berlin Dr.phil. 1899; Munich 1902 (Piloty; worked on diketohexamethylene). Later with Schering Berlin.

Steiner, Antal [1840–1905]. Dr.phil.; Munich 1878 (Baeyer; worked on dithymolethane). Later factory owner in Hungary.

Stenzl, Hans [1880–1980]. Munich Dr.phil. 1907 (Wieland; worked on addition of nitrogen oxides to unsaturated compounds). Later at Anstalt für anorg. Chem. Basle.

Sternitzki, Hermann [1869–?] (*RUSS*). Munich Dr.phil. 1892 (Bamberger; worked on dihydromethylketol). No further chemical publications 1896–1926; later activity not determined.

Stettenheimer, Ludwig [1866–1932]. Munich Dr.phil. 1889 (Bamberger; worked on tetrahydronaphthoquinoline). Later commercial chemist in Mannheim.

Stock, Josef [1888–1931]. Munich Dr.phil. 1915 (Piloty; worked on synthesis of pyrroles). Later with Meister, Lucius & Brüning Hoechst.

Stock, Robert [1858–?]. Erlangen Dr.phil. 1889; Munich 1891 (Claisen; worked on action of hydroxylamine on a benzoyl aldehyde). Later in metal industry.

Stockhausen, Ferdinand [1875–1949]. Munich Dr.phil. 1903 (Koenigs; worked on quinaldine derivatives). Later Prof. brewing technology Berlin.

Stokes, Henry Newlin [1859–1942] (*US*). Johns Hopkins Ph.D. 1884; Munich 1885 (Pechmann; worked on action of ammonia on acetone dicarboxylic acid ester). Chemist with U.S. Geological Survey 1889–1903; with Bureau of Standards 1903–09; later journal editor.

Stolz, Friedrich [1860–1936]. Munich 1885 (Baeyer; worked on iodopropiolic acid); Erlangen Dr.phil. 1886. With Meister Lucius & Brüning Hoechst 1890–26; I. G. Farben 1926–30. A gifted pharmaceutical chemist; introduced pyramidon, suprarenin.

Storch, Ludwig [1859–1938]. Munich Dr.phil. 1893 (Bamberger; worked on action of ferricyanide on diazobenzene). Prague TH ao.Prof. physical chem. 1900–04; o.Prof. 1904–34. Made valuable contributions to inorganic and analytical chemistry.

Strasser, Ludwig [1865–1933]. Munich Dr.phil. 1890 (Bamberger; worked on fichtelite, naphthoquinaldine). Later chemist with A. G. Hagen (electrical co.) Munich.

Straus, Fritz Ludwig [1877–1942]. Munich Dr.phil. 1901 (J. Thiele; worked on lactones of desylacetic and dihydrocorniculario acids). Strassburg Pv.Dz. chem. 1905; ao.Prof. 1917. Breslau TH o.Prof. 1923–34. Went to U.S.; with Ecusta Paper Corp. 1939–42. A productive organic chemist; many valuable papers on unsaturated compounds.

Strauss, Eduard [1876–1952]. Berlin Dr.phil. 1899 (Gabriel); Munich 1901 (Hofmann; worked on radioactive lead). Frankfurt a.M. chemist at city hospital 1907–22; Georg Speyer Haus 1922–36. Went to U.S.; worked in several scientific institutions in New York City.

Stützel, Ludwig [1873–?]. Munich Dr.phil. 1899 (Muthmann; worked on double thiosulfates of copper and potassium, spectroscopic analysis of rare earths). Later commercial chemist in Nürnberg.

Stylos, Nicolaos [1865–?] (*GREECE*). Munich Dr.phil. 1888 (Claisen; worked on reaction of oxalic acid ester with acetone, acetoacetic aldehyde). No further chemical publications 1892–1926; later activity not determined.

Süssenguth, Otto [?–1893]. Berlin 1865 (Baeyer; worked on linoleic acid). Dr.phil.?; later commercial chemist in Berlin.

Süsser, Artur [1879–?]. Munich Dr.phil. 1912 (Wieland; worked on tetraarylhydrazines). No further chemical publications 1917–36; later activity not determined.

Suida, Wilhelm [1853–1922]. Budapest Dr.phil. 1876; Munich 1878 (Baeyer; worked on isatin). Vienna TH Pv.Dz. chem. 1882–1902; Prof. 1902–22. Published papers on sterols.

Sulzberger, Nathan [1874–1954] (*US*). Munich Dr.phil. 1901 (J. Thiele; worked on unsaturated lactones). Later commercial chemist in New York, also writer and poet.

Tahara, Yoshisumi [1855–1935] (*JAP*). Tokyo Dr.pharm. 1881; Munich 1893 (Einhorn; worked on anhydroecgonine). Later pharmaceutical chemist at National Hygienic laboratory Tokyo; a productive organic chemist, best known for his isolation of tetrodoxin.

Taub, Ludwig [1877–?]. Tübingen Dr.phil. 1905; Munich 1906 (Dimroth; worked on a triazolone derivative). Later with I. G. Farben Elberfeld.

Thal, Alexander [1889–?] (*RUSS*). Munich Dr.phil. 1913 (Schlenk; worked on metal ketyls). No further chemical publications 1917–56; later activity not determined.

THANNHAUSER, SIEGFRIED JOSEPH [1885–1962]. Munich Dr.med. 1911; Dr.phil. 1912 (Piloty; worked on bile pigments, hemoglobin); ao.Prof. medicine 1917–27. o.Prof.med. Düsseldorf 1927–31; Freiburg i. B. 1931–34. Went to U.S.; Prof. Tufts Medical College 1935–. Made many notable contributions to the study of metabolic diseases.

Thiele, Edmund [1867–1927]. Munich Dr.phil. 1894 (Krüss; worked on spectroscopy of iodine solutions). Later active in artificial silk industry; then with Kohorn Co. Chemnitz.

THIELE, JOHANNES [1865–1918]. Halle Dr.phil. 1890; Munich ao.Prof. 1893–1902 (published 77 papers, had 31 research students). Strassburg o.Prof. 1902–18. Made many important contributions to theoretical and experimental organic chemistry (see TEXT).

Tietze, Hermann [1864–?]. Munich Dr.phil. 1894 (Bamberger; worked on hexahydroquinoline). No further chemical publications 1896–1936; later activity not determined.

Tingle, John Bishop [1866–1919] (*UK*). Munich Dr.phil. 1889 (Claisen; worked on action of ethyl oxalate on aliphatic ketones). After various teaching posts in UK (1889–96) and in US (1897–1907), became Prof. chem. McMaster Univ. Toronto 1907–22.

Tischbein, Robert. Munich Dr.phil. 1899 (J. Thiele; worked on angelica lactone). No further chemical publications 1907–36. Later in Garfield, N.J.; activity not determined.

Tönnies, Paul. Munich 1878–80 (Baeyer; worked on mucic acid, unsaturated hydrocarbons). Dr.phil.; later in organic chem. lab. Charlottenburg TH.

Treubert, Franz. Munich 1897–99 (Vanino; determination of mercury and bismuth); Munich TH Dr.phil. 1909. Continued as research asst. of Vanino.

Troschke, Hermann Oswald [1851–?]. Göttingen Dr.phil. 1875; Munich 1879 (E. Fischer; worked on ethyl hydrazine derivatives, amarine and lophine). Later at Agricultural Experiment Station Regenwalde.

Tschacher, Oswald. Munich 1886 (Baeyer; worked on condensation of nitrobenzaldehyde with hydrocarbons). Dr.phil.?; no further chemical publications 1892–1936; later activity not determined.

Tschirner, Frederick (*US*). Munich Dr.phil. 1899 (Bamberger; worked on oxidation of aromatic bases). Later with Zeolite Chem. Corp.

Tutein, Friedrich [1862–1930]. Erlangen Dr.phil. 1889; Munich 1889 (Baeyer; worked on reduction of hydroxyphthalic acid). Became pharmacist in Mannheim.

Uhlfeder, Emil [1871–1935]. Munich Dr.phil. 1896 (J. Thiele; worked on biuret and dicyanodiamidine derivatives). Remained in Munich chemical institute as a research and teaching assistant.

Ulrich, Carl. Marburg Dr.phil. 1858 (Kolbe); Berlin 1867 (Baeyer; worked on derivatives of ricinoleic acid). Later at Chem. Inst. Univ. Vienna.

Unger, Oskar [1870–1937]. Munich Dr.phil. 1894 (Krüss; worked on heavy metal bichromates). Later with I. G. Farben.

Vanino, Ludwig [1861–1944]. Erlangen Dr.phil. 1891; Munich 1893–95 (Pechmann; worked on acylperoxides, action of benzoyl chloride on urethane); Pv.Dz. 1895–1911; tit. Prof. 1911–20 (before 1915 published 57 papers, had 12 research students); Hauptkonservator 1920–. Published extensively on topics in inorganic chemistry.

Villiger, Victor [1868–1934]. Munich Dr.phil. 1893 (Baeyer; worked on hexahydroisophthalic acid). Research asst. of Baeyer 1892–1904 (28 papers). With BASF Ludwigshafen 1904–31 (after 1926 I. G. Farben).

Vogel, Wilhelm [1878–1943]. Munich Dr.phil. 1903 (Piloty; worked on porphyrexide). Later in leather industry; head tannery school Freiburg/Sa. 1919–38.

Volck, Conrad [1867–?]. Munich Dr.phil. 1892 (Krüss; worked on sulfur compounds of thorium). Later pharmacist in Bietingheim.

VOLHARD, JACOB [1834–1910]. Giessen Dr.phil. 1855; Munich (with Liebig 1863–73; Pv.Dz. 1863; ao.Prof. 1868; with Baeyer 1875–79). O.Prof. chem. Erlangen 1879–82; Halle 1882–1910. A productive organic chemist who took over many of Liebig's teaching duties in Munich; also wrote about the history of chemistry and prepared a major biography of Liebig (see TEXT).

Vongerichten, Eduard [1852–1930]. Erlangen Dr.phil. 1873; Pv.Dz. 1875; Munich 1878–81 (Baeyer; worked on phthalyl chloride, alkaloids). With Meister, Lucius & Brüning Hoechst 1883–93; Strassburg private lab. 1893–1902; Jena ao.Prof. chem. 1902–22. Published valuable papers on chemistry of alkaloids.

Wagner, Heinrich [1880–?]. Munich Dr.phil. 1910 (Hofmann; worked on metal cyanides). Later with I. G. Farben.

Wagstaffe, Ernest Arthur [1870–1928] (UK). Munich Dr.phil. 1893 (Koenigs; worked on condensation of chloral with ketones). Later consulting chemist in Manchester.

WALDEN, PAUL (PAVEL IVANOVICH) [1863–1957] (RUSS). Leipzig Dr.phil. 1891 (Ostwald); Munich 1893 (J. Thiele; worked on tetrazole). Prof. chem. Riga Polytechnic 1894–1919; Rostock 1919–34. Made important contributions to physical-organic chemistry and to stereochemistry; also wrote extensively on history of chemistry.

Wecker, Ernst [1884–1961]. Munich Dr.phil. 1910 (Wieland; worked on quinonoid compounds). Head of Hagenbucher oil factory Heilbronn 1918–61.

Wedekind, Edgar [1870–1938]. Munich Dr.phil. 1895 (Pechmann; worked on diacetyl, tetrazolium bases). After assistantships in Munich, Riga and Tübingen (Pv.Dz. 1899–1904; ao.Prof. 1904–09), and brief faculty appointments in Strassburg and Frankfurt a. M., became head of chem. dept. of the Forestry School Hamm-Münden 1920–38. Published extensively on chemistry of natural products and on topics in inorganic chemistry.

Wehsarg, Karl [1862–1887]. Munich Dr.phil. 1886 (Pechmann; worked on di-isonitrosoacetone, hydrazoximes). Died soon after his doctorate.

Weickel, Tobias Friedrich [1883–?]. Munich Dr.phil. 1910 (Schlenk; worked on biphenylmethyl compounds). No further chemical publications 1917–46; later activity not determined.

Weil, Friedrich Josef [1888–1917]. Munich Dr.phil. 1914 (Wieland; worked on cholic acid, bufotalin). Later research asst. in Kaiser-Wilhelm Institutes of Willstätter and of Haber in Berlin.

Weil, Hugo [1863–1942?]. Erlangen Dr.phil. 1885 (E. Fischer). Munich 1894 (Baeyer; worked on triphenylmethane dyes). With Meister, Lucius & Brüning Hoechst 1887–91; CIBA Basle 1894–1902. After 1904 owner of commercial lab. Dr. H. Weil; died in Nazi concentration camp near Minsk.

Weiler, Julius [1850–1904]. Strassburg Dr.phil. 1874 (Baeyer; worked on action of methylal on aromatic compounds). Later took over family chem. firm, which became Weiler & ter Meer A. G.

Weinberg, Artur von [1860–1943]. Munich Dr.phil. 1882 (Friedländer; worked on carbostyril). Later head of Cassella factory in Fechenheim, director of I. G. Farben in Frankfurt a. M.; friend and patron of Paul Ehrlich. Died in Theresienstadt concentration camp.

Welsh, William (UK). Munich 1884 (Pechmann; worked on coumarins, malic acid, pyridine derivatives). Dr.phil.?; no further chemical publications 1892–1936; later activity not determined.

Wertheimer, Peter. Munich 1914 (K. H. Meyer; worked on desmotropism of nitro compounds). Dr.phil.?; no further chemical publications 1917–36; later activity not determined.

Weuringh, Pierre Guillaume [1886–?] (*NETH*). Munich Dr.phil. 1912 (Dimroth; worked on carminic acid). No further chemical publications 1917–36; later activity not determined.

Wheeler, Henry Lord [1867–1914] (*US*). Yale Ph.D. 1893; Munich 1894 (J. Thiele; worked on conversion of hydrazines into diamines). Yale asst.Prof. 1895–1908, Prof. 1908–1914; published valuable papers on pyrimidine chemistry.

Wheelwright, Edwin Whitfield [1868–1916] (*UK*). Munich Dr.phil. 1893 (Bamberger; worked on action of diazobenzene on ethyl acetoacetate). Later at Albright & Wilson Oldbury (match company).

Widman, Oscar [1852–1930] (*SWED*). Uppsala Dr.phil. 1877; Munich 1880 (Baeyer; worked on toluidine). Uppsala Pv.Dz. anal. chem. 1877–85; Prof. 1885–1907; Prof. chem. 1905–17. Made valuable contributions in organic chemistry.

Wiede, Otto Fritz [1871–1905]. Munich Dr.phil. 1897 (Hofmann; worked on metal-ammonia compounds, iron-nitroso compounds). Later commercial chemist in Rosenthal.

WIELAND, HEINRICH [1877–1957]. Munich Dr.phil. 1901 (J. Thiele; worked on aromatic allenes); Pv.Dz. 1905–09; ao.Prof. 1909–17 (published 73 papers, had 18 research students). o.Prof. chem. Munich TH 1917–21; Freiburg i. B. 1921–24; Munich 1926–50. His outstanding work on organic nitrogen compounds, steroids, alkaloids and pterins, and his studies on biological oxidation, place him among the great organic chemists of the twentieth century (see TEXT).

Wilke, Karl [1887–?]. Munich Dr.phil. 1915 (Piloty; worked on pyrranthracene compounds). Later with I. G. Farben.

Will, Hanns [1891–?]. Munich 1913 (Piloty; condensation of oxalic ester with acetyl pyrroles). Dr.phil.?; later became pharmacist.

Williamson, Sidney [1867–1939] (*UK*). Munich Dr.phil. 1889 (Bamberger; worked on decahydroquinoline, diethylnaphthylamine). Later research asst. of Tilden, Purdie; then with Cooper Co. (sheep and cattle dips).

Willson, Francis George. (*UK*). Munich 1914 (K. H. Meyer; worked on keto-enol tautomerism). With Cooper Co. Watford 1919–21 and then Royal Arsenal Woolwich 1921–56.

WILLSTÄTTER, RICHARD [1872–1942]. Munich Dr.phil. 1894 (Einhorn; worked on cocaine, hydrogenated aromatic acids); Pv.Dz. 1896–1902; ao.Prof. 1902–05 (published 83 papers, had 16 research students). Zurich ETH o.Prof. chem. 1905–12; head of Kaiser-Wilhelm Institute for Chemistry Berlin 1912–15; Munich o.Prof

1915–24. Did outstanding work on alkaloids, quinones, chlorophyll and anthocyanins; after 1918 published extensively on enzymes (see TEXT).

Winter, Ernst [1874–?]. Munich Dr.phil. 1900 (J. Thiele; worked on oxidation of quinones with acetic acid anhydride and sulfuric acid). Later with A. G. Stickstoffdünger (fertilizer firm) in Köln.

Wirth, Ernst [1859–1927]. Munich 1894 (Baeyer; worked on indigo). Dr.phil.?; later factory owner in Wiesbaden (tar industry).

Wischin, Carl [1871–?]. Erlangen Dr.phil. 1890; Munich 1893 (Moraht; worked on osmium). No further chemical publications 1894–1936; later activity not determined.

Witter, Hugo [1862–1908]. Rostock Dr.phil. 1891; Munich 1893–95 (Buchner; worked on trimethylene carboxylic acids, pyrazole tricarboxylic acids). Later with Bayer & Co. Elberfeld.

Wittmack, Charles. (US). Munich Dr.phil. 1884 (O. Fischer; worked on quinoline carboxylic acid). No further chemical publications 1892–1936; later activity not determined.

Wleügel, Severin Segelke [1852–?] (NOR). Munich 1884 (Friedländer; worked on anthranil). Dr.phil.?; later head technical school in Trondhjem.

Wölfl, Valentin [1879–1932]. Munich Dr.phil. 1905 (Hofmann; worked on radioactive lead). Later consulting chemist in Munich.

Wolff, Fritz [1866–1931]. Munich Dr.phil. 1897 (Koenigs; worked on cinchomeronic and apophyllic acids). Later co-owner Chemische Fabrik Ludwig Meyer (weed killer, etc.).

Wulz, Paul [1867–1933]. Erlangen Dr.phil. 1890; Munich 1891 (Bamberger; worked on quinoline and toluidine derivatives). Later commercial chemist in Heidenheim.

Wurster, Casimir [1854–1913]. Zurich Dr.phil. 1875; Pv.Dz. 1874–76; Munich 1878 (published 11 papers, had 7 research students; worked largely on aniline dyes). Berlin Physiol. Institute 1886–89 (worked on oxidation of proteins). Later in paper industry in New York (De Jonge Co.) and in London (Dickinson Co.).

Zahn, Karl [1885–?] Marburg Dr.phil. 1909; Munich 1913 (K. H. Meyer; worked on anthracene derivatives). Later with I. G. Farben.

Zedel, Wilhelm [1863–?]. Munich Dr.phil. 1889 (Claisen; worked on derivatives of acetylacetone, ethyl acetoacetate and ethyl malonate). Continued as research asst. of Claisen.

Zedwitz, Armin von (Graf). [1886–?]. Munich Dr.phil. 1911 (Hofmann; worked on perchloric acid). No further chemical publications; later activity not determined.

Zeidler, Othmar [1859–1911]. Strassburg Dr.phil. 1873 (Baeyer; worked on compounds of chloral with bromo- and chlorobenzene). Became pharmacist in Vienna.

Zerban, Friedrich Wilhelm [1880–1957]. Munich Dr.phil. 1903 (Hofmann; worked on radioactive thorium). Went to U.S. in 1904; during 1906–20 was with various sugar experiment stations, and during 1920–55 was associated with the U.S. sugar industry.

Ziegler, Joseph [1853–?]. Munich 1880 (O. Fischer; worked on aniline dyes); Erlangen Dr.phil. 1888. Later became pharmacist.

Zimmermann, Clemens [1856–1885]. Munich Dr.phil. 1879 (Baeyer; worked on analysis and separation of heavy metals); Pv.Dz. 1882–85 (published 14 papers, mostly on the analysis and atomic weights of uranium, cobalt and nickel; had 5 research students). An outstanding inorganic chemist, whose promising career was cut short by early death, possibly caused by his excessive self-imposed teaching duties.

Zucker, Eugen [1883–?]. Munich Dr.phil. 1906 (Vanino; worked on reactions of metallic salts). No further chemical publications; later activity not determined.

Zumbro, E. A. (US?). Munich 1893 (Bamberger; worked on dihydromethylketol). Dr.phil.?; no further chemical publications 1894–1926; later activity not determined.

Zumbusch, Emilie [1880–1923]. Munich Dr.phil. 1911 (Vanino; worked on lumiphorin). Later science teacher in Odenwaldschule (Heppenheim a.d. Bergstrasse).

Appendix 6

The Fischer Research Group

ABDERHALDEN, EMIL [1877–1950] (*SWITZ*). Basle Dr.med. 1902 (Bunge); Berlin research asst. of Fischer 1902–08 (worked on amino acids. peptides, proteins, proteolytic enzymes; published 140 papers). Berlin veterinary school o.Prof. physiol. 1908–11; o.Prof. physiol. chem. Halle 1911–45; Zurich 1945–47. Published over 1200 papers on peptides, proteolytic enzymes, vitamins and other topics (see TEXT).

Ach, Friedrich [1864–1902]. Würzburg Dr.phil. 1888; research asst. of Fischer 1889–92 (worked on sugars and purines; 7 papers, 3 acknowledgments). With Boehringer Mannheim 1892–1902.

Ach, Lorenz [1868–1948]. Würzburg Dr.phil. 1892; Berlin research asst. of Fischer 1892–94 (worked on sugars and purines; 5 papers, 10 acknowledgments). With Boehringer Mannheim 1894–1930; in 1902 succeeded his deceased brother Friedrich as head of the scientific laboratory.

Alexander, Walter [1871–?] (*US*). Berlin Dr.phil. 1894 (worked on condensation of amino acetal with phthalyl chloride). Later consulting chemist (fats and oils) in New York City.

Andreae, Edward Philip [1879–1975] (*UK*). Berlin Dr.phil. 1905 (worked on chitonic acid, action of diethylmalonylchloride on diamines). Later consulting chemist in London.

Anger, Gerda. Berlin research asst. of Fischer 1915–19 (worked on synthesis of linamarin and other glucosides; 1 paper, 2 acknowledgments). No further chemical publications 1920–44; later activity not determined.

Antrick, Otto [1862–1942]. Erlangen Dr.phil. 1884 (worked on ben-
zylindole, reaction of diacetonamine with aldehydes). With Schering
Berlin 1896–1926.

Arheidt, Richard (?–1924). Würzburg Dr.phil. 1887 (worked on diphe-
nylene dihydrazine). Later with BASF Ludwigshafen.

Armstrong, Edward Frankland [1878–1945] (UK). Berlin Dr.phil. 1901
(van't Hoff); with Fischer 1900–02 (worked on ethyluric acid, synthesis of disaccharides). After 1905 was associated with several British industrial firms, but continued to work on enzymatic cleavage of
glycosides; later published valuable papers on catalysis.

Axhausen, Walter [1882–?]. Berlin Dr.phil. 1904; research asst. of
Fischer 1904–09 (worked on peptide synthesis; 10 acknowledgments). In poor health; died ca. 1910? (see TEXT).

Babkin, Boris Petrovich [1877–1950] (RUSS). St. Petersburg Dr.med.
1904; Berlin 1906 (Abderhalden; worked on amino acid analysis of
legumin, metabolism of leucylglycine in dog). Odessa o.Prof. physiol.
1915–22. Went to Canada; McGill Prof. physiol. 1928–47. Made important contributions to physiology of digestive glands.

Baerwind, Heinrich [1892–1968]. Berlin Dr.phil. 1920 (worked on synthesis of glycerides). Later with Carl Freudenberg Co. Weinheim.

Barker, Lewellys Franklin [1867–1943] (US). Toronto M.B. 1890; Berlin 1904 (Abderhalden; worked on amino acids in urine). Chicago
Rush Med. College Prof. anatomy 1900–05; Johns Hopkins Prof.
medicine 1905–21. Published extensively on clinical topics.

Bauman, Louis [1880–1954] (US). Columbia Univ. M.D. 1901; Berlin
1907 (Abderhalden; worked on amino acid analysis of hemoglobin,
tryptophan). Practiced medicine (Presbyterian Hosp.) in New York
City; published on biochemical subjects.

Beensch, Leo [1869–?]. Berlin Dr.phil. 1894; research asst. of Fischer
1894–96 (worked on synthesis of glucosides; 4 papers). Later with
Boehringer Mannheim.

Behringer, Carl. Würzburg Dr.phil. 1891 (worked on higher members
of the galactose series). Later with Wilke-Dörfurt Co.

Bergell, Peter [1875–?]. Breslau Dr.med. 1898; Berlin 1902–04
(worked on naphthalenesulfonyl amino acids, enzymatic hydrolysis
of dipeptides). Later in private medical practice in Berlin.

Berghausen, Oscar [1860–1959] (US). Univ. Cincinnati M.D. 1904;
Berlin 1906 (Abderhalden; worked on amino acid analysis of crystalline protein from pumpkin seeds). In private medical practice Cincinnati 1908–58.

BERGMANN, MAX [1886–1944]. Berlin Dr.phil. 1911 (I. Bloch;
worked on acyl polysulfides); research asst. of Fischer 1912–19
(worked on sugars, depsides, glycerides, amino acids; 12 papers, 8
acknowledgments). Dresden, Director Kaiser–Wilhelm Institute for

Leather Research 1921–33. Went to U.S.; with Rockefeller Institute New York 1934–44. Made significant contributions to sugar and peptide chemistry (see TEXT).

BESTHORN, EMIL [1858–1921]. Freiburg i.B. Dr.phil. 1882 (for work with Fischer in Munich on derivatives of phenylhydrazine). With Meister, Lucius & Brüning Hoechst 1882–90, then in Baeyer institute in Munich; see Appendix 5.

Bethmann, Fritz [1874–?]. Berlin Dr.phil. 1899 (Gabriel); research asst. of Fischer 1899–1901 (worked on diaminobutyric and diaminovaleric acids; 3 acknowledgments). With Hoechst 1901–11; later landowner in Schönberg bei Naumburg a.S.

Blank, Paul [1882–?]. Berlin Dr.phil. 1906 (worked on synthesis of phenylalanine peptides). No further chemical publications 1907–44; later activity not determined.

Blix, Martin [1877–1933] (SWED). Berlin Dr.phil. 1902 (worked on borimid). Went to U.S. 1904; with Franklin Sugar Refining Co. Philadelphia 1908–33.

Bloch, Bruno [1878–1933] (SWITZ). Basle Dr.med. 1902; Berlin 1907 (Abderhalden; worked on cleavage of tyrosine and phenylalanine peptides, protein metabolism in alcaptonuria). Basle Pv.Dz. dermatol. 1909–13; ao.Prof. 1913–16; Zurich o.Prof. 1916–33. Made valuable contributions to biochemical aspects of dermatology.

Blochmann, Richard Hermann [1877–?]. Rostock Dr.phil. 1902 (Gattermann); Berlin 1902 (worked on indazole derivatives). Later consulting chemist (photography).

Blumenthal, Herbert [1885–1952]. Berlin Dr.phil. 1907 (worked on synthesis of aminobutyric acid). Later head of inorganic divn. in govt. testing bureau Berlin.

Boeckler, August [1871–?]. Würzburg 1892 (worked on oxalosuccinic acid); Dr.phil. 1895. Later with BASF Waldhof-Mannheim.

Böhner, Reginald [1880–1945] (CAN). Berlin 1910 (worked on proline content of gelatin, conversion of glutamic acid into proline); McGill Ph.D. 1911. Syracuse Prof. chem. 1912–39.

Boelsing, Friedrich. Würzburg Dr.phil. 1892 (worked on formation of secondary hydrazides). Later with Naef & Co. Geneva.

Boesler, Magnus [1852–1924]. Freiburg i.B. Dr.phil. 1881 (for work with Fischer in Munich on cuminoin and anisoin); research asst. of Fischer 1881–82 (worked on caffeine and furfurol; 2 papers, 5 acknowledgments). Later consulting chemist in Munich; see Appendix 5.

Bosart, Louis William [1874–1947] (US). Berlin Dr.phil. 1898 (worked on sugar hydrazones). With Proctor & Gamble Co. 1901–44.

Brauns, Friedrich Emil [1890–1982]. Berlin Dr.phil. 1915 (worked on furoyl derivatives of formic and glycolic acids, isopropylmalonamic acid). In various posts 1920–30; went to Canada (at McGill 1930–35) and U.S. (at Paper Institute Appleton, Wis. 1935–55).

Brieger, Walter [1881–?]. Berlin Dr.phil. 1914 (worked on sulfur derivatives of unsaturated aromatic acids, allylpropylcyanoacetic acid). Later with hydrotherapeutic institute in Berlin; also wrote on history of chemistry.

Bromberg, Otto [1870–?]. Berlin Dr.phil. 1895 (worked on reactions of alloxan with semicarbazide); research asst. of Fischer 1896–97 (worked on caffeine and pentoses, 5 papers). Later consulting chemist (water purification).

Brüning, Gustav von [1864–1913]. Würzburg Dr.phil. 1889 (worked on hydrazine derivatives). See Appendix 5.

Brunck, Rudolph [1867–1942]. Würzburg Dr.phil. 1892 (worked on indole derivatives). Later with I. G. Farben.

Brunner, Arnold [1880–1940]. Berlin Dr.phil. 1905 (worked on synthesis of leucylglycine and alanylleucylglycine). With Meister, Lucius & Brüning Hoechst (later I. G. Farben) 1907–40.

Bülow, Karl [1857–1933]. Rostock Dr.phil. 1882; Erlangen 1885 (worked on derivatives of phenylhydrazine and of benzoylacetone). With BASF Ludwigshafen 1887–96; Tübingen Pv.Dz. 1897–1901; ao.Prof. 1901–17; hon. Prof. 1917–33. Published on various topics, especially dye chemistry.

Bunte, Karl [1878–1944]. Berlin Dr.phil. 1903 (worked on aromatic derivatives of uric acid). Karlsruhe TH ao.Prof. chem. 1919–34; o.Prof. 1934–44; specialized in gas technology.

Cahn, Heinz [1892–?]. Berlin Dr.phil. 1919 (worked on carbomethoxy derivatives of aliphatic hydroxy compounds). Later consulting chemist in Berlin.

Carl, Hans [1880–1966]. Berlin Dr.phil. 1907 (worked on resolution of bromoisocaproic and bromohydrocinnamic acids). Later became publisher.

Chaplin, Edward Mitchell [1868–1948] (*UK*). Würzburg Dr.phil. 1892 (worked on hydrazides of camphoric acid). Leeds City Analyst 1896–1929; later gas examiner for towns in Yorkshire.

Clarke, Hans Thacher [1887–1972] (*UK*). London D.Sc. 1910; Berlin 1911–13 (worked on thiazanes). Went to U.S.; with Eastman Kodak Co. Rochester 1914–28. Columbia Univ. Prof. biol. chem. 1928–56; at Yale Univ. 1956–64 (see TEXT).

Clemm, Hans [1872–1927]. Kiel Dr.phil. 1896; Berlin research asst. of Fischer 1897–98 (worked on purine derivatives; 3 papers). Director Zellstoff-Fabrik Waldhof-Mannheim 1902–27.

Colman, Harold Govett [1866–1954] (*UK*). Würzburg Dr.phil. 1888 (worked on methylindole). Chief chemist Birmingham Gas Dept. 1892–1903; then consulting chemist to gas industry.

Cone, Lee Holt [1880–1957] (*US*). Univ. Michigan Ph.D. 1905; Berlin 1907 (worked on synthesis of histidine peptides). Prof. chem. Michigan 1909–15; with Midland Chem. Co. 1915–17, National Aniline Co. 1918–29; DuPont Co. 1929–45.

Crossley, Arthur William [1869–1927] (*UK*). Würzburg Dr.phil. 1892 (worked on dulcite, oxidation of saccharic and mucic acids). Prof. chem. School of Pharmacy London 1904–14. After extended military service, Director Shirley Institute of the British cotton industry 1920–27.

Culmann, Julius [?–1930] (*US*). Würzburg Dr.phil. 1889 (worked on action of aromatic amines and hydrazines on bromoacetophenone). Later chemist with Seigle Corp.

Curtiss, Richard Sydney [1863–1944] (*US*). Würzburg Dr.phil. 1892 (worked on optically-active gulonic acids). Between 1894 and 1919, held brief academic appointments at Chicago, Hobart, Union, Illinois, Throop; then managed his ranch and his real estate interests.

Dangschat, Gerda. Dr.phil.; Berlin research asst. of Fischer 1919 (worked on synthesis of rhamnosides). Later research asst. of H. O. L. Fischer in Berlin.

Dangschat, Paul [1893–?]. Dr.phil.?; Berlin 1919 (Bergmann; worked on depsides). Later consulting chemist in Berlin (essential oils and pharmaceuticals).

Deetjen, Hermann [1867–1915]. Kiel Dr.med. 1894; Pv.Dz. physiol. 1900–03. Berlin 1907 (Abderhalden; worked on breakdown of peptides by erythrocytes). Heidelberg institute for cancer research 1908–14; died in active service World War I.

Degen, Joseph [1861–1942]. Würzburg Dr.phil. 1886 (worked on methylphenylhydrazine). In 1887 founded chemical firm Degen & Kuth in Düren.

Delbrück, Konrad [1884–1915]. Berlin Dr.phil. 1907 (Buchner); research asst. of Fischer 1908–09 (worked on thiophenylglucosides and disaccharides). With Bayer & Co. Elberfeld 1909–15; died in action World War I.

DIELS, OTTO [1876–1954]. Berlin Dr.phil. 1899 (worked on cyano compounds); Pv.Dz. 1904–15; ao.Prof. 1915–16. Kiel o.Prof. 1916–48. Made many important contributions to organic chemistry, most notably in his work on cholesterol and (with Alder) on the invention of the method of diene synthesis (see TEXT).

Dilthey, Alfred [1877–1915]. Würzburg Dr.phil. 1900 (Hantzsch); Berlin research asst. of Fischer 1902–03 (worked on action of ammonia on malonates, barbituric acid derivatives). Set up private chemical company; died in action World War I.

Dörpinghaus, Wilhelm Theodor [1878–?]. Berlin Dr.phil. 1902 (worked on amino acid analysis of keratin). Later consulting chemist (mining).

Dyckerhoff, August [1868–1947]. Berlin Dr.phil. 1894 (worked on hydroxy acids). Later joined family cement factory.

Easterfield, Thomas Hill [1866–1949] (*UK*). Würzburg Dr.phil. 1894 (worked on citrazonic acid). Prof. chem. and physics Victoria Univ. College, Wellington, New Zealand 1899–1920; Director Cawthorn Institute of Scientific Research, Nelson, N.Z. 1920–33.

Ebstein, Erich [1880–1931]. Heidelberg Dr.med. 1904; Berlin 1906 (Abderhalden; worked on amino acid analysis of hen's egg chorion). Leipzig chief physician hospitals 1913–31. Published extensively on clinical subjects and on history of medicine.

Ehrensberger, Emil [1858–1940]. Munich 1881 (worked on arsenic compounds). Later Director of Krupp Works.

Ehrhardt, Wilhelm. Dr.phil; Munich research asst. of Fischer 1878–79 (worked on derivatives of phenylhydrazine, 2 papers). No further chemical publications 1885–1934; later activity not determined.

Einbeck, Hans [1873–?]. Berlin Dr.phil. 1905 (worked on lactam condensations). Later with Salkowski at chem. divn. institute of pathology Berlin.

Elbers, Alfred [1861–1936]. Erlangen Dr.phil. 1884 (worked on compounds of hydrazines with aldehyde- and ketone-acids). Later joined family textile factory in Düsseldorf.

Elsinghorst, Gerhard [1858–1929]. Erlangen Dr.phil. 1884 (worked on halogen-substituted hydrazines). Later consulting chemist in Münster.

Ephraim, Fritz [1876–1935]. Berlin TH Dr.phil. 1899 (Liebermann); Berlin 1900 (worked on indacene derivatives). Berne Pv.Dz. 1902–11; ao.Prof. 1911–32; o.Prof. 1932–35. Made many important contributions in inorganic chemistry.

Evans, Thomas Brown [1863–?] (*US*). Erlangen Dr.phil. 1886 (worked on halogen derivatives of quinoline). No further chemical publications 1882–1926; later activity not determined.

Falta, Wilhelm [1875–1950]. Prague Dr.med. 1900; Berlin 1903 (Abderhalden; worked on amino acid analysis of blood proteins of alcaptonuric patients). Basle Pv.Dz. med. 1904–13; Vienna ao.Prof. 1913–19; o.Prof. 1919–47. Published extensively on metabolic diseases.

Fay, Irving Wetherbee [1861–1936] (*US*). Berlin Dr.phil. 1896 (worked on derivatives of idose). Prof. chem. Brooklyn Polytechnic Institute 1897–1932.

Felser, Heinrich [1881–?]. Berlin Dr.phil. 1908 (Leuchs; worked on synthesis of hydroxyproline). Later at Kaiser-Wilhelm Institute for Textile Research.

Fiedler, Albert [1886–1961] (*US*). Berlin Dr.phil. 1910 (worked on synthesis of aspartic acid peptides). Later consulting chemist (photography).

FISCHER, FRANZ [1877–1947]. Giessen Dr.phil. 1899; Freiburg i.B. Pv.Dz. 1903; Berlin 1904–11 (conducted independent research program). Charlottenburg TH o.Prof. electrochem. 1911–13; Director Kaiser-Wilhelm Institute for Coal Research 1913–1943. Famous for his invention (with Hans Tropsch) of method to prepare synthetic benzene.

FISCHER, HANS [1881–1945]. Marburg Dr.phil. 1904; Munich Dr.med. 1908. Berlin research asst. of Fischer 1909 (worked on derivatives of lactose and maltose, new glucosides; 1 paper). Munich, chem. lab. of medical clinic (Friedrich Müller) 1910–15. o.Prof. med. chem. Innsbruck 1915–18; Vienna 1918–21; Munich TH org. chem. 1921–45. Made decisive contributions in the synthesis of hemin, other porphyrin derivatives, and bile pigments (see TEXT).

FISCHER, HERMANN OTTO LAURENZ [1888–1960]. Jena Dr.phil. 1912 (Knorr); Berlin research asst. of his father 1912–14 (worked on derivatives of orsellinic and resorcylic acids; 4 papers); ao.Prof. chem 1922–32. Prof. chem. Basle 1932–37; Toronto 1937–48; Univ. California Berkeley (biochem.) 1948–56. Made many important contributions in the synthesis of trioses. quinic acid, inositol and carbohydrates.

Fischer, Wilhelm [1868–?]. Würzburg Dr.phil. 1892 (worked on glucoheptose). Later commercial chemist in Schweinfurt.

Flatau, Erich. Berlin 1909 (worked on conversion of optically active bromopropionic acid into active methylsuccinic acid). Dr.phil.?; no further chemical publications 1912–44; later activity not determined.

Fodor, Kalman von [1881–1931]. Szeged Dr.phil. 1906; Berlin 1912–14 (worked on theophylline rhamnoside, cellobial; 2 papers, 1 acknowledgment). Later head Royal Experiment Station Budapest; published extensively on alkaloids from paprika.

Fogh, Johan Bertil [1866–1925] (*DEN*). Jena Dr.phil. 1889; Berlin 1895 (no publication). With chem. lab. agricult. school Copenhagen 1897–1925.

Forster, Martin Onslow [1872–1945] (*UK*). Würzburg Dr.phil. 1892 (worked on methylketole, quinoline derivatives). London Roy. Coll. Science 1901–13; Director Salters' Institute of Industrial Chem. 1918–22. Bangalore, Director Indian Institute of Science 1922–33.

FOURNEAU, ERNEST [1872–1949] (*FR*). Berlin 1901 (worked on phenyladenine, derivatives of glycine). See Appendix 5.

Frank, Fritz [1869–1949]. Berlin 1896 (worked on theobromine); Dr.phil. Basle 1896 for this work. With Henriques Laboratory for

Commerce and Industry Berlin 1901–38 and head of coal and oil institute at Berlin TH 1922–38. Went to UK; set up Frank Laboratories Ltd. in London. Made many contributions to coal and oil technology.

FREUDENBERG, KARL [1886–1983]. Berlin Dr.phil. 1910; research asst. of Fischer 1910–14 (worked on carbomethoxy derivatives of phenolcarboxylic acids, tannins; 7 papers). Kiel Pv.Dz. chem. 1914–20; Freiburg i.B. ao.Prof. 1921–22. o.Prof. Karlsruhe TH 1922–26; Heidelberg 1926–56. Made valuable contributions to stereochemistry and to the chemistry of polysaccharides and tannins.

Fritz, Victor [1872–1926]. Berlin Dr.phil. 1896 (worked on derivatives of benzoylcarbinol). With Boehringer Mannheim 1897–1926.

Funk, Casimir [1884–1967] (POL). Berne Dr.phil. 1904; Berlin 1906–10 (Abderhalden; worked on utilization of dietary proteins). At Lister Institute London 1910–13; in U.S. 1915–23, 1939–67 associated with various commercial enterprises related to vitamins.

GABRIEL, SIEGMUND [1851–1924]. Heidelberg Dr.phil. 1874 (Bunsen). At Berlin 1874–1921 (ao.Prof. with A. W. Hofmann and Fischer; conducted independent research program). An outstanding organic chemist who developed valuable new synthetic methods (see TEXT).

Geiger, Walter [1866–?]. Berlin Dr.phil. 1908 (Leuchs; worked on synthesis of serine). No further chemical publications 1910–46; later activity not determined.

Gerlach, Ferdinand [1885–?]. Jena Dr.phil. 1911 (Knorr); Berlin 1912 (worked on pyrroline carboxylic acid). Later with Siemens-Schuckerts Werke A. G. Berlin.

Gerngross, Otto [1882–1966]. Berlin Dr.phil. 1905 (Gabriel; worked on synthesis of thymine); research asst. of Fischer 1908 (worked on synthesis of cystine peptides); Pv.Dz. 1920. Charlottenburg TH ao.Prof. 1923–33. Went to Turkey; o.Prof. chem. Ankara 1933–43. Wrote extensively on proteins and tannins.

Geserick, Arthur [1860–?]. Berlin Dr.phil. 1908 (worked on phloroglucinol). Later established private chemical factory in Berlin.

Gevekoht, Heinrich [1858–?]. Erlangen Dr.phil. 1883 (worked on isomeric nitroacetophenones). No further chemical publications 1890–1944; later activity not determined.

Giebe, George [1874–1899]. Berlin Dr.phil. 1896; research asst. of Fischer 1896–98 (worked on acetals).

Gigon, Alfred [1883–1975] (SWITZ). Basle Dr.med. 1906; Berlin 1907 (Abderhalden; worked on metabolism of amino acids). Basle Pv.Dz. med. 1911–21; ao.Prof. 1921–32; o.Prof. 1932–53. Published extensively on metabolic disorders.

Gluud, Wilhelm [1887–1936]. Berlin Dr.phil. 1909; research asst. of Fischer 1910–12 (worked on derivatives of alanine, leucine, phenylglycine). Head of research institute of coal technology Dortmund 1918–36.

Göddertz, Bernhard Albert [1888–1916]. Berlin Dr.phil. 1911 (worked on synthesis of aminohydroxybutyric acid). With Bayer & Co. 1911–14; died in action World War I.

Grävenitz, Richard von [1890–1918]. Berlin Dr.phil. 1914 (worked on Walden inversion, conversion of hydroxybenzoic acid into hydroxyphenylglyoxylic acid). Died in action World War I.

Groh, Reinhart. Berlin 1910 (worked on Walden inversion, aminovaleric acid, phenylhydrazones of keto acids). Dr.phil.?; no further chemical publications 1912–44; later activity not determined.

Guggenheim, Markus [1885–1970] (SWITZ). Basle Dr.phil. 1907; Berlin 1908 (Abderhalden; worked on peptides of diiodotyrosine). After 1910 associated with Hoffmann-LaRoche Basle; despite his blindness, caused by a laboratory accident in 1916, he effectively headed the pharmaceutical group of the company.

Gwinner, Hans von [1887–1959]. Berlin Dr.phil. 1912 (worked on optically active dialkylacetic acids). Later with Nitritfabrik A.G. Munich.

Haberland, Hermann (1865–?]. Würzburg Dr.phil. 1888 (worked on methylation of indoles). Later head of chemical factory Neu-Stassfurt, Bitterfeld.

Hähnel, Otto (1884–?]. Berlin Dr.phil. 1909 (worked on preparation of pure argon and nitrogen). After 1914, chief chemist govt. post office.

Hämäläinen, Yuho Heikki [1883–1915] (FIN). Berlin 1907 (Abderhalden; worked on amino acid analysis of avenine). Helsingfors Dr.med. 1912; Pv.Dz. med. chem. 1913–15.

Haga, Tamesama [1856–1914] (JAP). Dr.phil.; Berlin 1897 (Harries; worked on methylation of hydrazine hydrate). Tokyo Prof. chem.

Haenisch, Victor [1871–?]. Berlin Dr.phil. 1894 (worked on higher homologues of galactose). Later director sulfuric acid factory in Duisburg.

Hagenbach, Rudolf [1875–1927] (SWITZ). Basle Dr.phil. 1900; Berlin 1901 (worked on resolution of racemic amino acids). With Meister, Lucius & Brüning Hoechst 1902–1921; Director chemical factory Durand & Huguenin Basle 1921–27.

Harper, Charles Athiel [1868–?] (US). Berlin Dr.phil. 1896 (worked on isoquinoline and isocarbostyril). Later consulting chemist (water purification).

HARRIES, CARL [1866–1923]. Berlin Dr.phil. 1890 (Tiemann); Pv.Dz. 1897; tit. Prof. 1900–04 (conducted independent research program). Kiel o.Prof. 1904–16. With Siemens & Halske A.G. 1916–23. Made significant contributions to the study of the action of ozone on organic compounds and on rubber.

Hartmann, Gerhard [1868–1945]. Würzburg Dr.phil. 1891 (worked on mannoheptonic acid). Later with Griesheim-Elektron A.G.

Hauff, Friedrich [1863–1935]. Würzburg Dr.phil. 1888 (worked on derivatives of naphthylhydrazine). Later joined family pharmaceutical factory.

Hegel, Sigmund [1863–?]. Erlangen Dr.phil. 1885 (worked on synthesis of indole derivatives). Later in govt. patent office in Berlin.

HELFERICH, BURKHARDT [1887–1982]. Berlin Dr.phil. 1911; research asst. of Fischer 1911–13 (worked on synthesis of glucosides; 3 papers, 1 acknowledgment); Pv.Dz. 1920. o.Prof. chem. Greifswald 1925–30; Leipzig 1930–45; Bonn 1947–57. Made valuable contributions to the study of the specificity of glycoside-cleaving enzymes.

Heller, Gustav [1866–1946]. Würzburg Dr.phil. 1890; research asst. of Fischer 1890–93 (worked on fructose, mucic acid, isomaltose and on action of phosgene and ethylchloroformate on phenylhydrazine; 4 acknowledgments). At Leipzig 1904–44 (ao.Prof. 1911–44). Made many contributions in organic chemistry.

Heller, Hans [1870–?]. Berlin Dr.phil. 1894 (worked on derivatives of amino acetal). Until 1921 research asst. at Leipzig and Munich, then with Brinkmann & Thergell, oil factory in Harburg.

Henkel, Paul. Berlin 1910 (worked on oxidation of dimethyluracil). Dr.phil.?; no further chemical publications 1917–44; later activity not determined.

Herborn (Avenarius-Herborn), Heinrich [1873–1955]. Berlin Dr.phil. 1896 (worked on isorhamnose). After 1914 with Avenarius Co. (oil refinery) in Gau-Algesheim.

Herrick, James Bryan [1861–1954] (US). Rush Med. College M.D. 1888; Berlin 1905 (Abderhalden; worked on amino acid analysis of conglutin of lupin seeds). With Rush Med. College Chicago 1890–1927 (Prof. medicine 1900–27). A noted cardiologist.

Hertz, Johann Nicolaus [1869–1908]. Würzburg Dr.phil. 1892 (worked on reduction of mucic acid). Later became Director of Voelper Montanwachsfabrik.

Hess, Kurt [1888–1961]. Jena Dr.phil. 1911; Berlin 1912 research asst. of Fischer (worked on reaction of sugar derivatives with methyl magnesium iodide). Freiburg i.B. Pv.Dz. 1914–16; Prof. 1916–21; Kaiser-Wilhelm Institute for Chemistry 1921–59. Made many contributions in organic chemistry, especially on cellulose.

Hess, Otto Paul [1860–?]. Erlangen Dr.phil. 1884 (worked on indole derivatives, benzoyl derivatives of aromatic amines). With Meister, Lucius & Brüning Hoechst 1895–1925.

Hilpert, Siegfried [1883–1951]. Berlin Dr.phil. 1905 (worked on photochemical reactions of stilbene derivatives); research asst. of Fischer 1905–07 (worked on resolution of valine and on mercury dipropionic acid; 2 acknowledgments). Charlottenburg TH Pv.Dz. inorganic chem. 1914; in industry 1915–30; Braunschweig TH o.Prof. chem. 1930–39.

Hirschberger, Joseph [1866–1954] (US). Würzburg Dr.phil. 1889; research asst. of Fischer 1889–90 (worked on mannose and fructose, reduction of sugar acids; 4 papers, 4 acknowledgments). Later with Consolidated Color and Chemical Co., then Arkansas Co., both in Newark, N.J.

Hirszowski, Alfred [1882–1943] (POL). Berlin Dr.phil. 1908 (worked on peptides of glycine, alanine, leucine, tyrosine). Later director of chemical factories in Lodz and Warsaw.

Hoesch, Kurt [1882–1932]. Berlin Dr.phil. 1911 (worked on depsides). After World War I became owner of paper factory near Düren. Wrote biography of Emil Fischer.

Hoffa, Erwin Friedrich [1875–1967]. Berlin Dr.phil. 1897 (worked on aromatic acetals and aldehydes). After several brief assistantships, with Meister, Lucius & Brüning (later I. G. Farben) 1908–35.

Holzapfel, Julius [1883–1918] (UK). Heidelberg Dr.phil. 1909 (Curtius); Berlin 1911 (worked on optically active dialkylacetic acids). With International Paint Co. 1911–18 (died of influenza).

Houben, Josef [1875–1940]. Bonn Dr.phil. 1898 (Bredt). After assistantships in Aachen and Bonn, Berlin in Pv.Dz. 1908–14 (conducted independent research program). After service in World War I, ao.Prof. Berlin 1921–33. A gifted organic chemist; introduced valuable synthetic methods and wrote important treatise on the methods of organic chemistry.

Hübner, Friedrich [1876–1953]. Berlin Dr.phil. 1897; research asst. of Fischer 1897–99 (worked on resolution of racemic amino acids; 7 acknowledgments). Later factory director in Wiesbaden.

Hütz, Hugo [1871–?]. Berlin Dr.phil. 1895 (worked on phenylindoxyl, benzoin derivatives). With Meister, Lucius & Brüning Hoechst 1899–1904; later consulting chemist (rubber industry).

Hunsalz, Paul [1871–?]. Berlin Dr.phil. 1894; research asst. of Fischer 1894–98 (worked on hydrazinoacetaldehyde, purines; 2 papers, 13 acknowledgments). With Boehringer Mannheim 1898–1900 and Schering Berlin 1900–05; suicide soon afterward (date not determined).

Hunter, Andrew [1876–1969] (UK). Edinburgh M.D. 1901; Berlin 1906 (Abderhalden; worked on amino acid analysis of milk proteins, proteolytic enzymes). Cornell Univ. asst. Prof. biochem. 1908–14. Prof. biochem. Toronto 1915–29; Glasgow 1929–47. Published on many aspects of protein metabolism and on other biochemical topics.

Hutchinson, Arthur [1866–1937] (*UK*). Würzburg Dr.phil. 1892 (worked on reduction of aromatic amides). Became distinguished mineralogist at Cambridge 1892–1937 (Prof. 1926–31).

Ince, Walter Holinshed [1865–1907] (*UK*). Würzburg Dr.phil. 1889 (worked on phenylindoles). Held various brief academic, government and commercial appointments in England, Trinidad, Australia and Bolivia, with primary interest in mining technology.

Jacobi, Friedrich Wilhelm [1863–?]. Kiel Dr.phil. 1893; Berlin 1895 (no publication?). Later activity not determined.

Jacobi, Hermann [?–1901]. Würzburg Dr.phil. 1891 (worked on oximes and hydrazones of sugars). Later with Griesheim-Elektron A.G.; died in picric acid explosion.

JACOBS, WALTER ABRAHAM [1883–1967] (*US*). Berlin Dr.phil. 1907 (worked on resolution of racemic amino acids). At Rockefeller Institute for Medical Research, New York 1908–49. An outstanding organic chemist who made decisive contributions on nucleic acids, chemotherapeutic agents (Tryparsamide) and alkaloids (see TEXT).

Jänecke, Ernst [1875–1957]. Berlin Dr.phil. 1898 (worked on aminodiethylketone and its derivatives). Hannover TH Pv.Dz. 1905; with Kaliwerk Benthe 1905–20 and BASF Ludwigshafen 1920–35.

Jennings, Walter Louis [1866–1944] (*US*). Harvard Ph.D. 1892; Berlin 1893–94 (worked on constitution of hydrocyanorosaniline and fuchsin, compounds of sugars with polyphenols). Prof. chem. Worcester Polytechnic 1901–37.

Jessel, Henry Rose [1867–1933] (*US*). Berlin Dr.phil. 1900 (worked on purines). With Scaife Chem. Co. 1904–27; asst.Prof. Marquette 1927–32.

Joachimoglu, Georg [1887–1979] (*GREECE*). Berlin Dr.med. 1911; in Fischer lab. 1912 (no publication?); Pv.Dz. pharmacol. 1918–22; oa.Prof. 1922–27. Athens o.Prof. pharmacol. 1928–58. Published extensively on pharmacological topics.

Joedicke, Friedrich [1864–?]. Erlangen Dr.phil. 1887 (worked on action of ethylnitrobenzoylacetoacetate on phenylhydrazine). Later consulting chemist in asphalt industry.

Johlin, Jacob Martin [1884–1954] (*US*). Berlin Dr.phil. 1910 (worked on ketoses). Syracuse Univ. 1913–? (assoc.Prof. biochem. 1925–?). Published on physical chemistry of proteins.

Johnson, Theodore (1884–?] (*US*). Berlin Dr.phil. 1906 (worked on synthesis of glutamyl peptides). No further chemical publications 1907–44; later activity not determined.

Jourdan, Friedrich. Mainz Dr.phil. 1879; Erlangen 1883 (worked on hydrazine derivatives of pyruvic acid). Later industrial chemist in Rome.

Kaas, Carl [?–1915]. Berlin 1906 (worked on action of hippuryl chloride on methylindole). Dr.phil.?; later asst. in chem. Graz.

Kadisade, A. Refik [1892–?] (*TURKEY*). Berlin Dr.phil. 1918 (worked on synthesis of tannins). No further chemical publications 1920–46; later activity not determined.

Kalantharianz, Anuschawan [1867–?] (*RUSS*). Berlin Dr.phil. 1898 (worked on cleavage of polysaccharides by yeast enzymes). No further chemical publications 1907–46; later activity not determined.

Kall, Henry von der. Würzburg Dr.phil. 1890 (worked on derivatives of hydroxylamine). No further chemical publications 1900–46; later activity not determined.

Kametaka, Tokuhei [1872–1935] (*JAP*). Dr.phil.; Berlin 1909 (worked on derivatives of protocatechuic acid, reduction of alanine and phenylalanine esters). Later Prof. chem. in Tokyo; published on topics in pharmaceutical chemistry.

Kauffmann, Hugo [1890–1916]. Berlin Dr.phil. 1914 (worked on chloro- and bromo-oximinoacetic acid). Died in action World War I.

Kautzsch, Karl [1879–1920]. Leipzig Dr.phil. 1904; Berlin 1905 (worked on synthesis of alanylalanine). Later with Meister, Lucius & Brüning Hoechst.

Kay, Francis William [1883–1967] (*UK*). Berlin Dr.phil. 1908 (worked on resolution and derivatives of methylisoserine). At Univ. Liverpool 1910–47 (Senior Lecturer chem. 1920–47).

Kempe, Martin [1884–?]. Berlin Dr.phil. 1907 (worked on tryptophan peptides). No further chemical publications 1910–44; later activity not determined.

Kempf, Richard [1879–1935]. Berlin Dr.phil. 1903 (worked on para-substituted nitrobenzaldehydes). With Chem.-Techn. Reichsanstalt Berlin 1923–35.

Kleberg, Werner. Würzburg Dr.phil. 1891 (worked on higher members of glucose series, action of formaldehyde on phenols). No further chemical publications 1907–46; later activity not determined.

Klemperer, Georg [1865–1946]. Berlin Dr.med. 1886; Pv.Dz. med. 1889–1905; ao.Prof. 1905– (after 1906 chief physician Moabit Hosp.). In 1913 collaborated with Fischer on synthesis of lipoidal arsenic compounds. Went to U.S. in 1936; published extensively on clinical topics.

Klotz, Carl [1862–1938]. Tübingen Dr.phil. 1885; Würzburg 1887 research asst. of Fischer (worked on naphthylhydrazine and indole derivatives; 3 acknowledgments). With Meister, Lucius & Brüning Hoechst 1891–1930.

Knoevenagel, Oscar [1862–1944]. Würzburg Dr.phil. 1887 (worked on compounds of phenylhydrazine with acrolein, mesityl oxide and allyl bromide). Later with BASF Ludwigshafen.

KNORR, LUDWIG [1859–1921]. Erlangen Dr.phil. 1882; research asst. of Fischer 1882–85 (worked on purines; 2 acknowledgments). See TEXT and Appendix 5.

Koch, Hermann [1858–1939]. Erlangen 1883–86 (worked on ethyl-phthalylacetoacetate, trimethylene diamine; 5 papers); Berlin Dr.phil. 1890. With Kali-Chemie A.G. Heilbronn 1891–1929.

Koelker, Arthur Heinrich [1883–1911] (US). Berlin Dr.phil. 1908 (worked on synthesis of leucyl peptides). Later on staff of Roosevelt Hospital New York.

Koelker, Wilhelm Friedrich [1880–1911] (US). Berlin Dr.phil. 1905 (worked on synthesis of leucylisoserine). Later asst. Prof. chem. Univ. Wisconsin.

Köllisch, Anton [1888–1916]. Berlin Dr.phil. 1911 (worked on diacetyl). Later with Merck Darmstadt.

Koenigs, Ernst [1878–1945]. Berlin Dr.phil. 1903 (worked on synthesis of aspartyl peptides). Breslau Pv.Dz. org. chem. 1912–21; ao.Prof. 1921–45.

Körösy, Kornel von [1879–1948]. Budapest Dr.med. 1903; Berlin 1907 (Abderhalden; worked on digestion of proteins). Budapest Pv.Dz. physiol. 1911–16; ao.Prof. 1916–20; o.Prof. 1920–48.

Kötzle, Arthur [1865–1930]. Würzburg Dr.phil. 1889 (worked on deriv-atives of diketohydrindene). Later consulting chemist in leather industry.

Kohlrausch, Karl [1862–?]. Würzburg Dr.phil. 1889 (worked on action of methylphenylhydrazine on dialdehydes and diketones). No further chemical publications 1900–36; later activity not determined.

Kopisch, Friedrich [1867–?]. Berlin Dr.phil. 1894 (worked on hydroxy derivatives of phenylbutyric acid). Later landowner in Weizenroda bei Schweinitz.

Korczynski, Antoni [1879–1929] (POL). Erlangen Dr.phil. 1902; Berlin 1908–09 (worked on chromoisomeric salts of nitrophenols). Cracow Pv.Dz. 1909–19; Poznan o.Prof. chem. 1919–29. Published on vari-ous topics in organic chemistry, especially catalytic reactions.

Krämer, Adolf [1883–1914]. Berlin Dr.phil. 1907 (worked on oximi-noacetic acid, synthesis of aminodioxyvaleric acid). With Bayer & Co. 1909–14; died in action during World War I.

Kraus, Johann [1869–?]. Berlin Dr.phil. 1895 (worked on aminopropyl-aldehyde). No further chemical publications; later activity not deter-mined.

Kropp, Walther [1885–1939]. Berlin Dr.phil. 1908 (worked on deriva-tives of glutamic and aminostearic acids). With Bayer & Co. Elber-feld 1909–38.

Kühlewein, Malte von [1887–?]. Berlin Dr.med. 1913; Dr.phil. 1915 (worked on purine glucosides). Became pharmacist in Utrecht.

Kuzel, Hans [1859–1921]. Erlangen Dr.phil. 1883 (worked on hydrazides of cinnamic acid, benzoylacetone). With Meister, Lucius & Brüning Hoechst 1885–1900; then set up private lab. in Vienna for lamp technology.

La Forge, Frederick Burr [1882–1958] (US). Berlin Dr.phil. 1908 (Pschorr; worked on isomerism of carboethoxyglycylglycine ethyl ester). At Rockefeller Institute New York (with Levene) 1910–15; then with U.S. Dept. Agriculture 1915–52. Made valuable contributions to the study of insecticides.

Langstein, Leo [1876–1933]. Vienna Dr.med. 1899. Strassburg Dr.phil. 1902 (Hofmeister); Berlin 1903 (amino acid analysis of zein). Head Kaiserin-Augusta-Victoria hospital (pediatrics) 1908–33.

LANDSTEINER, KARL [1868–1943]. Würzburg 1891 (worked on glycolaldehyde). See TEXT and Appendix 5.

Langenwalter, Jacob [1868–1943]. Würzburg Dr.phil. 1892 (worked on glucoheptonic acid). Later with BASF Ludwigshafen.

Laubmann, Heinrich [1865–1951]. Würzburg Dr.phil. 1888 (worked on compounds of phenylhydrazine with ketoalcohols). With Meister, Lucius & Brüning Hoechst until 1912, when he became a geologist.

Lawrence, William Trevor [1870–1934] (UK). Berlin Dr.phil. 1896 (worked on compounds of sugars with mercaptans). Manchester Lecturer org. chem.; then turned to horticulture.

Lawson, Thomas Atkinson [1862–1903] (UK). Marburg Dr.phil. 1885; Würzburg 1890 (no publication?). Later with various British chemical firms.

Laycock, William Frederick [1866–1912] (UK). Würzburg Dr.phil. 1889 (worked on metacetone and isophorone). Later consulting chemist in Leeds (dyes).

Le Count, Edwin Raymond [1868–1935] (US). Rush Med. College M.D. 1892; Berlin 1905 (Abderhalden; worked on amino acid analysis of feather keratin). Prof. pathology Rush Med. College Chicago 1902–35.

Lehmann, Fritz Julius [1874–?]. Berlin Dr.phil. 1898; research asst. of Fischer 1898–99 (worked on action of hydroxylamine on phorone, resolution of racemic amino acids; 5 acknowledgments). Later with Bayer & Co. Elberfeld.

Lehnert, Hermann [?–1901]. Munich 1880 (no publication); Heidelberg Dr.phil. 1896 (worked on detection of mercury, derivatives of xylol). Later with govt. patent office Berlin.

Lemme, Georg [1865–1925]. Berlin Dr.phil. 1889 (Rammelsberg); research asst. of Fischer 1894 (worked on glucosamine). Later with Haarmann & Reiner Co. Holzminden.

Lepsius, Richard [1885–1969]. Berlin Dr.phil. 1911 (worked on synthesis of polydepsides). After World War I with various government agencies and scientific societies.

LEUCHS, HERMANN [1879–1945]. Berlin Dr.phil. 1902; research asst. of Fischer 1902–04 (worked on synthesis of hydroxyamino acids and peptides, glucosamine; 2 papers, 4 acknowledgments); Pv.Dz. 1910–14; ao.Prof. 1914–34; o.Prof. 1935–45. Made important contributions to the organic chemistry of amino acids, spiranes and strychnine alkaloids.

LEVENE, PHOEBUS AARON THEODORE [1869–1940] (US). St. Petersburg Dr.med. 1892; Berlin 1902 (worked on hydrolysis of gelatin). With Rockefeller Institute for Medical Research New York 1905–40. Led a productive research group that worked on many chemical and biochemical topics—nucleic acids, carbohydrates, proteins, stereochemistry (see TEXT).

Lipschitz (Lindley), Werner [1892–1948]. Berlin Dr.phil. 1915 (worked on optically active methyl derivatives of amino acids, synthesis of tannins); Leipzig Dr.med. 1916. Frankfurt a.M. Pv.Dz. pharmacol. 1920–24; o.Prof. 1926–33; Istanbul o.Prof. 1933–39. In U.S. with Lederle Co. 1939–48.

Loeb, Walther [1872–1916]. Berlin Dr.phil. 1894 (worked on action of aminoacetal on nitrobenzyl chloride). Aachen TH Pv.Dz. electrochem. 1896. Later at Virchow Hospital Berlin 1906–16. Wrote extensively on application of electrochemistry to the study of biological problems.

Loeben, Wolf von [1869–1913]. Leipzig Dr.phil. 1896; Berlin 1900 (worked on phenylpurine, heat of combustion of glucosides). Later with govt. testing bureau Berlin.

London, Efim Semenovich [1868–1939] (RUSS). Warsaw Dr.med. 1894; Berlin 1906 (Abderhalden; worked on protein digestion, intestinal cleavage of peptides). Head of section for general pathology at Leningrad institute for experimental medicine 1919–39.

Lorenz, Henry William Frederick [1871–?] (US). Berlin Dr.phil. 1897 (worked on urea derivatives of diacetonamine). Later consulting chemist (photography).

Lüppo-Cramer, Hinricus [1871–1943]. Berlin Dr.phil. 1894 (worked on derivatives of caffeine). With Schering Berlin 1895–1902, Schleussner Co. Frankfurt a.M. 1902–17, Deutsche Gelatinfabrik Schweinfurt 1922–32. Made valuable contributions to photography.

Luniak, Andrei Ivanovich [1881–1957] (RUSS). St. Petersburg Dr.med.1904; Berlin (worked on synthesis of peptides of proline and phenylalanine). Prof. chem. Perm 1916–24; Kazan 1924–52.

Lussana, Filippo (ITAL). Dr.med.; Berlin 1907 (Abderhalden; worked on cleavage of peptides by tissue extracts). Later activity not determined.

McLester, James Somerville [1877–1954] (US). Univ. Virginia Dr.med. 1904; Berlin 1907 (Abderhalden; worked on action of plasma on peptides). Prof. Birmingham Med. College 1908–12; Prof. med. Univ. Alabama 1919–50.

Macnair, Duncan Scott [1861–1937] (UK). Würzburg Dr.phil. 1889 (worked on derivatives of furil and furoin). Educational inspector Scotland 1893–1927.

Madelung, Walther [1879–1963]. Strassburg Dr.phil. 1905 (Thiele); Berlin 1911–13 (worked on indigo synthesis); Freiburg i.B. Pv.Dz. 1914; ao.Prof. 1921–46. Published on various topics in organic chemistry, especially dyes.

Manasse, Wilhelm [1881–?]. Berlin Dr.phil. 1907 (Leuchs; worked on isomerism of carboethoxyglycylglycine ethyl ester). No further chemical publications 1907–46; later activity not determined.

Malengreau, Fernand [1880–1958] (BELG). Louvain Dr.med. 1903; Dr.rer.nat. 1905; Berlin 1906 (Abderhalden; worked on amino acid analysis of gluten). Louvain 1908–50 (Prof. physiol. chem. 1913–50); published on nucleoproteins, histones, vitamins.

Manwaring, Wilfrid Hamilton [1871–1960] (US). Johns Hopkins M.D. 1904; Berlin 1908 (Abderhalden; worked on peptide metabolism in erythrocytes). At Rockefeller Institute New York 1910–13; Prof. bacteriol. and experim. pathol. Stanford Univ. 1913–37.

Marko, Dmitri Militiadov [1878–?] (RUSS). Kazan Dr.phil. 1900?; Berlin 1908 (worked on resolution of aminocaproic acid). Later at chemical-technical institute Perm.

Marx, Max [1868–?]. Würzburg Dr.phil. 1891 (worked on reduction of trimethylgallamide). Went to U.S.; became consulting chemist in Newark.

Matsubara, Koichi [1872–1955] (JAP). Tokyo Dr.phil. 1906; Berlin 1906 (worked on resolution of racemic formylaminoisovaleric acid). Tokyo Prof. organic chem. 1909–33.

Mauritz, Alfred [1867–1938]. Würzburg Dr.phil. 1891 (Tafel; worked on phenacyl derivatives). Director Dortmund Actien-Brauerei 1892–1938.

Max, Jules [1882–?] (FR). Berlin Dr.phil. 1906 (worked on amino acid chlorides). Joined family chemical company in Paris.

Mechel, Lucas von [1893–1956]. Berlin Dr.phil. 1917 (worked on synthesis of phenylglucosides). With CIBA Basle 1918–56.

Medigreceanu, Florentin [?–1917] (RUM). Bucharest Dr.med. 1907; Berlin 1908 (Abderhalden; worked on peptide-cleaving enzymes in intestinal tract). At Rockefeller Institute New York 1910–13, then opened private lab. in Bucharest. Died in active service World War I.

Meyer, Jacob [1867–?]. Würzburg Dr.phil. 1889 (worked on oxidation of lactose, methylation of indoles). Went with Fischer to Berlin, and helped in administration of the institute; also did independent technical work (dye patents). After 1920, consulting chemist in Berlin.

Meyer, Paul [?–1914]. Würzburg Dr.phil. 1892 (worked on bromination of phenylhydrazine). With Kalle & Co. Biebrich; died in action World War I.

Michaelis, Ludwig [1869–1942]. Berlin Dr.phil. 1893 (worked on bromination of hydrazines and amines). Later associated with various chemical firms. Died in Nazi concentration camp near Lodz.

Minovici, Stefan [1867–1935] (*RUM*). Berlin Dr.phil. 1897 (worked on oxazoles). Prof. chem. Bucharest 1911–35. Published extensively on topics in organic chemistry, biochemistry and forensic chemistry.

Moreschi, Annibale [1886–1931] (*ITAL*). Dr.phil.; Berlin 1912 (worked on Walden inversion). Later Director Stacchini Powder Factory Rome; killed during explosion test.

Morrell, Robert Selby [1867–1946] (*UK*). Würzburg Dr.phil. 1893 (worked on configuration of rhamnose and galactose). Chief chemist Manders Bros. Ltd. (paint and varnish) Wolverhampton 1904–30.

Mouneyrat, Antoine [1870–1952] (*FR*). Paris D.Sc. 1899; Dr.med. 1902; Berlin 1900 (worked on resolution of racemic amino acids, conversion of amino acids into phenylhydantoins). Became assoc. Prof. medical faculty Paris; published on use of arsenicals in treatment of syphilis and later on sulfa drugs; promoted development of French pharmaceutical industry.

Müller, Franz [1874–?]. Berlin 1894 (worked on action of hydrocyanic acid on phenylhydrazine); Erlangen Dr.phil. 1898. Later with Geigy Basle.

Müller, Hermann [1867–?]. Berlin Dr.phil. 1894 (worked on hydrazinodiphenyl). Later activity not determined.

Mylo, Bruno [1884–1915]. Berlin Dr.phil. 1906 (Koenigs; worked on amino acid amides). Asst. chem. Danzig 1907–14; died in action World War I.

Nastvogel, Oscar [1863–1910]. Würzburg Dr.phil. 1887 (worked on reaction of hydrazines with dibromopyruvic acid). With Bayer & Co. Elberfeld 1889–1907.

Neufeld, Karl Albert [1866–1914]. Würzburg Dr.phil. 1888 (worked on halogen derivatives of phenylhydrazine). With state testing bureau Munich 1890–1911 and state nutrition bureau Würzburg 1911–14.

Neymann, Hans von Splawa [1885–1916]. Kiel Dr.phil.1910 (Harries); Berlin 1913–14 (worked on chloromethylfurfurol). With Wolfener Farbenfabrik Bielefeld 1914; died in action World War I.

Niebel, Wilhelm [?–1901]. Berlin 1896 (worked on action of animal fluids and tissues on polysaccharides).

Noth, Hartmut [1892–?]. Berlin Dr.phil. 1917 (worked on partially acylated sugars). Research asst. of Hans Pringsheim 1920; later activity not determined.

Nouri, Osman [1885–?] (*TURKEY*). Berlin Dr.phil. 1917 (worked on nitriles of phenol carboxylic acids and on phloretin). No further chemical publications 1918–44; later activity not determined.

Oetker, Rudolf [1889–1916]. Berlin Dr.phil. 1914 (worked on acyl derivatives of glucose and mannose). Died in action World War I.

Olinger, Josef [1881–?]. Berne Dr.med. 1908; Berlin 1908 (Abderhalden; worked on metabolism of protein cleavage products). No further chemical publications 1907–46; later activity not determined.

Oppenheimer, Carl [1874–1941]. Berlin Dr.phil. 1894; Dr.med. 1898; in Fischer lab. 1904 (Abderhalden; worked on occurrence of albumoses in blood). Prof. agricult. school Berlin 1908–36; also associated with Scheidemandel Chemical Co. The Hague 1938–41. A prolific writer of biochemical treatises.

Oppler, Bernhard [1871–1943]. Munich Dr.med. 1895; Berlin 1907 (Abderhalden; worked on action of blood plasma on peptides). Later physician in Hannover.

Osa, Adolphe de [1873–?] (*FR*). Berlin Dr.phil. 1904 (worked on oxidation by ozone, phenylbutene ozonide). Later chemist in French iron and steel industry.

Ostmann, Paul [1893–1965]. Berlin Dr.phil. 1920 (worked on derivatives of glucose bromohydrin). No further chemical publications 1920–46; later activity not determined.

Otto, Erich [?–1902]. Berlin 1902 (worked on peptides of glycine and alanine). Died during course of this work.

Ou, Ching-Ko [1885–?] (*CHINA*). Berlin Dr.phil. 1911 (worked on dipeptides of aminocaproic acid). No further chemical publications 1917–46; later activity not determined.

Passmore, Francis William [1867–1921] (*UK*). Würzburg Dr.phil. 1890 (worked on higher homologues of mannose, formation of phenylhydrazides, formation of acrose from formaldehyde). Consulting chemist London 1902–21.

Pfähler, Ernst [1890–1981]. Berlin Dr.phil. 1913; research asst. of Fischer 1913–14, 1919 (worked on reaction of phthalimidoacyl chlorides with sodium malonic ester, synthesis of glucosides, glucal, glycerides; 1 paper, 5 acknowledgments). Later became consulting chemist in Buenos Aires.

Pfeffer, Otto [1885–1918]. Leipzig Dr.phil. 1911; Berlin 1912–13 (worked on methylation of polyphenol carboxylic acids). Son of botanist Wilhelm Pfeffer; after brief service in testing dept. of govt. finance ministry, died in action World War I.

Pfülf, August [1862–1921]. Würzburg Dr.phil. 1887 (worked on hydrazinobenzene sulfonic acids, indoles). With Buckau chemical factory in Ammendorf 1898–1921.

Pickel, Max [1861–1933]. Erlangen Dr.phil. 1886 (worked on derivatives of phenylhydrazine). Later commercial chemist in Nürnberg.

Piloty, Oskar [1866–1915]. Würzburg Dr.phil. 1890; research asst. of Fischer 1889–96 (worked on higher homologues of rhamnose, reduction of saccharic acid; 4 papers, 3 acknowledgments). See Appendix 5.

Pinkus, Georg [1870–1943]. Berlin Dr.phil. 1893 (A. W. Hofmann); research asst. of Fischer 1893–98 (worked on acetaldehyde phenylhydrazone, various sugars and their derivatives; 13 acknowledgments). Later with Noelting dye company in Mulhouse and with A.G. Anilinfabrikation in Berlin, before becoming consulting chemist. Went to U.S. in 1933; died in Miami.

Plimmer, Robert Henry Aders [1877–1955] (*UK*). London D.Sc. 1902; Berlin 1902 (worked on amino acid analysis of gelatin). London University College 1904–19 (Reader physiol. chem. 1912–19); St. Thomas's Hosp. Med. School Prof. chem. 1922–43. Published on chemistry of proteins and was active in the development of biochemistry in Britain.

Poppenberg, Otto [1876–1956]. Berlin Dr.phil. 1900 (worked on pyridazine). At Charlottenburg TH 1922–45 (o.Prof. techn. chem. 1933–45); specialized in explosives.

PREGL, FRITZ [1869–1930]. Graz Dr.med. 1894; Berlin 1904 (Abderhalden; worked on amino acid analysis of crystalline egg albumin). Graz Pv.Dz. med. chem. 1905–10; o.Prof. 1913–30 (Innsbruck o.Prof. 1910–13). Inventor of methods of organic microanalysis (see TEXT).

Pribram, Hugo [1882–1943]. Prague Dr.med. 1905; Berlin 1907 (Abderhalden; worked on amino acid analysis of lactalbumin). Prague Pv.Dz. pathology 1912–21; ao.Prof. 1921–38?; died in Theresienstadt concentration camp.

Prym, Oskar [1873–?]. Bonn Dr.med. 1898; Berlin 1907 (Abderhalden; worked on autolysis of liver, absorption of amino acids in intestinal tract). Bonn Pv.Dz. med. 1907; ao.Prof. 1921–33.

Pschorr, Robert [1868–1930]. Jena Dr.phil. 1894; Berlin 1895–13 (Pv.Dz. 1901–09; ao.Prof. 1909–13; conducted independent research program); Charlottenburg TH o.Prof. chem. 1914–30. Published valuable organic-chemical papers, especially on alkaloids.

Pulvermacher, Georg [1866–?]. Berlin Dr.phil. 1887 (Gabriel); with Fischer 1892–94 (no publication?). Later active in chemical societies; for a time General Secretary of International Chemical Congress.

Rabe, Arthur [1889–?]. Berlin Dr.phil. 1920 (worked on derivatives of rhamnose). Later with Bayer & Co. Elberfeld.

Rahnenführer, Carl [1859–1921]. Königsberg Dr.phil. 1884; Würzburg research asst. of Fischer 1885–88 (worked on phenylhydrazine derivatives of sugars; 7 acknowledgments). With Weiler-ter Meer Uerdingen 1890–1921.

Ramsay, Henrik [1886–1951] (*FIN*). Berlin 1908 (worked on glycocyamines); Helsingfors Dr.phil. 1909. Later in sugar industry, then manager of Finnish Steamship Co., finally with various govt. agencies (Minister of Foreign Affairs 1943–44).

Rapaport, Max [1889–1915]. Berlin Dr.phil. 1914 (worked on carbomethoxy derivatives of phenol carboxylic acids). Died in action World War I.

Raschen, Julius [1863–?] (*UK*). Würzburg Dr.phil. 1887 (worked on synthesis of indole derivatives). Later with United Alkali Co.

Raske, Karl August Heinrich [1863–?]. Berlin Dr.med. 1886; Dr.phil. 1905; research asst. of Fischer 1903–09 (worked on synthesis of aminobutyryl peptides, glucosides, conversion of serine to alanine and serine; 8 papers). Later returned to private medical practice.

Reese, Ludwig. Erlangen 1883 (worked on caffeine, xanthine and guanine); Leipzig Dr.phil. 1887. Later activity not determined.

Rehländer, Paul [1869–?]. Berlin Dr.phil. 1893; research asst. of Fischer 1892–95 (worked on various topics in sugar chemistry; 10 acknowledgments). Later with Boehringer Mannheim, Griesheim-Elektron, Schering Berlin; specialized in electrochemistry.

Reif, Johann Georg [1880–1964]. Freiburg i.B. Dr.phil. 1906; Berlin 1908 (worked on derivatives of proline). With govt. institute of health 1920–45 and Koch Institute Berlin 1945–49.

Reinbold, Bela [1875–1927]. Klausenberg Dr.med. 1900; Berlin 1905 (Abderhalden; worked on amino acid analysis of edestin). Klausenberg Pv.Dz. med. chem. 1907–09; ao.Prof. 1909–11; o.Prof. 1911–21; Szeged o.Prof. 1921–27.

Reinbrecht, Otto. Würzburg Dr.phil. 1891 (worked on carboxylic acids of lactose and maltose). No further chemical publications 1896–1939; later activity not determined.

Reisenegger, Hermann [1861–1930]. Erlangen Dr.phil. 1883 (worked on hydrazine compounds of phenol and anisole). With Meister, Lucius & Brüning Hoechst 1884–1915; o.Prof. chem. technology Charlottenburg TH 1915–26. Made valuable contributions to dye technology.

Reitzenstein, Friedrich [1868–1940]. Würzburg Dr.phil. 1892 (worked on diketohydrindene). Later ao.Prof. chem. Würzburg 1906–13; then retired.

Renouf, Edward [1848–1934] (*US*). Munich 1880 (worked with E. Fischer on dimethylhydrazine; took Dr.phil. for this work at Freiburg i.B.); see Appendix 5.

Reuter, Ferdinand [1879–1942]. Berlin Dr.phil. 1903; research asst. of Fischer 1904–06 (worked on synthesis of peptides by halogen acyl chloride method; 4 acknowledgments). Later with Didier-Werke A.G.

Richter, Ernst [1861–1917]. Berlin Dr.phil. 1887 (A. W. Hofmann); with Fischer 1892–93 (no publication?). Later commercial chemist in Stettin.

Rilliet, Auguste [1880–?] (*SWITZ*). Geneva Dr.phil. 1907; Berlin 1907 (Abderhalden; worked on cleavage of peptides by mushroom extracts). Geneva Pv.Dz. chem. 1910–.

Ristenpart, Eugen [1873–1953]. Berlin Dr.phil. 1896 (Gabriel; worked on action of ammonia and alkylamines on bromoethylphthalimide). With several textile dyeing firms 1899–1937.

Roder, Anton [1864–1931]. Würzburg Dr.phil. 1887 (worked on synthesis of indoles). Later head of private chemical factory in Munich.

Roeder, Georg [1874–?]. Berlin Dr.phil. 1899; research asst. of Fischer 1900–01 (worked on pulegon, synthesis of uracil, thymine, phenyluracil; 2 papers). Later chemist at pharmacol. institute Naples, then with Gerber Milk Co.

Roesner, Hans [1886–?]. Berlin Dr.phil. 1911 (worked on serine peptides, reactions of hydroxyisovaleric acid). Later with Ilse Bergbau A.G. (petroleum technology).

Rohde (Davis), Alice [1882–1933] (*US*). Johns Hopkins M.D. 1910; Berlin 1912 (worked on reactions of ethylisopropylcyanoacetic ester and ethylisopropylmalonamic acid). During 1913–30, at Carnegie Tech, Johns Hopkins, Univ. California, Pittsburgh, Cornell.

Rohmer, Martin [1878–1941]. Berlin Dr.phil. 1900 (worked on rearrangement of methylene dinitroaniline); research asst. of Fischer 1900–02 (worked on arsenic analysis). With Meister, Lucius & Brüning Hoechst 1902–33.

Rona, Peter [1871–1945]. Vienna Dr.med. 1895; Dr.phil. 1903; Berlin 1904 (Abderhalden; worked on hydrolysis of thymus histone, metabolism of glycyltyrosine); Pv.Dz. med.chem. 1905; ao.Prof. 1920–33 (head of chem. divn. pathol. institute Charité Hosp. 1921–33). Published extensively on biochemical topics.

Rose, Frederick [1867–1932] (*UK*). Berlin Dr.phil. 1894 (worked on action of cumylhydrazine on sugars, ketones and aldehydes). Later active on UK government educational commissions.

Rostoski, Otto [1872–1962]. Würzburg Dr.med. 1896; Pv.Dz. med. 1902; Berlin 1905 (Abderhalden; worked on amino acid analysis of edestin and Bence Jones protein). ao.Prof. and head of medical clinic Dresden municipal hospital 1907–? Published extensively on clinical topics.

Rudolph, Otto [1864–?]. Würzburg Dr.phil. 1888 (worked on phenylhydrozones). Later consulting chemist in Berlin.

RUFF, OTTO [1871–1939]. Berlin Dr.phil. 1897 (Piloty); Pv.Dz. 1901 (worked on various problems in sugar chemistry). O.Prof. inorganic chem. Danzig 1904–16; Breslau 1916–39. A versatile experimenter who made important contributions on the chemistry of sugars and of fluorine.

Rund (Schmidt-Rund), Charlotte [1890–?]. Berlin Dr.phil. 1916 (worked on partial acylation of polyalcohols and sugars); Dr.med. 1918. Later in private medical practice.

Sachs, Franz [1875–1919]. Berlin Dr.phil. 1898 (Gabriel); Pv.Dz. 1903; tit. Prof. 1908–19. Published on various topics in organic chemistry.

Samuely, Franz [1879–1913]. Strassburg Dr.med. 1903; Berlin 1905 (Abderhalden; worked on amino acid analysis of gliadin, metabolism of cystine and cystine peptides). Freiburg i.B. ao.Prof. med. 1910–13.

Sasaki, Takaoki [1878–1966] (*JAP*). Tokyo Dr.med. 1903; Berlin 1907 (Abderhalden; amino acid analysis of muscle protein). Prof. med. Kyoto 1910–17; head of private hospital Tokyo 1917–?. Published extensively on topics in biochemistry and bacteriology.

Sattler, Wilhelm [1859–?]. Würzburg Dr.phil. 1890 (worked on reaction of ethyl oxalate with anilides). Later owner of dye factory in Schweinfurt.

SCHEIBLER, HELMUTH [1882–1966]. Berlin Dr.phil. 1909; research asst. of Fischer 1908–11 (worked on synthesis of valine peptides, Walden inversion; 6 papers). Charlottenburg TH Pv.Dz. 1915; o.Prof. 1924–55. Made many valuable contributions in organic chemistry, especially on compounds of divalent carbon.

Schenkel, Julius Wilhelm [1878–1967]. Berlin Dr.phil. 1906 (worked on synthesis of valine peptides). Later Pv.Dz. chem. technol. Dresden TH.

Schittenhelm, Alfred [1874–1954]. Tübingen Dr.med. 1898; Berlin 1904 (Abderhalden; worked on amino acid analysis of edestin, excretion of tyrosine and leucine in cystinuria). O.Prof. medicine Königsberg 1912–15; Kiel 1916–34; Munich 1934–45. Published extensively on clinical problems, notably purine metabolism and infectious diseases.

Schlieper, Adolf [1865–1945]. Würzburg Dr.phil. 1887 (worked on synthesis of naphthylindoles). With family chemical firm (Schlieper & Baum) in Wuppertal 1889–1945.

Schlotterbeck, Fritz [1876–1940]. Tübingen Dr.phil. 1902; Berlin 1903–04 (worked on conversion of sorbic acid to amino acids). Würzburg Pv.Dz. 1903–15; head of Stockhausen chemical factory Krefeld 1915–35; with institute of plant chem. forestry school Tharandt 1935–40.

Schmidlin, Julius [1880–1962]. Berlin 1905 (worked on peptides of phenylglycine). See Appendix 5.

Schmidmer, Eduard [1861–1933]. Würzburg Dr.phil. 1892 (worked on uptake of salt solutions by filter paper). Later owner chemical factory in Nürnberg.

Schmidtmann, Hermann [1868–1919]. Berlin Dr.phil. 1896 (worked on nitrogen derivatives of malononitrile). Later owner chemical factory Kaliwerke Aschersleben.

Schmitt, Theodor Friedrich [1859–?]. Würzburg Dr.phil. 1889 (worked on phenylindoles). Later with govt. agricult. station in Vienna.

Schmitz, Wilhelm [1879–1939]. Göttingen Dr.phil. 1904; Berlin 1906 (worked on synthesis of amino acids from bromo fatty acids). With BASF Ludwigshafen 1906–39.

Schneider, Wilhelm [1882–1939]. Berlin Dr.phil. 1909 (worked on carboxymethyl derivatives of phenol carboxylic acids). Jena Pv.Dz. chem. 1914–22; ao.Prof. 1922–33; o.Prof. 1933–39.

Schoeller, Walter [1880–1965]. Berlin Dr.phil. 1906 (worked on synthesis of phenylalanine peptides); Pv.Dz. chem. 1915. Freiburg i.B. o.Prof. med. chem. 1919–23; with Schering Berlin 1923–56, where his research group made valuable contributions to the chemistry of sex hormones.

Schotte, Herbert [1897–1950]. Berlin Dr.phil. 1920 (worked on glucal). With Schering Berlin 1922–50.

Schottmüller, Arnold [1879–?]. Berlin Dr.phil. 1908 (Houben; worked on aromatic amino acids). Later at state agricult. exper. station Dresden.

Schrader, Hans [1887–1982]. Berlin Dr.phil. 1910 (worked on carbomethoxy derivatives of phenol carboxylic acids). Köln Pv.Dz. chem. 1921; with T. Goldschmidt & Co. Essen 1922–45; consulting chemist in Seeheim 1945–.

Schrauth, Walther [1881–1939]. Berlin Dr.phil. 1906 (worked on cleavage of diketopiperazines). With Deutsche Hydrierwerke A.G. Berlin 1916–39.

Schuler, Josef [1883–1963]. Berlin Dr.phil. 1910 (worked on synthesis of isoleucine peptides). Later with CIBA Basle.

Schulze (Schulze-Forster), Arnold [1882–1946]. Berlin Dr.phil. 1907 (worked on synthesis of alanine peptides). With govt. institute of hygiene Berlin 1912–45.

Sembritzki, Kurt [1878–1943]. Berlin Dr.phil. 1897 (worked on malonyldiethylurea and diethyluric acid). Later consulting chemist in Berlin.

Seuffert, Otto [1875–1952]. Munich Dr.phil. 1900; Berlin 1901 (worked on indazole). Later with Merck Darmstadt; see Appendix 5.

Severin, Josef [1866–?]. Bonn Dr.med. 1891; Berlin 1912 (worked on glucosides). Later chief physician hospital in Breslau.

Skita, Aladar [1876–1953]. Heidelberg Dr.phil. 1900; Berlin 1901–02 (worked on amino acid analysis of silk fibroin). With Meister, Lucius & Brüning 1903–05; Karlsruhe TH Pv.Dz. 1906–11; ao.Prof. 1911–14; Freiburg i.B. ao.Prof. 1914–22; Kiel ao.Prof. 1922–24; Hannover TH o.Prof. org. chem. 1924–48. Made valuable contributions in organic chemistry, especially in studies on catalytic hydrogenation.

Slimmer, Max Darwin [1877–?] (US). Berlin Dr.phil. 1902 (worked on aminovaleric acid, asymmetric synthesis). Later became consulting chemist in Chicago.

Smith, William Stanley [1863–?] (UK). Würzburg Dr.phil. 1891 (worked on higher sugars related to mannose). Later commercial chemist in rubber industry.

Sonn, Adolf [1882–1957]. Berlin Dr.phil. 1907 (worked on reduction of glycine ethyl ester). Königsberg Pv.Dz. chem. 1913–20; o.Prof. 1921–45.

Speier, Arthur. Berlin 1895 (worked on preparation of esters, compounds of acetone with polyalcohols); Rostock Dr.phil. 1896. Later with BASF Ludwigshafen.

Sperling, Rudolf [1888–1914]. Berlin Dr.phil. 1913 (worked on synthesis of condensed benzene derivatives). Died in action World War I.

SPIRO, KARL [1867–1932]. Würzburg Dr.phil. 1889 (worked on oxaloacetic acid ester); Leipzig Dr.med. 1893. Strassburg 1894–1919 (with Hofmeister 1896–1919); Basle, with Sandoz 1919–21; o.Prof. physiol. chem. 1923–32 (see TEXT).

Stähler, Arthur [1877–1950]. Berlin Dr.phil. 1902 (worked on carvone, santonic acid); Pv.Dz. chem. 1908–17; tit. Prof. 1917–19. Published extensively in inorganic and physical chemistry.

Stahel, Rudolf [1866–?]. Würzburg Dr.phil. 1891 (worked on xylose, derivatives of diphenylhydrazine and methylphenylhydrazine). Later director Sydowsawe factory Stettin.

Stahlschmidt, Alex [1882–1966]. Berlin Dr.phil. 1910 (worked on synthesis of peptides of glutamic and aminovaleric acids, dihydrofurane dicarboxylic acid). Later with Bayer & Co. Elberfeld.

Steche, Albert [1862–?]. Würzburg Dr.phil. 1887 (worked on methylation of indoles). Later with Heine & Co. Leipzig (perfumes).

Steingroever, Joseph [1884–?]. Berlin Dr.phil. 1907 (worked on synthesis of leucine, alanine and glycine peptides). Later with Deutsche Oxyhydriergesellschaft Düsseldorf.

Stewart, Anthony (US). Würzburg Dr.phil. 1893 (worked on aromatic sugars). Later manager of family sugar plantation in Cuba.

STOCK, ALFRED [1876–1946]. Berlin Dr. phil. 1899 (Piloty; worked on monobromoacrolein and tribromopropionaldehyde, separation of antimony and arsenic); Pv.Dz. 1900–09. Breslau o.Prof. chem. 1909–16; Kaiser-Wilhelm Institute of Chemistry Berlin 1916–26; Karlsruhe TH 1926–36. Made outstanding contributions in inorganic chemistry, especially in relation to compounds of silicon and boron.

Strauss, Eduard [1876–1952]. Berlin 1906 (Abderhalden; worked on amino acid analysis of egg albumin). See Appendix 5.

Strauss, Hermann [1884–1942]. Strassburg Dr.med. 1908; Berlin 1912 (worked on carbomethoxy derivatives of phloroglucinic acid, synthesis of phenylglucosides). Later physician in Berlin; died in Nazi concentration camp at Drancy.

Suzuki, Umetaro [1874–1943] (JAP). Tokyo Dr.agr. 1896; Berlin 1904–05 (worked on synthesis of cystine, proline, histidine and phenylalanine peptides). Became Prof. chem. Tokyo; made valuable contributions to study of vitamins and other natural products.

Täuber, Ernst [1861–1944]. Breslau Dr.phil. 1882; Erlangen 1884 (worked on uric acid, triacetonamine). With govt. patent office Berlin 1901–18; later consulting chemist, especially on art conservation.

Tafel, Julius [1862–1918]. Erlangen Dr.phil. 1884 (worked on indazole); Würzburg research asst. of Fischer 1885–89 (worked on many aspects of sugar chemistry, 9 papers); Pv.Dz. 1888; ao.Prof. 1902–03; o.Prof. 1903–06. Because of illness retired to Munich. Published important papers on the structure of strychnine and brucine.

Tappen, Hans [1879–1969]. Berlin Dr.phil. 1904 (worked on synthesis of phenanthrenes); research asst. of Fischer 1905–07 (worked on peptide synthesis, Walden inversion; 4 acknowledgments). Later with Goertz Photochemische Werke, then Director Zeiss-Ikon Göttingen.

Techow, Walter [1870–?]. Berlin Dr.phil. 1894 (worked on derivatives of dimethylalloxan). No further chemical publications 1900–39; later activity not determined.

Teru-uchi, Yukata [1873–1936] (JAP). Tokyo Dr.med. 1900; Berlin 1905 (Abderhalden; worked hydrolysis of proteins from pine seeds). With Kitasato Institute for Infectious Diseases Tokyo 1915–36.

Thieme, Bruno. Würzburg Dr.phil. 1891 (worked on salts and derivatives of phenylhydrazine). Later with Pulverfabrik Troisdorf.

Torrey, Henry Augustus [1871–1910] (*US*). Harvard Ph.D. 1897; Berlin 1898 (worked on allocaffeine). Asst.Prof. organic chem. Harvard 1900–10.

Traube, Wilhelm [1866–1942]. Berlin Dr.phil. 1888 (Hofmann); Pv.Dz. 1896–1902; ao.Prof. 1902–29; o.Prof. 1929–34. Published important papers in organic and inorganic chemistry. Murdered by Nazis in his home.

Trenkler, Bruno. Würzburg Dr.phil. 1887 (worked on indoles). No further chemical publications 1890–1939; later activity not determined.

Troschke, Hermann Oswald [1851–?]. Munich research asst. of Fischer 1879; see Appendix 5.

Tüllner, Hermann [1879–?]. Berlin Dr.phil. 1903 (worked on conversion of isouric acid to uric acid and thioxanthine). No further chemical publications 1905–39; later activity not determined.

Unna, Eugen [1885–1958]. Berlin Dr.phil. 1911 (worked on synthesis of galactosides). With Beiersdorf & Co. Hamburg 1913–48.

VAN SLYKE, DONALD DEXTER [1883–1971] (*US*). Univ. Michigan Ph.D. 1907 (Gomberg); Berlin 1911 (worked on pyrrole carboxylic acid). At Rockefeller Institute for Medical Research New York 1909–10, 1912–48; Brookhaven Natl. Lab. 1948–71. A gifted chemist who made decisive contributions to the development of modern clinical chemistry, in addition to important work on protein chemistry and acid-base equilibria (see TEXT).

Voegtlin, Carl [1879–1960] (*SWITZ*). Freiburg i.B. Dr.phil. 1902; went to U.S. in 1904. Berlin 1906 (Abderhalden; worked on cleavage of casein by pancreatic juice). Johns Hopkins assoc.Prof. pharmacol. 1906–13; with U.S. Public Health Service 1913–43. Wrote extensively on biochemical and pharmacological topics.

Voitinovici, Arthur [1881–?] (*RUM*). Berlin Dr.phil. 1907 (Abderhalden; worked on amino acid analysis of keratin); later activity not determined.

Voss, Arthur [1882–1940]. Berlin Dr.phil. 1908 (worked on derivatives of phloroglucinol, ring closure in naphthalene series). With Meister, Lucius & Brüning Hoechst (I. G. Farben) 1910–40.

Wacker, Leonhard [1864–1936]. Erlangen Dr.phil. 1888 (O. Fischer); with BASF Ludwigshafen 1888–93; in Buffalo, N.Y. 1902–07. Berlin 1908 (Abderhalden; worked on metabolic cleavage of diketopiperazines). Munich head of chem. divn. institute of pathology 1910–34.

Wagner, Philipp [1863–?]. Würzburg Dr.phil. 1887 (worked on methylketole, rosindole). Later owner of chemical factory in Worms.

Wagner, Theodore Brentano [1867–1936] (*US*). Würzburg 1891 (Dr.phil.?; no publication?). With Corn Products Co. 1900–19; U.S. Food Products Corp. 1919–22; consulting chemist New York 1922–36.

WARBURG, OTTO [1883–1970]. Berlin Dr.phil. 1906 (worked on resolution of racemic leucine, synthesis of peptides); Heidelberg Dr.med. 1911; Kaiser-Wilhelm Institute Berlin 1914–70. Made striking contributions to the study of biological oxidations and glycolysis; perhaps the greatest biochemist of the twentieth century (see TEXT).

Weichhold, Oskar [1883–1965]. Berlin Dr.phil. 1909 (worked on resolution of phenylglycine and synthesis of phenylglycine peptides). No further chemical publications 1910–39; later activity not determined.

Weigert, Fritz [1876–1947]. Berlin Dr.phil. 1899 (Gabriel); research asst. of Fischer 1901–02 (worked on synthesis of lysine); Pv.Dz. chem. 1908. Leipzig o.Prof. photochemistry 1914–35; went to U.K. where he headed physical-chem. dept. cancer research institute Northwood 1936–47.

Weil, Albert [1867–?]. Freiburg i.B. Dr.phil. 1889; Berlin research asst. of Fischer 1892 (no publication?). Later with Schuchardt Co. Görlitz.

Weil, Hugo [1863–1942?]. Erlangen Dr.phil. 1885 (worked on diacetonamine). With Meister, Lucius & Brüning Hoechst 1887–91, then had private consulting lab. in Munich; died in Nazi concentration camp near Minsk.

Weller, Heinrich [1853–1923]. Freiburg i.B. Dr.phil. 1882; Erlangen 1883 (worked on triphenylmethane dyes). Director state chemical and bacteriological bureau Darmstadt 1886–1923.

WELLS, HARRY GIDEON [1875–1943] (US). Rush Med. College M.D. 1898; Berlin 1905 (Abderhalden; worked on amino acid analysis of keratin). Univ. Chicago Dept. Pathology 1901–43 (Prof. 1913–43). Published extensively on problems of tissue autolysis, immunology and cancer.

Wenzing, Max. Würzburg Dr.phil. 1887 (worked on derivatives of methylindoles). Later Pv.Dz. chem. Graz.

Wheeler, Alvin Sawyer [1866–1940] (US). Harvard Ph.D. 1900; Berlin 1910 (worked on Walden inversion). Univ. North Carolina Prof. org. chem. 1912–36.

WINDAUS, ADOLF [1876–1959]. Freiburg i.B. Dr.phil. 1899 (Kiliani); Berlin 1900 (worked on formation of quaternary ammonium compounds from aniline homologues). Pv.Dz. Freiburg 1903–13; o.Prof. med. chem. Innsbruck 1913–15; o.Prof. chem. Göttingen 1915–44. One of the leading investigators of the chemistry of sterols and their relation to vitamin D; also made valuable contributions to chemistry of imidazoles.

Wirthle, Ferdinand [1863–1936]. Würzburg 1890 (worked on mannosaccharic acid); Erlangen Dr.phil. 1890. With state food research institute Würzburg 1911–28.

Wislicenus, Wilhelm [1861–1922]. Würzburg Dr.phil. 1885 (worked on esters of keto acids, action of cyanide on phthalide, isoglucosamine); Pv.Dz. 1888–90; ao.Prof. 1890–1902. Tübingen o.Prof. chem. 1902–22.

Wohl, Alfred [1863–1939]. Berlin Dr.phil. 1886 (A. W. Hofmann); Pv.Dz. 1891–1904 (worked on compounds of hexamethylenetetramine with aldehydes, chemistry of sugars). Danzig o.Prof. chem. 1904–33; then went to Stockholm. Made important organic-chemical contributions, especially in relation to the fermentation of sugars and oxidation reactions.

Wolfes, Otto [1877–1942]. Berlin Dr.phil. 1899; research asst. of Fischer 1900–02 (worked on resolution of racemic amino acids, peptide synthesis, preparation of amino acid esters, hydrolysis of casein; 6 acknowledgments). With Merck Darmstadt 1903–42.

Wrede, Franz [1877–1946]. Berlin Dr.phil. 1903; research asst. of Fischer 1903–06 (worked on determination of heats of combustion of organic compounds; 3 papers). Later with Schering Berlin.

Zach, Karl [1888–1968]. Berlin Dr.phil. 1911 (worked on derivatives of glucosides). In 1922 became Director of the chemical works Schuster & Wilhelmy A.G.

ZEMPLÉN, GÉZA [1883–1956]. Budapest Dr.phil. (physics) 1904; Berlin research asst. of Fischer 1909–12 (worked on cellobiose, synthesis of diaminovaleric acid, proline and hydroxyamino acids, 9 papers). Budapest o.Prof. chem. 1913–56; made important contributions to carbohydrate chemistry.

Zöllner, Clemens [1884–?]. Berlin Dr.phil. 1910 (worked on isoquinoline derivatives). Later with Schering Berlin.

Appendix 7

The Hofmeister Research Group

Alexander, Franz [1872–?]. Freiburg i.B. Dr.med. 1895; Strassburg 1898 (worked on casein and its cleavage by pepsin); was asst. in otolaryngology 1897–1901. Settled in Frankfurt a.M., where he had extensive practice in this specialty.

Almagia, Marco [1876–?] (*ITAL*). Dr.med.; Strassburg 1906 (worked on metabolism of uric acid). Later Prof. physiology Rome.

Alzona, Federico [1888–1958] (*ITAL*). Pavia Dr.med. 1912; Strassburg 1914 (worked on chondroitin sulfate). Chief physician municipal hospital Bologna 1922–53.

Auer, Aloys [1888–1948]. Strassburg Dr.med. 1918 (worked on dietary insufficiency). After 1924 head of medical clinic at Hoechst Hospital. Published extensively on clinical problems.

Baer, Julius [1876–1961?]. Strassburg Dr.med. 1899; worked with Embden on cystine metabolism (1906) and with Parnas on carbohydrate metabolism (1912); Pv.Dz. med. 1907. After World War I in private practice in Frankfurt a.M. until 1935, when he went to Palestine.

Barrenscheen, Hermann Karl [1887–1958]. Vienna Dr.phil. 1910; Strassburg 1913 (worked on glycogen and glucose formation). Göttingen Dr.med. 1923; Vienna Pv.Dz. physiol. chem. 1924–39; ao.Prof. 1939–47; chief physician Salzburg hospital 1947–58. Published extensively on carbohydrate metabolism.

Bassermann, Heinrich [1886–1965]. Strassburg Dr.phil. 1912 (worked with Spiro on food preservation). Later became owner of family preserves factory in Mannheim.

Bauer, Freidrich Alfred [1883–1957]. Strassburg Dr.med. 1907 (worked on structure of inosinic acid and the pentose in muscle). Chief physician at sanatoria in Davos 1916–51.

Baum, Fritz [1872–1939]. Heidelberg Dr.phil. 1896; Strassburg 1903 (worked on pancreatic autolysis). Director Wülfing-Dahl & Co. chemical factory 1907–19.

Baumann, Arno [1877–?]. Leipzig Dr.med. 1904; Strassburg 1913 (worked on cephalin). Later in private medical practice in Dresden.

Bayer, Kurt [1888–?]. Strassburg Dr.med. 1915 (worked on lipids of intestinal mucosa). Later in private medical practice in Frankfurt a.M.

BERGMANN, GUSTAV von [1878–1955]. Strassburg Dr.med. 1903 (worked on metabolic conversion of cystine to taurine). After clinical experience in Berlin (1903–12), became o.Prof. medicine Marburg 1916–20, Frankfurt a.M. 1920–27, Berlin 1927–55 (Charité Hosp.). One of the leading German clinicians of his time; published extensively on metabolic disorders.

Bernert, Richard. Strassburg 1898 (worked on oxidation of proteins by permanganate). Dr.med. Vienna 1904; later in private medical practice (cardiology) in Berlin.

BETHE, ALBRECHT [1872–1954]. Munich Dr.phil. 1895; Strassburg Dr.med. 1898; asst. physiology 1896–1911 (worked in Hofmeister lab. on staining of tissues). O.Prof. physiology Kiel 1911–15; Frankfurt a.M. 1915–37, 1945–54 (see TEXT).

BLUM, LÉON [1878–1930] (FR). Strassburg Dr.med. 1901; asst. of Hofmeister 1901–04 (worked on metabolism of cystine, nutritional value of albumoses of fibrin and casein); o.Prof. medicine 1918–25. Made valuable contributions to study of metabolic disorders.

Blumenthal, Franz [1878–1971]. Strassburg Dr.med. 1903 (worked on carbohydrate metabolism). Berlin Pv.Dz. dermatol. 1919; o.Prof. 1921–32. Went to U.S.; Prof. dermatology Univ. Michigan 1936; private practice in Ann Arbor, Mich. 1937–.

Bonanni, Attilio [1869–1938] (ITAL). Dr.med.; Strassburg 1902 (worked on borneol and menthol glucuronides). After teaching at Sassari and Pavia, became Prof. pharmacology Rome.

Brion, Albert [1874–1936] (FR). Strassburg Dr.med. 1898 (worked on metabolism of tartaric acid). Chief physician Strassburg hospitals 1904–36; known for his studies on typhoid and paratyphoid.

Bürger, Max [1885–1966]. Würzburg Dr.med. 1911; Strassburg 1916 (worked on lipids of tubercle bacilli). Pv.Dz. med. Kiel 1918; o.Prof. medicine Bonn (1931–37); Leipzig (1937–59). Published on carbohydrate metabolism and various clinical topics.

Comessatti, Guiseppe [1880–1964] (*ITAL*). Strassburg 1907 (worked on carbohydrate metabolism in muscle); Padua Dr.med. 1915. Chief physician hospital Palmanova 1915–50.

Conradi, Heinrich [1876–?]. Strassburg Dr.med. 1899 (worked in Hofmeister lab. 1902 on blood coagulation). Pv.Dz. Dresden TH 1913–21 and chief bacteriologist Bureau of Public Health Dresden, then at hospital in Zwickau.

CZAPEK, FRIEDRICH [1868–1921]. Prague Dr.med. 1891 (worked on effect of selenium and tellurium on animal organisms). Vienna Dr.phil. (botany) 1894; Pv.Dz. 1895. Prague ao.Prof. botany 1896–1902; o.Prof. 1909–21. Made many important contributions to plant biochemistry and physiology (see TEXT).

CZERNY, ADALBERT [1863–1941]. Prague Dr.med. 1888; Pv.Dz. pediatrics 1893; worked in Hofmeister lab. 1894 on blood pathology. Breslau ao.Prof. ped. 1894–1906; o.Prof. 1906–19; Berlin o.Prof. ped. 1913–35. Published extensively on pediatric problems, especially those related to nutrition.

Dauwe, Ferdinand [1881–1948] (*BELG*). Strassburg 1905 (worked on adsorption of enzymes by colloids); Ghent Dr.med. 1914. Later Prof. medicine Ghent; specialized in gastroenterology.

Dietrich, M. (*RUSS*). Strassburg 1909 (worked on casein peptones). Dr.med.?; later activity not determined.

Dittrich, Paul [1859–1936]. Prague Dr.med. 1883; worked in Hofmeister lab. 1891 on formation of methemoglobin by poisons. Vienna Pv.Dz. 1891; Innsbruck ao.Prof. legal med. 1892–93; Prague o.Prof. 1895–1929.

Ducceschi, Virgilio [1871–1952] (*ITAL*). Florence Dr.med. 1895; Strassburg 1902 (worked on aromatic groups in proteins). Successively Prof. physiology Cordoba (Argentina), Sassari, Pavia, Padua, Rome.

Ebbecke, Ulrich [1883–1960]. Strassburg 1907 (worked on urinary excretion of non-dialyzable substances); Kiel Dr.med. 1909. Göttingen Pv.Dz. physiol. 1913–21; ao.Prof. 1921–24. Bonn o.Prof. physiol. 1924–53.

Ehrmann, Rudolf [1879–1955]. Dr.med. Strassburg 1903 (worked on peroxy acids). Berlin Pv.Dz. med. 1912–15; ao.Prof. 1915–35; chief physician municipal hospital.

Elias, Herbert [1885–1975]. Vienna Dr.med. 1909; Strassburg 1913 (worked on muscle metabolism). Vienna ao.Prof. medicine 1929–33; hospital physician 1933–38. Went to U.S.; with N.Y. Medical College 1939–60.

ELLINGER, ALEXANDER [1870–1923]. Berlin Dr.phil. 1892 (Marckwald); Strassburg 1895–97 (worked with Spiro on blood coagulation). Königsberg Dr.med. 1898; Pv.Dz. med. chem. and pharmacol. 1899–1906; ao.Prof. 1906–11; o.Prof. 1911–14. Frankfurt a.M.

o.Prof. pharmacol. 1914–23. A gifted experimenter, who made significant contributions in biochemistry and pharmacology (see TEXT).

EMBDEN, GUSTAV [1874–1933]. Strassburg Dr.med. 1899; with Hofmeister 1899–1903 (worked on glucuronides, albumoses, adrenaline, glucose formation in perfused liver). Frankfurt a.M. head of physiol. chem. institute 1906–08; ao.Prof. 1909–14; o.Prof. physiology 1914–33. One of the leading biochemists of his time; made outstanding contributions in studies on the metabolic formation of ketone bodies, the intermediary metabolism of amino acids and fatty acids, and on the biochemistry of muscular contraction (see TEXT).

Emerson, Robert Leonard [1872–1951] (*US*). Harvard M.D. 1900; Strassburg 1901–02 (worked on formation of tyramine in pancreatic digestion). Later in private practice, specialized in forensic medicine.

Eppinger, Hans [1879–1946]. Graz Dr.med. 1903; Strassburg 1905 (worked on formation of allantoin and urea). Vienna Pv.Dz. med. 1909–18; ao.Prof. 1918–26; o.Prof. med. Freiburg i.B. 1926–30; Köln 1930–33; Vienna 1933–45. Stated to have participated in medical experiments on inmates of Dachau concentration camp; committed suicide upon being summoned to appear before Nuremberg war crimes tribunal.

Falk, Fritz [1878–1912]. Berlin Dr.med. 1896; Strassburg 1908 (worked on cephalin). Later physician at Vienna medical clinic.

Filosofov, Petr Ivanovich [1879–1935] (*RUSS*). St. Petersburg Dr.med. 1905; Strassburg 1910 (worked on formation of uraminic acids). Later head of hospital in Leningrad.

Fränkel, Sigmund [1868–1939]. Vienna Dr.med. 1892; Pv.Dz. med. chem. 1896; Strassburg 1903 (worked on histidine). Vienna ao.Prof. med. chem. 1916–39 (head of laboratory of Spiegler Foundation).

Frank, Armando [1885–1951]. Erlangen Dr.med. 1910; Strassburg 1913 (worked on liver triglycerides and phosphatides). Chief physician pediatric clinic Leipzig 1920–45.

Freise, Eduard [1882–1921]. Göttingen Dr.med. 1906; Strassburg 1913 (worked on CO_2 formation in liver). Leipzig Pv.Dz. pediatrics 1918–20; head of children's hospital in Essen 1920–21.

FREUDENBERG, ERNST [1884–1967]. Munich Dr.med. 1910; Strassburg 1912 (worked on lipid metabolism). Heidelberg Pv.Dz. pediatrics 1917; o.Prof. Marburg 1922–37; Basle 1938–54. Made important contributions to biochemical aspects of pediatrics.

FRIEDMANN, ERNST [1877–1956]. Strassburg Dr.phil. 1902; Dr.med. 1905; Pv.Dz. physiol. chem. 1906; asst. of Hofmeister 1905–07 (worked on constitution of cystine, sulfur metabolism, adrenaline). Berlin ao.Prof. med. chem. Charité Hospital 1907–30. Went to England; at Cambridge 1932–56.

FÜRTH, OTTO von [1867–1938]. Vienna Dr.med. 1894; with Hofmeister in Prague and Strassburg 1894–1905; Pv.Dz. physiol. chem. 1899 (worked on muscle proteins, adrenaline, tyrosinase; published 21 papers). Vienna ao.Prof. 1906–17; o.Prof. physiol. 1917–29; med. chem. 1929–38. Published extensively on many topics in biochemistry (see TEXT).

Fuld, Ernst [1873–1955]. Strassburg Dr.med. 1895; with Hofmeister 1900–02 (worked on blood coagulation, reaction of proteins with metaphosphoric acid). Later in private and hospital medical practice in Berlin, specialized in gastrointestinal disorders; published on enzymes. Went to U.S. in 1934; in private medical practice New York City.

Githens, Thomas Stotesbury [1878–1966] (US). Univ. Penna. M.D. 1899; Strassburg 1904 (worked on effect of diet and blood loss on composition of plasma). At Rockefeller Institute 1910–19; Mulford Co. 1919–32; Sharp & Dohme Co. 1933–50.

Glässner, Karl (Charles) [1876–1944]. Prague Dr.med. 1900; Strassburg 1902 (worked on digestive enzymes, kynurenic acid). Vienna Pv.Dz. med. 1909; ao.Prof. 1923–38; chief physician at Vienna hospitals. Went to U.S. 1939; in private medical practice (gastroenterology).

Goldschmidt, Franz [1869–?]. Strassburg Dr.med. 1898 (worked on action of acids on proteins). Later activity not determined; probably died before 1920.

Goldschmidt, Max [1884–1972]. Strassburg Dr.med. 1910 (worked on chemistry of adrenal gland). Leipzig Pv.Dz. ophthalmology 1916–22; ao.Prof. 1922–; went to U.S. ca. 1933; private medical practice in New York City.

Goodman, Edward Harris [1880–1939] (US). Univ. Penna. M.D. 1902; Strassburg 1906 (worked on influence of diet on metabolism of cholesterol and bile acids). Later in private and hospital medical practice in Philadelphia.

Granström, Eduard Andreevich [1879–?] (RUSS). Dr.med.; Strassburg 1908 (worked on metabolism of glyoxylic acid). Later published on tuberculosis.

Gross, Alfred [1876–1904]. Strassburg Dr.med. 1899 (worked on ovovitellin); Kiel Pv.Dz. medicine 1903.

Gümbel, Theodor [1879–1938]. Strassburg Dr.med. 1903 (worked on chloroma). Later in private medical practice in Berlin.

Haake, Bruno [1874–1942]. Strassburg 1902 (worked with Spiro on diuretic action of salt solutions); Leipzig Dr.med. 1905. Later in private medical practice in Berlin.

Haas, Georg [1886–1971]. Freiburg i.B. Dr.med. 1911; Strassburg 1912–16 (worked on metabolism of glyoxylic acid and glycine). Giessen Pv.Dz. medicine 1916–24; ao.Prof. 1924–50; o.Prof. 1950–54.

Halsey, John Taylor [1870–1951] (*US*). Columbia Univ. M.D. 1893; Strassburg 1898 (worked on urea formation). Tulane Prof. pharmacology 1904–40.

Hanssen, Olav Mikal [1878–1965] (*NOR*). Stavanger Dr.med.; Strassburg 1908 (worked on amyloid, CO_2 formation in organ minces). Chief physician Bergen hospital 1909–48.

Hausmann, Walther [1877–1938]. Strassburg 1899 (worked on analysis of proteins). Vienna Dr.med. 1901; Pv.Dz. pharmacol. 1909–20; ao.Prof. 1920–38 (head of state institute of photobiology and photopathology).

Hebting, Josef [1876–1932]. Freiburg i.B. Dr.med. 1905; Strassburg 1910–14 (worked on chondroitin sulfate). Later in private medical practice in Freiburg i.B.

Heller, Friedrich [1883–1963]. Berlin Dr.med. 1911; Strassburg 1917 (no publication?). Later in private medical practice in Berlin.

HENDERSON, LAWRENCE JOSEPH [1878–1942] (*US*). Harvard M.D. 1902; Strassburg 1902–04, 1908 (worked on analysis of hemoglobin, acid-base equilibria in blood). At Harvard 1904–42 (Prof. biol. chem. 1919–42). Made important contributions to study of blood chemistry; wrote extensively on philosophical aspects of biology (see TEXT).

Hildebrandt, Paul [1877–?]. Berlin Dr.med. 1900; Strassburg 1904 (worked on milk formation). Later in private medical practice in Erfurt.

Hirsch, Rahel [1870–1953]. Strassburg Dr.med. 1903 (worked on glycolysis in liver). Berlin Charité Hospital 1903–13; ao.Prof. med. 1913; then in private medical practice 1913–38. Went to England in 1938. In 1905–07 discovered passage of food particles through lymphatic system.

Hoesslin, Heinrich von [1878–1955]. Munich Dr.med. 1902; Strassburg 1906 (worked on metabolism of choline). Halle Pv.Dz. med. 1909–13; ao.Prof. 1913–16. Chief physician Ziethen Hospital in Berlin 1916–. Published many papers on clinical topics.

Hohlweg, Hermann [1879–1941]. Munich Dr.med. 1903; Strassburg 1908 (worked on blood analysis, urochrome). Giessen Pv.Dz. med. 1910–13; ao.Prof. 1913–15; then chief physician in hospitals in Duisburg 1915–25 and Köln 1925–41.

Inada, Ryokichi [1874–1950] (*JAP*). Tokyo Dr.med. 1900; Strassburg 1906 (worked on metabolism of glyoxylic acid). Prof. medicine Kyushu 1906–18; Tokyo 1918–. Made valuable contributions in bacteriology.

Isaac, Simon [1881–1942]. Strassburg Dr.med. 1910 (worked on formation of purines during autolysis). Frankfurt a.M. Pv.Dz. med. 1916–21; ao.Prof. 1921– (Head of Jewish Hospital).

Ishimori, Kuniomi [1874–1955] (*JAP*). Strassburg 1910 (worked on metabolism of glycogen); Tokyo Dr.med. 1916. Kyoto Prof. physiol. and physiol. chem. 1912–20.

Jacoby, Martin [1872–1941]. Berlin Dr.med. 1895; Strassburg 1900 (worked on aldehyde oxidase, cleavage of proteins by liver tissue; autolysis). Heidelberg Pv.Dz. pharmacol. 1901–06; tit. Prof. 1906; Berlin head of biochem. divn. Moabit Hospital 1907–33. Published extensively on enzymes and other biochemical topics.

Jochem, Emil [1873–1943]. Strassburg 1900 (worked on hydrolysis of proteins, deamination of amino acids); Würzburg Dr.phil. (org. chem.) 1901. Later became pharmacist in Metz.

Kauder, Gustav. Prague Dr.med. 1886 (worked on serum proteins). Later activity not determined.

Kelly, Agnes [1875–?] (*AUSTRALIA*). Munich Dr.phil. 1901; Strassburg 1904 (worked on ethereal sulfates). Later activity not determined.

KNOOP, FRANZ [1875–1946]. Freiburg i.B. Dr.med. 1900; Strassburg 1901–04 (worked on albumoses, metabolic oxidation of fatty acids). Freiburg i.B. Pv.Dz. 1904–08; ao.Prof. 1908–20; o.Prof. 20–28. Tübingen o.Prof. physiol. chem. 1928–45 (see TEXT).

Kondo, Kura [1876–?] (*JAP*). Tokyo Dr.med. 1903; Strassburg 1910 (worked on chondroitin sulfate, urinary excretion of organic phosphates). Tokyo chief physician railway hospital 1920–?

Koppel, Max [1890–1916]. Strassburg Dr.med. 1914 (worked with Spiro on buffers in biological fluids). Died in action World War I.

KRAUS, FRIEDRICH [1858–1936]. Prague Dr.med. 1882; with Hofmeister 1882–1885 (worked on post-mortem changes in tissues, effect of erythrocyte breakdown on alkalinity of blood); Pv.Dz. med. 1888. o.Prof. medicine Graz 1894–1902; Berlin (Charité Hospital) 1902–26. One of the leading German clinicians of his time.

Krieger, Hans Theodor [1874–?]. Strassburg Dr.med. 1899 (worked on crystalline proteins). Later activity not determined; probably died before 1920.

Kuraev, Dmitri Ivanovich [1869–1908] (*RUSS*). St. Petersburg Dr.med. 1896; Strassburg 1898 (worked on iodination of crystalline serum albumin and ovalbumin). Prof. physiol. chem. Kharkov 1902–08; published on plastein, protamines, iodoproteins.

Landolt, Hans Robert Georg [1865–1932]. Kiel Dr.med. 1886; Strassburg 1899 (worked on melanin of optic membranes). Later ao.Prof. ophthalmology Frankfurt a.M.

Lang, Sandor. Budapest Dr.med.; Prague 1894–96; Strassburg 1896–1904 (worked on metabolism of acetonitrile, cyanide antidotes, metabolism of sulfur and amino compounds). Later at Institute of Physiology Budapest.

Langer, Josef [1866–1937]. Prague Dr.med. 1892 (worked on bee venom); Pv.Dz. pediatrics 1903–06. O.Prof. Graz 1906–14; Prague 1915–36.

Langstein, Leo [1866–1933]. Strassburg Dr.phil. 1902 (worked on carbohydrate of crystalline ovalbumin, coagulable component of egg white, kynurenic acid, end products of pepsin action; 7 papers). See Appendix 6.

Lewith, Siegmund. Prague Dr.med. 1887 (worked on salting out of serum proteins). No further chemical publications 1900–36; later activity not determined.

Liebermeister, Gustav [1879–1943]. Tübingen Dr.med. 1902; Strassburg 1906 (worked on serum nucleoprotein). Later in private medical practice; specialized in tuberculosis.

Lieblein, Victor [1869–1939]. Prague Dr.med. 1893 (worked on nitrogen excretion in liver disease); Pv.Dz. surgery 1903; ao.Prof. 1907–39.

Limbeck, Rudolf von [1861–1900]. Prague Dr.med. 1884; with Hofmeister 1888–92 (worked on diuretic action and toxicity of salts); Pv.Dz. med. 1893. Vienna chief physician Rudolf Hospital 1894–1900 (o.Prof. med. 1898–1900).

Löffler, Karl Wilhelm [1887–1972] (SWITZ). Basle Dr.med. 1911; Strassburg 1912–13 (worked on urea formation in liver). Basle Pv.Dz. med. 1917–21; Zurich ao.Prof. 1921–37; o.Prof. 1937–57.

LOEWE, WALTER SIEGFRIED [1884–1963]. Strassburg Dr.med. 1908 (worked on action of pepsin on casein and serum globulin, tetanus toxin). Göttingen Pv.Dz. pharmacol. 1912–18; ao.Prof. 1918–21; Dorpat o.Prof. 1921–28. Director lab. municipal hospital Mannheim 1928–34. Went to U.S.; at Cornell Med. College 1936–46, Univ. Utah 1946–63 (see TEXT).

LOEWI, OTTO [1873–1961]. Strassburg Dr.med. 1896; with Hofmeister 1897 (worked on urea formation in liver). Marburg Pv.Dz. pharmacol. 1900–05; Vienna ao.Prof. 1905–09; Graz o.Prof. 1909–38. Went to U.S.; at New York Univ. College of Med. 1940–61. Discovered action of acetyl choline and made other important contributions in physiology and pharmacology (see TEXT).

Lopez-Suarez, Juan [1884–?] (SPAIN). Madrid Dr.med. 1912; Strassburg 1912–13 (worked on gastric HCl production, mucin). Later Prof. public health Madrid.

Lotmar, Fritz [1878–1964]. Strassburg Dr.med. 1904 (worked on albumoses from crystalline serum albumin). Later in private medical practice in Berne.

Luzzatto, Riccardo [1876–1922] (ITAL). Padua Dr.med. 1898; Strassburg 1905–06 (worked on pentosuria, metabolic oxidation of fatty acids). Later Prof. pharmacology Modena.

Maas, Otto [1871–1942?]. Strassburg Dr.med. 1898; with Hofmeister 1900 (worked on products of cleavage of proteins by alkali). Later in neurological divn. Moabit Hospital Berlin; died in Nazi concentration camp near Minsk.

Magnus-Alsleben, Ernst [1879–1936]. Strassburg Dr.med. 1903 (worked on adenomyomas of pylorus). Basle Pv.Dz. med. 1909; Würzburg ao.Prof. 1915–28; o.Prof. 1928–33; after dismissal went to Turkey and died soon afterward.

MAGNUS-LEVY, ADOLF [1865–1955]. Heidelberg Dr.med. 1890; Erlangen Dr.phil. 1893; Strassburg 1897–1901 (worked in Hofmeister lab. on Bence Jones protein). Berlin chief physician at municipal hospital 1910–22; at Charité Hospital 1922–40. Went to U.S. (see TEXT).

Mancini, Stefano (*ITAL*). Dr.med.; Strassburg 1908–09 (worked on urochrome, blood chemistry). Later chief physician Riuniti Hospital Livorno.

Meyer, Hans [1877–1964]. Kiel Dr.med. 1902; Strassburg 1906 (worked on blood analysis). Kiel Pv.Dz. radiology 1911; director X-ray divn. municipal hospital Bremen 1920–46; hon.Prof. Marburg 1946–64. A leading German radiologist.

Meyer, Kurt [1882–1942?]. Strassburg Dr.med. 1905 (worked on diffusion in gels). Later director bacteriol. divn. Virchow Hospital Berlin; died in Majdanek concentration camp.

Meyer, Max [1890–1954]. Strassburg Dr.med. 1914 (worked on analysis of blood of epileptics). Würzburg Pv.Dz. otolaryngol. 1923–27; ao.Prof. 1927–35; o.Prof. Ankara 1935–41; Teheran 1941–47.

Mochizuki, Jun-Ichi [1859–?] (*JAP*). Tokyo Dr.med. 1891; Strassburg 1902 (worked on cleavage of proteins by trypsin). Later in hospital and private medical practice in Kyoto.

Morawitz, Paul [1879–1936]. Jena Dr.med. 1901; Strassburg 1904 (worked on blood coagulation). Heidelberg Pv.Dz. med. 1907; o.Prof. Greifswald 1913–21; Würzburg 1921–26; Leipzig 1926–36. Published extensively on clinical subjects, especially hematology.

Münzer, Egmont [1865–1924]. Prague Dr.med. 1887; Pv.Dz. med. 1892; with Hofmeister 1895 (worked on physiological action of salts); ao.Prof. med. 1907–24.

Oseki, Sakaye [1881–?] (*JAP*). Osaka Dr.med. 1906; Strassburg 1914 (worked on nutritional deficiencies). Later in hospital and private medical practice in Osaka.

Oswald, Adolf [1870–1956] (*SWITZ*). Zurich Dr.phil. (zool.) 1893; Freiburg i.B. Dr.med. 1897; Strassburg 1897–99 (worked on thyroglobulin). Zurich hon.Prof. physiol. chem. 1919–. Continued the work of Eugen Baumann on thyroxine; made important contributions to thyroid endocrinology.

PARNAS, JACOB KAROL [1884–1949] (*POL*). Munich Dr.phil. 1908 (Willstätter); Strassburg 1908–15 (ao.Prof. physiol. chem. 1913; worked on aldehyde mutase, metabolism of lactic acid and glucose, cephalin, muscle biochemistry; 8 papers). Prof. physiol. chem. Warsaw 1916–19; Lwow 1920–41. A leading contributor to study of muscle biochemistry (see TEXT).

Pascucci, Olinto (*ITAL*). Dr.med.; Strassburg 1905–06 (worked on hemolysis). Later activity not determined.

PAULI (PASCHELES), WOLFGANG [1869–1955]. Prague Dr.med. 1893 (worked on metabolic conversion of cyano compounds). Vienna Pv.Dz. med. 1899–1913; ao.Prof. 1913–22; o.Prof. colloid chem. 1922–39; then lived in Zurich. Published extensively on colloid chemistry of proteins (see TEXT).

Pemsel, Wilhelm [1873–?]. Heidelberg Dr.phil. 1888 (Bamberger); Strassburg 1898 (worked with Spiro on acid-base equilibria of blood and proteins). No further chemical publications 1900–39; later activity not determined.

Petry, Eugen [1873–1945]. Graz Dr.med.; Strassburg 1900 (worked on excretion of sulfur compounds). Later became radiologist in Graz; published on metabolic effects of X-rays.

Pfaundler, Meinhard von [1872–1947]. Graz Dr.med. 1896; Pv.Dz. pediatrics 1900; Strassburg 1900 (worked on determination of urinary amino-nitrogen, products of action of pepsin); ao.Prof. ped. Graz 1902–06; Munich 1906–11; o.Prof. 1911–39.

Pfeiffer, Wilhelm [1879–1937]. Tübingen Dr.med. 1903; Strassburg 1906 (worked on metabolism of uric acid). Later in private medical practice (otolaryngology) in Frankfurt a.M.

Philippson, Paula [1874–1949]. Strassburg 1902 (worked on dialysis of biological fluids); Breslau Dr.med. 1904. In 1907 set up private medical practice (pediatrics) in Frankfurt a.M.; later became Greek scholar in Basle.

Picard, Martin [1879–1945]. Strassburg 1911 (worked on determination of CO); Munich TH Dr.phil. 1911. Later ao.Prof. chem. technol. Charlottenburg TH.

PICK, ERNST PETER [1872–1960]. Prague Dr.med. 1896; Strassburg 1897–99 (worked on albumoses and peptones, blood coagulation). Vienna Pv.Dz. pharmacol. 1904–17; ao.Prof. 1917–24; o.Prof. 1924–38. Went to U.S.; clin. Prof. Columbia Univ. 1939–60. Made valuable contributions to immunology, protein chemistry and pharmacology.

Pick, Friedrich Gottfried [1867–1926]. Prague Dr.med. 1890 (worked on role of liver in carbohydrate metabolism); Pv.Dz. med. 1895; ao-.Prof. otolaryngol. 1909–26. Also wrote about history of medicine.

POHL, JULIUS [1861–1942]. Prague Dr.med. 1884; Pv.Dz. pharmcol. 1892; asst. of Hofmeister 1894–96 (worked on serum proteins, metabolic oxidation of alcohols and fatty acid; 15 papers); o.Prof. 1897–1911. Breslau o.Prof. pharmacol. 1911–28 (see TEXT).

Pollak, Leo [1878–1946]. Prague Dr.med. 1903; Strassburg 1905–06 (worked on pancreatic trypsin, oxidation of glycylglycine). Vienna Pv.Dz. med. 1914–32; ao.Prof. 32–38. Went to England 1938; with White Lodge Hosp. Newmarket 1943–46.

Pons, Charles [1876–1952] (*BELG*). Ghent Dr.med. 1903; Strassburg 1907 (worked on excretion of chondroitin sulfate). Later in private medical practice in Ghent; also participated in Belgian medical work in the Congo.

Porges, Otto [1879–1968]. Prague Dr.med. 1903; Strassburg 1903 (worked with Spiro on serum globulins). Vienna Pv.Dz. med. 1910–20; ao.Prof. 1920–38. Went to U.S. 1938; with Northwestern Univ. Med. School 1941–68.

Przibram, Hans Leo [1874–1944]. Vienna Dr.phil. 1899; Strassburg 1900–02 (worked on muscle proteins). Vienna Pv.Dz. exper. biol. 1903–13; ao.Prof. 1913–. Died in Theresienstadt concentration camp.

RAPER, HENRY STANLEY [1882–1951] (*UK*). Strassburg 1906–07 (worked on peptones); Leeds M.D. 1910. Toronto Lecturer pathol. chem. 1910–23; Leeds Lecturer physiol. chem. 1913–23; Manchester Prof. physiol. 1923–46. Made valuable contributions to study of tyrosine metabolism and action of tyrosinase.

Raudnitz, Robert Wolf [1856–1921]. Prague Dr.med. 1881; with Hofmeister 1893 (worked on resorption of rare earths in intestinal tract); Pv.Dz. pediat. 1888–1906; ao.Prof. 1906–21.

Reach, Felix [1872–?]. Prague Dr.med. 1895; Strassburg 1904 (worked on gastric digestion). Vienna Pv.Dz. physiology 1909–20; ao.Prof. 1921–.

Reh, Alfred [1878–1959] (*FR*). Strassburg Dr.med. 1904 (worked on autolysis of lymph glands, paranuclein). Later in private medical practice (pediatrics) in Strassburg; also wrote on history of Alsace.

Reichel, Heinrich [1876–1943]. Vienna Dr.med. 1901; Strassburg 1905–06 (worked with Spiro on enzymes, milk coagulation). Vienna Pv.Dz. hygiene 1910–14; ao.Prof. 1914–33; Graz o.Prof. 1933–43.

Reiss, Emil [1878–1923]. Strassburg Dr.med. 1902; with Hofmeister 1904 (worked on refractive index of serum proteins). Frankfurt a.M. municipal hospital 1906–23 (chief physician 1910–23).

Renall, Montague Henry [1888–?] (*UK*). Strassburg 1913 (worked on cephalin). No further chemical publications 1914–36; later activity not determined.

Reye, Wilhelm [1871–1916]. Strassburg Dr.med. 1898 (worked on fibrinogen). Later with Eppendorf Hospital Hamburg.

Rogozinski, Felix [1879–1940] (*POL*). Dr.med.; Strassburg 1908 (worked on peptones). Later Prof. physiology Cracow; died in Sachsenhausen concentration camp.

Rosell, Max [1876–1938]. Strassburg Dr.med. 1901 (worked on intracellular enzymes). Later in private medical practice in Ballenstadt.

Rothera, Arthur Cecil Hamel [1880–1915] (*UK*). Strassburg 1904 (worked on linkages in proteins). Melbourne Lecturer biochem. 1906–15; died in active service World War I.

Rücker, Hans [1874–?]. Strassburg Dr.med. 1901 (worked on hematoporphyrin). Later in private medical practice in Aachen.

Rupprecht, Paul [1887–?]. Kiel Dr.med. 1911; Strassburg 1917 (no publication?). Later in private medical practice in Magdeburg.

Salomonsen, Knud Ejnar [1883–1950] (*SWED*). Dr.med.; Strassburg 1908 (worked on urochrome). Later specialized in military medicine.

Samuely, Franz [1879–1913]. Strassburg Dr.med. 1903 (worked on formation of melanins from proteins). See Appendix 6.

Sasaki, Takaoki [1878–1966] (*JAP*). Strassburg 1907 (worked on nondialyzable constituents of urine, benzoyl peptide of asparagine). See Appendix 6.

Savaré, Mario (*ITAL*). Strassburg 1907–08 (worked on placental enzymes, nucleoproteins). Later pharmaceutical chemist in Milan.

Schickelé, Gustave [1875–1927] (*FR*). Strassburg Dr.med. 1900; with Hofmeister 1911 (worked on internal secretion of ovary); at obs. and gyn. clinic 1906–1927 (Prof. 1919–27).

Schlesinger, Eugen [1869–?]. Strassburg Dr.med. 1893; with Hofmeister 1904 (worked on tissue autolysis). Later in private medical practice (pediatrics) in Frankfurt a.M.

Schloss, Ernst [1882–1918]. Strassburg Dr.med. 1906 (worked on metabolism of glyoxylic acid). Later in private and hospital practice (pediatrics) in Rummelsberg.

Schmidt-Nielsen, Sigval [1877–1956] (*NOR*). Basle Dr.phil. (chem.) 1901; Strassburg 1902–03 (worked on autolysis of fish muscle). With Norwegian dept. of health 1908–13; Prof. chem. Trondheim TH 1913–46.

Schneider, Hugo. Dr.med.; Strassburg 1902 (worked with von Fürth on tyrosinases); Pv.Dz. pediatrics 1902; later activity not determined.

Schroeder, Henry [1873–1945]. Bonn Dr.phil. (botany) 1904; Strassburg 1906 (worked on enzymes in plants). Bonn Pv.Dz. bot. 1907–11; Kiel 1911–22 (o.Prof. 1921–22); o.Prof. Hohenheim 1922–39.

Schrumpf, Pierre [1882–?] (*FR*). Strassburg Dr.med. 1905 (worked on pepsin). Later in private medical practice (cardiology) in Mulhouse.

Schütz, Emil [1853–1941]. Prague Dr.med. 1877; with Hofmeister 1886–90 (worked on gastric motility). Vienna Pv.Dz. med. 1894–1912; ao.Prof. 1912–33.

Schütz, Julius [1876–1923]. Vienna Dr.med. 1899; Strassburg 1900 (worked on kinetics of pepsin action). Vienna Pv.Dz. med.; physician at several hospitals.

Schulz, Friedrich Nikolaus [1871–1956]. Bonn Dr.med. 1894; Strassburg 1897 (worked on hemoglobin). Jena ao.Prof. physiol. chem. 1900–23; o.Prof. 1923–36. Published on many biochemical topics.

Schwarz, Leopold [1877–1962]. Strassburg 1899–1903 (worked on reaction of proteins with aldehydes, formation of gastric hydrochloric acid); Berlin Dr.med. 1904. Hamburg Pv.Dz. hygiene 1919–23; ao. Prof. 1923–.

Schwarz, Oswald [1883–?]. Strassburg 1905 (worked on antipepsins); Vienna Dr.med. 1906; Pv.Dz. urology 1919–.

Schwarzschild, Moritz [1881–1933]. Strassburg Dr.med. 1904 (worked on trypsin). Later in private medical practice in Köln.

Seyderhelm, Richard [1888–1940]. Strassburg Dr.med. 1913; with Hofmeister 1918 (worked on pernicious anemia); Pv.Dz. med. 1918. In private medical practice in Frankfurt a.M.; after 1928 director of hospital clinic. Published extensively on topics in hematology.

Siegert, Ferdinand [1865–1946]. Strassburg Dr.med. 1889; Pv.Dz. pediatrics 1896–1904; with Hofmeister 1902 (worked on liver autolysis, lipid metabolism in infants). Köln o.Prof. pediatrics 1904–35.

SPIRO, KARL [1867–1932]. Strassburg 1894–1919 (with Hofmeister 1896–1919; Pv.Dz. physiol. chem. 1897–1912; ao.Prof. 1912–19; published 38 papers, largely on physical-chemical aspects of biochemistry). See TEXT and Appendix 6.

STEPP, WILHELM [1882–1963]. Munich Dr.med. 1907; Strassburg 1909 (worked on effect of lipid-free diets). Giessen med. clinic 1916–24; o.Prof. medicine Jena 1924–26; Breslau 1926–34; Munich 1934–45. A pioneer in the study of lipid-soluble vitamins.

Stolte, Karl [1881–1951]. Strassburg Dr.med. 1904 (worked on metabolism of amino acids and glucosamine). Berlin Pv.Dz. pediatrics 1913–16; o.Prof. Breslau 1916–45; Greifswald 1945–48; Rostock 1948–51.

Stookey, Lyman Brumbaugh [1878–1940] (US). Yale Ph.D. (physiol. chem.) 1902; Strassburg 1904–05 (worked on peptones). Prof. physiol. Univ. Southern California 1905–40; published extensively on various biochemical topics.

Strada, Ferdinando [1872–1969] (ITAL). Pavia Dr.med. 1897; Strassburg 1906 (worked on nucleoproteins). Prof. pathol. anat. Cordoba (Argentina) 1912–45.

Swain, Robert Eckles [1875–1961] (*US*). Strassburg 1902 (worked on skatole derivatives); Yale Ph.D. (physiol. chem.) 1904. Prof. chem. Stanford 1912–41.

Tachau, Paul [1887–1967]. Heidelberg Dr.med. 1912; Strassburg 1914 (worked on nutritional deficiencies). Later in private medical practice (dermatology); in 1935 went to U.S. and continued practice in Chicago.

Takagi (Takaki), Kenji [1881–1919] (*JAP*). Strassburg 1908 (worked on hemolysis; binding of tetanus toxin by brain tissue); Tokyo Dr.med. 1910. Chief physician Tokyo hospital 1910–19.

Tanaka, Masahiko [1876–?] (*JAP*). Nagasaki Dr.med. 1902; Strassburg 1911 (worked on' calcium metabolism). Later head of private hospital in Nagasaki.

Tauber, Siegfried. Prague 1895 (worked on effect of sulfur compounds on poisoning by phenol). No further chemical publications 1900–36; later activity not determined.

Tikhmenev, N. (*RUSS*). Strassburg 1914 (worked on protein storage in the liver). No further information found.

Umber, Friedrich [1871–1946]. Heidelberg Dr.med. 1896; Strassburg 1898–99 (worked on action of pepsin on crystalline ovalbumin and serum albumin, glycolysis). Berlin Pv.Dz. med. 1903. Chief physician at municipal hospital Hamburg 1903–11; Berlin 1912–41.

Urano, Fumihiko [1870–?] (*JAP*). Munich Dr.med. 1907; Strassburg 1907 (worked on creatine metabolism in muscle); later activity not determined.

Vogt, Hans [1874–1963]. Marburg Dr.med. 1898; Strassburg 1901 (worked with Spiro on experimental glycosuria). Marburg Pv.Dz. 1906; at hospitals in Breslau and Berlin 1907–24. Münster o.Prof. pediatrics 1924–43.

Wagner, Richard [1887–1974]. Vienna Dr.med. 1912; Strassburg 1912–14 (worked with Parnas on carbohydrate metabolism, adrenal lipids). Vienna Pv.Dz. pediatrics 1924–38. Went to U.S.; Prof. pediatrics Tufts School of Med. 1942–58. Published extensively on disorders of carbohydrate metabolism in children.

Wallerstein, Saly [1878–?]. Strassburg Dr.med. 1902 (worked on determination of globulins in biological fluids). Later in private medical practice in Frankfurt a.M.

Weil, Josef. Prague 1893 (worked with Czapek on effect of selenium and tellurium on animal organism). Later activity not determined.

Wieland, Hermann [1885–1929]. Strassburg Dr.med. 1909 (worked on lipid constituents of intestinal mucosa); Pv.Dz. pharmacol. 1915. o.Prof. Königsberg 1921–25; Heidelberg 1925–29.

Winternitz, Rudolf [1859–?]. Prague Dr.med. 1883; with Hofmeister 1889–93 (worked on uptake and release of mercury in the liver, ab-

sorption through the skin, locally sensitizing agents); Pv.Dz. dermatol. 1894–06; tit.Prof. 1906–29.

Würtz, Adolf [1873–?]. Strassburg Dr.med. 1898; with Hofmeister 1912
(worked on excretion of phosphate in urine and feces). Later in private medical practice in Strassburg, then in Stuttgart.

Yokota, Kotaro (*JAP*). Strassburg 1904 (worked on excretion of phloridzin). Later in chem divn. Virchow Hospital Berlin.

Zunz, Edgard [1874–1939] (*BELG*). Brussels Dr.med. 1897; Strassburg
1899 (worked on fractionation of cleavage products and kinetics of
pepsin action). Brussels Dr.phil. (physiol.) 1901; Prof. pharmacol.
1919–39. Published extensively on blood coagulation, anaphylactic
shock and other topics in biochemistry and pharmacology.

Abbreviations of Serial Publications

Adv. Carbohydrate Chem.	Advances in Carbohydrate Chemistry
Am. J. Physiol.	American Journal of Physiology
Am. Phil. Soc. Year Book	American Philosophical Society Year Book
Anat. Anz.	Anatomischer Anzeiger
Ann.	Annalen der Pharmacie (1832–39) Annalen der Chemie und Pharmacie (1840–73) Justus Liebigs Annalen der Chemie (1873–)
Ann. Acad. Roy. Belg.	Annuaire de l'Académie Royale des Sciences, des Lettres et des Beaux-Arts de Belgique
Ann. N.Y. Acad. Sci.	Annals of the New York Academy of Sciences
Ann. Rev. Biochem.	Annual Review of Biochemistry
Ann. Rev. Phys. Chem.	Annual Review of Physical Chemistry
Ann. Sci.	Annals of Science
Arch. exp. Path. Pharm.	Archiv für experimentelle Pathologie und Pharmakologie
Arch. ges. Physiol.	(Pflügers) Archiv für die gesamte Physiologie des Menschen und der Tiere

Arch. Int. Pharm. Ther.	Archives Internationales de Pharmaco-dynamie et de Thérapie
Arch. path. Anat.	(Virchows) Archiv für pathologische Anatomie und Physiologie und für klinische Medizin
Ber. Bot. Ges.	Berichte der deutschen botanischen Gesellschaft
Ber. chem. Ges.	Berichte der deutschen chemischen Gesellschaft
Biochem. J.	Biochemical Journal
Biochem. Z.	Biochemische Zeitschrift
Biog. Mem. FRS	Biographical Memoirs of Fellows of the Royal Society of London
Biog. Mem. NAS	Biographical Memoirs of the National Academy of Sciences U.S.A.
Brit. J. Hist. Sci.	British Journal of the History of Science
Bull. Hist. Med.	Bulletin of the History of Medicine
Bull. Johns Hopkins Hosp.	Bulletin of the Johns Hopkins Hospital
Bull. Soc. Chim.	Bulletin de la Société Chimique de France
Bull. Soc. Chim. Biol.	Bulletin de la Société de Chimie Biologique
Chem. Ber.	Chemische Berichte
Chem. Wkbl.	Chemisch Weekblad
Chem. Z.	Chemiker-Zeitung
Cold Spring Harbor Symp.	Cold Spring Harbor Symposia on Quantitative Biology
Comp. Rend.	Comptes Rendus Hebdomadaires des Séances de l'Académie des Sciences, Paris
DSB	Dictionary of Scientific Biography
Deutsche med. Wchschr.	Deutsche medizinische Wochenschrift
Erg. Physiol.	Ergebnisse der Physiologie
Helv. Chim. Acta	Helvetica Chimica Acta
Hist. Phil. Life Sci.	History and Philosophy of the Life Sciences
Hist. Sci.	History of Science
Hist. Stud. Phys. Sci.	Historical Studies in the Physical Sciences

J. Am. Chem. Soc.	Journal of the American Chemical Society
J. Chem. Ed.	Journal of Chemical Education
J. Chem. Soc.	Journal of the Chemical Society, London
J. Clin. Chem.	Journal of Clinical Chemistry and Clinical Biochemistry
J. Gen. Microbiol.	Journal of General Microbiology
J. Gen. Physiol.	Journal of General Physiology
J. Hist. Biol.	Journal of the History of Biology
J. Hist. Med.	Journal of the History of Medicine and Allied Sciences
J. prakt. Chem.	Journal für praktische Chemie
J. Soc. Chem. Ind.	Journal of the Society of Chemical Industry
Klin. Wchschr.	Klinische Wochenschrift
Med.-chem. Unt.	Medicinisch-chemische Untersuchungen
Med. Hist.	Medical History
Med.-hist. J.	Medizinhistorisches Journal
Med. Klin.	Medizinische Klinik
Mem. Acad. Roy. Belg.	Mémoires de l'Académie Royale de Belgique, Classe des Sciences
Münch. med. Wchschr.	Münchener medizinische Wochenschrift
NTM	Schriftenreihe für Geschichte der Naturwissenschaften, Technik und Medizin
Naturw. Rund.	Naturwissenschaftliche Rundschau
Naturwiss.	Naturwissenschaften
Notes Rec. RS	Notes and Records of the Royal Society, London
Obit. Not. FRS	Obituary Notices of Fellows of the Royal Society, London
Persp. Biol. Med.	Perspectives in Biology and Medicine
Physiol. Revs.	Physiological Reviews
Proc. Am. Phil. Soc.	Proceedings of the American Philosophical Society
Proc. Chem. Soc.	Proceedings of the Chemical Society, London

Proc. Roy. Soc.	Proceedings of the Royal Society, London
Proc. Welch Found.	Proceedings of the Robert A. Welch Foundation
Rev. Hist. Pharm.	Revue de l'Histoire de Pharmacie
Samml. chem. Vort.	Sammlung chemischer und chemisch-technischer Vorträge
Sitzber. Pr. Akad.	Sitzungsberichte der Preussischen Akademie der Wissenschaften zu Berlin
Sudhoffs Arch.	Sudhoffs Archiv für Geschichte der Medizin und der Naturwissenschaften
TIBS	Trends in Biochemical Sciences
Unt. Physiol. Heidelberg	Untersuchungen aus dem physiologischen Institut der Universität Heidelberg
Verhandl. Nat. Heidelberg	Verhandlungen des Naturhistorisch-medicinischen Vereins zu Heidelberg
Wiener klin. Wchschr.	Wiener klinische Wochenschrift
Yale J. Biol. Med.	Yale Journal of Biology and Medicine
Z. ang. Chem.	Zeitschrift für angewandte Chemie
Z. Biol.	Zeitschrift für Biologie
Z. Chem.	Zeitschrift für Chemie
Z. physiol. Chem.	(Hoppe-Seylers) Zeitschrift für physiologische Chemie

Bibliography

ABDERHALDEN, E. (1922). Franz Hofmeister. *Med. Klin.* 18:1167–1168.

ABEL, F. A. (1896). The history of the Royal College of Chemistry and reminiscences of Hofmann's professorship. *J. Chem. Soc.* 69:580–596.

ABERNETHY, J. L. (1967). Franz Hofmeister, the impact of his life and research on chemistry. *J. Chem. Ed.* 44:177–180.

ACKERKNECHT, E. H. (1953). *Rudolf Virchow, Doctor, Statesman, Anthropologist.* Madison, Wis.: University of Wisconsin Press.

ADAMS, R. (1952). William Albert Noyes. *Biog. Mem. NAS* 27:179–208.

ANDREWS, F. M., ed. (1979). *Scientific Productivity: The Effectiveness of Research Groups in Six Countries.* Cambridge University Press.

ANON. (1839). Das enträthelte Geheimnis der geistiger Gährung. *Ann.* 29:100–104.

ANON. (1890). *Verzeichnis der an der Kaiser-Wilhelms-Universität Strassburg vom Sommer-semester 1872 bis Ende 1884 erschienen Schriften.* Strassburg: Heitz.

ANON. (1909). Arthur Gamgee. *Nature* 80:194–196.

ANON. (1917). Thomas Lauder Brunton. *Proc. Roy. Soc.* B89:xliv–xlviii.

ANON. (1935). Herbert Eugene Smith. *National Cyclopedia of American Biography* 24:429–430.

ANON. (1938). Sigmund Pollitzer. *Archives of Dermatology and Syphilology* 37:499–503.

ANON. (1965). *Wie die Ersten Heilmittel nach Hoechst Kamen; Dokumente aus Hoechster Archiven No.8.* Frankfurt a.M.: Hoechst.

ANSCHÜTZ, R. (1929). *August Kekulé.* Berlin: Verlag Chemie.

ANSCHÜTZ, R. (1936). Ludwig Claisen, ein Gedenkblatt. *Ber. chem. Ges.* 69A:97–170.

ARMSTRONG, H. E. (1896). Notes on Hofmann's scientific work. *J. Chem. Soc.* 69:580–732.

ARRHENIUS, S. (1915). *Quantitative Laws in Biological Chemistry.* London: Bell.

ASHER, L. (1932). Karl Spiro. *Erg. Physiol.* 34:1–17.

BADDILEY, J. (1943). Arthur George Green. *Obit. Not. FRS* 4:251–270.

BAEYER, A. (1866). *Ueber den Kreislauf des Kohlenstoffs in der organischen Natur.* Berlin: Lüderitz.

BAEYER, A. (1870). Ueber die Wasserentziehung und ihre Bedeutung für das Pflanzenleben und die Gährung. *Ber. chem. Ges.* 3:63–75.

BAEYER, A. (1878). *Die chemische Synthese.* Munich: K. B. Akademie.

BAEYER, A. (1880). Über die Beziehungen der Zimtsäure zu der Indigogruppe. *Ber. chem. Ges.* 13:2254–2263.

BAEYER, A. (1885). Clemens Zimmermann. *Ber. chem. Ges.* 18(3):826–833.

BAEYER, A. (1892). Über die Konstitution des Benzols. Siebente Abhandlung. *Ann.* 269:145–206.

BAEYER, A. (1900). Zur Geschichte des Indigos. *Ber. chem. Ges.* 33 (Sonderheft): li–lxviii.

BAEYER, A. (1905). Erinnerungen aus meinem Leben 1835 bis 1905. In *Adolf von Baeyers Gesammelte Werke,* vol. 1, pp. vii–xx, xxviii–lv. Braunschweig: Vieweg.

BAEYER, A. and GEUL. A. (1880). *Das neue chemische Laboratorium der Akademie der Wissenschaften zu München.* Munich: Ackermann.

BAEYER, A. and KNOP, C. A. (1866). Untersuchungen über die Gruppe des Indigblaus. *Ann.* 140:1–38.

BAILAR, J. C., Jr. (1970). Moses Gomberg. *Biog. Mem. NAS* 41:141–173.

BARY, J. de (1866). Untersuchungen über die Verdauung von Eiweissstoffen. *Med.-chem. Unt.,* pp. 76–87.

BAUER, H. (1954). Paul Ehrlich's influence on chemistry and biochemistry. *Ann. N. Y. Acad. Sci.* 59:150–167.

BAUMANN, E. (1878). *Ueber die synthetische Prozesse im Thierkörper.* Berlin: Hirschwald.

BAUMANN, E. (1882). Ueber das von O. Loew und T. Bokorny erbrachten Nachweis der chemischen Ursache des Lebens. *Arch. ges. Physiol.* 29:400–421.

BAUMANN, E. and KOSSEL, A. (1895). Zur Erinnerung an Felix Hoppe-Seyler. *Z. physiol. Chem.* 21:i–lxi.

BAUMANN, H. U. (1972). *Über Mehrfachentdeckungen.* Münster: Institut für Geschichte der Medizin.

BAYLISS, L. E. (1961). William Maddock Bayliss 1860–1924, life and scientific work. *Persp. Biol. Med.* 4:460–479.

BAYLISS, W. M. (1908). *The Nature of Enzyme Action.* London: Longmans Green.

BAYLISS, W. M. (1924). *Principles of General Physiology.* Fourth edition. London: Longmans Green.

BEER, J J. (1959). *The Emergence of the German Dye Industry.* Urbana: University of Illinois Press.

BENEDUM, J. (1982). Georg Haas. In *Giessener Gelehrte in der ersten Hälfte des 20. Jahrhunderts* (H. G. Gundel et al., eds.), pp. 357–364. Marburg: Elwert.

BENTLEY, J (1978). The chemical department of the Royal School of Mines, its origins and development under A. W. Hofmann. *Ambix* 17:153–181.

BERGMANN, G. von (1936). Friedrich Kraus. *Deutsche med. Wchschr.* 62:482–484.

BERGMANN, M. (1930). Emil Fischer. In *Das Buch der Grossen Chemiker* (G. Bugge, ed.), vol. 2, pp. 408–420. Berlin: Verlag Chemie.

BERGMANN, M. and ZERVAS, L. (1932). Über eine neue Methode der Peptidsynthese. *Ber. chem. Ges.* 65:1192–1201.

BERL, E. (1931). *Liebig und die Bittersalz- und Salzsäurefabrik zu Salzhausen.* Berlin: Verlag Chemie.

BERLIN (1899). *Verzeichnis der Berliner Universitätsschriften 1810–1885.* Berlin: Weber.

BERNARD, C. (1856). *Mémoire sur le Pancreas.* Paris: Ballière.

BERNHARD, C. G., CRAWFORD, E. and SÖRBOM, P., eds. (1982). *Science, Technology and Society in the Time of Alfred Nobel.* Oxford: Pergamon.

BERNTHSEN, A. (1912). Heinrich Caro. *Ber. chem. Ges.* 45:1987–2042.

BERZELIUS, J. J. (1806). *Föreläsningar i Djurkemien.* Stockholm: Delen.

BICKEL, M. H. (1972). *Marceli Nencki 1847–1901.* Berne: Huber.

BLANGEY, L. (1933). Eugen Bamberger. *Helv. Chim. Acta* 16:644–685.

BLINKS, L. R. (1974). Winthrop John Vanleuven Osterhout. *Biog. Mem. NAS* 44:213–249.

BLUME, S. S. and SINCLAIR, R. (1973). Chemists at British universities: A study in the reward system in science. *American Sociological Review* 38:126–138.

BOCHALLI, R. (1948). Die Gesellschaft deutscher Naturforscher und Ärzte als Spiegelbild der Naturwissenschaften und der Medizin. *Naturw. Rund.* 1:275–278.

BOCK, K. D. (1972). *Strukturgeschichte der Assistentur.* Düsseldorf: Bertelmann.

BODENSTEIN, M. (1936). Robert Wilhelm Bunsens Stellung zur organischen Chemie. *Naturwiss.* 24:193–196.

BODLÄNDER, G. (1895). Moritz Traube. *Ber. chem. Ges.* 28(4):1085–1108.

BOK, T. A. (1974). The history of physical chemistry in Denmark. *Ann. Rev. Phys. Chem.* 25:1–10.

BONNER, T. N. (1963). *American Doctors and German Universities.* Lincoln, Neb.; University of Nebraska Press.

BORESCH, K. (1922). Friedrich Czapek. *Ber. bot. Ges.* 39:(97)–(114).

BORSCHEID, P. (1976). *Naturwissenschaft, Staat und Industrie in Baden 1848–1914.* Stuttgart: Klett.

BORUTTAU, H. (1922). *Emil du Bois-Reymond.* Vienna: Rikola.

BOYDE, T. R. C. (1980). *Foundation Stones of Biochemistry.* Hong Kong: Voile et Aviron.

BRAND, K. (1931). Der Einfluss von Justus von Liebig auf die Entwicklung der pharmazeutischen Chemie. *Archiv der Pharmazie* 269:477–505.

BRANDT, P. (1980). Le rôle des Koechlin dans la chimie textile au XIXe siècle. *Bulletin du Musée Historique des Sciences Humaines de Mulhouse* 87:87–100.

BREDERECK, H. (1957). Zur Entwicklung der Chemie der Kohlenhydrate und der Glucosidspaltenden Enzyme. *Angewandte Chemie* 69:405–412.

BRIGHTMAN, R. (1956). Perkin and the dyestuffs industry in Britain. *Nature* 177:815–821.

BROCK, W. H. (1967). An attempt to establish the first principles of the history of chemistry. *Hist. Sci.* 6:156–169.

BROCK, W. H. (1981). Liebigiana: old and new perspectives. *Hist. Sci.* 19:201–218.

BROCK, W. H., ed. (1984). *Justus von Liebig und August Wilhelm Hofmann in ihren Briefen 1841–1873.* Weinheim: Verlag Chemie.

BROCKE, B. von (1988). Von der Wissenschaftsverwaltung zur Wissenschaftspolitik. Friedrich Althoff (19.2.1839–20.10.1908). *Berichte zur Wissenschaftsgeschichte* 11:1–26.

BROOKE, J. H. (1968). Wöhler's urea and its vital force? A verdict from the chemists. *Ambix* 15:84–114.

BROWNE, C. A. (1926). *A Half-Century of Chemistry in America 1876–1926.* Easton, Pa.: American Chemical Society.

BROWNE, C. A. (1932). The history of chemical education in America between the years 1820 and 1870. *J. Chem. Ed.* 9:696–728.

BRÜCKE, E. T. (1928). *Ernst Brücke.* Vienna: Springer.

BRUNTON, T. L. (1896). On a probable glycolytic ferment in muscle, on raw meat and the treatment of diabetes. *Z. Biol.* 34:487–489.

BUCHNER, R. (1963). Die politische und geistige Vorstellungswelt Eduard Buchners. *Zeitschrift für Bayerische Landesgeschichte* 26:631–645.

BUCHS, H. (1977). Ernst Freudenberg. In *Marburger Gelehrte in der ersten Hälfte des 20. Jahrhunderts* (I. Schnack, ed.), pp. 64–74. Marburg: Elwert.

BURCHARDT, L. (1975). *Wissenschaftspolitik im Wilhelminischen Deutschland.* Göttingen: Vandenhoeck & Ruprecht.

BURCHARDT, L. (1978). Die Ausbildung des Chemikers im Kaiserreich. *Tradition* 25:31–50.

BURCHARDT, L. (1980). Professionalisierung oder Berufskonstruktion? Der Beispiel des Chemikers im Wilhelminischen Deutschland. *Geschichte und Gesellschaft* 6:326–348.

BUSCH, A. (1959). *Die Geschichte der Privatdozenten.* Stuttgart: Enke.

BUSCH, A. (1963). The vicissitudes of the *Privatdozent:* Breakdown and adaptation in the recruitment of the German university teacher. *Minerva* 1:319–341.

BUTENANDT, A. (1961). The Windaus memorial lecture. *Proc. Chem. Soc.*, pp. 131–138.

CALVIN, M. (1977). Gilbert Newton Lewis. *Proc. Welch Found.* 20:116–149.

CANNON, W. B. (1943). Lawrence Joseph Henderson. *Biog. Mem. NAS* 23:31–58.

CARO, H. (1892). Ueber die Entwicklung der Teerfarben Industrie. *Ber. chem. Ges.* 25(3):955–1105.

CARRIÈRE, J., ed. (1893). *Berzelius und Liebig: Ihre Briefe von 1831–1845.* Munich: Lehmann.

CHALLENGER, F. (1950). Frederick Stanley Kipping. *Obit. Not. FRS* 7:183–219.

CHARGAFF, E. (1978). *Heraclitean Fire.* New York: Rockefeller University Press.

CHEVREUL, M. E. (1823). *Recherches Chimiques sur les Corps Gras de l'Origine Animale.* Paris: Levrault.

CHEVREUL, M. E. (1824). *Considérations sur l'Analyse Organique et ses Applications.* Paris: Levrault.

CHICK, H., HUME, M. and MACFARLANE, M. (1971). *War on Disease: A History of the Lister Institute.* London: Deutsch.

CHITTENDEN, R. H. (1894). Some recent chemico-physiological discoveries regarding the cell. *American Naturalist* 28:97–117.

CHITTENDEN, R. H. (1930). *The Development of Physiological Chemistry in the United States.* New York: Chemical Catalog Co.

CHITTENDEN, R. H. (1937). Lafayette Benedict Mendel. *Biog. Mem. NAS* 18:123–155.

CHITTENDEN, R. H. (1945). *The First Twenty-five Years of the American Society of Biological Chemists.* New Haven: American Society of Biological Chemists.

CLAESSON, S. and PEDERSEN, K. O. (1972). The Svedberg. *Biog. Mem. FRS* 18:595–627.

CLARK, G. and KASTEN, F. H. (1983). *History of Staining.* Third edition. Baltimore: Williams and Wilkins.

CLARK, W. M. (1950). Leonor Michaelis 1875–1949. *Science* 111:55.

CLARK, W. M. (1960). *Oxidation-Reduction Potentials of Organic Systems.* Baltimore: Williams and Wilkins.

CLARKE, H. T. (1956). William John Gies. *Am. Phil. Soc. Year Book*, pp. 111–115.

CLARKE, H. T. (1958). Impressions of an organic chemist in biochemistry. *Ann. Rev. Biochem.* 27:1–14.

CLOW, A. and CLOW, N. (1952). *The Chemical Revolution.* London: Butterworth.

COHEN, E. (1912). *Jacobus Henricus van't Hoff. Sein Leben und Wirken.* Leipzig: Akademische Verlagsgesellschaft.

COHN, E. J. (1941). Introduction to the conference on crystalline protein molecules. *Ann. N. Y. Acad. Sci.* 41:79–86.

COHN, E. J. and EDSALL, J. T. (1943). *Proteins, Amino Acids and Peptides as Ions and Dipolar Ions.* New York: American Chemical Society.

COHNHEIM, J. (1863). Zur Kenntnis der zuckerbildenen Fermente. *Arch. path. Anat.* 28:241–261.

COHNHEIM, O. (1898). *Ueber die Resorption im Dünndarm und der Bauchhöhle.* Munich: Oldenburg.

COHNHEIM, O. (1901). Die Umwandlung des Eiweisses durch die Darmwand. *Z. physiol. Chem.* 33:451–465.

COLE, S. and COLE, J. R. (1967). Scientific output and recognition: A study on the reward system in science. *American Sociological Review* 32:377–390.

COLLINS, K. D. and WASHABAUGH, M. W. (1985). The Hofmeister effect and the behavior of water at interfaces. *Quarterly Review of Biophysics* 18:323–422.

COLMAN, J. and ALBERT, A. (1926). Siegmund Gabriel. *Ber. chem. Ges.* 59A:7–26.

CONRAD, W. (1985). *Justus von Liebig und sein Einfluss auf die Entwicklung des Chemiestudiums und des Chemieunterrichts an Hochschulen und Schulen.* Dr. phil. dissertation, Darmstadt Technische Hochschule.

CORI, C. F. (1983). Embden and the glycolytic pathway. *TIBS* 8:257–259.

CORNER, G. W. (1964). *A History of the Rockefeller Institute 1901–1953.* New York: Rockefeller University Press.

COSTA, A. B. (1962). *Michel Eugène Chevreul, Pioneer of Organic Chemistry.* Madison, Wis.: University of Wisconsin Press.

COSTA, A. B. (1971). Arthur Michael (1853–1942): The meeting of thermodynamics and organic chemistry. *J. Chem. Ed.* 48:243–246.

COSTA, A. B. (1974). Victor Meyer. *DSB* 9:354–358.

CRAIG, J. E. (1984). *Scholarship and Nation Building: The University of Strasbourg and Alsatian Society 1870–1939.* University of Chicago Press.

CRANEFIELD, P. F., ed. (1982). *Two Great Scientists of the Nineteenth Century.* Baltimore: Johns Hopkins University Press.

CRAWFORD, E. (1984a). *The Beginnings of the Nobel Institution: The Science Prizes 1901–1915.* Cambridge University Press.

CRAWFORD, E. (1984b). Arrhenius, the atomic hypothesis, and the Nobel prizes in physics and chemistry. *Isis* 75:503–522.

CRAWFORD, E. and FRIEDMAN, R. M. (1982). The prizes in physics and chemistry in the context of Swedish science. In BERNHARD et al. (1982), pp. 311–331.

CRESCITELLI, F. (1977). Friedrich Wilhelm Kühne. *Vision Research* 17:1317–1323.

CURTIUS, T. (1886). *Diazoverbindungen der Fettreihe, eine neue Klasse von organischen Körpern.* Munich: Straub.

CURTIUS, T. (1904). Verkettung von Aminosäuren. *J. prakt. chem.* [2]70:57–72.

CURTIUS, T. (1906). *Robert Bunsen als Lehrer in Heidelberg.* Heidelberg: Hörnung.

CURTIUS, T. and BREDT, J. (1912). Wilhelm Koenigs. *Ber. chem. Ges.* 45:3781–3830.

DALE, H. H. (1955). Edward Mellanby. *Biog. Mem. FRS* 1:193–222.

DALE, H. H. (1962). Otto Loewi. *Biog. Mem. FRS* 8:67–90.

DALE, H. H. (1963). Otto Loewi. *Erg. Physiol.* 52:1–19.

DANE, E. et al. (1942). Die Arbeiten H. Wielands. *Naturwiss.* 30:333–373.

D'ANS, J. (1952). Erinnerungen an J. H. van't Hoff 1904–1905. *Chem. Wkbl.* 48:640–643.

DANZER, K. (1972). *Robert W. Bunsen und Gustav R. Kirchhoff: die Begründer der Spectralanalyse.* Leipzig: Teubner.

DARAPSKY, A. (1930). Theodor Curtius zum Gedächtnis. *J. prakt. Chem.* NF125:1–22.

D'ARCY THOMPSON, R. (1974). *The Remarkable Gamgees.* Edinburgh: Ramsay Head.

DAVENPORT, H. W. (1982). Physiology 1850–1923: The view from Michigan. *Physiologist* 25 Suppl.:20–44.

DEBUS, H. (1901). *Erinnerungen an Robert Wilhelm Bunsen und seine wissenschaftliche Leistungen.* Cassel: Fisher.

DECHEND, H. von (1963). *Justus von Liebig: in eigenen Zeugnissen und solche seiner Zeitgenossen.* Second edition. Weinheim: Verlag Chemie.

DE KRUIF, P. (1923). Jacques Loeb, the mechanist. *Harper's Monthly* 146:182–190.

DE KRUIF, P. (1962). *The Sweeping Wind.* New York: Harcourt Brace.

DE LA RUE, W. (1868). Walter Crum. *J. Chem. Soc.* 21:xvii–xviii.

DELÉPINE, M. (1950). Notice sur la vie et les travaux de Ernest Fourneau. *Bull. Soc. Chim.* [5]17:953–982.

DE MILT, C. (1948). Carl Weltzien and the congress at Karlsruhe. *Chymia* 1:153–169.

DE MILT, C. (1951). Auguste Laurent, guide and inspiration of Gerhardt. *J. Chem. Ed.* 28:198–204.

DENNSTEDT, M. (1899). Die Entwicklung der organischen Elementaranalyse. *Samml. chem. Vort.* 4:1–114.

DEUTICKE, H. J. (1933). Gustav Embden. *Erg. Physiol.* 35:32–49.

DIECKMANN, W. (1915). Adolf von Baeyers Arbeiten über die Harnsäuregruppe. *Naturwiss.* 3:569–573.

DIEPGEN, P. (1960). *Unvollendente von Leben und Wirken.* Stuttgart: Thieme.

DIMROTH, O. (1915). Adolf von Baeyers Arbeiten über die Konstitution des Benzols. *Naturwiss.* 3:582–587.

DIVERS, E. (1907). Alexander William Williamson. *Proc. Roy. Soc.* A78:xxiv–xliv.

DIXON, M. and TATE, P. (1966). David Keilin. *J. Gen. Microbiol.* 45:159–185.

DOENECKE, D. (1983). Hundert Jahre Biochemie des Chromatins. *Naturw. Rund.* 36:432–438.

DOLBY, R. G. A. (1976). Debates over the theory of solution. *Hist Stud. Phys. Sci.* 7:297–404.

DOLMAN, C. E. (1971). Paul Ehrlich. *DSB* 4:295–305.

DONNAN, F. G. (1942). Herbert Freundlich. *Obit. Not. FRS* 4:27–50.

DONNAN, F. G. (1948). Ernst Julius Cohen. *Obit. Not. FRS* 5:667–687.

DOUNCE, A. L. and ALLEN, P. Z. (1988). Fifty years later: recollections of the early days of protein crystallization. *TIBS* 13:317–320.

DRABKIN, D. (1958). *Thudichum: Chemist of the Brain*. Philadelphia: University of Pennsylvania Press.

DRÖSSMAR, F. (1964). Das publizistische Wirken Justus von Liebigs. Dr.phil. dissertation Berlin.

DU BOIS-REYMOND, R. (1927). *Zwei Grosse Naturforscher des 19. Jahrhunderts*. Leipzig: Barth.

DUBOS, R. J. (1976). *The Professor, the Institute and DNA*. New York: Rockefeller University Press.

DUCLAUX, E. (1898–1899). *Traité de Microbiologie*. Paris: Masson.

DUDEN, P. and KAUFMANN, H. P. (1927). Ludwig Knorr zum Gedächtnis. *Ber. chem. Ges.* 60A:1–27.

DUDEN, P. and DECKER, P. (1928). Nachruf auf Carl Graebe. *Ber. chem. Ges.* 61A:9–46.

DUISBERG, C. (1896). The education of chemists. *J. Soc. Chem. Ind.* 15:427–432.

DUISBERG, C. (1919). Emil Fischer und die Industrie. *Ber. chem. Ges.* 52A:149–164.

DUISBERG, C. (1933). *Meine Lebenserinnerungen*. Leipzig: Reclam.

DU VIGNEAUD, V. (1952). *A Trail of Research*. Ithaca: Cornell University Press.

EBERHARD, A. (1838). Liebig als Apothekenvisitator etc. *Süddeutsche Apotheker-Zeitung* 78:866–868.

EDLBACHER, S. (1928). Albrecht Kossel zum Gedächtnis. *Z. physiol. Chem. 177:1–14.*

EDSALL, J. T. (1950). *University Laboratory of Physical Chemistry Related to Medicine and Public Health*. Cambridge, Mass.: Harvard University.

EDSALL, J. T. (1955). Edwin J. Cohn. *Erg. Physiol.* 48:23–48.

EDSALL, J. T. (1961). Edwin J. Cohn. *Biog. Mem. NAS* 35:47–84.

EDSALL, J. T. (1981). Edwin J. Cohn and the physical chemistry of proteins. *TIBS* 6:335–337.

EDSALL, J. T. (1982). George Scatchard, John G. Kirkwood and the electrical interactions of amino acids and proteins. *TIBS* 7:414–416.

EDSALL, J. T. (1985). Carbon dioxide transport in blood: Equilibrium between red cells and plasma. The work of D. D. Van Slyke and L. J. Hendeerson 1920–1928. *Hist. Phil. Life Sci.* 7:105–120.

EDSALL, J. T. and STOCKMAYER, W. H. (1980). George Scatchard. *Biog. Mem. NAS* 52:335–377.

EHRLICH, P. (1910). *Studies in Immunity.* Second edition. New York: Wiley.

ELDERFIELD, R. C. (1980). Walter Abraham Jacobs. *Biog. Mem. NAS* 51:247–278.

ELIAS, N., MARTINS, H. and WHITLEY, R. (1982). *Scientific Establishments and Hierarchies.* Dordrecht: Riedel.

ELIEL, E. L. (1975). Conformational analysis—the last 25 years. *J. Chem. Ed.* 52:762–767.

ELLINGER, P. (1924). Alexander Ellinger. *Erg. Physiol.* 23:139–179.

ELLIOTT, T. R. (1934). Walter Morley Fletcher. *Obit. Not. FRS* 1:153–163.

EMBDEN, G. (1922). Franz Hofmeister. *Klin. Wchschr.* 1:1974–1975.

ENGELHARDT, W. von and DECKER-HAUFF, H. (1963). *Quellen zur Gründungsgeschichte der Naturwissenschaftliche Fakultät in Tübingen 1859–1863.* Tübingen: Mohr (Paul Siebeck).

ENKVIST, T. (1972). *The History of Chemistry in Finland 1828–1918.* Helsinki: Finnish Academy of Sciences.

EPSTEIN, B. (1937). Professor Josef Langer. *Jahrbuch der Kinderheilkunde* 149:265–266.

EULER, H. von (1952). In van't Hoff's laboratory in Berlin 1899 and 1900. *Chem. Wkbl.* 48:644–645.

EULNER, H. H. (1970). *Die Entwicklung der medizinischen Spezialfächer an der Universitäten des deutschen Sprachgebietes.* Stuttgart: Enke.

EVE, A. S. and CREASEY, C. H. (1945). *Life and Work of John Tyndall.* London: Macmillan.

FARRAR, W. V. (1977). Edward Schunck F.R.S.: A pioneer in natural product chemistry. *Notes Rec. RS* 31:273–296.

FARRAR, W. V., FARRAR, K. R. and SCOTT, E. L. (1977). The Henrys of Manchester. Part 6. *Ambix* 24:1–26.

FELDBERG, W. (1970). Henry Hallett Dale. *Biog. Mem. FRS* 16:77–174.

FELDMAN, G. D. (1973). A German scientist between illusion and reality: Emil Fischer 1909–1919. In *Deutschland in der Weltpolitik des 19. und 20. Jahrhunderts* (I. Geiss and B. J. Wendt, eds.), pp. 341–362. Düsseldorf: Bertelsmann.

FERBER, C. von (1956). *Die Entwicklung des Lehrkörpers der deutschen Universitäten und Hochschulen.* Göttingen: Vandenhoek & Ruprecht.

FICHTER, F. (1911). Rudolf Fittig. *Ber. chem. Ges.* 44:1339–1401.

FIESER, L. F. (1975). Arthur Michael. *Biog. Mem. NAS* 46:331–366.

FINDLAY, A. (1953). Wilder Dwight Bancroft. *J. Chem. Soc.* pp. 2506–2514.

FINLEY, K. T. (1965). The synthesis of carbocyclic compounds. *J. Chem. Ed.* 42:536–540.

FISCHER, E. (1874). *Uber Fluorescein und Phthalein-orcin.* Bonn: Neusser.

FISCHER, E. (1893). Antrittsrede als Mitglied der Akademie der Wissenschaften in Berlin. *Sitzber. Pr. Akad.* pp. 632–636.

FISCHER, E. (1894). Einfluss der Konfiguration auf die Wirkung der Enzyme I. *Ber. chem. Ges.* 27:2985–2993.

FISCHER, E. (1895). Felix Hoppe-Seyler. *Ber. chem. Ges.* 23:2333–2336.

FISCHER, E. (1899). Synthesen in der Puringruppe. *Ber. chem. Ges.* 32:435–504.

FISCHER, E. (1901). Uber die Hydrolyse des Caseins durch Salzsäure. *Z. physiol. Chemie* 33:151–176.

FISCHER, E. (1902). Uber die Hydrolyse der Proteinstoffe. *Chem. Z.* 26:939–940.

FISCHER, E. (1905). Erinnerungen aus der Strassburger Studien Zeit 1872 bis 1875. In *Adolf von Baeyers Gesammelte Werke,* vol. 1, pp. xxi–xxvi. Braunschweig: Vieweg.

FISCHER, E. (1906). Untersuchungen über Aminosäuren, Polypeptide und Proteine *Ber. chem. Ges.* 39:530–610.

FISCHER, E. (1907a). Synthese von Polypeptiden XVII. *Ber. chem. Ges.* 40:1754–1767.

FISCHER, E. (1907b). Synthetical chemistry in its relation to biology. *J. Chem. Soc.* 91:1749–1765.

FISCHER, E. (1907c). Proteine und Polypeptide. *Z. ang. Chem.* 20:913–917.

FISCHER, E. (1913). Synthese von Depside, Flechtenstoffe und Gerbstoffe. *Ber. chem. Ges.* 46:3253–3288.

FISCHER, E. (1916). Isomerie der Polypeptide. *Sitzber. Pr. Akad.,* pp. 990–1008.

FISCHER, E. (1922). *Aus meinem Leben.* Berlin: Springer.

FISCHER, E. (1955). Justus von Liebig und Wilhelm Ostwald. *Naturw. Rund.* 8:49–53.

FISCHER, E. (1958). Meister, Lucius und Brüning, die Gründer der Farbwerke Hoechst AG. *Tradition* 3:65–78.

FISCHER, E. (1987). *Aus meinem Leben.* Reprint of 1922 edition, with additions by B. Witkop. Berlin: Springer.

FISCHER, E. and FOURNEAU. E. (1901). Über einige Derivate des Glykocolls. *Ber. chem. Ges.* 34:2868–2877.

FISCHER, E. and GUTH, M. (1901). *Der Neubau des 1. Chemischen Institutes der Universität Berlin.* Berlin: Hirschwald.

FISHER, N. W. (1974). Kekulé and organic classification. *Ambix* 21:29–52.

FITTIG, R. (1895). *Ziele und Erfolge wissenschaftlicher chemischer Forschung.* Strassburg: Trübner.

FLEMMING, H. W. (1965). Ludwig Baist, der Gründer der chemischen Fabrik Griesheim. *Tradition* Beiheft 4.

FLEMMING, H. W. (1967). *Dr. Sells Teerdistillation in Offenbach.* Dokumente aus Hoechster Archiven No. 26. Frankfurt a.M.: Hoechst.

FLEMMING, H. W. (1968). *Ludwig Knorr, Begründer Hoechster wissenschaftlicher Tradition.* Dokumente aus Hoechster Archiven No. 31. Frankfurt a.M.: Hoechst.

FLETCHER, W. M. (1926). John Newport Langley. In memoriam. *J. Physiol.* 61:1–15.

FLETCHER, W. M. and HOPKINS, F. G. (1907). Lactic acid in amphibian muscle. *J. Physiol.* 35:247–309.

FLEXNER, S. and FLEXNER, J. T. (1941). *William Henry Welch and the Heroic Age of American Medicine.* New York: Viking Press.

FLORKIN, M. (1943). *Léon Fredericq et les Débuts de la Physiologie en Belgique.* Brussels: Office de Publicité.

FLORKIN, M. (1960). *Naissance et Déviation dans l'Oeuvre de Théodore Schwann.* Paris: Hermann.

FLORKIN, M. (1972). *A History of Biochemistry. Part I. Protobiochemistry. Part II. From Protobiochemistry to Biochemistry.* Amsterdam: Elsevier.

FLORKIN, M. (1975). *A History of Biochemistry. Part III. History of the Identification of the Sources of Free Energy in Organisms.* Amsterdam: Elsevier.

FLORKIN, M. (1977). *A History of Biochemistry. Part IV. Early Studies on Bioenergetics.* Amsterdam: Elsevier.

FLORKIN, M. (1979). *A History of Biochemistry. Part V. The Unravelling of Biosynthetic Pathways.* Amsterdam: Elsevier.

FOLLEY, S. J. (1955). Henry Stanley Raper. *J. Chem. Soc.,* pp. 2987–2988.

FORSTER, M. O. (1920). Emil Fischer memorial lecture. *Trans. Chem. Soc.* 117:1157–1201.

FOURNEAU, J. P. (1987). Ernest Fourneau, fondateur de la chimie thérapeutique française: feuillets d'album. *Rev. Hist. Pharm.* 34:335–355.

FREDERICQ, L. and MASSART, L. (1908). Notice sur L. Errera. *Ann. Acad. Roy. Belg.* 74:131–279.

FREETH, F. A. (1957). Frederick George Donnan. *Biog. Mem. FRS* 3:23–39.

FREUDENBERG, K. (1963). Theodor Curtius. *Chem. Ber.* 96:i–xxv.

FREUDENBERG, K. (1966). Emil Fischer and his contribution to carbohydrate chemistry. *Adv. Carbohydrate Chem.* 21:1–38.

FREUDENBERG, K. (1967). Von Emil Fischer zur molekularen Konstitution der Cellulose und Stärke. *Chem. Ber.* 100:clxii-clxxviii.

FRIEDEL, C. (1885). Notice sur la vie et les travaux de Charles-Adolphe Wurtz. *Bull. Soc. Chim.* 43:i–lxxx.

FRIEDLÄNDER, P. (1915). Die Bedeutung der Baeyerschen Indigoarbeiten. *Naturwiss.* 3:573–576.

FRUTON, J. S. (1938). Protein structure and proteolytic enzymes. *Cold Spring Harbor Symp.* 6:50–55.

FRUTON, J. S. (1949). The synthesis of peptides. *Advances in Protein Chemistry* 6:1–82.

FRUTON, J. S. (1951). The place of biochemistry in the university. *Yale J. Biol. Med.* 23:305–310.

FRUTON, J. S. (1966). The Rockefeller Institute for Medical Research. *J. Hist. Med.* 21:71–77.

FRUTON, J. S. (1972). *Molecules and Life: Historical Essays on the Interplay of Chemistry and Biology.* New York: Wiley.

FRUTON, J. S. (1976). The emergence of biochemistry. *Science* 192:327–334.

FRUTON, J. S. (1977). Willstätter lectures on enzymes. *TIBS* 2:210–211.

FRUTON, J. S. (1979). Early theories of protein structure. *Ann. N. Y. Acad. Sci.* 325:1–18.

FRUTON, J. S. (1982a). The education of a biochemist. In *Of Oxygen, Fuels and Living Matter.* Part 2 (G. Semenza, Ed.), pp. 315–360. Chichester: Wiley.

FRUTON, J. S. (1982b). *A Bio-bibliography for the History of the Biochemical Sciences since 1800.* Philadelphia: American Philosophical Society.

FRUTON, J. S. (1982c). The carbobenzoxy method of peptide synthesis. *TIBS* 7:37–39.

FRUTON, J. S. (1982d). The interplay of chemistry and biology at the turn of the century. In BERNHARD et al. (1982), PP. 74–96.

FRUTON, J. S. (1985a). *A Supplement to a Bio-bibliography for the History of the Biochemical Sciences since 1800.* Philadelphia: American Philosophical Society.

FRUTON, J. S. (1985b). Contrasts in scientific style. Emil Fischer and Franz Hofmeister: Their research groups and their theory of protein structure. *Proc. Am. Phil. Soc.* 129:313–370.

FRUTON, J. S. (1988a). The Liebig research group—a reappraisal. *Proc. Am. Phil. Soc.* 132:1–66.

FRUTON, J. S. (1988b). Energy-rich bonds and enzymatic peptide synthesis. In *The Roots of Modern Biochemistry* (H. Kleinkauf. H. von Döhren and L. Jaenicke, eds.), pp. 165–180. Berlin: de Gruyter.

FÜRTH, O. von (1912). *Probleme der physiologischen und pathologischen Chemie.* Leipzig: Vogel.

FÜRTH, O. von (1919). Emil Fischer. *Wiener klin. Wchschr.* 32:828–829.

FUOSS, R. M. (1971). Charles August Kraus. *Biog. Mem. NAS* 42:119–159.

GAMGEE, A. (1877). On the photochemical processes in the retina. *Nature* 15:196, 477–478.

GAMGEE, A. (1893). *A Text-book of the Physiological Chemistry of the Animal Body.* London: Macmillan.

GAMGEE, A. (1895). The late Professor Hoppe-Seyler. *Nature* 52:575–576, 623–625.

GANSS, G. A. (1937). *Geschichte der pharmazeutischen Chemie an der Universität Göttingen.* Marburg: Euker.

GASILOV, V. B. (1977). Analyse der Interpretationen des Terminus "wissenschaftliche Schule". In MIKULINSKIJ et al. (1977), vol. 1, pp. 291–321.

GEDENKBUCH (1986). *Opfer der Verfolgung der Juden unter die Nationalsozialistische Gewaltschaft in Deutschland 1933–1945.* Koblenz: Bundesarchiv.

GEISON, G. L. (1973). John Newport Langley. *DSB* 8:14–19.

GEISON, G. L. (1978). *Michael Foster and the Cambridge School of Physiology: The Scientific Enterprise in Late Victorian Society.* Princeton University Press.

GEISON, G. L. (1981). Scientific change, emerging specialties and research schools. *Hist. Sci.* 19:20–40.

GEMMILL, C. L. (1966). Recollections of Professor Otto Meyerhof. *Medical College of Virginia Quarterly* 2:141–142.

GERBER, G. and SAUER, G. (1985). Die Entdeckung des Adenins durch Albrecht Kossel in Berlin—sein Leben und Wirken. *Charité-Annalen* NF 5:355–365.

GETMAN, F. H. (1940). *The Life of Ira Remsen.* Easton, Pa.: American Chemical Society.

GILLIS, J. (1960). Lettres d'Adolf Baeyer à son ami Jean Servais Stas. *Mem. Acad. Roy. Belg.* 32:no. 2.

GILLIS, J. (1966). Auguste Kekulé et son oeuvre, realisée à Gand de 1858 à 1867. *Mem. Acad. Roy. Belg.* 37:no. 1.

GOLDNER, M. G. (1955). Adolf Magnus-Levy. *Diabetes* 4:422–424.

GOOD, H. G. (1936). On the early history of Liebig's laboratory. *J. Chem. Ed.* 13:557–562.

GOODMAN, D. C. (1972). Chemistry and the two kingdoms of nature in the nineteenth century. *Med. Hist.* 16:113–130.

GRAEBE, C. (1915). Zum achtzigsten Geburtstag von Adolf von Baeyer. *Z. ang. Chem.* 28:433–437.

GREEN, A. G. (1902). The relative progress of the coal-tar industry in England and Germany during the past fifteen years. *Science* 15:7–13.

GREENAWAY, A. J. (1932). *The Life and Work of William Henry Perkin.* London: Chemical Society.

GRIMAUX, E. (1882). Synthèse des colloides azotés. *Bull. Soc. Chim.* [2]38:64–69.

GRIMAUX, E. and GERHARDT, C. (1900). *Charles Gerhardt, sa Vie, son Oeuvre, sa Correspondance.* Paris: Masson.

GRMEK, M. D. (1973). *Raisonnement Expérimental et Recherches Toxicologiques chez Claude Bernard.* Paris: Droz.

GROSSMANN, H. (1917). Justus von Liebig und die Engländer, eine zeitgemässige Betrachtung. *Chem. Z.* 41:429–431.

GROTE, L. R. (1923). *Die Medizin der Gegenwart in Selbstdarstellungen.* Vol. 2. Leipzig: Meinel.

GRUBER, G. B. (1958). Friedrich von Müller—zum Gedächtnis seines 100. Geburtstages. *Med. Klin.* 53:1589–1595.

GRUMBACH, A. (1956). Adolf Oswald-Honegger. *Vierteljahrschrift der Naturforschenden Gesellschaft in Zürich* 101:226–231.

GÜNTHER, R. (1980). Vertrauen auf die Weisheit einer hohen Staatsregierung. Die Berufung des Nobelpreisträgers Emil Fischer. *Würzburg-Heute* 29(3):21–22.

GUGGENHEIM, E. A. (1960). The Niels Bjerrum memorial lecture. *Proc. Chem. Soc.,* pp. 104–114.

GUNDEL, H. G. (1973). Liebig als Dekan der philosophischen Fakultät der Universität Giessen 1846 und 1851. *Giessener Universitätsblätter* 6:58–80.

GUSTIN, B. H. (1975). The Emergence of the German Chemical Profession 1790–1867. Ph.D. dissertation, University of Chicago.

GUTFREUND, H. (1976). Wilhelm Friedrich Kühne. *FEBS Letters* 62 Suppl., E1–E12.

HABER, L. F. (1958). *The Chemical Industry during the Nineteenth Century.* Oxford: Clarendon Press.

HAFNER, K. (1979). August Kekulé—the architect of chemistry, commemorating the 150th anniversary of his birth. *Angewandte Chemie (Int. Ed).* 18:641–651.

HAGGARD, H. (1944). Yandell Henderson. *Am. Phil. Soc. Year Book,* pp. 369–374.

HALDANE, J. B. S. (1930). *Enzymes.* London: Longmans Green.

HANNAWAY, O. (1976). The German model of chemical education: Ira Remsen at Johns Hopkins. *Ambix* 23:145–164.

HANSON, H. (1970). Emil Abderhalden. *Nova Acta Leopoldina* NS36:257–317.

HARDIE, D. W. F. (1955). The Muspratts and the British chemical industry. *Endeavour* 14:29–33.

HARINGTON, C. R. (1945). Max Bergmann. *J. Chem. Soc.*, pp. 716–718.

HARMS, F., CRIEGEE, R. and EBERT, L. (1941). Otto Dimroth. *Ber. chem. Ges.* 74A:1–23.

HARRIES, C. (1915). Adolf von Baeyer und sein Einfluss auf die Entwicklung der Chemie der hydroaromatischen Verbindungen und Terpenkörper. *Naturwiss.* 3:587–594.

HARRIES, C. (1917). Eduard Buchner. *Ber. chem. Ges.* 50:1843–1876.

HARRIES, C. (1919a). Emil Fischers wissenschaftliche Arbeiten. *Naturwiss.* 7:843–860.

HARRIES, C. (1919b). Emil Fischer. *Die Umschau* 23:609–613.

HARRIES, C. (1920). Oskar Piloty. *Ber. chem. Ges.* 53A:153–168.

HARTECK, P. (1960). Physical chemists in Berlin 1919–1933. *J. Chem. Ed.* 37:462–466.

HARTLEY, P. (1953). Henry Stanley Raper. *Obit. Not. FRS* 8:567–582.

HARTMANN, H. and LONGUET-HIGGINS, H. C. (1982). Erich Hückel. *Biog. Mem. FRS* 28:153–162.

HASTINGS, A. B. (1976). Donald Dexter Van Slyke. *Biog. Mem. NAS* 48:309–360.

HAUPT, H. (1934). Friedrich Bopp. *Hessische Biographie* 3:261–264.

HEIDELBERGER, M. (1969). Karl Landsteiner. *Biog. Mem. NAS* 40:177–210.

HEINIG, K. (1960). Das chemische Institut der Berliner Universität unter der Leitung von August Wilhelm Hofmann und Emil Fischer. In *Forschen und Wirken, Festschrift der 150-Jahrfeier der Humboldt-Universität zu Berlin 1810–1960*. Vol. 1, pp. 339–357. Berlin: VEB.

HELFERICH, F. (1969). Max Bergmann. *Chem. Ber.* 102:i–xxvi.

HELLER, J. and MOZOLOWSKI, W. (1958). Jakub Karol Parnas. *Postepy Biochemii* 4:5–65.

HENRICH, R. (1908). *Neuere theoretische Ansichten auf dem Gebiete der organischen Chemie*. Braunschweig: Vieweg.

HENRY, T. A. (1952). Ernest Fourneau. *J. Chem. Soc.*, pp. 261–266.

HERNECK, F. (1970). Emil Fischer als Mensch und Forscher. *Z. Chem.* 10:41–48.

HERRICK, J. B. (1949). *Memories of Eighty Years*. University of Chicago Press.

HERRIOTT, R. M. (1962). A biographical sketch of John Howard Northrop. *J. Gen. Physiol.* 45 Suppl.:1–16, 255–265.

HERRIOTT, R. M. (1978). Moses Kunitz. *Nature* 275:351–352.

HERRIOTT, R. M. (1983). John H. Northrop: The nature of enzymes and bacteriophage. *TIBS* 8:296–297.

HERRIOTT, R. M. (1989). Moses Kunitz. *Biog. Mem. NAS* 58:305–317.

HESSE, F. (1976). *Professor Dr.med. Julius Eugen Schlossberger (1819–1860).* Düsseldorf: Triltsch.

HEYNS, K. (1951). Emil Abderhalden. *Arch. ges. Physiol.* 253:229–237.

HICKEL, E. (1978). Der Apothekerberuf als Keimzelle naturwissenschaftlicher Berufe in Deutschland. *Med.-hist. J.* 13:259–276.

HICKEL, E. (1979). Die organische Elementaranalyse. *Pharmazie in unserer Zeit* 8:1–10.

HIEBERT, E. N. (1978). Hermann Walther Nernst. *DSB* 15:432–453.

HIEBERT, E. N. and KÖRBER, H. G. (1978). Friedrich Wilhelm Ostwald. *DSB* 15:455–469.

HILDEBRAND, J. H. (1948). Gilbert Newton Lewis. *Biog. Mem. NAS* 31:210–235.

HILGETAG, G. and PAUL, H. (1970). Zur wissenschaftlichen Leistung Emil Fischers. *Z. Chem.* 10:281–289.

HILL, A. V. (1912). The heat production of surviving amphibian muscles during rest, activity and rigor. *J. Physiol.* 44:466–513.

HILL, A. V. (1932). The revolution in muscle physiology. *Physiol. Revs.* 12:56–67.

HILL, A. V. (1965). *Trails and Trials in Physiology.* London: Arnold.

HILL, A. V. (1966). Why biophysics? *Science* 124:1233–1237.

HILL, A. V. (1969). Bayliss and Starling and the happy fellowship of physiologists. *J. Physiol.* 204:1–13.

HILL, A. V. (1970). Autobiographical sketch. *Persp. Biol. Med.* 14:27–42.

HIMSWORTH, H. and PITT-RIVERS, R. (1972). Charles Robert Harington. *Biog. Mem. FRS* 18:267–308.

HJELT, E. (1916). *Geschichte der organischen Chemie von älterster Zeit bis zur Gegenwart.* Braunschweig: Vieweg.

HOESCH, K. (1921). *Emil Fischer, sein Leben und sein Werk.* Berlin: Verlag Chemie.

HOFMANN, A. W., ed. (1888). *Aus Justus Liebigs und Friedrich Wöhlers Briefwechsel in den Jahren 1829–1873.* Braunschweig: Vieweg.

HOFMANN, K. (1987). Vincent du Vigneaud. *Biog. Mem. NAS* 56:543–595.

HOFMEISTER, F. (1889). Ueber die Darstellung von kristallisierten Eieralbumin und die Kristallisierbarkeit kolloider Stoffe. *Z. physiol. Chem.* 14:165–172.

HOFMEISTER, F. (1894). Ueber Methylierungen im Thierkörper. *Arch. exp. Path. Pharm.* 33:198–215.

HOFMEISTER, F. (1896). Ueber die Bildung des Harnstoffs durch Oxydation. *Arch. exp. Path. Pharm.* 37:426–444.

HOFMEISTER, F. (1900). Willy Kühne. *Ber. chem. Ges.* 33:1875–1880.

HOFMEISTER, F. (1901). *Die chemische Organisation der Zelle.* Braunschweig: Vieweg.

HOFMEISTER, F. (1902a). Ueber den Bau der Eiweisskörper. *Naturw. Rund.* 17:529–533, 545–549.

HOFMEISTER, F. (1902b). Ueber Bau und Gruppierung der Eiweisskörper. *Erg. Physiol.* 1:759–802.

HOFMEISTER, F. (1908). Einiges über die Bedeutung und den Abbau der Eiweisskörper. *Arch. exp. Path. Pharm.* Suppl., pp. 273–281.

HOFMEISTER, F. (1912). *Chemische Steuerungsvorgänge im Tierkörper.* Strassburg: Trübner.

HOFMEISTER, F. (1913). *Der Kohlenhydratstoffwechsel der Leber.* Vienna: Nothnagel Stiftung.

HOFMEISTER, F. (1914). Vom chemisch-morphologischen Grenzgebiet. *Zeitschrift für Morphologie und Anthropologie* 18:717–724.

HOFMEISTER, F. (1918). Ueber qualitativ unzureichende Ernährung. *Erg. Physiol.* 16:1–39, 510–589.

HOLLEMAN, A. F. (1915). Fünfzigjähriges Benzolstudium. *Janus* 20:459–488.

HOLMES, F. L. (1963). Elementary analysis and the origin of physiological chemistry. *Isis* 54:50–81.

HOLMES, F. L. (1964). Introduction to facsimile edition of J. Liebig *Animal Chemistry.* New York: Johnson Reprint Corp.

HOLMES, F. L. (1973). Justus von Liebig. *DSB* 8:332–342.

HOLMES, F. L. (1974). *Claude Bernard and Animal Chemistry.* Cambridge, Mass: Harvard University Press.

HOLMES, F. L. (1984). Lavoisier and Krebs. The individual scientist in the near and deeper past. *Isis* 75:131–142.

HOLMES, F. L. (1985). *Lavoisier and the Chemistry of Life.* Madison, Wis.: University of Wisconsin Press.

HOLMES, F. L. (1986a). Patterns of scientific creativity. *Bull. Hist. Med.* 60:19–35.

HOLMES, F. L. (1986b). The complementarity of teaching and research in Liebig's laboratory. Paper presented at the annual meeting of the History of Science Society, 25 October 1986.

HOLMES, F. L. (1987). The intake-output method of quantification in physiology. *Hist. Stud. Phys. Sci.* 17:235–270.

HOLMES, F. L. (1988). The formation of the Munich school of metabolism. In *The Investigative Enterprise: Experimental Physiology in Nineteenth-Century Medicine* (W. Coleman and F. L. Holmes, eds.), pp. 179–210. Berkeley: University of California Press.

HOLMES, F. L. (1989). The complementarity of teaching and research in Liebig's laboratory. *Osiris* 2nd ser., 5:121–164.

HOLTER, H. (1976). K. U. Linderstrøm-Lang. In *The Carlsberg Laboratory 1876–1976* (H. Holter and K. M. Moller, eds.), pp. 88–117. Copenhagen: Rhodos.

HOPFF, H. (1959). Kurt H. Meyer. *Chem. Ber.* 92:cxxi–cxxxvi.

HOPKINS, B. S. (1944). William Albert Noyes. *J. Am. Chem. Soc.* 66:1045–1066.

HOPKINS, F. G. (1914). The dynamic side of biochemistry. *Report of the Eighty-third Meeting of the British Association for the Advancement of Science,* pp. 652–668.

HOPKINS, F. G. (1924). Biochemistry: its present position and outlook. *Lancet* (I):1247–1252.

HOPKINS, F. G. (1926). On current views concerning the mechanisms of biological oxidation. *Skandinavisches Archiv für Physiologie* 49:33–59.

HOPPE-SEYLER, F. (1866). Beiträge zur Kenntnis des Blutes des Menschen und der Wirbelthiere. *Med.-chem. Unt.,* pp. 169–208.

HOPPE-SEYLER, F. (1867). Zur Chemie des Blutes und seiner Bestandteile. 1. Ueber Oxydationsprocesse im lebenden Blute. *Med.-chem. Unt.,* pp. 293–297.

HOPPE-SEYLER, F. (1869). Ueber die chemische Zusammensetzung des Eiters. *Med.-chem. Unt.,* pp. 486–501.

HOPPE-SEYLER, F. (1870). Ueber Fäulnissprocesse und Desinfection. *Med.-chem. Unt.,* pp. 561–581.

HOPPE-SEYLER, F. (1876). Ueber die Processe der Gährungen und ihre Beziehung zum Leben der Organismen. *Arch. ges. Physiol.* 12:1–17.

HOPPE-SEYLER, F. (1877). Ueber die Stellung der physiologischen Chemie zur Physiologie im Allgemeinen. *Z. physiol. Chem.* 1:270–273.

HOPPE-SEYLER, F. (1878). Ueber Gährungsprocesse. *Z. physiol. Chem.* 2:1–28.

HOPPE-SEYLER, F. (1881). *Physiologische Chemie.* Berlin: Hirschwald.

HOYTINCK, G. J. (1970). Physical chemistry in The Netherlands after van't Hoff. *Ann. Rev. Phys. Chem.* 21:1–16.

HUBBARD, R. (1977). Preface to English translation of Boll, *On the Anatomy and Physiology of the Retina* and of Kühne, *Chemical Processes in the Retina*. *Vision Research* 17:1247–1248.

HUDSON, C. S. (1941). Emil Fischer's discovery of the configuration of glucose. *J. Chem. Ed.* 18:353–357.

HUDSON, C. S. (1948). Historical aspects of Emil Fischer's fundamental conventions for writing stereo-formulas in a plane. *Adv. Carbohydrate Chem.* 3:1–22.

HÜCKEL, E. (1975). *Ein Gelehrtenleben*. Weinheim: Verlag Chemie.

HÜCKEL, W. (1940). Otto Ruff. *Ber. chem. Ges.* 73A:125–156.

HÜCKEL, W. (1947). Windaus' Bedeutung für die theoretische organische Chemie. *Angewandte Chemie* 59:185–188.

HÜCKEL, W. (1949). Heinrich Kiliani. *Chem. Ber.* 82:i–ix.

HÜCKEL, W. (1958). Paul Walden. *Chem. Ber.* 91:xix–lxv.

HUHLE-KREUTZER, G. (1989). *Die Entwicklung arzneilichen Produktionsstätten aus Apotheklaboratorien*. Stuttgart: Deutscher Apotheker Verlag.

HUISGEN, R. (1958). Heinrich Wieland. *Proc. Chem. Soc.*, pp. 210–219.

HUISGEN, R. (1961). Richard Willstätter. *J. Chem. Ed.* 38:8–15.

HUISGEN, R. (1986). Adolf von Baeyer's scientific achievements. *Angewandte Chemie (Int. Ed.)* 25:297–311.

HURWIC, J. (1985). Badania Kazimierza Fajansa. *Kwartalnik Historii Nauki i Techniki* 30:215–245.

IHDE, A. J. (1961). The Karlsruhe Congress: a centennial retrospect. *J. Chem. Ed.* 38:83–86.

IHDE, A. J. (1974). Early American studies on respiration calorimetry. *Molecular and Cellular Biochemistry* 5:11–16.

IPATIEFF, V. N. (1946). *Life of a Chemist*. Stanford University Press.

JACOBSON, P. (1918). Carl Liebermann. *Ber. chem. Ges.* 51:1135–1204.

JACOBSON, P. (1921). Emil Knoevenagel. *Ber. chem. Ges.* 54A:269–271.

JAGTENBERG, T. (1983). *The Social Construction of Science*. Dordrecht: Reidel.

JAMES, F. A. J. L. (1985). The creation of a Victorian myth: The historiography of spectroscopy. *Hist. Sci.* 23:1–24.

JAPP, F. R. (1898). Kekulé memorial lecture. *J. Chem. Soc.* 73:97–138.

JAQUET, A. (1944). Professor Friedrich Miescher. *Helvetica Physiologica et Pharmacologica Acta* Suppl. II, pp. 5–43.

JARAUSCH, K. H. (1982). *Students, Society and Politics in Imperial Germany*. Princeton University Press.

JOHNSON, J. A. (1985a) Academic self-regulation and the chemical profession in imperial Germany. *Minerva* 23:241–271.

JOHNSON, J. A. (1985b) Academic chemistry in imperial Germany. *Isis* 76:500–524.

JONES, M. E. (1953). Albrecht Kossel, a biographical sketch. *Yale J. Biol. Med.* 26:80–97.

JONES, P. G. (1984). Crystal structure determination: A critical view. *Chemical Society Reviews* 13:157–172.

JONES, P. R. (1983). *Bibliographie der Dissertationen Amerikanischer und Britischer Chemiker an Deutschen Universitäten 1840–1914.* Munich: Forschungsinstitut des Deutschen Museums.

JORISSEN, W. P. and REICHER, L. T. (1912). *J. H. van't Hoffs Amsterdamer Periode 1877–1895.* Helder: De Boer.

JOSEPHSON, L. (1933). The story of the isolation of crystalline vitamin D. *Journal of the American Pharmaceutical Association* 22:309–312.

KAHANE, E. (1958). La vie et l'oeuvre scientifique de Charles Gerhardt. *Bull. Soc. Chim.*, pp. 4733–4742.

KALCKAR, H. (1987). Gerhard Schmidt. *Biog. Mem. NAS* 57:399–429.

KAPOOR, S. C. (1973). Auguste Laurent. *DSB* 8:54–61.

KARLSON, P. (1977). 100 Jahre Biochemie im Spiegel von Hoppe-Seyler's Zeitschrift für physiologische Chemie. *Z. physiol. Chem.* 358:717–752.

KARLSON, P. (1986). Wie und warum entstehen wissentschaftliche Irrtümer. *Naturw. Rund.* 39:380–389.

KATSCH, G. (1955). Nachruf für Gustav von Bergmann. *Münch. med. Wchschr.* 97:1398–1400.

KATSOYANNIS, P. G., ed. (1973). *The Chemistry of Polypeptides.* New York: Plenum.

KATTERMANN, R. (1984). Walter Siegfried Loewe. Sein Beitrag zur Analytik, Biologie und Pharmakologie der Sexualhormone. *J. Clin. Chem.* 22:505–514.

KATZ, B. (1978). Archibald Vivian Hill. *Biog. Mem. FRS* 24:71–149.

KAUFFMAN, G. B. (1908). Niels Bjerrum (1879–1958): A centennial evaluation. *J. Chem. Ed.* 57:779–782, 863–867.

KAUTSCH, K. (1906). Emil Fischers Forschungen auf dem Gebiete der Eiweisschemie. *Die Umschau* 10:129–131.

KAY, L. E. (1988). Laboratory technology and biological knowledge: The Tiselius electrophoresis apparatus. *Hist. Phil. Life Sci.* 10:51–72.

KEIBEL, F. (1916). Gustav Albert Schwalbe. *Anat. Anz.* 49:210–211.

KEILIN, D. (1966). *The History of Cell Respiration and Cytochrome.* Cambridge University Press.

KEKWICK, R. A. and PEDERSEN, K. O. (1974). Arne Tiselius. *Biog. Mem. FRS* 20:401–428.

KELLERMANN, H. (1915). *Der Krieg der Geister.* Weimar: Duncker.

KENDAL, L. P. (1952). Henry Stanley Raper. *Biochem. J.* 52:353–356.

KENDALL, J. (1932). Alexander Smith as an educator. *J. Chem. Ed.* 9:254–260.

KENNAWAY, E. (1952). Some recollections of Albrecht Kossel. *Ann. Sci.* 8:393–397.

KERKER, M. (1976). The Svedberg and molecular reality. *Isis* 67:190–216.

KERKER, M. (1986). The Svedberg and molecular reality. An autobiographical postscript. *Isis* 77:278–282.

KERSAINT, G. (1958). Sur une correspondance inédite de Nicolas Vauquelin. *Bull. Soc. Chim.*, pp. 1603–1619.

KERSAINT, G. (1966). *Antoine François Fourcroy, sa Vie et son Oeuvre.* Paris: Editions du Museum.

KLEMPERER, G. (1919). Erinnerungen an Emil Fischer. *Berliner Tageblatt* 20 July 1919, pp. 2–3.

KLENK, E. (1931). Hans Thierfelder. *Z. physiol. Chem.* 203:1–9.

KLETZINSKY, V. (1858). *Compendium der Biochemie.* Vienna: Braumüller.

KLOSTERMAN, L. J. (1985). A research school of chemistry in the nineteenth century: Jean Baptiste Dumas and his research students. *Ann. Sci.* 42:1–40.

KLUCKER, C. (1930). *Erinnerungen eines Bergführers.* Zurich: Rentsch.

KNOOP, F. (1904). *Der Abbau aromatischen Fettsäuren im Tierkörper.* Freiburg: Kuttruff.

KOCH-WESER, J. and SCHACHTER, P. J. (1978). Schmiedeberg in Strassburg 1872–1918: The making of modern pharmacology. *Life Sciences* 22:1361–1371.

KÖNIGSBERGER, L. (1902–1903). *Hermann von Helmholtz.* Braunschweig: Vieweg.

KOERTING, W. (1968). *Die Deutsche Universität in Prag. Die letzten hundert Jahre ihrer Medizinischen Fakultät.* Munich: Bayerische Landesärztekammer.

KÖSSLER, F. (1970). *Verzeichnis der Doktorpromotionen an der Universität Giessen von 1801–1884.* Giessen: Universitätsbibliothek.

KÖSSLER, F. (1971). *Katalog der Dissertationen und Habilitationen der Universität Giessen von 1801–1884.* Giessen: Universitätsbibliothek.

KÖSSLER, F. (1976). *Register zu den Matrikeln und Inscriptionsbüchern der Universität Giessen WS1807–WS1850.* Giessen: Universitätsbibliothek.

KOHLER, R. E. (1971). The background to Eduard Buchner's discovery of cell-free fermentation. *J. Hist. Biol.* 4:35–61.

KOHLER, R. E. (1972). The reception of Eduard Buchner's discovery of cell-free fermentation. *J. Hist. Biol.* 5:327–353.

KOHLER, R. E. (1973). The enzyme theory and the origins of biochemistry. *Isis* 64:181–196.

KOHLER, R. E. (1977). Rudolf Schoenheimer, isotopic tracers, and biochemistry in the 1930s. *Hist. Stud. Phys. Sci.* 8:257–298.

KOHLER, R. E. (1978). Walter Fletcher, F. G. Hopkins, and the Dunn Institute of Biochemistry: A case history in the patronage of science. *Isis* 69:331–355.

KOHLER, R. E. (1982). *From Medical Chemistry to Biochemistry: The Making of a Biomedical Discipline.* Cambridge University Press.

KOHUT, A. (1904). *Justus von Liebig.* Giessen: Roth.

KOLBE, H. (1874). Zur Erinnerung an Justus von Liebig. *J. prakt. Chem.* NF8:428–458.

KOLBE, H. (1878). Die chemische Synthese, ein chemischer Traum. *J. prakt. Chem.* 126:432–455.

KOPPEL, M. and SPIRO, K. (1914). Über die Wirkung von Moderatoren (Puffern) bei der Verschiebung des Säure-Base-Gleichgewichts in biologischen Flüssigkeiten. *Biochem. Z.* 65:409–439.

KORNBERG, A. (1987). The two cultures: chemistry and biology. *Biochemistry* 26:6888–6891.

KORNBERG, H. and WILLIAMSON, D. H. (1984). Hans Adolf Krebs. *Biog. Mem. FRS* 30:351–384.

KOSSEL, A. (1876). Ein Beitrag zur Kenntnis der Peptone. *Arch. ges. Physiol.* 13:309–320.

KOSSEL, A. (1878). Über die chemische Wirkungen der Diffusion. *Z. physiol. Chem.* 2:158–176.

KOSSEL, A. (1881). *Untersuchungen über die Nucleine und ihre Spaltprodukte.* Strassburg: Trübner.

KOSSEL, A. (1884). Ueber einen peptonartigen Bestandteil des Zellkerns. *Z. physiol. Chem.* 8:511–515.

KOSSEL, A. (1897). Zur Erinnerung an Eugen Baumann. *Z. physiol. Chem.* 23:1–22.

KOSSEL, A. (1901). Ueber den gegenwärtigen Stand der Eiweisschemie. *Ber. chem. Ges.* 34:3214–3245.

KOSSEL, A. (1928). *The Protamines and Histones.* London: Longmans Green.

KRAGH, H. (1987). *An Introduction to the Historiography of Science.* Cambridge University Press.

KRAUS, C. A. (1958). The present state of the electrolyte problem. *J. Chem. Ed.* 35:324–337.

KREBS, H. A. (1967). The making of a scientist. *Nature* 215:1441–1445.

KREBS, H. A. (1972). Otto Heinrich Warburg. *Biog. Mem. FRS* 18:629–699.

KREBS, H. A. and LIPMANN, F. (1974). Dahlem in the late nineteen twenties. In *Energy, Regulation and Biosynthesis in Molecular Biology* (D. Richter, ed.), pp. 7–27. Berlin: de Gruyter.

KREHL, L. von (1936). Paul Morawitz. *Münch. med. Wchschr.* 83:1397–1398.

KRÖHNKE, F. (1952). Hermann Leuchs. *Chem. Ber.* 85:lv–lxxxix.

KRÖHNKE, F. (1961). Organische Chemiker der Universität Berlin. *Nachrichten aus Chemie und Technik* 9:240–241, 255–257.

KRÖNER, P. (1983). *Vor fünfzig Jahren. Die Emigration deutschsprachiger Wissenschaftler 1933–1939.* Münster: Heckner.

KRONECKER, H. (1907). Ein eigenartiger deutscher Naturforscher. *Deutsche Revue* 32:99–112.

KÜHNAU, J. (1964). Wilhelm Stepp. *Deutsche med. Wchschr.* 89:1911–1912.

KÜHNE, W. (1860). *Myologische Untersuchungen.* Leipzig: Voit.

KÜHNE, W. (1864). *Untersuchungen über das Protoplasma und die Contractilität.* Leipzig: Engelmann.

KÜHNE, W. (1866–1868). *Lehrbuch der physiologischen Chemie.* Leipzig: Engelmann.

KÜHNE, W. (1867). Ueber die Verdauung der Eiweissstoffe durch den Pankreassaft. *Arch. path. Anat.* 39:130–174.

KÜHNE, W. (1876). Ueber das Verhalten verschiedener organisierter und sog. ungeformte Fermente. *Verhandl. Nat. Heidelberg* NF1:190–193.

KÜHNE, W. (1877). Nachtrag zur Geschichte des Trypsins. *Unt. Physiol. Heidelberg* 1:325–326.

KÜHNE, W. (1878a). *The Photochemistry of the Retina and on Visual Purple.* Edited by Michael Foster. London: Macmillan.

KÜHNE, W. (1878b). Erfahrungen und Bemerkungen über Enzyme und Fermente. *Unt. Physiol. Heidelberg* 1:291–324.

KÜHNE, W. (1878c). Erwiderung auf einem Angriff des Herrn Hoppe-Seyler. *Unt. Physiol. Heidelberg* 2:62–68.

KÜHNE, W. (1880). Ueber die Verbreitung einiger Enzyme im Tierkörper. *Verhandl. Nat. Heidelberg* NF2:1–6.

KÜHNE, W. (1882). Bemerkungen zu Herrn Hoppe-Seyler's Darstellung der Optochemie. *Unt. Physiol. Heidelberg* 2:488–492.

KÜHNE, W. (1888). On the origin and causation of vital movement. *Proc. Roy. Soc.* 44:427–448.

KÜHNE, W. (1891). Bemerkung zu der Mitteilung von C. A. Pekelharing. *Z. Biol.* 28:571–573.

KÜHNE, W. (1898). Ueber die Bedeutung des Sauerstoffs für die vitale Bewegung. *Z. Biol.* 36:425–522.

KUHN, R. (1949). Richard Willstätter. *Naturwiss.* 36:1–5.

KUHN, T. S. (1963). The essential tension: tradition and innovation in scientific research. In TAYLOR and BARRON (1963), pp. 341–354.

KUHN, W. (1962). Georg Bredig. *Chem. Ber.* 95:xlii–lxii.

KUNZMANN, T. (1930). Die Bedeutung der wissenschaftlichen Tätigkeit von Justus von Liebig, Friedrich Wöhler und Christ. Fr. Schönbein für die Entwicklung der deutschen chemischen Industrie. Dr.phil. dissertation Berlin.

KURTI, N. (1958). Franz Eugen Simon. *Biog. Mem. FRS* 4:225–256.

KURZER, F. (1956). Biuret and related compounds. *Chemical Reviews* 56:95–197.

LA BARRE, J. (1939). Edgard Zunz. *Bull. Soc. Chim. Biol.* 21:1040–1042.

LAIDLER, K. J. (1984). The development of the Arrhenius equation. *J. Chem. Ed.* 61:494–498.

LAMSON, P. D. (1941). John Jacob Abel—A portrait. *Bull. Johns Hopkins Hosp.* 68:119–157.

LANGLEY, J. N. (1917). Arthur Sheridan Lea. *Proc. Roy. Soc.* B89:xxv–xxvii.

LAUBENDER, W. (1949). Werner Lipschitz. *Arch. exp. Path. Pharm.* 207:243–255.

LEA, A. S. (1890). A comparative study of natural and artificial digestion. *J. Physiol.* 11:226–263.

LEA, A. S. (1893). *The Chemical Basis of the Animal Body: An Appendix to Foster's Text-book of Physiology, Sixth edition.* New York: Macmillan.

LEBER, T. (1903). Willy Kühne. In *Heidelberger Professoren aus dem 19. Jahrhundert* (K. Friedrich, ed.), vol. 2, pp. 209–220. Heidelberg: Winter.

LEEGWATER, A. (1986). The development of Wilhelm Ostwald's chemical energetics. *Centaurus* 29:314–337.

LEICESTER, H. M. (1974). *Development of Biochemical Concepts from Ancient to Modern Times.* Cambridge, Mass.: Harvard University Press.

LEMBECK, F. and GIERE, W. (1968). *Otto Loewi: ein Lebensbild in Dokumenten.* Berlin: Springer.

LENZ, M. (1910). *Geschichte der königlichen Friedrich-Wilhelms-Universität zu Berlin.* Halle: Buchhandlung des Waisenhauses.

LEUTHARDT, F. (1932). Karl Spiro. *Kolloid Zeitschrift* 59:257–263.

LEWIN, L. (1919). Eine toxicologische Erinnerung an Emil Fischer. *Naturwiss.* 7:878–892.

LEWIS, G. N. and RANDALL, M. (1923). *Thermodynamics and the Free Energy of Chemical Substances.* New York: McGraw-Hill.

LICHTENTHALER, F. W. (1987). Karl Freudenberg, Burckhardt Helferich, Hermann O. L. Fischer. *Carbohydrate Research* 164:1–22.

LIEBEN, F. (1935). *Geschichte der physiologischen Chemie.* Vienna: Deuticke.

LIEBEN, F. (1948). Otto von Fürth—Ein Gedenkblatt. *Wiener klin. Wchschr.* 60:377–379.

LIEBERMANN, C. (1897). Victor Meyer. *Ber. chem. Ges.* 30:2157–2168.

LIEBIG, J. (1836). Preisverteilung in dem chemischen Laboratorium in Giessen. *Ann.* 17:119–121.

LIEBIG, J. (1838). Der Zustand der Chemie in Oesterreich. *Ann.* 25:339–347.

LIEBIG, J. (1839). Ueber die Erscheinungen der Gährung, Fäulnis und Verwesung und ihre Ursachen. *Ann.* 30:250–288.

LIEBIG, J. (1840). *Traité de Chimie Organique.* Paris: Masson.

LIEBIG, J. (1843). *Familiar Letters on Chemistry.* London: Taylor & Walton.

LIEBIG, J. (1844). *Chemische Briefe.* Heidelberg: Winter.

LIEBIG, J. (1845). Herr Gerhardt und die organische Chemie. *Ann.* 57:93–118; (French translation) *Revue Scientifique et Industrielle* 23:422–439.

LIEBIG, J. (1870). Ueber die Gährung und die Quelle der Muskelkraft. *Ann.* 153:1–47, 137–228.

LIEBIG, J. (1874). *Reden und Abhandlungen.* Leipzig & Heidelberg: Winter.

LIEBIG, J. (1890). Eigenhändige biographische Aufzeichnungen. *Ber. chem. Ges.* 23:785–816.

LINDERSTRØM-LANG, K. (1939). S. P. L. Sørensen. *Comptes Rendus des Travaux du Laboratoire Carlsberg* 23:i–xxi.

LIPMAN, T. O. (1964). Wöhler's preparation of urea and the fate of vitalism. *J. Chem. Ed.* 41:452–458.

LIPMANN, F. (1975). Reminiscences of Embden's formulation of the Embden-Meyerhof cycle. *Molecular and Cellular Biochemistry* 6:171–175.

LOEB, J. (1922). *Proteins and the Theory of Colloidal Behavior.* New York: McGraw-Hill.

LOEWI, O. (1902). Über Eiweisssynthese im Tierkörper. *Arch. exp. Path. Pharm.* 48:303–330.

LOEWI, O. (1960). An autobiographical sketch. *Persp. Biol. Med.* 4:3–25.

LOGAN, S. R. (1982). The origin and status of the Arrhenius equation. *J. Chem. Ed.* 59:279–281.

LONG, E. R. (1949). Harry Gideon Wells. *Biog. Mem. NAS* 26:233–261.

LONGUET-HIGGINS, H. C. and FISCHER, M. E. (1978). Lars Onsager. *Biog. Mem. FRS* 24:443–471.

LOTZE, S. (1986). Die Chemie in Kurhessen von 150 Jahren. Robert Wilhelm Bunsens 175. Geburtstag. *Zeitschrift des Vereins für Hessische Geschichte und Landeskunde* 91:105–121.

LOWNDES, J. (1956). Robert Henry Aders Plimmer. *Biochem. J.* 62:353–357.

LUSK, G. (1928). *Elements of the Science of Nutrition.* Fourth edition. Philadelphia: Saunders.

LUTWAK-MANN, C. and MANN, T. (1981). The Parnas school. *TIBS* 6:309–310.

LYNEN, F. (1965). Hans von Euler-Chelpin. *Bayerische Akademie der Wissenschaften Jahrbuch*, pp. 206–212.

LYONS, P. (1981). Lars Onsager. *Am. Phil. Soc. Year Book*, pp. 485–495.

McBRIDE, J. M. (1974). The hexaphenylethane riddle. *Tetrahedron* 30:2009–2022.

McCARTY, M. (1985). *The Transforming Principle.* New York: Norton.

McCOLLUM, E. V. (1956). Camille Méhu. *J. Chem. Ed.* 33:507.

MacINNES, D. A. and GRANICK, S. (1958). Leonor Michaelis. *Biog. Mem. NAS* 31:282–321.

McKEE, R. H. (1932). Alexander Smith, the investigator. *J. Chem. Ed.* 9:246–253.

MacMUNN, C. A. (1914). *Spectrum Analysis Applied to Biology and Medicine.* London: Longmans Green.

MacNIDER, W.deB. (1946). John Jacob Abel. *Biog. Mem. NAS* 24:231–257.

MALLOIZEL, G. (1886). *Oeuvres Scientifiques de Michel Eugène Chevreul.* Paris: Lecerf.

MANI, N. (1956). Das Werk von Friedrich Tiedemann und Leopold Gmelin: "Die Verdauung nach Versuchen" und seine Bedeutung für die Entwicklung der Ernährungslehre in der ersten Hälfte des 19. Jahrhunderts. *Gesnerus* 13:190–214.

MANI, N. (1976). Die wissenschaftliche Ernährungslehre im 19. Jahrhundert. In *Ernährung und Ernährungslehre im 19. Jahrhundert* (E. Heischkel-Artelt, ed.), pp. 22–95. Göttingen: Vandenhoek & Ruprecht.

MANN, T. (1964). David Keilin. *Biog. Mem. FRS* 10:183–197.

MARGOLIASH, E. and SCHEJTER, A. (1984). . . . and 70 years ago: myohematins and histohematins (cytochromes). *TIBS* 9:364–367.

MARSEILLE, J. (1968). Das physiologische Lebenswerk von Emil du Bois-Reymond mit besonderer Berücksichtigung seiner Schüler. Dr. med. dissertation, Münster.

MARTIUS, E. W. (1847). *Erinnerungen aus meinem neunzigjährigen Leben.* Leipzig: Voss.

MASON, S. F. (1987). From molecular morphology to universal dissymmetry. In *Essays on the History of Organic Chemistry* (J. G. Traynham, ed.), pp. 35–53. Baton Rouge: Louisiana State University Press.

MATHEWS, A. P. (1927). Professor Albrecht Kossel. *Science* 66:293.

MATTHEWS, D. M. (1977). Protein absorption then and now. *Gastroenterology* 73:1267–1279.

MATTHEWS, D. M. (1978). Otto Cohnheim, the forgotten physiologist. *British Medical Journal* II:618–619.

MAULITZ, R. C. (1978). Rudolph Virchow, Julius Cohnheim and the program in pathology. *Bull. Hist. Med.* 52:162–182.

MAYNARD, L. A. (1958). James Batcheller Sumner. *Biog. Mem. NAS* 31:376–396.

MAZUMDAR, P. (1976). *Karl Landsteiner and the Problem of Species.* Baltimore: Johns Hopkins University Press.

MEINEL, C. (1978). *Die Chemie an der Universität Marburg seit Beginn des 19. Jahrhunderts.* Marburg: Elwert.

MELTZER, S. J. (1914). Professor Hugo Kronecker. *Science* 40:411–414.

MENDELSOHN, E. (1964). The biological sciences in the nineteenth century: Some problems and sources. *Hist. Sci.* 3:39–59.

MENDELSSOHN, K. (1973). The World of Walther Nernst. London: Macmillan.

MERTON, R. K. (1961). Singletons and multiples in scientific discovery. *Proc. Am. Phil. Soc.* 105:470–486.

MERTON, R. K. (1968). The Matthew effect in science. *Science* 59:56–63.

MERTON, R. K. (1973). *The Sociology of Science.* University of Chicago Press.

MEURON-LANDOT, M. de (1965). Friedrich Miescher, l'homme qui a découvert les acides nucléiques. *Histoire de la Médecine* 15:2–25.

MEYER, H. H. (1922). Schmiedebergs Werk. *Arch. exp. Path. Pharm.* 92:i–xxvii.

MEYER, R. (1904). Friedrich Knapp. *Ber. chem. Ges.* 37:4777–4814.

MEYER, R. (1915). Die Phthaleine. *Naturwiss.* 3:576–582.

MEYER, R. (1917). *Victor Meyer.* Leipzig: Akademische Verlagsgesellschaft.

MICHAELIS, L. (1898). *Kompendium der Entwicklungsgeschichte des Menschen, mit Berücksichtigung der Wirbeltiere.* Berlin: Boas & Hesse.

MICHAELIS, L. (1914). *Die Wasserstoffionenkonzentration: Ihre Bedeutung für die Biologie und die Methode ihrer Messung.* Berlin: Springer.

MICHAELIS, L. (1929). *Oxidation-Reduktions-Potentiale, mit besondere Berücksichtigung ihrer physiologischen Bedeutung.* Berlin: Springer.

MICHAELIS, L. and MENTEN, M. L. (1913). Die Kinetik der Invertasewirkung. *Biochem. Z.* 49:333–369.

MIESCHER, F. (1869). Ueber die chemische Zusammensetzung der Eiterzellen. *Med.-Chem. Unt.,* pp. 441–460.

MIKULINSKIJ, S. R., JAROSEVSKIJ, M. G., KROBER, G. and STEINER, H., eds. (1977, 1979). *Wissenschaftliche Schulen.* Berlin: Akademie-Verlag.

MILES, W. D., ed. (1976). *American Chemists and Chemical Engineers.* Washington, D.C.: American Chemical Society.

MILNE-EDWARDS, H. (1862). *Leçons sur la Physiologie et l'Anatomie Comparée de l'Homme et des Animaux.* Paris: Masson.

MIRSKY, A. (1967). The discovery of DNA. *Scientific American* 248:78–88.

MITCHELL, J. S. (1956). Prof. E. J. Friedmann. *Nature* 178:397.

MOCHNACKA, I. (1956). Prace Jakuba Karola Parnasa. *Acta Biochimica Polonica* 3:3–39.

MÖLLER, H. (1984). Wissenschaft in der Emigration—Quantitative und geographische Aspekte. *Berichte zur Wissenschaftsgeschichte* 7:1–9.

MOLITOR, H. (1961). Ernst Peter Pick. *Arch. Int. Pharm. Therap.* 132:205–221.

MOORE, B. (1898). Chemistry of the digestive process. In E. A. Schäfer, *Textbook of Physiology,* vol. 1, pp. 312–474. Edinburgh: Pentland.

MORGAN, N. (1980). The development of biochemistry in England through botany and the brewing industry. *Hist. Phil. Life Sci.* 2:141–166.

MORGAN, N. (1983). William Dobinson Halliburton F.R.S. (1860–1931). Pioneer of British biochemistry? *Notes Rec. RS* 38:129–145.

MORRELL, J. B. (1972). The chemist breeders: The research schools of Liebig and Thomas Thomson. *Ambix* 19:1–46.

MORTON, R. A. (1972). Biochemistry at Liverpool 1902–1971. *Med. Hist.* 16:321–353.

MOULTON, F. R., ed. (1942). *Liebig and after Liebig: A Century of Progress in Agricultural Chemistry.* Washington, D.C.: American Association for the Advancement of Science.

MÜLLER, K. (1970). *Moritz Traube (1826–1894) und seine Lehre der Fermente*. Zurich: Juris.

MULDER, G. J. (1846). *Liebig's Question to Mulder, Tested by Morality and Science*. London: Blackwood. A translation by P. F. H. Fromberg from the Dutch edition (1846); a German translation *Liebig's Frage, sittlich und wissenschaftlich geprüft* (Frankfurt: Schmerber) also appeared in 1846.

MURALT, A. von (1950). Leon Asher. *Erg. Physiol.* 46:1–5.

MURALT, A. von (1952). Otto Meyerhof. *Erg. Physiol.* 47:i–xx.

MURNAGHAN, J. H. and TALALAY, P. (1967). John Jacob Abel and the crystallization of insulin. *Persp. Biol. Med.* 10:334–380.

NACHMANSOHN, D. (1979). *German-Jewish Pioneers in Science 1900–1933*. Berlin: Springer.

NAGEL, W. (1924). Carl Dietrich Harries. *Z. ang. Chem.* 37:105–106.

NEEDHAM, D. M. (1971). *Machina Carnis: The Biochemistry of Muscular Contraction and its Historical Development*. Cambridge University Press.

NEEDHAM, J. (1962). Frederick Gowland Hopkins. *Persp. Biol. Med.* 6:2–46.

NEEDHAM, J. and BALDWIN, E., eds. (1949). *Hopkins and Biochemistry*. Cambridge: Heffer.

NENCKI, M. (1904). *Opera Omnia*. Braunschweig: Vieweg.

NERNST, W. (1896). Das Institut für physikalische Chemie und besonders Elektrochemie an der Universität Göttingen. *Zeitschrift für Elektrochemie* 2:629–636.

NEUBERG, C. (1923). Ernst Leopold Salkowski. *Biochem. Z.* 138:1–4.

NEURATH, H. and ZWILLING, R. (1986). Willy Kühne und die Anfänge der Enzymologie. In *Semper Apertus* (W. Doerr, ed.), vol. 2, pp. 361–374. Berlin: Springer.

NORTHROP, J. H. (1961). Biochemists, biologists and William of Occam. *Ann. Rev. Biochem.* 30:1–10.

NORTHROP, J. H., KUNITZ, M. and HERRIOTT, R. M. (1948). *Crystalline Enzymes*. Second edition. New York: Columbia University Press.

NOVAK, J. (1968). Zur Erinnerung an Professor Dr. Otto Porges. *Wiener klin. Wchschr.* 80:559–560.

NOYES, W. A. (1941). Julius Stieglitz. *Biog. Mem. NAS* 21:275–314.

NOYES, W. A. and NORRIS, J. F. (1932). Ira Remsen. *Biog. Mem. NAS* 14:207–257.

OCKLITZ, H. W. (1951). Prof. Dr.med. Karl Stolte. *Archiv für Kinderheilkunde* 143:161–162.

O'CONNOR, W. J. (1988). *Founders of British Physiology: A Biographical Dictionary*. Manchester University Press.

OESPER, R. E. (1937). Emil Abderhalden. *J. Chem. Ed.* 14:237.

OHLMEYER, P. (1948). Probleme des Zwischenstoffwechsels. Gedenkblatt für Franz Knoop. *Angewandte Chemie* 60:29–35.

OLBY, R. (1974). Friedrich Miescher. *DSB* 9:380–381.

OLSEN, S. (1962). Otto Diels. *Chem. Ber.* 95:v–xlvi.

OPITZ, H. (1963). In Memoriam Adalbert Czerny. *Deutsche med. Wchschr.* 88:723–725.

OPPENHEIMER, E. (1929). Hermann Wieland. *Klin. Wchschr.* 8:1286–1287.

ORTEN, J. M. (1956). Eugen Baumann. *Journal of Nutrition* 58:3–10.

OSBORNE, T. B. (1911). Samuel William Johnson. *Biog. Mem. NAS* 7:204–222.

OSBORNE, T. B. (1913). Heinrich Ritthausen. *Biological Bulletin* 2:335–346.

OSBORNE, T. B. and GUEST, H. H. (1911). Hydrolysis of casein. *J. Biol. Chem.* 9:333–353.

OSBORNE, T. B. and JONES, D. B. (1910). A consideration of the sources of loss in analyzing the products of protein hydrolysis. *Am. J. Physiol.* 26:305–328.

OSTERHOUT, W. J. V. (1930). Jacques Loeb. *Biog. Mem. NAS* 13:318–401.

OSTROWSKA, T. (1980). Jakub Karol Parnas. *Polski Slownik Biograficzny* 25:218–221.

OSTWALD, W. (1926–1927). *Lebenslinien.* Berlin: Klasing.

O'SULLIVAN, H. D. (1934). *The Life of C. O'Sullivan F.R.S.* Guernsey: Star & Gazette Co.

PAOLINI, C. (1968). *Justus von Liebig, eine Bibliographie sämtlicher Veröffentlichungen.* Heidelberg: Winter.

PARASCANDOLA, J. (1971). The L. J. Henderson papers at Harvard. *J. Hist. Biol.* 4:115–118.

PARASCANDOLA, J. (1981). The theoretical basis of Paul Ehrlich's chemotherapy. *J. Hist. Med.* 36:19–43.

PARASCANDOLA, J. (1982). John J. Abel and the early development of pharmacology at the Johns Hopkins University. *Bull. Hist. Med.* 56:512–527.

PARNAS, J. (1981). Thorvald Madsen, leader in international public health. *Danish Medical Bulletin* 28:82–86.

PARTINGTON, J. R. (1964). *A History of Chemistry.* Volume 4. London: Macmillan.

PAUL, H. W. (1972). *The Sorcerer's Apprentice: The French Image of German Science.* Gainesville: University of Florida Press.

PAULY, P. J. (1987). *Controlling Life.* New York: Oxford University Press.

PELLING, M. (1978). *Cholera, Fever and English Medicine 1825–1865.* Oxford University Press.

PELZ, D. C. and ANDREWS, F. M. (1966). *Scientists in Organizations: Productive Climates for Research and Development.* New York: Wiley.

PENZOLDT, F. (1917). Erinnerungen an Adolf v. Baeyer. *Münch. med. Wchschr.* 64:1331–1332.

PERKIN, W. H. (1896). The origin of the coal-tar colour industry, and the contributions of Hofmann and his pupils. *J. Chem. Soc.* 69:596–637.

PERKIN, W. H. (1923). Baeyer memorial lecture. *J. Chem. Soc.* 123:1520–1546.

PEYER, U. (1972). *Rudolf Schoenheimer.* Zurich: Juris.

PFLÜGER, E. (1877). Die Physiologie und ihre Zukunft. *Arch. ges. Physiol.* 15:361–365.

PHILIPPI, E. (1962). Fritz Pregl. *Microchemical Journal* 6:5–16.

PHILLIPS, J. P. (1966). Liebig and Kolbe, critical editors. *Chymia* 11:89–97.

PICCO, C. (1981). *Das Biochemische Institut der Universität Zürich.* Aarau: Sauerländer.

PICHLER, H. (1967). Franz Fischer. *Chem. Ber.* 100:cxxvii–clviii.

PIRIE, N. W. (1983). Sir Frederick Gowland Hopkins. In *Selected Topics in the History of Biochemistry* (G. Semenza, ed.), pp. 103–128. Amsterdam: Elsevier.

PLANTEFOL, L. (1968). Le genre du mot enzyme. *Comp. Rend.* 266C:41–46.

PLATT, B. S. (1956). Sir Edward Mellanby. *Ann. Rev. Biochem.* 25:1–28.

PLAYFAIR, L. (1896). Personal reminiscences of Hofmann etc. *J. Chem. Soc.* 69:575–579.

PLIMMER, R. H. A. (1908). *The Chemical Constitution of the Proteins.* Part II. London: Longmans Green.

POGGENDORFF, J. C. (1971). *Biographisch-literärisches Handwörterbuch zur Geschichte der exacten Naturwissenschaften.* Vol. 7a Supplement. Berlin: Akademie-Verlag.

POHL, J. (1893). Ueber die Oxydation des Methyl- und Aethylalcohols im Thierkörper. *Arch. exp. Path. Pharm.* 31:281–302.

POHL, J. (1896). Ueber den oxydativen Abbau der Fettsäuren im thierischen Organismus. *Arch. exp. Path. Pharm.* 37:413–425.

POHL, J. and SPIRO, K. (1923). Franz Hofmeister, sein Leben und Wirken. *Erg. Physiol.* 22:1–50.

PORTUGAL, F. H. and COHEN, J. S. (1977). *A Century of DNA.* Cambridge, Mass.: M.I.T. Press.

PRANDTL, W. (1949). Das chemische Laboratorium der Bayerischen Akademie der Wissenschaften in München. *Chymia* 2:81–97.

PRANDTL, W. (1952). *Die Geschichte des chemischen Laboratoriums der Bayerischen Akademie der Wissenschaften in München.* Weinheim: Verlag Chemie.

PRELOG, V. and JEGER, O. (1980). Leopold Ruzicka. *Biog. Mem. FRS* 26:411–501.

PRELOG, V. and JEGER, O. (1983). Leopold Ruzicka. *Helv. Chim. Acta* 66:1307–1342.

PRÉVOST, J. L. and DUMAS, J. B. (1823). Examen du sang et de son action dans les divers phénomènes de la vie. *Annales de Chimie* [2]23:90–104.

PROSS, H. (1955). *Die Deutsche Akademische Emigration nach den Vereinigten Staaten 1933–1941.* Berlin: Duncker & Humblot.

RAMSAY, O. B. (1987). The early history and development of conformational analysis. In *Essays on the History of Organic Chemistry* (J. G. Traynham, ed.), pp. 54–77. Baton Rouge: Louisiana State University Press.

RANDLE, A. B. (1934). Sydney Howard Vines. *Journal of Botany* 72:139–141.

RAPER, H. S. (1935). Julius Berend Cohen. *J. Chem. Soc.,* pp. 1331–1337.

RATNAM, C. V. (1961). Depsides: Synthesis by Emil Fischer. *J. Chem. Ed.* 38:93–94.

RAVETZ, J. (1971). *Scientific Knowledge and its Social Problems.* Oxford: Clarendon Press.

REINGOLD, N. (1981). Science, scientists and history of science. *Hist. Sci.* 19:274–283.

REINKE, J. (1925). *Mein Tagewerk.* Freiburg i.B.: Herder.

REITZENSTEIN, L. (1915). Ludwig Medicus. *Ber. chem. Ges.* 48:1744–1748.

REMANE, H. (1984). *Emil Fischer.* Leipzig: Teubner.

REMSEN, I. (1911). Rudolf Fittig. *American Chemical Journal* 45:210–215.

RESNECK, S. (1970). The European education of an American scientist and its influence in 19th-century America: Eben Norton Horsford. *Technology and Culture* 11:366–388.

RHEINBOLDT, H. (1950). Bunsens Vorlesung ueber allgemeine Experimentalchemie. *Chymia* 3:224–241.

RIDEAL, E. (1953). James William McBain. *Obit. Not. FRS* 8:529–547.

RIESENFELD, E. H. (1931). *Svante Arrhenius.* Leipzig: Akademische Verlagsgesellschaft.

RIESSER, O. (1931). Julius Pohl zum 70. Geburtstag. *Deutsche med. Wchschr.* 57:1869.

RIETSCHEL, H. (1933). Leo Langstein. *Med. Klin.* 29:963–964.

RITTER, H. and ZERWECK, W. (1956). Arthur von Weinberg. *Chem. Ber.* 89:xix–xli.

ROBERTS, G. R. (1976). The establishment of the Royal College of Chemistry: An investigation of the social context of early Victorian chemistry. *Hist. Stud. Phys. Sci.* 7:437–485.

ROBINSON, B. (1983). *The Fischer Indole Synthesis.* Chichester: Wiley.

ROBINSON, R. (1947). Arthur Lapworth. *Obit. Not. FRS* 5:555–572.

ROBINSON, R. (1953). Richard Willstätter. *J. Chem. Soc.,* 999–1026.

ROBINSON, R. (1976). *Memoirs of a Minor Prophet. 70 Years of Organic Chemistry.* Amsterdam: Elsevier.

ROBINSON, T. (1960). Michael Tswett. *Chymia* 6:146–161.

ROCKE, A. J. (1981). Kekulé, Butlerov and the historiography of the theory of chemical structure. *Brit. J. Hist. Sci.* 14:27–57.

ROCKE, A. J. (1985). Hypothesis and experiment in the early development of Kekulé's benzene theory. *Ann. Sci.* 42:355–381.

ROCKE, A. J. (1987). Kolbe versus the "transcendental" chemists: The emergence of classical organic chemistry. *Ambix* 34:156–168.

ROE, D. A. (1981). William Cumming Rose. *Journal of Nutrition* 111:1313–1320.

ROOS, A. and BARON, W. F. (1980). The buffer value of weak acids and bases etc. *Respiration Physiology* 40:1–32.

ROSCOE, H. E. (1906). *The Life and Experiences of Henry Enfield Roscoe D.C.L., LL.D., F.R.S., Written by Himself.* London: Macmillan.

ROSE, W. C. and COON, M. J. (1974). Howard Bishop Lewis. *Biog. Mem. NAS* 44:139–174.

ROSEN, G. (1936). Carl Ludwig and his American students. *Bulletin of the Institute of the History of Medicine* 4:609–650.

ROSENBERG, C. E. (1970). Wilbur Olin Atwater. *DSB* 1:325–326.

ROSSITER, M. W. (1975). *The Emergence of Agricultural Science: Justus Liebig and the Americans.* New Haven: Yale University Press.

ROTH, W. (1898). Justus von Liebig, ein Gedenkblatt zu seinem 25jährigen Todestage 18. April 1898. *Samml. chem. Vort.* 3:165–200.

ROTHSCHUH, K. E. (1971). Emil Heinrich du Bois-Reymond. *DSB* 4:200–205.

ROUSSEL, P. A. (1953). The Fischer indole synthesis. *J. Chem. Ed.* 30:122–125.

ROY, A. B. (1976). Eugen Baumann and sulphate esters. *TIBS* 1:N233–N234.

RUBIN, L. P. (1980). Styles in scientific speculation: Paul Ehrlich and Svante Arrhenius on immunochemistry. *J. Hist. Med.* 35:397–425.

RUPE, H. (1932). *Adolf von Baeyer als Lehrer und Forscher.* Stuttgart: Enke.

RUSSELL, C. A. (1983). *Science and Social Change in Britain and Europe 1700–1900.* New York: St. Martin's Press.

RUSSELL, C. A. (1987). The changing role of synthesis in organic chemistry. *Ambix* 34:169–180.

RUZICKA, L. (1973). In the borderland between bioorganic chemistry and biochemistry. *Ann. Rev. Biochem.* 42:1–20.

SACHSE, A. (1928). *Friedrich Althoff und sein Werk.* Berlin: Mittler.

SACHSE, H. (1890). Ueber die geometrische Isomerien der Hexamethylenderivate. *Ber. chem. Ges.* 23:1365–1366.

SACHTLEBEN, R. (1958). Nobel prize winners descended from Liebig. *J. Chem. Ed.* 35:73–75.

SALOMON-BAYET, C. (1982). Bacteriology and Nobel prize selections. In BERNHARD et al. (1982), pp. 377–400.

SALTZMAN, M. D. (1980). The Robinson-Ingold controversy. *J. Chem. Ed.* 57:484–488.

SALTZMAN, M. D. (1986). The development of physical organic chemistry in the United States and the United Kingdom 1919–1939, parallels and contrasts. *J. Chem. Ed.* 63:588–593.

SARTON, G. (1938). The scientific basis of the history of science. In *Cooperation in Research.* Carnegie Institution of Washington, Publication No. 501, pp. 465–481.

SAUER, G., RAPAPORT, S. M. and ROST, G. (1961). Zur Geschichte der Biochemie in Berlin. *NTM* 1:119–147.

SCATCHARD, G. (1960). John Gamble Kirkwood. *Journal of Chemical Physics* 33:1279–1281.

SCATCHARD, G. (1969). Edwin J. Cohn and protein chemistry. *Vox Sanguinis* 17:37–44.

SCHALCK, A. (1940). *Das Leben und Wirken des Heidelberger Physiologen Willy Kühne.* Düsseldorf: Nolte.

SCHARRER, K. (1949). Justus von Liebig and today's agricultural chemistry. *J. Chem. Ed.* 26:515–518.

SCHEFFER, F. E. C. (1930). Dr. J. J. van Laar, 70 Jaar. *Chem. Wkbl.* 27:418–427.

SCHIFF, E. (1956). Adalbert Czerny. *Journal of Pediatrics* 48:391–399.

SCHILLER, J. (1967). *Claude Bernard et les Problèmes Scientifiques de son Temps.* Paris: Cèdre.

SCHLENK, W. (1915). Adolf von Baeyers Stellung zur Problem der basischen Natur des Kohlenstoffs. *Naturwiss.* 3:596–599.

SCHMAUDERER, E. (1969). Die Stellung des Wissenschaftlers zwischen chemischer Forschung und chemischer Industrie im 19. Jahrhundert. *Technikgeschichte in Einzeldarstellungen* 11:37–93.

SCHMAUDERER, E. (1971a). Leitmodelle im Ringen der Chemiker um eine optimale Ausformung des Patentwesens auf die besonderen Bedürfnisse der Chemie während der Gründerzeit. *Chemie Ingenieur Technik* 43:531–540.

SCHMAUDERER, E. (1971b). Der Einfluss der Chemie auf die Entwicklung des Patentwesens in der zweiten Hälfte des 19. Jahrhundert. *Tradition* 16:144–176.

SCHMIDT, C. F. (1971). Alfred Newton Richards. *Biog. Mem. NAS* 42:271–318.

SCHMIDT, O. T. (1959). Géza Zemplén. *Chem. Ber.* 92:i-xix.

SCHMITZ, R. (1978). *Die Naturwissenschaften an der Philipps-Universität Marburg 1527–1977.* Marburg: Elwert.

SCHMORL, K. (1952). *Adolf von Baeyer.* Stuttgart: Wissenschaftliche Verlagsgesellschaft.

SCHNABEL, F. (1953). Friedrich Theodor Althoff. *Neue Deutsche Biographie* 1:222–224.

SCHOLZ, R. (1980–1981). Zur Geschichte synthetischer organischer Farbstoffe in der zweiten Hälfte des 19. Jahrhunderts. *NTM* 17(2):84–104, 18(1)62–83, 18(2)29–48.

SCHREIBER, R. (1957). Der Einfluss Justus von Liebigs auf die Entwicklung der englischen Landwirtschaft. *Bericht der Oberhessischen Gesellschaft für Natur- und Heilkunde* NF28:141–154.

SCHRÖER. H. (1967). *Carl Ludwig.* Stuttgart: Wissenschaftliche Verlagsgesellschaft.

SCHÜTT, H. W. (1973). Zum Berufsbild des Chemikers im Wilhelminischen Zeitalter. In *Der Chemiker im Wandel der Zeiten* (E. Schmauderer, ed.), pp. 55–71. Weinheim: Verlag Chemie.

SCHULZ, F. N. (1903). *Die Grösse des Eiweissmoleküls.* Jena: Fischer.

SCHWAB, G. M. (1976). Kasimir Fajans. *Bayerische Akademie der Wissenschaften Jahrbuch,* pp. 227–229.

SCHWABE, K. (1969). *Wissenschaft und Kriegsmoral.* Göttingen: Musterschmidt.

SCHWARZACHER, W. (1958). Wolfgang Pauli. *Almanach der Akademie der Wissenschaften in Wien* 108:458–460.

SCHWEIZER, H. G. (1986). Otto Meyerhof. In *Semper Apertus* (W. Doerr, ed.), vol. 2, pp. 359–375. Berlin: Springer.

SCRIVEN, E. F. V. and TURNBULL, K. (1988). Azides: their preparation and synthetic uses. *Chemical Reviews* 88:297–368.

SECHENOV, I. M. (1952). *Avtobiograficheskie Zapiski.* Moscow: Akademia Nauk.

SECHENOV, I. M. (1965). *Autobiographical Notes* (D. B. Lindsay, ed.; K. Hanes, translator). Washington, D.C.: American Institute of Biological Sciences.

SEGEL, I. H. (1975). *Enzyme Kinetics.* New York: Wiley-Interscience.

SERVOS, J. W. (1982). A disciplinary program that failed: Wilder D. Bancroft and the Journal of Physical Chemistry 1896–1913. *Isis* 73:207–232.

SHAMIN, A. N. (1977). *Istoria Khimii Belka.* Moscow: Nauka.

SHARPEY-SHAFER, E. (1927). *History of the Physiological Society during the First Fifty Years 1876–1926.* Cambridge University Press.

SHEMIN, D. (1974). Hans Thacher Clarke. *Am. Phil. Soc. Year Book,* pp. 134–137.

SHOPPEE, C. W. (1972). Christopher Kelk Ingold. *Biog. Mem. FRS* 18:349–411.

SIEGFRIED, M. (1916). *Über partielle Eiweisshydrolyse.* Berlin: Borntraeger.

SIMMER, H. (1955). Aus den Anfängen der physiologischen Chemie in Deutschland. *Sudhoffs Arch.* 39:216–236.

SIMMS, G. R. (1963). *The Scientific Work of Karl Landsteiner.* Biberstein: Freundesdienst.

SMEATON, W. A. (1962). *Fourcroy, Chemist and Revolutionary.* Cambridge: Heffer.

SNELDERS, H. A. M. (1984). J. H. van't Hoff's research school in Amsterdam 1877–1895. *Janus* 71:1–30.

SOCIÉTÉ INDUSTRIELLE DE MULHOUSE (1902). *Histoire Documentale de l'Industrie de Mulhouse et ses Environs aux XIXe Siècle.* Mulhouse: Bader.

SØRENSEN, S. P. L. (1909). Enzymstudien II. Über die Messung und Bedeutung der Wasserstoffionenkonzentration bei enzymatischen Prozessen. *Biochem. Z.* 21:131–200.

SOWDEN, J. C. (1962). Hermann Otto Laurenz Fischer. *Adv. Carbohydrate Chem.* 17:1–14.

SPAUDE, M. (1973). *Eugen Albert Baumann.* Zurich: Juris.

SPEE, W. von (1906). Walther Flemming. *Anat. Anz.* 28:41–59.

SPEISER, P. (1961). *Karl Landsteiner.* Vienna: Hollinek.

SPIRO, K. (1915). Dr. Max Koppel. *Maly's Jahresbericht der Thierchemie* 45:i-iv.

SPIRO, K. (1922). Franz Hofmeister. *Arch. exp. Path. Pharm.* 95:i–vii.

SPIRO, K. (1930). Léon Blum. *Schweizerische medizinische Wochenschrift* 60:379–380.

STANLEY, W. M. and HASSID, W. Z. (1969). Hermann Otto Laurenz Fischer. *Biog. Mem. NAS* 40:91–112.

STAUDINGER, H. (1961). *Arbeitserinnerungen.* Heidelberg: Hüthig.

STEINER, H. (1977). Soziale und kognitive Bedingungen wissenschaftlicher Schulen in Geschichte und Gegenwart. In MIKULINSKIJ et al. (1977), pp. 82–156.

STEPHENS, M. D. and RODERICK, G. W. (1962). The Muspratts of Liverpool. *Ann. Sci.* 29:287–311.

STOCK, A. (1935). Carl Duisberg. *Ber. chem. Ges.* 68A:111–148.

STÖRRING, F. K. (1941). Friedrich Umber zum 70. Geburtstag. *Deutsche med. Wchschr.* 67:1185–1186.

STRAUS, F. (1927). Johannes Thiele. *Ber. chem. Ges.* 60A:75–132.

ŠTRBÁŇOVÁ, S. (1981). Biochemical journals and their profile. *Acta Historiae Rerum Naturalium necon Technicarum,* Special issue 16, pp. 150–195.

SURREY, A. R. (1954). *Name Reactions in Organic Chemistry.* New York: Academic Press.

SWAZEY, J. P. (1974). Rudolf Magnus. *DSB* 9:19–21.

SZABADVÁRY, F. (1966). *Geschichte der analytischen Chemie.* Budapest: Akademia Kiadó.

SZWEIJCEROWA, A. and GROSZYŃSKA, J. (1956). *Marceli Nencki: Materialy Biograficzne i Bibliograficzne.* Warsaw: Panstwowe Wydawnictwo Naukowe.

TARBELL, D. S. and TARBELL, A. T. (1981). *Roger Adams, Scientist and Statesman.* Washington, D.C.: American Chemical Society.

TARBELL, D. S. and TARBELL, A. T. (1982). Roger Adams. *Biog. Mem. NAS* 53:3–47.

TAYLOR, C. W. and BARRON, F., eds. (1963). *Scientific Creativity: its Recognition and Development.* New York: Wiley.

TEICH, M. (1981). Ferment or enzyme: what's in a name? *Hist. Phil. Life Sci.* 3:193–215.

TENDERLOO, H. J. C. et al. (1952). Jacobus Henricus van't Hoff. *Chem. Wkbl.* 48:621–663.

THAUER, R. (1955). Albrecht Bethe. *Arch. ges. Physiol.* 261:i–xiv.

THENARD, P. (1950). *Un Grand Français: le Chimiste Thenard.* Dijon: Jobard.

THIERFELDER, H. (1926). Felix Hoppe-Seyler. *Tübinger Naturwissenschaftliche Abhandlungen* No. 10.

THOMAS, K. (1948). Franz Knoop zum Gedächtnis. *Z. physiol. Chem.* 283:1–8.

THOMPSON, A. L. (1973). *Half a Century of Medical Research.* London: HMSO.

THOMSON, W. (1885). Scientific laboratories. *Nature* 31:409–413.

THORPE, T. E. (1900). The Victor Meyer memorial lecture. *J. Chem. Soc.* 77:169–206.

THUNBERG, T. (1933). Olof Hammarsten. *Erg. Physiol.* 35:13–31.

TIEDEMANN, F. and GMELIN, L. (1827). *Die Verdauung nach Versuchen.* Heidelberg: Groos.

TIEMANN, F. (1896). Eugen Baumann. *Ber. chem. Ges.* 29:2675–2580.

TIFFENEAU, M. (1916). Conférence sur l'oeuvre de Charles Gerhardt. *Bull. Soc. Chim.* Supplement, pp. 13–103.

TIFFENEAU, M. (1918). *Correspondance de Charles Gerhardt.* Paris: Masson.

TODD, A. R. and CORNFORTH, J. W. (1976). Robert Robinson. *Biog. Mem. FRS* 22:415–527.

TODD, A. R. and CORNFORTH, J. W. (1981). Robert Burns Woodward. *Biog. Mem. FRS* 27:629–695.

TRAUBE, M. (1858). *Theorie der Fermentwirkungen.* Berlin: Dümmler.

TRAUBE, M. (1899). *Gesammelte Abhandlungen.* Berlin: Mayer & Müller.

TREIBS, H. (1971). *Das Leben und Wirken von Hans Fischer.* Munich: Technische Universität.

TRENDELENBURG, E. (1919). Emil Fischer in seiner Betätigung für die deutsche Wissenschaftspflege. *Naturwiss.* 7:873–876.

TURNER, R. S. (1982). Justus Liebig versus Prussian chemistry. Reflections on early institute building in Germany. *Hist. Stud. Phys. Sci.* 13:129–162.

UHLFEDER, E. (1917). Alfred Einhorn. *Ber. chem. Ges.* 50:668–670.

VALENTIN, J. (1949). *Friedrich Wöhler.* Stuttgart: Wissenschaftliche Verlagsgesellschaft.

VALKO, E. I. (1957). In memoriam: Wolfgang Pauli, Sr. *Journal of Colloid Science* 12:241–242.

VAN KLOOSTER, H. S. (1944). Friedrich Wöhler and his American students. *J. Chem. Ed.* 21:158–170.

VAN KLOOSTER, H. S. (1962). J. J. van Laar, pioneer in chemical thermodynamics. *J. Chem. Ed.* 39:74–76.

VAN SLYKE, D. D. and JACOBS, W. A. (1944). Phoebus Aaron Theodore Levene. *Biog. Mem. NAS* 23:75–126.

VICKERY, H. B. (1931). Thomas Burr Osborne. *Biog. Mem. NAS* 14:261–304.

VICKERY, H. B. (1942). Liebig and proteins. *J. Chem. Ed.* 19:73–79.

VICKERY, H. B. (1945). Russell Henry Chittenden. *Biog. Mem. NAS* 24:59–104.

VICKERY, H. B. (1967). William Mansfield Clark. *Biog. Mem. NAS* 39:1–26.

VICKERY, H. B. (1975). Hans Thacher Clarke. *Biog. Mem. NAS* 46:3–20.

VICKERY, H. B. and OSBORNE, T. B. (1928). A review of hypotheses of the structure of proteins. *Physiol. Revs.* 8:393–446.

VINES, S. H. (1915). Joseph Reynolds Green. *Proc. Roy. Soc.* B88:xxxvi–xxxviii.

VIS, G. N. (1900). Adolf Claus. *J. prakt. Chem.* 62:127–133.

VOEGTLIN, C. (1939). John Jacob Abel. *Journal of Pharmacology and Experimental Therapeutics* 67:373–406.

VÖLKER, O. (1984). Carl Friedrich Krukenberg. *Neue Deutsche Biographie* 13:118–119.

VOGT, C. (1896). *Aus meinem Leben: Erinnerungen und Rückblicke.* Stuttgart: Nägele.

VOIT, C. (1900). Willy Kühne. *Z. Biol.* 40:i-viii.

VOLHARD, J. (1909). *Justus von Liebig.* Leipzig: Barth.

VOLHARD, J. and FISCHER, E. (1902). *August Wilhelm Hofmann, ein Lebensbild.* Berlin: Friedländer.

VOSS, H. E. (1959). Herrn Professor Dr.med. W. S. Loewe . . . zum 75. Geburtstag. *Arzneimittelforschung* 9:533–535.

VOSS, H. E. (1964). In Memoriam Prof. Dr. W. S. Loewe. *Deutsche med. Wchschr.* 89:93–94.

WAGEMANN, A. (1917). Theodor Leber. *Archiv der Ophthalmologie* 93:i-vi.

WALDEN, P. (1941). *Geschichte der organischen Chemie seit 1880.* Berlin: Springer.

WALKER, J. (1928). Arrhenius memorial lecture. *J. Chem. Soc.*, pp. 1380–1401.

WALLING, C. (1977). Moses Gomberg. *Proc. Welch Found.* 20:72–84.

WANKMÜLLER, A. (1967). Ausländische Studierende der Pharmazie und Chemie bei Liebig in Giessen. *Deutsche Apotheker Zeitung* 107:463–467.

WANKMÜLLER, A. (1980). Die Professoren und Dozenten der physiologischen Chemie in Tübingen. In *Physik, Physiologische Chemie und Pharmazie an der Universität Tübingen* (A. Hermann and A. Wankmüller, eds.), pp. 41–77. Tübingen: Mohr (Paul Siebeck).

WANKMÜLLER, A. (1981–1982). Studenten der Pharmazie und Chemie an der Universität Giessen von 1800 bis 1852. *Beiträge zur Württembergischen Apotheken Geschichte* 13:54–64, 95–96, 121–128, 148–160.

WANZLIK, H. W. (1969). Helmuth Scheibler. *Chem. Ber.* 102:xxvii–xxxix.

WARBURG, O. (1908). Beobachtungen über die Oxydationsprozessen im Seeigelei. *Z. physiol. Chem.* 57:1–16.

WATERMANN, R. (1960). *Theodor Schwann, Leben und Werk.* Düsseldorf: Schwann.

WATSON, C. J. (1965). Reminiscences of Hans Fischer and his laboratory. *Persp. Biol. Med.* 8:419–435.

WEBB, G. B. and POWELL, D. (1946). *Henry Sewall.* Baltimore: Johns Hopkins Press.

WEIL, H. and WILLIAMS, T. I. (1953). Der Ursprung der Papierchromatographie. *Naturwiss.* 40:1–7.

WEINBERG, A. (1917). Paul Friedländer. *Ber. chem. Ges.* 57A:13–29.

WEINBERG, A. (1919). Emil Fischer's Tätigkeit während des Krieges. *Naturwiss.* 7:868–873.

WEINLAND, E. (1906). Nachruf auf Richard Neumeister. *Z. Biol.* 48:141–143.

WEISSER, U. (1984). *Das erste Hormon aus der Retorte. Arbeiten am synthetischen Adrenalin (Suprarenin) bei Hoechst 1900–1908.* Dokumente aus Hoechst-Archiven No. 52. Frankfurt a.M.: Hoechst.

WELTE, E. (1968). Die Bedeutung der mineralischen Düngung und die Düngemittelindustrie in den letzten 100 Jahren. *Technikgeschichte* 35:37–65.

WENDEL, G. (1975). *Die Kaiser-Wilhelm Gesellschaft 1911–1914.* Berlin: Akademie Verlag.

WHITAKER, M. A. B. (1984). Science, scientists and history of science. *Hist. Sci.* 22:421–424.

WIBERG, K. B. (1986). The concept of strain in organic chemistry. *Angewandte Chemie (Int. Ed.)* 25:312–322.

WIDMANN, H. (1973). *Exil und Bildungshilfe: Die deutschsprachige akademische Emigration in die Türkei nach 1933.* Berne: Lang.

WIELAND, H. (1915). Adolf von Baeyers Untersuchungen über Peroxyde und Oxonium-Verbindungen. *Naturwiss.* 3:594–596.

WIELAND, H. (1950). Hans Fischer und Otto Hönigschmid zum Gedächtnis. *Angewandte Chemie* 62:1–4.

WILLIAMS, T. I. and WEIL, H. (1952). The phases of chromatography. *Arkiv för Kemi* 5:283–299.

WILLIAMSON, A. (1885). Charles Adolphe Wurtz. *Proc. Roy. Soc.* 38:xxiii–xxxiv.

WILLSTÄTTER, R. (1926). Carl Dietrich Harries. *Ber. chem. Ges.* 59A:123–157.

WILLSTÄTTER, R. (1956). *Aus meinem Leben.* Second edition. Weinheim: Verlag Chemie. An English translation (1965) *From my Life* was prepared by L. S. Hornig. New York: Benjamin.

WILLSTÄTTER, R. and ROHDEWALD, M. (1940). Die enzymatische Systeme der Zucker-umwandlung im Muskel. *Enzymologia* 8:1–63.

WITEBSKY, E. (1954). Ehrlich's side-chain theory in the light of present immunology. *Ann. N. Y. Acad. Sci.* 59:168–181.

WITKOP, B. (1981). Paul Ehrlichs Leitgedanken und lebendiges Werk. *Naturw. Rund.* 34:361–379.

WITKOWSKI, J. A. (1985). The magic of numbers. *TIBS* 10:139–141.

WOHL, A. (1919). Emil Fischer. *Die Chemische Industrie* 42:269–275.

WOLFENDEN, J. H. (1976). The anomaly of strong electrolytes. *Ambix* 19:175–196.

WOLFROM, M. L. (1960). John Ulric Nef. *Biog. Mem. NAS* 34:204–227.

WOTIZ, J. H. and RUDOFSKY, S. (1987). The unknown Kekulé. In *Essays on the History of Organic Chemistry* (J. G. Traynham, ed.), pp. 21–34. Baton Rouge: Louisiana State University Press.

YARSLEY, V. E. (1967). Hermann Staudinger—His life and work. *Chemistry and Industry,* pp. 250–271.

ZADEK, I. (1955). Magnus-Levy. *Münch. med. Wchschr.* 97:834–835.

ZEYNEK, R. von (1904). Hugo Huppert. *Prager medizinische Wochenschrift* 29:593–596.

ZEYNEK, R. von (1908). Gustav Hüfner. *Z. physiol. Chem.* 58:1–38.

ZIMAN, J. M. (1983). The collectivization of science. *Proc. Roy. Soc.* B219:1–19.

ZUPAN, P. (1987). *Der Physiologe Carl Ludwig in Zurich 1849–1855.* Zurich: Juris.

Index of Personal Names

www.ingramcontent.com/pod-product-compliance
Lightning Source LLC
Chambersburg PA
CBHW081340190326
41458CB00018B/6055